PLANETARY
SATELLITES

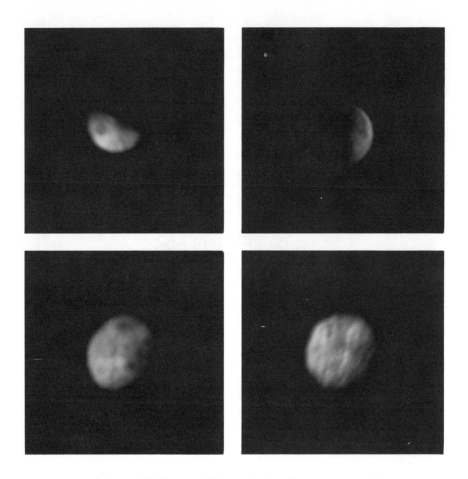

Images of Galilean satellites made from Pioneer spacecraft:

Io (3 December 1974, 17:34 U.T.; distance, 715,000 km; phase, 68°); Europa (3 December 1973, 18:55 U.T.; distance, 325,000 km; phase, 80°); Ganymede (2 December 1974, 19:39 U.T.; distance, 770,000 km; phase, 43°); Callisto (1 December 1974, 23:31 U.T.; distance, 1,041,000 km; phase, 37°).

(NASA/University of Arizona photographs)

PLANETARY SATELLITES

JOSEPH A. BURNS, editor

with 33 collaborating authors

THE UNIVERSITY OF ARIZONA PRESS
TUCSON, ARIZONA

About the Editor ...
Joseph A. Burns joined the faculty of Cornell University in 1966, the same year
in which he completed his Ph.D. in space mechanics, applied mathematics, and
physics at that institution. His B.S. degree in naval architecture was obtained in
1962 from Webb Institute, where he was also awarded the Lewis Nixon Prize
upon graduation. He served as a National Academy of Sciences postdoctoral
fellow in the Theoretical Studies Division of NASA's Goddard Space Flight
Center in 1967–68. From 1975 to 1976 he was a senior investigator at the
Theoretical and Planetary Studies Branch of NASA's Ames Research Center.
During 1973 Burns traveled, lectured, and conducted research in the Soviet
Union and Czechoslovakia under an exchange fellowship granted by the National Research Council. His publications include research papers on turbulence
stimulation, on charged particle dynamics, on heuristic studies in classical
mechanics, on kinetic art, on the orbital evolution of satellites, and on the
rotational dynamics and strength of pulsars, planets, and asteroids.

CONTENTS

Part I.   INTRODUCTION

Part II.   ORBITS AND DYNAMICAL EVOLUTION

[v]

Part III. PHYSICAL PROPERTIES

Part IV. OBJECTS

Part V. SATELLITE ORIGIN

REFERENCE SECTION

FIGURES

TABLES

COLLABORATING AUTHORS

Kaare Aksnes
Hannes Alfvén
Leif E. Andersson
Gustaf Arrhenius
Joseph A. Burns
John Caldwell
A. G. W. Cameron
Guy J. Consolmagno
Allan F. Cook
Dale P. Cruikshank
Bibhas R. De
Thomas C. Duxbury
Fraser P. Fanale
Fred A. Franklin
Tom Gehrels
Richard Greenberg
Donald M. Hunten

Torrence V. Johnson
Jean Kovalevsky
John S. Lewis
Dennis L. Matson
David Morrison
Nancy D. Morrison
Brian O'Leary
Dan Pascu
Stanton J. Peale
Carl B. Pilcher
James B. Pollack
E. L. Ruskol
V. S. Safronov
Jean-Louis Sagnier
P. Kenneth Seidelmann
Joseph Veverka
John A. Wood

PREFACE

Interest in the solar system has grown dramatically since the mid-1960s. Planetary satellites, once the neglected children of the solar system community, have shared this growth of interest and now are being seriously reinvestigated for the first time in nearly a hundred years. Toward the end of the nineteenth century, scientists were principally concerned with discovering new satellites and then properly describing their orbits, whereas today—during the latter part of the twentieth century—research is progressing on numerous fronts: celestial mechanicians are improving classical satellite theories for space age applications and devising new observational schemes to test these theories; observational astronomers are using more powerful instruments and techniques, including those from spacecraft, to understand the composition of the satellites' outer layers and the physical processes occurring therein; and theoreticians are developing detailed models for the origin and subsequent evolution of the companions to the planets.

These modern investigations have recognized for the first time that each of the satellites is an entity with its own character. At the same time scientists have started to view the satellites as a generic whole, as objects that give valuable clues to the origin and formation of their parent bodies, the planets, and of the solar system itself. There is much excitement, considerable understanding—and even some mystery—engendered by contemporary satellite research. All of this we have tried to capture in the chapters of this book.

This volume summarizes quite completely man's knowledge of planetary satellites in the mid-1970s. It also describes some of the techniques used to obtain this knowledge. The information presented, the data tabulated, and the bibliography will make the book valuable as a reference tool to planetary astronomers and spacecraft engineers for years to come. The book discusses all planetary satellites, including Saturn's rings. Consideration of Earth's Moon, however, is limited to one chapter because contemporary scientific concerns about the Moon are quite distinct from those of other satellites which are less well known.

Two chapters serve as an introduction to the volume. Chapter 1, "Introducing the Satellites," by David Morrison, Dale Cruikshank, and Joseph Burns, describes the gross characteristics of the natural satellites, gives a little history, and

tabulates physical properties. Chapter 2, "An Overview," outlines the book, giving a précis of each chapter so as to help the reader find what interests him.

Most of these chapters originated as invited review papers presented to the International Astronomical Union (IAU) Colloquium No. 28, "Planetary Satellites." The colloquium, sponsored by IAU Commissions 16 (Physical Study of Planets and Satellites) and 20 (Positions and Motions of Minor Planets, Comets and Satellites) and also by the Committee on Space Research (COSPAR), was held at Cornell University in Ithaca, New York, on 18–21 August 1974. It was partially supported by NASA, New York State Science and Technology Foundation, the IAU, and Cornell University. The colloquium was the first ever held on planetary satellites; the attendance at the meeting, the number and quality of the presentations, and the enthusiasm of the attendees all point to the fact that satellite studies have reached the stage of vigorous youth, but not sober maturity.

The contributed papers to IAU Colloquium No. 28 show the important problems being attacked in the study of planetary satellites as of 1974. They themselves are a significant part of the research summarized in this volume. Most of them have been published in special "Planetary Satellites" issues of *Icarus* (Volume 24, No. 4, April 1975 and Volume 25, No. 3, July 1975) and *Celestial Mechanics* (Volume 12, No. 1, August 1975). These issues can be purchased separately and bound to form a supplement to this book. All colloquium papers are indicated in the bibliography by a note and the appropriate reference given there.

Preliminary drafts of most invited papers were received by November 1974. They were refereed by at least two scientists and then edited during late 1974 or early 1975. The book as a whole was re-edited in June 1975 and again in Fall 1975 at which time cross-referencing was given between the various articles, a few new references added and the numbering of all tables and figures made sequential for the entire book.

It is intended that the chapters should be complete in themselves, and thus there is occasional duplication or overlap of material. The points where this occurs are cross-referenced so that the reader can appreciate the different points of view.

Authors and topics for the articles of this volume were selected by the colloquium's Organizing Committee (Kaare Aksnes, Edward Anders, Joseph Burns, Audouin Dollfus, Tom Gehrels, Peter Gierasch, Thomas Gold, A. D. Kuzmin, John Lewis, Thomas McCord, David Morrison, V. S. Safronov, Carl Sagan, Joseph Veverka and G. A. Wilkins).

Appreciation for their fine and substantial effort should go to the referees of "Planetary Satellites," who included, among others, K. Alfriend, K. Aksnes, L. Andersson, E. Bilson, G. Born, F. Briggs, R. Brinkman, J. Burns, J. Caldwell, A. Cameron, R. Carlson, C. Chapman, A. Cook, D. Cruikshank, A. Dollfus, T. Duxbury, J. Elliot, I. Ferrin, F. Franklin, O. Franz, T. Gehrels, P.

Gierasch, T. Gold, R. Greenberg, T. Greene, A. Harris, W. Hartmann, P. Herget, D. Hunten, W. Irvine, T. Johnson, W. Kaula, J. Kovalevsky, J. Lewis, J. Lieske, W. Macy, B. Marsden, D. Matson, D. McAdoo, T. McCord, T. McDonough, R. Millis, D. Morrison, R. Murphy, B. O'Leary, T. Owen, D. Pascu, S. Peale, C. Pilcher, J. Pollack, R. Rand, V. Safronov, C. Sagan, J. Sagnier, E. Salpeter, P. Seidelmann, A. Sinclair, S. Singer, W. Smythe, S. Soter, G. Taylor, O. Toon, L. Trafton, J. Veverka, W. Ward, L. Wasserman, G. Wilkins, C. Yoder, and B. Zellner.

Tom Gehrels, David Morrison, and Joseph Veverka gave especially valuable advice and assistance on the composition and the scientific contents of this book. The Frontispiece was made from Pioneer photos of the Galilean satellites produced by John Fountain and William Swindell. Judith Klein Burns designed the cover of *Planetary Satellites*. The organization of the book was improved by comments from Mildred Matthews and by the efforts of JoAnn Cowan, Maida Gierasch, Ann Lemley, and especially Harriet Moore, who devoted much of her time to the detailed work. Many others at Cornell University and elsewhere helped make the colloquium and this book possible. Judith Klein Burns gave artistic, editorial, and technical assistance, but most of all emotional support through the various crises, real and imaginary, associated with *Planetary Satellites*. Finally, appreciation is due to the University of Arizona Press and its editors for bringing this work to publication. Thank you all for your help.

J.A.B.

GLOSSARY

(Commonly used symbols, unless defined differently by individual authors)

a   semimajor axis of a satellite's orbit

aeon 10^9 years

arcsec seconds of arc

A   semimajor axis of the planet's orbit; Bond albedo; minimum principal moment of inertia

AU   astronomical unit $= 1.496 \times 10^{13}$ cm

Å   angstrom $= 10^{-10}$m

B   intermediate principal moment of inertia

B(1,0) see UVB

C   maximum principal moment of inertia

cm   centimeter

e   eccentricity of the orbit

E   emissivity; energy

f   true anomaly

$F(\alpha)$ phase function

g   absolute magnitude, $g \approx B(1,0) - 0.10$; local gravitational acceleration

G   universal constant of gravitation

i   inclination of the orbit

I   intensity; moment of inertia

IAU   International Astronomical Union

J_2 oblateness coefficient

k   thermal conductivity

k_2 Love number of the second kind

km   kilometer

°K   degrees Kelvin

ℓ longitude

L   angular momentum

m   mass

mag   astronomical magnitude proportional to $-2.5 \log_{10}(I/I_o)$

m.e.   mean error

M   mean anomaly

n   mean motion

N   number

p   geometric albedo; pressure

p.e.   probable error

P P_r, period of rotation or revolution; P_o, orbital period; polarization

q   photometric phase integral

Q   specific dissipation function

r   variable distance of an object from the planet's center

R   radius of a planet

$\mathscr{R}$ disturbing potential

sec   second of time

t   time

t_0 epoch

T   temperature

UBV   the photometric system described in *Basic Astronomical Data* (K. Aa. Strand, ed.; U. of Chicago Press, 1963). The observed magnitude B is related to the absolute magnitude B(1,0) at unit distance and zero phase by $B = B(1,0) + \log r\triangle$

v   linear velocity

V   volume

V(1,0) visual magnitude at zero phase reduced to 1 AU distance from both Earth and Sun

α solar phase angle, the angle at the object between the radius vectors to the Earth and Sun; right ascension

δ declination

$\triangle$ distance from Earth

ϵ obliquity, the angle between the rotational axis and the perpendicular to the orbital plane; tidal lag angle

λ wavelength; ecliptic longitude

μ micron $= 10^{-6}$m; rigidity

μm   micron $= 10^{-6}$m

ω argument of pericenter

Ω longitude of ascending node

$\tilde{\omega}$ longitude of pericenter $= \Omega + \omega$; single particle scattering albedo

ϕ latitude

ρ radius of the satellite; mass density

σ thermal conductivity; cross-sectional area; standard deviation

τ optical depth

θ orbital phase angle, measured prograde from superior conjunction

$^{\circ}$ angular degrees

PART I

INTRODUCTION

INTRODUCING THE SATELLITES

David Morrison, University of Hawaii
Dale P. Cruikshank, University of Hawaii
and
Joseph A. Burns, Cornell University

The physical properties and orbits of the known satellites in the solar system are briefly described and tabulated. The evaluation of satellite sizes and masses is presented to calculate mean densities.

In order to set the stage for the information presented in this volume, it is desirable to introduce the 33 known natural satellites of the solar system and to tabulate some basic data on their orbits, sizes, and masses. Brief mention of some of the history of their discoveries also is included.

Only one satellite—Earth's Moon—is visible from Earth without telescopic aid. The Galilean satellites of Jupiter are bright enough to be seen by the naked eye, were it not for their close proximity to Jupiter, an object several hundred times brighter. As is well known, the discovery of these four large satellites by Galileo in 1610 was one of the most significant early results of the application of telescopes to astronomy. In the second half of the seventeenth century, the five largest satellites of Saturn were found, and the nature of the rings of Saturn was deduced. Four more satellites were discovered in the eighteenth century, eight in the nineteenth, and eleven so far in the twentieth (as of June 1975). Table 1.1 lists the 33 known satellites, together with some particulars about their discoveries.

NATURE OF THE SATELLITE SYSTEMS

On the basis of their orbits, the satellites are conveniently separated into two classes. The 19 *regular* satellites move prograde in nearly circular orbits in the equatorial plane of the parent body. The 13 *irregular* satellites move either prograde (7 satellites) or retrograde (6 satellites) in orbits that may be both elliptic and inclined at a considerable angle to the equatorial plane. Because of its anomalously large mass ratio to its parent, Earth's satellite sometimes is considered to be one of the irregular satellites.

Satellites can exist only within a specific distance range from their primaries, limited, at the inner boundary, by the disruptive influence of tidal forces for large satellites and, at the outer boundary, by the stability of their orbits under the action of the Sun. Within this region, many searches for satellites have been

[3]

TABLE 1.1
Satellite Discoveries

Year	Satellite		Pronunciation*	Discoverer	Country
1610	J1	Io	ī'ō	Galileo	Italy
1610	J2	Europa	yo͝o rō' pə	Galileo	Italy
1610	J3	Ganymede	gan' ə mēd'	Galileo	Italy
1610	J4	Callisto	kə lis' tō	Galileo	Italy
1655	S6	Titan	tīt' ən	Huygens	Holland
1671	S8	Iapetus	ī ap' i təs	Cassini	France
1672	S5	Rhea	rē' ə	Cassini	France
1684	S3	Tethys	tē' this	Cassini	France
1684	S4	Dione	dī o' nē	Cassini	France
1787	U3	Titania	ti tā' nē ə	Herschel	England
1787	U4	Oberon	ō' bə ron'	Herschel	England
1789	S1	Mimas	mī' mas	Herschel	England
1789	S2	Enceladus	en sel' ə dəs	Herschel	England
1846	N1	Triton	trīt' ən	Lassell	England
1848	S7	Hyperion	hī pēr' ē ən	Bond/Lassell	U.S./England
1851	U1	Ariel	âr' ē əl	Lassell	England
1851	U2	Umbriel	um' brē el'	Lassell	England
1877	M1	Phobos	fō' bos	Hall	U.S.
1877	M2	Deimos	dī' mos	Hall	U.S.
1892	J5	Amalthea	am' əl thē' ə	Barnard	U.S.
1898	S9	Phoebe	fē' bē	Pickering	U.S. (Peru)
1904/5	J6	Himalia	him' ə li ə	Perrine	U.S.
1904/5	J7	Elara	ē' lər ə	Perrine	U.S.
1908	J8	Pasiphae	pə sif' ə ē'	Melotte	England
1914	J9	Sinope	si nō' pē	Nicholson	U.S.
1938	J10	Lysithea	lis' i thē' ə	Nicholson	U.S.
1938	J11	Carme	kär' mē	Nicholson	U.S.
1948	U5	Miranda	mi ran' də	Kuiper	U.S.
1949	N2	Nereid	nēr' ē id	Kuiper	U.S.
1951	J12	Ananke	ə' nan kē	Nicholson	U.S.
1966	S10	Janus	jā' nəs	Dollfus	France
1974	J13	Leda	lē' də	Kowal	U.S.

Note: Data are from Porter (1960), with revision and additions.

* Pronunciation is based on the unabridged *Random House Dictionary of the English Language* where possible. Markings include *â* as in air and *th* as in thin or path, as well as the usual long vowels, marked *ā* for example, and the short vowels which are unmarked. The schwa, written ə, has the sound of *a* in above, of *e* in system, of *i* in easily, of *o* in gallop and of *u* in circus (after Newburn and Gulkis, 1973).

carried out (see Kuiper, 1961a, who also reviews earlier surveys, and Chapt. 5, Pascu). Since the publication of Kuiper's paper, additional unsuccessful searches have been made for new satellites of Uranus by Sinton (1972), of Mars by

Pollack *et al.* (1973a), and of Mercury by Murray *et al.* (1974). One additional satellite of Saturn has been found by Dollfus (1967) and another outer Jovian satellite by Charles Kowal of Hale Observatories (Kowal *et al.*, 1975). Kowal also found a probable fourteenth satellite of Jupiter having 21st magnitude in fall 1975 but it has been lost (Marsden, 1975).

We now consider briefly the history and the geometrical aspects of each of the satellite systems following Porter (1960), who gives extensive references on the mechanics of the satellite motions. The orbital parameters of the satellites are summarized in Table 1.2.

Mars

Asaph Hall began his search for satellites of Mars in August 1877 and described his discovery of two the next year (Hall, 1878). He called them Phobos (fear) and Deimos (panic) after the attendants of the war god Mars in the *Iliad*. The motion of both satellites (Burns, 1972) is prograde, with the period of Phobos being less than the rotation period of Mars, a unique condition in the solar system. They are described in Chapter 14 by Pollack and in Chapter 15 by Duxbury.

Kuiper (1961a) used special photographic techniques to search for fainter satellites in 1954 and 1956, setting an upper limit for the diameter of any object outside the orbit of Phobos of about 1.5 km for an assumed albedo of 0.05. An additional search for new satellites, primarily near the orbit of Phobos, was made by Pollack *et al.* (1973a) on 19 preinsertion television images from Mariner 9. The upper limit in diameter for satellites of albedo 0.05 ranges from 1.6 to 0.25 km for the first to last picture, respectively, although only a fraction of the possible space-time domain of an unknown satellite was surveyed, particularly near the lower limit of size.

Jupiter

The extensive system of Jupiter has been a profitable hunting ground for six satellite searchers, including Galileo and Nicholson, who each discovered four. The innermost five Jovian satellites move prograde in nearly circular orbits. The four Galilean satellites are lunar-sized objects whose masses are sufficient to induce substantial mutual perturbations in their orbital motions (see Chapt. 16, Morrison and Morrison, for photometry, and Chapt. 8, Greenberg, for the orbital resonance). They are known both by numerical designations and by names from mythology: J1, Io; J2, Europa; J3, Ganymede; and J4, Callisto. They were named by Simon Marius—who had discovered them at nearly the same time as did Galileo—for illicit loves of Zeus (Jupiter). The innermost satellite, J5 (also called Amalthea, the nymph who nursed Jupiter), is very close to the oblate planet, and as a result, its line of nodes regresses rapidly (Chapt. 3, Aksnes).

TABLE 1.2
Orbital Data for the Satellites

Planet	Satellite	Orbital radius [10^3km]	Orbital radius (planetary radii)	Period (days)	Eccentricity	Inclination*(°)
Earth	Moon	384.4	(60.2)	27.3217	0.05490	18.2-28.6†
Mars	1 Phobos	9.37	(2.76)	0.3189	0.0150	1.1
	2 Deimos	23.52	(6.90)	1.262	0.0008	0.9-2.7‡
Jupiter	5 Amalthea	181.3	(2.55)	0.489	0.003	0.4
	1 Io	421.6	(5.95)	1.769	0.000	0.0
	2 Europa	670.9	(9.47)	3.551	0.000	0.5
	3 Ganymede	1070.	(15.1)	7.155	0.001	0.2
	4 Callisto	1880.	(26.6)	16.689	0.01	0.2
	13 Leda	11110.	(156.)	240.	0.146	26.7
	6 Himalia	11470.	(161.)	250.6	0.158	27.6
	10 Lysithea	11710.	(164.)	260.	0.130	29.0
	7 Elara	11740.	(165.)	260.1	0.207	24.8
	12 Ananke	20700.	(291.)	617.	0.17	147.
	11 Carme	22350.	(314.)	692.	0.21	164.
	8 Pasiphae	23300.	(327.)	735.	0.38	145.
	9 Sinope	23700.	(333.)	758.	0.28	153.
Saturn	10 Janus	159.5	(2.65)	0.74896	0.	0.
	1 Mimas	186.	(3.09)	0.942	0.0201	1.5

Planet	Satellite	Orbital radius [10^3km]	Orbital radius (planetary radii)	Period (days)	Eccentricity	Inclination*(°)
Saturn	2 Enceladus	238.	(3.97)	1.370	0.0044	0.0
	3 Tethys	295.	(4.91)	1.888	0.0000	1.1
	4 Dione	377.	(6.29)	2.737	0.0022	0.0
	5 Rhea	527.	(8.78)	4.518	0.0010	0.4
	6 Titan	1222.	(20.4)	15.95	0.0289	0.3
	7 Hyperion	1481.	(24.7)	21.28	0.1042	0.4
	8 Iapetus	3560.	(59.3)	79.33	0.0283	14.7‡
	9 Phoebe	12930.	(216.)	550.4	0.1633	150.
Uranus	5 Miranda	130.	(5.13)	1.4135	0.017	3.4
	1 Ariel	192.	(7.54)	2.520	0.0028	0.
	2 Umbriel	267.	(10.5)	4.144	0.0035	0.
	3 Titania	438.	(17.2)	8.706	0.0024	0.
	4 Oberon	586.	(23.0)	13.46	0.0007	0.
Neptune	1 Triton	354.	(14.6)	5.877	0.00	160.0
	2 Nereid	5510.	(227.)	365.2	0.75	27.6

Note: Data are taken from Morrison and Cruikshank (1974), except for: the eccentricities of Phobos and Deimos (Born, 1974); the elements of Nereid (Rose, 1974); the preliminary elements of J13 (Kowal *et al.*, 1975). Proper eccentricities and inclinations are given for the Galilean satellites. The orbital elements of the outer Jovian satellites vary considerably.

* The inclinations of the inner satellites are relative to the planet's equator, while for the outer satellites (J6-J13, S9, N2) the inclinations are relative to the planet's orbital plane. Thus the inclination of the outer satellites relative to the planet's equator changes as the planet precesses and as the satellite's orbital plane precesses, both about the pole of the planet's orbit plane (see Chapt. 3, Aksnes).

† 5.145° to ecliptic

‡ variable

The eight outer satellites of Jupiter are located at relative distances from Jupiter similar to those of the terrestrial planets from the Sun. They are designated both by numerals corresponding to the order of their discovery and by names of lovers of Jupiter. The irregular Jovian satellites are, according to the orbital elements displayed in Table 1.2, clustered in two groups. The inner group, consisting of Himalia (J6), Elara (J7), Lysithea (J10) and Leda (J13), have similar prograde orbits of moderate inclination (25° to 30°) and moderate eccentricity (0.15 to 0.20) with semi-major axes of 1.1 to 1.2 × 10⁷ km. The outer group—Pasiphaë (J8), Sinope (J9), Carme (J11), and Ananke (J12)—have orbits similar in eccentricity (0.17 to 0.38) but retrograde with inclinations 18° to 35° above the Jovian orbital plane and about twice as large (2.1 to 2.4 × 10⁷ km) as the inner group. The small satellites on prograde orbits have names ending in a, whereas those in retrograde orbits have names ending in e.

The grouping by distance has led to the suggestions that the outer satellites were captured (cf. Kuiper, 1956; Bailey, 1971; Heppenheimer, 1975) or formed by two separate collisions (Colombo and Franklin, 1971). These suggestions are discussed in Chapter 7 by Burns. Little information is available concerning the physical nature of any of the Jovian satellites other than the Galilean ones. The Jovian satellite system is comprehensively reviewed by Morrison and Burns (1976).

Saturn

All the satellites of Saturn have names, most of them suggested by Sir John Herschel early in the nineteenth century. In Greek mythology a Titan was one of a family of giants born to Uranus and Gaea; Tethys, Dione, Rhea, and Phoebe were Titans and sisters of Saturn, while Hyperion and Iapetus were his brothers. Mimas was a giant, and Enceladus a giant or a Titan. Janus is the two-faced ancient deity of all beginnings.

The satellite system of Saturn consists of a regular and an irregular group. The seven inner satellites comprise the regular group, having nearly circular orbits which lie close to the ring plane and thus the equatorial plane of Saturn. Janus, the innermost, is the most recently discovered (Dollfus, 1967). Titan, the largest of all and the outermost of the regular group, has a well-determined atmosphere and has received great attention as a possible object for study by automated space probes and a possible abode for primitive life (see Chapt. 20, Hunten; Chapt. 21, Caldwell; and Chapt. 22, Andersson).

The irregular group, consisting of three satellites, is quite mixed. Hyperion is locked in a strong orbital commensurability with Titan, resulting in part in a variable orbital inclination (see Chapt. 8, Greenberg). Iapetus has a large and nearly circular orbit of high inclination, which is variable because of the combined action of the Sun and Saturn's oblateness; Iapetus is best known for the puzzling large amplitude of its light variations (Morrison et al., 1975). The

TABLE 1.3
Dimensions of Saturn's Ring Elements at 9.5388 AU

Feature	Radius (arcsec)	Extreme value (arcsec)
Outer A	19.82	20.30
Inner A	17.57	17.38
Outer B	16.87	17.09
Inner B	13.21	12.81
Inner C	10.5	10.2

Note: Data are from Cook *et al.* (1973).

outermost satellite, Phoebe, moves retrograde in a large orbit at high inclination.

The rings are a collection of a great many small particles of high albedo moving in circular orbits in Saturn's equatorial plane; that is, they are a collection of small satellites. The distribution of the ring particles apparently is affected dynamically by the satellite masses, which cause the well-known gaps and regions of low particle density in the rings (Chapt. 19, Cook and Franklin). Table 1.3 (from Cook *et al.,* 1973) gives the dimensions of the principal rings: A, B, and C. Rings A and B contribute more than 99% of the light from the ring system, and nearly all studies of the physical properties of the rings refer to these two main components. Two more recently discovered features are the interior ring D (Guérin, 1970) and a broad exterior ring, tentatively called D' or Z, apparently extending nearly to the orbit of Dione (Feibelman, 1967; Kuiper, 1972; Smith *et al.,* 1975). The radiometric results for the rings are reviewed in Chapter 12 by Morrison.

Uranus and Neptune

Uranus is distinguished from the other planets in that its rotation axis lies nearly in the plane of its orbit, being tilted at 98° to the normal to that plane. It is perhaps surprising then that the companions of Uranus, excluding Miranda, form the most regular satellite family in the solar system; they revolve in the same sense as their primary rotates on almost perfectly circular orbits of nearly zero inclination (Greenberg, 1975). Long after their discovery, the first four Uranian satellites were named by Sir John Herschel after fairies of Pope and Shakespeare; Kuiper continued the tradition with the name Miranda. Sinton (1972) has searched for satellites of Uranus interior to Miranda and down to the Roche limit by taking photographs through a filter centered on the strong methane absorption band at 0.89 μ, thereby greatly reducing the glare from the planet, but he has found no satellite brighter than magnitude 17.

By comparison, the two satellites of Neptune have irregular orbits. Triton, the largest in the Neptune system and the second largest relative to its primary in the solar system, has a highly inclined circular retrograde orbit of a relatively short projected lifetime (McCord, 1966), while Nereid moves prograde in the most eccentric orbit of any known satellite in the solar system. In mythology, Tritons and Nereids are the attendants to Neptune, the son of Saturn.

DIAMETERS, MASSES AND DENSITIES

One of the most fundamental bulk properties of a satellite is its mean density, which can be calculated only when both the diameter and mass are known. Masses have been derived for eleven of the satellites, most by knowing the mutual perturbations. Data on sizes, masses, and densities are summarized in Table 1.4. First, however, we address the problem of determining satellite sizes. Only the Galilean satellites and Titan are large enough and near enough to present reliably measurable disks; a number of indirect techniques have been applied to obtain diameters of other satellites.

Visual Measurements of Size

The direct measurements of the disks of the larger satellites have been reviewed by Dollfus (1970); our discussion is derived largely from his comprehensive work. Three instruments have been used to determine the angular diameters: (1) the filar micrometer, used in the second half of the nineteenth century with telescopes of up to 100-cm aperture; (2) the double-image micrometer, used by Dollfus and his collaborators, largely at the Pic-du-Midi 60-cm refractor; and (3) the diskmeter, developed by Camichel, which he used at the same 60-cm telescope, and which subsequently was employed by Kuiper at the Hale 5-m telescope. Each of these instruments is capable, with care, of yielding diameters reproducible to within between 0.05 and 0.10 arcsec. Systematic errors of at least 0.1 arcsec are possible, however, and these errors can be expected to depend on the size of the satellite, the size of the telescope, and the seeing, as well as on the limb darkening of the satellite and the personal equation of the observer. At best, visual techniques cannot be relied upon to yield radii of even the largest satellites with a precision greater than $\pm$ 10%.

A fourth classical technique is optical interferometry, employed by Michelson, Hamy, and Danjon to measure the diameters of the Galilean satellites but not of Titan (Dollfus, 1970). In the absence of limb darkening, the diameters measured in this way should be accurate to about $\pm$ 0.10 arcsec. A more recent interferometric approach proposed by KenKnight (1972) could, in principle, yield diffraction-limited measurements of the sizes of small objects, but it has not yet been demonstrated in practice. Recent advances in speckle interferometry

1. INTRODUCING THE SATELLITES   11

suggest that this technique will also be capable of yielding diameters for the brighter satellites, but here also no results are yet available.

Table 1.5, adapted from Dollfus (1970), summarizes the radii of the five large satellites as determined by each of these methods. Systematic differences are apparent. Dollfus concludes that the most reliable of these radii are those obtained with the double-image micrometer, and he estimates that the uncertainties for the Galilean satellites are ± 75 km, or about ± 5%.

Stellar Occultation Diameters of Io and Ganymede

A powerful method of measuring satellite diameters is to time the passage of the satellite in front of a bright star. Suitable events are rare (O'Leary, 1972, and Chapt. 13, herein), and only recently have satellite ephemerides been predicted. For even the largest satellites, the band from which a given occultation can be seen is only a few thousand kilometers wide; in order to determine the radius, the occultation must be observed photometrically from at least two stations. The effort is worthwhile, however, because occultations provide the only ground-based technique by which to measure radii accurately to within a few kilometers.

In 1971 Io occulted the 5th magnitude star β Scorpii C. Photoelectric observations from four observing parties gave a diameter, on the assumption that Io is spherical, of 3,660 ± 4 km (Taylor, 1972). If the satellite has the shape expected for a fluid in hydrostatic equilibrium, the mean diameter is 3,636 ± 10 km (O'Leary and van Flandern, 1972). We adopt a diameter for this satellite of 3,640 km. In 1972, Ganymede occulted the 8th magnitude star SAO 186800; the analysis by Carlson et al. (1973) yields a diameter of 5,270(+30,−200) km. We choose instead a radius of 2,635(± 25) km after reviewing the original data, so as to have a symmetric error.

The occultation diameter of Io is 4% larger than the value preferred by Dollfus (1970), and that of Ganymede is 5% smaller. A comparison with the radii in Table 1.5 suggests that there is no major systematic error in the classical determinations by filar micrometer or double-image micrometer but that the radii obtained with the interferometer and diskmeter are systematically low.

Diameters From Lunar Occultations

Radii of satellites also can be determined from high time-resolution photometry of occultations by the dark limb of the Moon, as pointed out for the specific example of Titan by Veverka (1974). Elliot et al. (1975) applied this method during the 1974 occultation of Saturn to derive the diameters of Titan, Iapetus, Rhea, Dione, and Tethys with a precision of 100 to 200 km. They deduced a large diameter (5,800 km) and substantial limb-darkening for Titan. On the basis of this measurement, Titan is the largest satellite in the solar system.

TABLE 1.4
Basic Physical Data for the Satellites

Satellite	Mean Opposition Visual Magnitude	V(1,0)	p_V	R(km)
Moon		+ 0.21	0.12	1738
1 Phobos	11.6	+11.9	0.06	(13.5±0.5; 10.8±0.7; 9.4±0.7)
2 Deimos	12.7	+13.0	0.07	$(7.5^{+3}_{-1}; 6.1±1.0; 5.5±1.0)$
1 Io	5.0	− 1.68	0.63	1820±10
2 Europa	5.3	− 1.41	0.64	1500±100
3 Ganymede	4.6	− 2.09	0.43	2635±25
4 Callisto	5.6	− 1.05	0.17	2500±150
5 Amalthea	13.0	+ 6.3	0.10	120±30
6 Himalia	14.8	+ 8.0	0.03	85±10
7 Elara	16.4	+ 9.3	0.03	40±10
8 Pasiphae	17.7	+11.0‡		~18
9 Sinope	18.3	+11.6‡		~14
10 Lysithea	18.4	+11.7‡		~12
11 Carme	18.0	+11.3‡		~15
12 Ananke	18.9	+12.2‡		~10
13 Leda	20	+13.3‡		~ 5
1 Mimas	12.9	+ 3.3		~170
2 Enceladus	11.8	+ 2.2		~250
3 Tethys	10.3	+ 0.7		~500
4 Dione	10.4	+ 0.88	0.60	575±100
5 Rhea	9.7	+ 0.16	0.60	800±125
6 Titan	8.4	− 1.20	0.21	2900±200
7 Hyperion	14.2	+ 4.6		110*
8 Iapetus	10.2-11.9	+ 1.6	0.12	800±100
9 Phoebe	16.5	+ 6.9		40
10 Janus	14	+ 4		~110
Rings				
1 Ariel	14.4	+ 1.7		400*
2 Umbriel	15.3	+ 2.6		275*
3 Titania	14.0	+ 1.3		500*
4 Oberon	14.2	+ 1.5		450*
5 Miranda	16.5	+ 3.8		150*
1 Triton	13.6	− 1.2		1600*
2 Nereid	18.7	+ 4.0‡		150*

Note: Most data in this table are taken from Morrison and Cruikshank (1974). Estimated radii preceded by ~ are from Morrison and Cruikshank (1974); estimated radii of Titan and Iapetus are from Elliot *et al.* (1975); of J5 (Amalthea) from Rieke (1975a); of the outer Jovian satellites from Andersson (1974) or Cruikshank (1976b); of Mimas from Koutchmy and Lamy (1975) and Franz (1975).

* indicates radii estimates based on an albedo of 0.5 (not listed by Morrison and Cruikshank).

R_{max}	R_{min}	M (10^{23}g)	Inverse Mass	ρ (g cm^{-3})
		735	1.23×10^{-2}	3.34
			2.7×10^{-8}†	
			4.8×10^{-9}†	
		891	$(4.696 \pm 0.06) \times 10^{-5}$	3.52 ± 0.10
		487	$(2.565 \pm 0.06) \times 10^{-5}$	3.45 ± 0.75
		1490	$(7.845 \pm 0.08) \times 10^{-5}$	1.95 ± 0.08
		1065	$(5.603 \pm 0.17) \times 10^{-5}$	1.62 ± 0.34
			$\sim 100 \times 10^{-10}$	
			$\sim 40 \times 10^{-10}$†	
			$\sim 4 \times 10^{-10}$†	
23	4		$\sim 0.4 \times 10^{-10}$†	
18	3		$\sim 0.2 \times 10^{-10}$†	
16	3		$\sim 0.1 \times 10^{-10}$†	
20	4		$\sim 0.2 \times 10^{-10}$†	
14	3		$\sim 0.1 \times 10^{-10}$†	
7	1		$\sim 0.01 \times 10^{-10}$	
950	160	0.37 ± 0.01	6.59×10^{-8}	~ 1.4
1000	240	0.85 ± 0.03	1.48×10^{-7}	~ 1.3
1000	480	6.26 ± 0.11	1.09×10^{-6}	~ 1.2
		11.6 ± 0.3	2.04×10^{-6}	1.45 ± 0.80
		18.2 ± 31.8	3.2×10^{-6}	
		1401 ± 2	2.46×10^{-4}	1.37 ± 0.60
460	80		1.5×10^{-8}†	
		22.4 ± 10.9	$\sim 3.94 \times 10^{-6}$	
160	30		7.1×10^{-10}†	
600	100		1.5×10^{-8}†	
			3.5×10^{-8}	
1700	300		$\sim 15 \times 10^{-6}$	Approximate
1100	200		$\sim 6 \times 10^{-6}$	inverse
2000	360		$\sim 50 \times 10^{-6}$	masses
1900	330		$\sim 29 \times 10^{-6}$	from
650	110		$\sim 1 \times 10^{-6}$	perturbations
6500	1100	3400 ± 2000	$\sim 3.3 \times 10^{-3}$	
600	100		2×10^{-7}†	

† indicates estimated inverse masses assuming $\rho = 1.5$ g cm^{-3} for Saturnian and Uranian satellites and $\rho = 3.0$ g cm^{-3} for Martian and outer Jovian satellites. Masses of Galilean satellites are Pioneer 10 values from Anderson et al. (1974b). All other masses are from Duncombe et al. (1973a); those based on perturbations are listed under M, while values based exclusively on estimated radii and densities are given only under Inverse Mass.

‡ indicates values based on a mean color index of +0.8. V(1,0) is the visual magnitude of zero phase reduced to 1 AU distance from both Earth and Sun. It is generally derived from observations at phase angles of 1° to 5°, so it does not necessarily include the opposition surge in brightness shown by a few satellites.

TABLE 1.5
Visually Measured Radii (in km)
of the Galilean Satellites and Titan (from Dollfus, 1970)

Method	Io	Europa	Ganymede	Callisto	Titan
Filar micrometer	1890	1660	2720	2520	2440
Interferometer	1780	1590	2380	2270	——
Diskmeter	1650	1420	2460	2290	2430
Double-image micrometer	1760	1550	2780	2500	2430
Occultation values	1820		2635		2900

Diameters From Satellite-Satellite Occultations

Approximately every six years, as the Earth passes through the orbital plane of the Galilean satellites, there takes place a sequence of partial and total occultations of one satellite by another (Brinkmann and Millis, 1973; Brinkmann, 1973; Aksnes, 1974a). When the size of one satellite is well-known (as is the case for Io and Ganymede), it is possible from high-quality photometry of the occultations to derive the size of the other satellite. In the mutual occultations of 1973/74, a series of occultations of Europa by Io has proven particularly useful as a means of determining the size of Europa. Based on preliminary reductions of a part of the data (Aksnes and Franklin, 1974; Vermilion *et al.*, 1974), we adopt a diameter for Europa of 3,000 km rather than 3,100 km, as suggested from the classical measurements alone (Dollfus, 1970).

Diameters Based on Estimates of Albedo

The visual magnitude of a satellite is given in terms of $V_\odot$, the magnitude of the Sun, and p_V, the geometric visual albedo, by:

$$V = V_\odot - 2.5 \log p_V - 5 \log r + 5 \log (R.\Delta),$$

where r is the satellite radius and R and Δ are the distances from the satellite to the Sun (in AU) and to Earth (in the same units as r), respectively, and $V_\odot = -26.8$. If V is measured and p_V can be estimated, it is possible to derive a photometric radius, r. This method has at various times been applied to all of the satellites with essentially arbitrary choice of albedo; its real usefulness, however, occurs in cases where p_V can be estimated with some reliability.

Perhaps the simplest way to estimate the albedo is from visual or photometric measurements of contrast as the satellite transits the planet. Of course, such estimates can be made only if the disk is nearly resolved. Transit observations of the Galilean satellites yield less precise radii than do other visual methods. Dollfus (1970) used this technique, nevertheless, to estimate an albedo of 0.50 ± 0.05 for Tethys from the transit of 12 June 1966. The corresponding diameter is 1,350 ± 60 km; however, Morrison (1974a) has noted that this diameter indicates an anomalously low density for Tethys and suggests that the albedo may be substantially higher than the value reported by Dollfus.

It has been shown by Widorn, KenKnight, Veverka, Zellner and their co-workers that both the depth of the minimum and the slope of the rising branch of the curve of visual polarization as a function of phase angle depend on the albedo of the reflecting surface (see Chapt. 10, Veverka). Calibration of the dependence from laboratory measurements of pulverized terrestrial rock, of lunar soil, and of crushed meteorites (Veverka and Noland, 1973; Bowell and Zellner, 1974) shows that the slope of the rising branch is a function primarily of the albedo alone, while the depth of the minimum is a more ambiguous indicator of albedo. Since the maximum range of phase angles for the satellites of the outer planets is not sufficient to define the slope from Earth, the only recourse is to obtain rough values for the albedos from the minimum polarization. These estimates are consistent with the albedos known for the Galilean satellites, and Zellner (1972b) and Bowell and Zellner (1974) have applied the method to Rhea and Iapetus, suggesting diameters of about 1,500 km for each of these satellites.

Another method of estimating the albedos and sizes of satellites is from combined visual photometry and infrared radiometry, as discussed in Chapter 12 by Morrison. If both the directly reflected radiation and the energy absorbed at the surface and re-radiated in the thermal infrared are measured, and appropriate modeling assumptions are made, the albedo can be determined from the ratio of the two fluxes. Diameters of three of Saturn's satellites have been measured by this photometric/radiometric technique: Iapetus, with d = 1,700 ± 200 km; Rhea, with d = 1,600 ± 250 km; and Dione, with d = 1,150 ± 200 km (Murphy *et al.*, 1972; Morrison, 1973a, 1974a; Morrison *et al.*, 1975; Morrison, Chapt. 12, herein). These agree fairly well with the lunar occultation values. For this summary, we adopt radii of the satellites of Saturn that are consistent (within the stated errors) with results of both the radiometric and the occultation measurements. Cruikshank (1975, 1976b) has also found sizes for J6 and J7 in this way. Rieke (1975a) used an infrared color temperature to define the size of J5.

In cases where no disk is measurable and where no polarimetric or radiometric data allow the albedo to be estimated, the only recourse is to assume a plausible range of geometric albedos and derive the corresponding range of radii from the visual magnitudes, those summarized by Morrison and Cruikshank in 1974. A choice of albedo for this purpose is not easy. The geometric albedos of the Galilean satellites range from 0.2 to 0.7. Typical albedos of asteroids are ~0.1, while those of Phobos and Deimos are ~0.06 (Zellner and Capen, 1974). The small irregular satellites are thought to have dark surfaces. At the other extreme, some of Saturn's inner satellites (Rhea, Dione, and Tethys) have albedos $\geqslant 0.5$, suggesting ice-covered surfaces; Enceladus, Mimas, and Janus most likely have similar albedos. To estimate the possible range in size, we compute a lower limit based on a geometric albedo of 1.0. The upper limit corresponds to the smaller of (1) a diameter of ⅓ arcsec, which would be detectable visually, or (2) an albedo of 0.03, the lowest known in the solar system (Matson, 1971). While the resulting range in size is large, we cannot justify restricting it further in our present state of ignorance. We do, however, list in Table 1.4 a best guess in order to permit a single estimate of the satellite mass.

Diameters of Phobos and Deimos

The satellites of Mars are far too small to be resolved from Earth, but as a result of close-range imagery by the Mariner spacecraft (Pollack et al., 1972, 1973a), they are among the most thoroughly studied of the satellites (Chapt. 14, Pollack). The best Mariner 9 pictures resolve features as small as 1% the linear size of the satellites (Chapt. 15, Duxbury). Because of their very irregular shapes, it is most meaningful to speak of their radii in the sense of the radius of a sphere whose projected area equals the average projected area of the satellite. Using this definition, Pollack et al. (1972) obtain diameters of 11.4 ± 1.0 km for Deimos and 21.8 ± 3.0 km for Phobos. Combining these diameters with the photometry by Pascu (1973a) and Zellner and Capen (1974), we find geometric visual albedos of 0.06 to 0.07 for these two satellites. Similar results were obtained for Phobos by Smith (1970), based on a Mariner 7 photo that showed the satellite in projection on the disk of Mars.

Phobos and Deimos are better described as triaxial ellipsoids, where the sizes listed are, respectively, the diameters in kilometers of the axes pointing towards Mars, lying in the orbit plane and being normal to the orbit plane. Duxbury (Chapt. 15, herein) gives (27 ± 1.0, 21.6 ± 1.4, 18.8 ± 1.4) for Phobos and (15.0^{+6}_{-2}, 12.2 ± 2.0, 11.0 ± 2.0) for Deimos. It appears as though Phobos' shape could be governed by tidal distortion (Soter and Harris, 1976).

Masses of the Satellites

Ground-based determinations of the masses of the satellites have been reviewed by Brouwer and Clemence (1961b), Kovalevsky (1970), Duncombe et al. (1973a; 1973b), and Aksnes (Chapt. 3, herein, see Table 3.2). In addition, Pioneer 10 measurements have improved the values for the masses of Galilean satellites (Anderson et al., 1974a; 1974b). No estimates of mass are available for the smaller satellites, and none are in prospect; our attention is thus confined to those satellites of mass greater than about 10^{22} g.

Ground-based measurements of the masses of the Galilean satellites yield values with formal errors of less than 10%; however, a listing (Duncombe et *al.*, 1973a) of the individual values suggests that the true uncertainty for Io and Callisto may be larger than these formal values, while, in contrast, different mass determinations for Europa and Ganymede are quite consistent. The Pioneer 10 results indicate that the masses of both Io and Callisto are larger (by about one standard error) than had been thought previously, while the ground-based mass determinations for Europa and Ganymede are confirmed. Ganymede, with a mass of 1.49 × 10²⁶ g, is the most massive satellite in the Galilean system.

The masses of several of the satellites of Saturn have been determined, despite their small size, from the orbital commensurabilities within this satellite system (see Chapt. 8, Greenberg). The value of Titan (1.91 times the lunar mass) is known with an uncertainty of only about 1% from its influence on the motion of Hyperion. The masses of members of the pairs Dione-Enceladus and Mimas-Tethys have formal uncertainties of about 5 and 20%, respectively. The masses of Rhea and Iapetus have been estimated, but since neither of these satellites possesses an orbital commensurability with another satellite, these masses are poorly known and should not be taken as definitive.

The only other satellite with a measured mass is Triton, found from its effect on the position of Neptune, similar to the way in which Earth's Moon affects the Earth's motion. The uncertainty in Triton's mass is, however, quite large.

Mean Densities

There are 11 satellites with measured masses and 12 with measured radii, but for only 8 have both quantities been determined: the Moon, the Galilean satellites, Titan, and, with substantial uncertainties, Tethys and Dione. These satellites, except for the Moon, Io and Europa, have densities much lower than those of the terrestrial planets, suggesting a basic difference in composition, consistent with the ice-and-rock models discussed by Lewis (1971a, 1973, 1974a) and others (cf. Chapt. 25 by Consolmagno and Lewis).

One can only guess at the probable densities of the other satellites. If it is assumed that the inner satellites of Saturn all have densities near unity, it is possible to derive a mass for Rhea that is consistent with the very uncertain dynamical value. Neither the mass nor the radius of Triton is known with sufficient accuracy to set any useful constraints on its density. In view of the apparent rocky nature of Phobos and Deimos and of the possible capture of the irregular Jovian satellites from the asteroid belt, it is reasonable to expect that these are silicate objects with densities near 3 g cm⁻³, but such a speculation cannot be verified at this time.

OVERVIEW

Joseph A. Burns
Cornell University

The investigation of satellites is very much a multi-disciplinary subject, a subject with no focus other than the objects themselves, and the objects are quite disparate. As might be expected, the approaches used to study planetary satellites differ from moon to moon and from investigator to investigator: techniques that are of value in exploring Earth's Moon likely will not be useful in investigating the thirteenth satellite of Jupiter!

Planetary astronomers have always taken pride in the fact that it is necessary for them to have a catholic scientific understanding in attacking most problems in their field. As our knowledge of planets expands—particularly through the detailed information being returned from spacecraft—planetary scientists are becoming more specialized. The same is not yet true for researchers of the moons of planets. Our picture of satellites is still given with the broadest of brush strokes. We must preserve this broad perspective in order to understand the gross features of the satellites and to find where they belong in the solar system hierarchy. It is with such a view that the chapters of this book were written.

The goal of virtually all natural satellite research, whether completed through telescopes or by computers, is to understand the origin of the solar system. This book has the same intent. Information on planetary satellites will provide some critical clues to our understanding of the solar system, because the satellites are such diverse bodies, existing in so many different environments, and because most are so much less processed than are their parents, the planets. Undoubtedly, comprehending the origin and subsequent evolution of planetary satellites will give more data on how the solar system originated and, thereby, ultimately on who we are.

The text of this book is divided into five parts. Part I, of which this Overview is part, gives some overall characteristics of satellites; it places the chapters that follow in historical and scientific perspective. Part II, *Orbits and Dynamical Evolution*, describes the orbital and rotational motions of satellites based on their present configuration. Part III, *Physical Properties*, summarizes photometric, polarimetric, spectrophotometric and radiometric measurements of the satellites and what these measurements say about surface composition and processes occurring therein. Part IV, *Objects*, reviews our knowledge of those planetary companions that have been most extensively studied: the Martian satellites, the

Galilean satellites, Saturn's rings and Titan. Part V, *Satellite Origin,* discusses current theories of the origin of satellites within the context of solar system cosmogony. In addition, an Appendix, a complete Bibliography, a Glossary, and an Index are included.

The chapters of Part II are concerned with celestial mechanics and dynamics, which are used to discern the original configurations of the satellite systems from the current orbital motions and rotations; such studies also permit the evaluation of many average physical properties of the planet-satellite systems (e.g., mass or higher gravitational coefficients).

Part II, *Orbits and Dynamical Evolution,* describes the motions of planetary satellites and shows how this motion may have changed since the time of the origin of the solar system. Chapter 3, "Properties of Satellite Orbits: Ephemerides, Dynamical Constants, and Satellite Phenomena," by Kaare Aksnes, geometrically describes a satellite orbit. It uses the orbital elements, and their variation with time, to derive some dynamical constants of the planet-satellite system. It nicely describes how the analysis of satellite phenomena (eclipses and occultations) tells of the surface reflectance of the involved satellites and suggests changes in the hypothesized satellite orbits. J. Kovalevsky and J.-L. Sagnier, in Chapter 4, give a more theoretical treatment of satellite celestial mechanics in "Motions of Natural Satellites." They present a qualitative introduction to satellite perturbation theory, showing how several classes of motion can be distinguished, depending on the type of disturbing force. The French celestial mechanicians then classify the satellites according to those for which solar perturbations are most important, those for which higher order planetary terms dominate, and those for which satellite interactions control. Much of their final discussion of individual satellites closely parallels that of Aksnes in Chapter 3.

A lengthy presentation of the historical development of astrometry in satellite studies is given by Dan Pascu in Chapter 5, "Astrometric Techniques for the Observations of Planetary Satellites." The mean errors of various methods are discussed and the quality of photography possible through the use of masking filters is illustrated. The probability of discovering further satellites is briefly mentioned.

Longer range, evolutionary problems in dynamics are considered in the last three chapters of Part II. Joseph Burns's "Orbital Evolution," Chapter 7, describes in terms of elementary physics many of the forces acting on satellites. The chapter views not only the moons of today, whose orbits are most influenced by tides and gravitational interactions with other satellites and planets, but also small dust particles with radii near a micron, for which radiation pressure and Poynting-Robertson drag can be predominant perturbations. The effects of these forces on long-time orbital evolution is considered, primarily in terms of how changes in orbital energy and angular momentum influence the variations of the orbital parameters. Solid body tides also affect the rotations of satellites as described in Chapter 6, "Rotation Histories of Natural Satellites," by S. J. Peale. This comprehensive report is the first attempt to apply the many recent advances in planetary rotation research to satellite problems. It shows that the rotations of all inner satellites are synchronously locked to their orbital revolutions and that all satellites rotate around their maximum axes of inertia. The tidal evolution of satellite obliquities is evaluated and two final obliquity states are identified. Richard Greenberg, in Chapter 8, gives a clear physical exposition of orbital resonances in "Orbit-Orbit Resonances among Natural Satellites." He characterizes the resonances commonly seen in satellite systems of the outer solar system and describes how tidal processes may produce some of the resonances.

The standard techniques of observational astronomy as applied to planetary satellites are the topic of Part III: *Physical Properties*. From them we attempt to learn of the surface properties and character of satellites with the hope that this knowledge will tell of the satellite's interior composition and will show the effect of external agents on these properties. J. Veverka's Chapter 9, "Photometry of Satellite Surfaces," and Chapter 10, "Polarimetry of Satellite Surfaces," first disclose the basic principles underlying these disciplines before reviewing the results that have been obtained primarily by Earth-based telescopes for individual satellites. Veverka shows how these techniques help to identify the composition and physical properties of the surface layers. Veverka finally points out areas for future research in satellite photometry and polarimetry. These chapters are a valuable and needed updating of the classical paper by Harris (1961). Torrence Johnson and Carl Pilcher in Chapter 11, "Satellite Spectrophotometry and Surface Compositions," present a very complete summary of the existing results, principally for the range of 0.3 to 3 μm, where satellite spectra are dominated by reflected sunlight. They compare these findings versus the spectral reflectivity curves of cosmologically abundant substances in order to identify the materials on the surfaces of the planetary companions. They show that materials such as water ice and other frosts are common in the outer solar system but that most surface compositions cannot yet be unambiguously defined. David Morrison, Chapter 12, succinctly describes the observational results available in the wavelength region of the spectrum longward of about 5μm, where thermal emission dominates reflected light, in "Radiometry of Satellites and of the Rings of Saturn." He points out how radiometric brightnesses, in conjunction with photometric results, can be employed to derive satellite and asteroid sizes (cf., Chapt. 1, Morrison *et al.*) and gives extensive discussions of the puzzling infrared and radio results for Titan and Saturn's rings. Morrison also summarizes those surface properties determined by eclipse cooling curves. Brian O'Leary's Chapter 13, "Stellar Occultations by Planetary Satellites," tells that satellite sizes as well as the presence of a satellite atmosphere can be ascertained by carefully monitoring the manner in which a star's light is extinguished during an occultation. It describes what has been already learned from occultations of Io and Ganymede and tabulates how common stellar occultations should be.

Whereas the chapters of Parts II and III were organized on the basis of scientific disciplines, those in Part IV, *Objects*, try to bring to bear information from diverse sources for a specific subject. As previously stated, it is necessary to comprehend the interrelationship of a variety of phenomena in order to perceive *what* a particular satellite is. The satellites that have been studied in the most detail—occasionally because they are the easiest to work on, but often because they are the most intriguing—are the Martian satellites, the Galilean satellites (especially Io), Titan, and Saturn's rings.

The Martian satellites are the subject of J. B. Pollack's thorough review in Chapter 14, "Phobos and Deimos." Brought together in this paper are the description of the rotation of satellites given by Peale (Chapt. 6) and the results of the striking Mariner 9 photographs shown and discussed in T. Duxbury's Chapter 15, "Phobos and Deimos: Geodesy." Pollack gives special emphasis to the surface composition, using Veverka's discussion of Chapts. 9 and 10 as a basis, and concludes that a regolith of basalt or carbonaceous chondritic material is most likely. He discusses questions of cratering, internal strength, and composition. Pollack uses orbital arguments of Burns (Chapt. 7) to develop a scenario for the origin of the companions to Mars in which the satellites form about their planet at the same time as the planet itself.

The Galilean satellites were the first satellites, excluding the Moon, to be scientifically investigated, and they have remained the most popular planetary satellites for study. In part this is because their relative nearness to Earth and their size make them obvious objects to investigate; but also it is because the increase in information has not led to a concomitant growth in understanding (cf. Morrison and Burns, 1976). David and Nancy Morrison (Chapt. 16) have provided a useful service in compiling "The Photometry of the Galilean Satellites," in which they reduce the photometric results of earlier workers to a common photometric system using the V magnitude. Their presentation, which expands on Veverka (Chapt. 9), derives the dependence of the magnitudes and colors of the Galilean satellites on solar and orbital phase angles. The Morrisons point out the difficulty in interpreting these results in terms of a complex albedo distribution. The most intriguing Galilean satellite, Io, becomes more puzzling as more is learned of it. Its unusual surface characteristics and other properties (Lyman α torus and sodium emission, to mention just two) have provoked numerous explanations. The most comprehensive of these theories is the surface evaporite model of F. Fanale, T. V. Johnson, and D. Matson, who in Chapter 17, "Io's Surface and the Histories of the Galilean Satellites," give their most complete presentation. They rely on the observations given in Part III and cosmochemical constraints in building their theory. The JPL model postulates that Io's exterior is covered by evaporite salts which were carried from Io's fluid interior and left on its surface after evaporation of the transporting water. Interaction of this material with the surrounding environment may account for many of Io's unique features. The hypothesis is tested against what is known about the surfaces and histories of the other Galilean satellites. The surface of the third Galilean satellite is seen in Chapter 18, "Picture of Ganymede," by Tom Gehrels. This remarkable image by Pioneer 10 in red and blue light illustrates surface detail at a resolution of a few hundred kilometers. Saturn's rings, composed of small particles, can be viewed as representative of a different satellite class. Allan Cook and Fred Franklin (Chapt. 19) employ recent radiometry, photometry and spectrophotometry (cf. Morrison, Chapt. 12) to tighten the constraints placed on particle size and composition during their earlier review (Cook et al., 1973). They suggest that the particles probably are about 7 cm in size and composed of water ice, possibly mixed with a clathrated hydrate of methane and ammonia hydrate. However, large particles with nodules about 7 cm cannot be excluded.

Titan, the first satellite to have its atmosphere recognized (Kuiper, 1944), has continued to be an object of considerable interest. Its unusual thermal emission spectrum (cf. Morrison, Chapt. 12) has prompted several explanations, which are being more narrowly confined by better photometry and polarimetry (cf. Veverka, Chapts. 9 & 10). The two most popular models of Titan are propounded in this book by Donald Hunten ("Titan's Atmosphere and Surface," Chapt. 20) and by John Caldwell ("Thermal Radiation from Titan's Atmosphere," Chapt. 21). Basing his discussion on a workshop he edited (Hunten, 1974), Hunten first sketches our current view of Titan: an ammonia/water interior overlain by a cloudy or hazy atmosphere of methane, and perhaps molecular hydrogen. He then attempts to explain Titan's unusual thermal emission spectrum as produced by a greenhouse effect, plus a warm stratosphere of CH_3 and C_2H_6, and suggests that Titan's surface temperature is near 125° K. Caldwell simply extends the thermal inversion model of Danielson et al. (1973), in which Titan's low ultraviolet albedo is used to imply high altitude heating by "dust," to include the thermal emission from acetylene. This model has a much less dense atmosphere than the greenhouse model and a low surface temperature (~73°K). Leif Andersson, in "Variability of Titan: 1896–1974" (Chapt. 22), summarizes four clusters of visual and photoelectric photometric observations extending over seventy-five years. Because of its atmosphere, Titan's brightness does not vary on a short time scale; Andersson points out, however, that it has changed by nearly 0.2 mag over longer periods and that Titan seems to be brightening.

To answer correctly the question of how satellites were born requires even more breadth than most problems in satellite studies. The cosmogonist should understand the celestial mechanics results of Part II of this book so that he knows where the satellites were when they originated, and so that his origin scenarios are dynamically permitted. The measurements of physical properties outlined in Part III must be correctly interpreted to determine what properties are intrinsic to the satellite and what are caused by the milieu in which it resides. The difficulty of making such interpretations for several specific satellites is illustrated in Part IV. Cosmogonic models must satisfy all these constraints as well as be chemically sound and reasonable in terms of stellar evolution theory. Some aspects of these issues are dealt with in Part V, *Satellite Origin*.

A. G. W. Cameron's "Formation of the Outer Planets and Satellites" (Chapt. 23) places the origin of satellites in the context of commonly accepted solar system cosmogony as a natural outgrowth of the development of the planets in the solar nebula. He is principally concerned about the birth of the satellites of the outer planets for which the processes of origin are likely to be similar to that of the solar system as a whole. Chapter 24, "The Critical Velocity Phenomenon and the Origin of the Regular Satellites," by B. De, H. Alfvén and G. Arrhenius, too regards the satellite systems of Jupiter, Saturn and Uranus as crucial to any origin theory. These authors believe that the same processes are important for the origins of both the planets and the satellites, and they feel that the emphasis should be placed on theories of satellite origin since several extant satellite systems are known, whereas there is only one known planetary system. This group of cosmogonists sees remarkable similarities in planetary, satellite and atomic systems which they contend can be explained as the result of a plasma instability operating in the early solar system. Guy Consolmagno and John Lewis's "Preliminary Thermal History Models of Icy Satellites" (Chapt. 25) limns the thermal consequences of some of the chemically plausible satellite origins which were developed, principally by Lewis, in the early 1970s. The internal melting and differentiation of large icy satellites ($R \gtrsim 500$ km) as a function of size is considered for several compositional models. "The Accumulation of Satellites" (Chapt. 26) by V. S. Safronov and E. L. Ruskol lays out in considerable mathematical detail the processes which they consider to be significant in the growth of satellites. The satellites are taken to accumulate from a circumplanetary swarm which is fed by captured heliocentric particles. Expressions are derived by the Soviet cosmogonists for the characteristic time scales of the important steps in this accretion process. John Wood in "Origin of Earth's Moon" (Chapt. 27) clearly summarizes the different hypothesis of lunar origin: fission of Earth, capture from heliocentric orbit, and binary accretion of the Earth and Moon together. He outlines the difficulties facing each class of theory and argues for binary accretion (cf. Safronov and Ruskol, Chapt. 26) without disregarding the chemical problems of such a scenario.

The Reference Section contains a Glossary of the symbols most frequently used in the book. The Bibliography compiles about one thousand items, including all those cited in the chapters of the book, plus a number of other pertinent works. This Bibliography will be useful in its own right to researchers. The works generally are those for which a complete reference was available by about December 1975. Single author entries are listed alphabetically by author and then in order of publication; multiple author entries are alphabetical, regardless of publication date. An Index of subjects and authors will increase the book's value as a text.

This, then, is what *Planetary Satellites* is about.

PART II
ORBITS AND DYNAMICAL EVOLUTION

PROPERTIES OF SATELLITE ORBITS: EPHEMERIDES, DYNAMICAL CONSTANTS, AND SATELLITE PHENOMENA

Kaare Aksnes
Harvard College Observatory and
Smithsonian Astrophysical Observatory

A short account is given of the history of the observation of the Galilean satellites, with an emphasis on early orbital work and practical applications thereof to world mapping and navigation. The general character of satellite motion is described from a geometric and physical point of view without reference to mathematics. This is followed by a similar discussion of ordinary and mutual satellite phenomena. The accuracies of published ephemerides are discussed on the basis of the observations and theories available for the various satellites. The ephemeris tables need to be revised and, perhaps, partly replaced by well-documented computer programs. The determination of physical parameters (planetary masses and oblatenesses; masses, sizes, and albedo maps of satellites) from positional observations of satellites and photometric observations of ordinary and mutual satellite phenomena is discussed.

HISTORICAL NOTES

Galileo Galilei's discovery of the four bright moons of Jupiter in 1610 and his realization that they were orbiting a planet other than the Earth had profound and far-reaching cosmological and philosophical consequences. Even before that time, Galileo had been a firm believer in Copernicus' beautifully simple but unproved heliocentric model of the solar system. Now he had proof that the Earth did not enjoy a preferred central position about which everything else revolved.

Galileo quickly realized that his satellites, as they move regularly into and out of eclipses in Jupiter's shadow, provided a clock in the sky that could be used to solve a vexing problem of his time—that of determining the longitude on land and at sea. (At the time, available clocks were not capable of keeping reliable time on long voyages, but they did suffice for determining local time through frequent observations of the Sun.) If the times of the eclipses could be predicted, the longitude of an eclipse-observer would be given simply as the difference between the observed local time of the eclipse and the local time predicted for zero longitude.

To this end, Galileo made regular timings of the eclipses, from which he deduced the periods of the satellites. It was, however, left to one of his countrymen, Giovanni Domenico Cassini, to devise a practical scheme for mapping the world by means of the Galilean satellites. Cassini had been invited to work on this problem at the French Royal Academy of Sciences as the result of the publication of his tables of the satellite eclipses for the year 1668. Under his direction, a worldwide program of eclipse observations was started, from which the first accurate map of the world was produced. However, the method never proved practicable for the navigation of ships. The solution to that problem had to await the invention of the marine chronometer about 100 years later.

For two centuries, these practical applications inspired astronomers such as Römer, Bradley, Lagrange, Laplace, and Souillart to work on the problem of the motion of the Galilean satellites.This led to Römer's famous determination of the speed of light and Laplace's discovery of the so-called libration condition, according to which the mean longitudes of the three inner satellites obey the simple relationship $\lambda_1 - 3\lambda_2 + 2\lambda_3 = \pi$ (see Chapt. 8, Greenberg). Souillart's theory of 1880 remains the most complete, though not the most accurate, analytical theory of the motions of the satellites to date.

The other 29 natural satellites also have interesting histories, but not as colorful and diverse as those for the Galilean satellites. The reader is referred to two popular books, *The Moons of Jupiter* by M. K. Wetterer (1971) and *Satellites of the Solar System* by W. Sandner (1965). Most recently discovered is a 20th magnitude object—the thirteenth satellite of Jupiter with an orbit similar to the orbits of J6, J7, and J10 (Aksnes and Marsden, 1974; Kowal *et al.*, 1975)— found by Charles Kowal on plates taken by him with the 48-inch Schmidt reflector at Palomar Mountain in the autumn of 1974.

The motivation for much of the early work on the Galilean satellites has an interesting parallel today. These and the other satellites of the outer planets may yet come to play a role in the navigation of ships—not at sea, but in space! The satellites are of interest not only as targets for observation on space missions, but also, through a technique called approach guidance, as a means of spacecraft navigation, during the critical planetary flyby phases on the Mariner-Jupiter-Saturn (MJS) missions. The narrow-angle television cameras on the MJS spacecraft have a field of view of one degree or less. Therefore, the requirements of camera pointing and navigation place stringent demands on the accuracies of the ephemerides of the satellites. The feasibility of approach guidance and the potential power of spacecraft-centered observations of satellites to improve their orbits have been demonstrated on the Mariner 9 mission to Mars and its two companions, Phobos and Deimos (Born and Duxbury, 1975). Furthermore, from the gravitational perturbations felt by Pioneer 10, as revealed by the highly accurate radio-tracking data on the spacecraft during its Jupiter flyby of December 1973, the masses of the Galilean satellites have been determined to an unprecedented accuracy (Null *et al.*, 1974). Improved satellite ephemerides are thus both a prerequisite for, and a by-product of, space missions to the planets. This has resulted in a revival of interest in the orbits of the natural satellites (see Appendix by Seidelmann) which also are worthy of renewed investigations for reasons entirely of their own.

GENERAL CHARACTER OF SATELLITE MOTION

We shall be concerned here only with the geometric and physical aspects of satellite motion. A mathematical discussion of the satellite theories can be found in Chapter 4, by Kovalevsky and Sagnier (see also Brouwer and Clemence, 1961a).

The purely elliptic motion that a satellite would have in the absence of disturbing forces will be perturbed by (1) the oblateness of the primary, (2) the attraction of the Sun, (3) the attractions of any other satellites revolving about the same planet, and (4) the attractions of other planets. The relative importance of the resulting perturbations will depend on the closeness of the satellite to the primary, on the distance to the Sun, on the masses and periods of neighboring satellites, and finally on the distances and masses of nearby planets (see Chapt. 4, by Kovalevsky and Sagnier, who classify the satellites in terms of which perturbations are important).

When acting alone, the oblateness of the primary will cause the orbit of the satellite to rotate in the orbital plane and to precess about the pole of the primary's equator, the rates of rotation $\dot{\omega}$ and of precession $\dot{\Omega}$ being proportional to $J_2(R/a)^2$ (see Fig. 3.1 and eqn. 4). Here J_2 is the dynamical oblateness of the primary of equatorial radius R, and a is the semimajor axis of the satellite's orbit. Thus, the closer to the primary the satellite is, the faster the precession; for Jupiter V it amounts to 2.5 rev/year. The Sun will have a precisely similar effect, but the precession will now take place about the pole of the primary's orbital plane at a rate proportional to $(n'/n)^2$, where n and n' are the mean motions of the satellite and the primary, respectively. Although one effect usually dominates, both the primary's oblateness and the Sun will, of course, be active at the same time. The result is a combined precession about the pole of the so-called Laplacian plane, which lies between the plane of the equator and that of the orbit of the primary, all three planes having a common node.

The periodic perturbations due to the primaries' oblatenesses are too small to be detectable from the Earth for any of the satellites, while the Sun gives rise to sizable periodic perturbations for Jupiter's outermost satellites, for which the ratio $(n'/n)^2$ may be as large as 3×10^{-2}.

Except for the massive Galilean satellites, the short-period terms arising from satellite interactions are too small to be observable. However, large long-period perturbations are present in the motions of both the Galilean satellites and Saturn's satellites. This is due to commensurabilities in their mean motions, which can cause the coefficients of periodic terms to be greatly magnified through the phenomenon of resonance (Chapt. 8, Greenberg). An example is furnished by the interaction in the Saturnian system between Mimas and Tethys, which are in a 2:1 resonance, causing an impressive 44° libration with a period of 70 years in the longitude of Mimas. Since the three innermost Galilean satellites

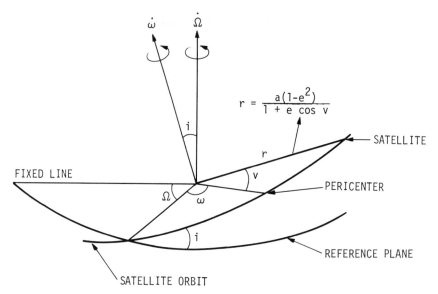

Fig. 3.1. The semimajor axis a, *eccentricity* e, *and true anomaly* v *define the shape of, and the satellite's position in, the orbit whose orientation on the reference plane (equatorial or orbital plane of the primary, ecliptic, etc.) is given by the inclination* i, *the longitude of the ascending node* Ω, *and the argument of the pericenter* ω.

partake in a triple resonance, their motions are by far the most complex of any of the bodies in the solar system. Not only are the mean motions of Io, Europa, and Ganymede nearly in the ratio 4:2:1, but their mean longitudes satisfy the aforementioned Laplace libration condition, $\lambda_1 - 3\lambda_2 + 2\lambda_3 = \pi$, to within the accuracy of the observations. It is a somewhat curious fact that although Callisto is second among the Galilean satellites in terms of mass, its motion is not commensurate with that of the other three satellites and therefore does not enter into this equation.

The most characteristic feature of the motions of the Galilean satellites is the regularity with which they pass in front of (transit) or behind (occult) Jupiter's disk and into or out of Jupiter's shadow (eclipse). In addition, the passage of a satellite's shadow across Jupiter's disk is called a shadow transit. These phenomena are referred to collectively as *ordinary satellite phenomena*. An interesting consequence of the libration condition is that all three inner satellites can never exhibit similar phenomena at the same time; for instance, if Ganymede and Europa are seen in transit across Jupiter's disk, Io will be occulted by the planet. The times of disappearance and reappearance for the eclipses and occultations, and the times of ingress and egress for the transits and shadow transits, are predicted in the *American Ephemeris and Nautical Almanac* (A.E.) to the nearest minute.

Saturn's satellites also exhibit ordinary satellite phenomena during a limited period about every fifteenth year, when the planet passes through one of the nodes of its ring plane. Predictions, based on a method due to Comrie (1931), are published in the *BAA* (British Astronomical Association) *Handbook* (cf. Peters, 1975).

The Galilean satellites have sizes comparable to that of the Moon, and they all move nearly in the equatorial plane of Jupiter. They thus occult and eclipse each other for a period of several months about every 6 years (half of Jupiter's orbital period), when the Earth and the Sun are nearly coplanar with the orbits of the satellites. These *mutual satellite phenomena* can last from a fraction of a minute to several hours, depending on how central the alignment of one satellite with another or its shadow is and on whether the apparent motions of the satellites are in the same or in opposite directions as viewed from the Earth. A mutual eclipse is classified as umbral or penumbral according to whether the satellite enters the umbra or only the penumbra of the other satellite's shadow. The terminology *partial, annular,* or *total* is also used to describe the manner in which the umbra or the occulting disk covers a satellite at midevent.

Since 1931, the mutual phenomena of Jupiter's satellites have been predicted regularly, except for the apparition of 1943–44, in the *BAA Handbook* based on a method due to Levin (1931). The predictions give the beginning and ending times of the occultations and, for the eclipses, the times of entry into and exit from the umbra and the penumbra. The *magnitude* of an umbral eclipse, that is, the fraction of the satellite's diameter covered by the umbra, is also given. For the 1973–74 series of mutual satellite phenomena, more detailed predictions, based on electronic computer searches, were computed by Brinkmann (1973) and by Aksnes (1974a). The latter set of predictions also included estimates of the light decreases accompanying the phenomena. Predictions for Saturn's satellites are given by Peters (1975).

APPLICATIONS: DETERMINATION OF DYNAMICAL CONSTANTS

Constants From Analyses of Orbits

The following gives the foundation for the evaluation of the dynamical constants; see also Chapter 4, by Kovalevsky and Sagnier, and Duncombe *et al.* (1973a) for further discussions.

In the absence of disturbing forces, Kepler's third law for the sum of the satellite's mass m_s and the primary's mass m_p yields

$$G(m_s + m_p) = a^3 n^2 \quad , \tag{1}$$

where a is the semimajor axis of the satellite, n is its sidereal mean motion, and G is the universal constant of gravitation. Because of perturbations, a and n will vary. Their mean values, a_0 and n_0, can be determined from satellite observations

TABLE 3.1
Inverse Planetary Masses (Sun = 1)

Planet	1/m	Author	Method
Mars	3094000 ± 3000	Wilkins (1966)	Deimos
	3097000 ± 3000	Sinclair (1972)	Phobos + Deimos
	3098714 ± 5	Null (1969)	Mariner 4
Jupiter	1047.400 ± 0.045	de Sitter (1915)	Jupiter I-IV + Planets
	1047.335 ± 0.077	Herget (1968b)	Jupiter VIII
	1047.342 ± 0.020	Null et al. (1974)	Pioneer 10
Saturn	3494.8 ± 1.1	Jeffreys (1954)	Saturn I-VI, VIII
	3501.47 ± 1.75	Garcia (1972)	Saturn III-VI
Uranus	22934 ± 9	Harris (1949)	Uranus I-V
	22945 ± 15	Dunham (1971)	Uranus I-V
Neptune	19331 ± 31	Eichelberger and Newton (1926)	Triton
	19296 ± 13	Gill and Gault (1968)	Triton
	19438 ± 116	Rose (1974)	Nereid

after removal of the periodic perturbations obtained from a theory of the motion. Furthermore, as explained earlier, the satellite's motion will also be affected by secular terms arising from various perturbations, so that it is necessary to modify eqn. (1) to:

$$G(m_s + m_p) = a_0^3 n_0^2 \left(1 - \frac{\sigma}{2}\right) + O(\sigma^2) , \qquad (2)$$

where (see the preceding section and also Chapter 4, by Kovalevsky and Sagnier)

$$\sigma = 3J_2 \frac{R^2}{a^2} - \frac{15}{4} J_4 \frac{R^4}{a^4} - \left(\frac{n'}{n}\right)^2 - \sum_{s'} \frac{m_{s'}}{m_p} a^2 \frac{\partial A_0}{\partial a}. \qquad (3)$$

Here, A_0 is the constant term of $(a^2 + a_s^2 - 2aa_{s'} \cos \psi)^{-1/2}$ developed as a Fourier series in ψ, the difference between the orbital longitudes of the perturbed s and the perturbing s' satellite. This formula was first used by Laplace in the theory of the Galilean satellites. For a derivation, see Struve (1888) or Tisserand (1896).

Before the preceding formula can be used to determine the mass of the primary, it is necessary to know the zonal harmonic coefficients J_2 and J_4 of the primary and the masses $m_{s'}$ of the perturbing satellites. The first two parameters can be calculated from the observed secular motions of the satellite's pericenter and node by means of

$$\frac{d\omega}{dt} = n \left(3J_2 \frac{R^2}{a^2} - \frac{9}{2} J_2^2 \frac{R^4}{a^4} - \frac{15}{2} J_4 \frac{R^4}{a^4} \right) ,$$

$$\frac{d\Omega}{dt} = -n \left(\frac{3}{2} J_2 \frac{R^2}{a^2} - \frac{27}{8} J_2^2 \frac{R^4}{a^4} - \frac{15}{4} J_4 \frac{R^4}{a^4} \right) , \qquad (4)$$

in which the squares of the eccentricity e and the inclination i have been ne-glected (Brouwer, 1946, 1959). This is a sufficient approximation, except for Triton, since all the other satellites that move close enough to their primaries to give reliable estimates for the rates $d\omega/dt$ and $d\Omega/dt$ have nearly circular and nearly equatorial orbits. In fact, for some satellites, e and i are so small (see Table 1.2) that it is difficult to observe these rates because of the ill-determined positions of the pericenters and the nodes. For this reason, J_2 and J_4 of Jupiter are rather weakly determined from its satellites, while Uranus' satellites (see Greenberg, 1975) give an inconclusive result about their primary's dynamical oblateness (see Table 3.3).

The J_4 terms in eqn. (4) are very small and rather strongly correlated with the J_2^2 terms. Only through a combined analysis of the motions of two or more satellites of the same planet is it possible to solve for both J_2 and J_4. From these coefficients, assuming hydrostatic equilibrium, the planet's dynamical flattening ϵ, that is, $(R - \text{polar radius})/R$, can be obtained from the relation

$$\epsilon = \left[\frac{3}{2} J_2 + \frac{1}{2} \nu^2 \frac{R^3(1-\epsilon)}{Gm_p} \right] \left(1 + \frac{3}{2} J_2 \right) + \frac{5}{8} J_4 , \qquad (5)$$

where ν is the rate of rotation of the planet (de Sitter, 1938).

Until quite recently, the mass of a satellite could be estimated only if it produces observable periodic or secular perturbations in the motions of nearby satellites (see Chapt. 8, by Greenberg, and Chapt. 4, by Kovalevsky and Sagnier). The Moon and Triton are exceptions, since their masses were obtained from the oscillations they induce in the motions of the primaries (see Chapt. 1, Morrison et al.). Only the Galilean satellites and most of the Saturnian satellites are large enough and close enough to each other to reveal their masses in this manner. The existence of resonant pairs of satellites, which result in large mutual perturbations of very long periods, has made it possible to determine very accurate masses for most of Saturn's satellites. The inner three Galilean satellites suffer even larger mutual perturbations owing to the remarkable triple resonance mentioned before. However, as a consequence, their motions are so complex that it is a formidable task to derive expressions for the multitude of mutual perturbations from which the masses must be disentangled through comparison with observations.

TABLE 3.2
Satellite Masses (Primary = 1)

Satellite	m	Author	Method
Io (JI)	4.50×10^{-5}	Sampson (1921)	Mutual perturbations
	$(3.81 \pm 0.44) \times 10^{-5}$	de Sitter (1931)	Nodal motion of III
	$(4.696 \pm 0.06) \times 10^{-5}$	Null et al. (1974)	Pioneer 10
Europa (JII)	2.54×10^{-5}	Sampson (1921)	Mutual perturbations
	$(2.48 \pm 0.07) \times 10^{-5}$	de Sitter (1931)	Nodal motion of III
	$(2.565 \pm 0.06) \times 10^{-5}$	Null et al. (1974)	Pioneer 10
Ganymede (JIII)	7.99×10^{-5}	Sampson (1921)	Mutual perturbations
	$(8.17 \pm 0.15) \times 10^{-5}$	de Sitter (1931)	Nodal motion of III
	$(7.845 \pm 0.08) \times 10^{-5}$	Null et al. (1974)	Pioneer 10
Callisto (JIV)	4.50×10^{-5}	Sampson (1921)	Mutual perturbations
	$(5.09 \pm 0.59) \times 10^{-5}$	de Sitter (1931)	Nodal motion of III
	$(5.603 \pm 0.17) \times 10^{-5}$	Null et al. (1974)	Pioneer 10
Mimas (SI)	$(6.69 \pm 0.20) \times 10^{-8}$	Jeffreys (1953)	Mutual perturbations
	$(6.59 \pm 0.15) \times 10^{-8}$	Kozai (1957)	Mutual perturbations
Enceladus (SII)	$(1.27 \pm 0.53) \times 10^{-7}$	Jeffreys (1953)	Mutual perturbations
	$(1.48 \pm 0.61) \times 10^{-7}$	Kozai (1957)	Mutual perturbations
Tethys (SIII)	$(1.141 \pm 0.030) \times 10^{-6}$	Jeffreys (1953)	Mutual perturbations
	$(1.095 \pm 0.022) \times 10^{-6}$	Kozai (1957)	Mutual perturbations
Dione (SIV)	$(1.825 \pm 0.061) \times 10^{-6}$	Jeffreys (1953)	Mutual perturbations
	$(2.039 \pm 0.053) \times 10^{-6}$	Kozai (1957)	Mutual perturbations
Rhea (SV)	$(3.2 \pm 3.8) \times 10^{-6}$	Jeffreys (1954)	Mutual perturbations
Titan (SVI)	$(2.411 \pm 0.018) \times 10^{-4}$	Jeffreys (1954)	Mutual perturbations
Iapetus (SVIII)	$(2.5 \pm 1.9) \times 10^{-6}$	Struve (1933)	Mutual perturbations
	$(3.94 \pm 1.93) \times 10^{-6}$	Kozai (1957)	Mutual perturbations
Triton (NI)	$(1.34 \pm 0.23) \times 10^{-3}$	Alden (1943)	Neptune

Before the advent of planetary probes and orbiters, the determination of the dynamical constants just discussed rested almost entirely on analyses of satellite orbits. The same techniques can now be applied to spacecraft, whose orbital perturbations in the gravitational fields of nearby planets or satellites can be evaluated extremely accurately by means of radio tracking from the Earth (Null et al., 1974). Planetary probes and orbiters also offer another advantage: Their orbits can be chosen so as to maximize the perturbations from which the relevant dynamical constants can be deduced. For example, Uphoff et al. (1974) have found that by utilizing the attractions of Jupiter's satellites, it is possible to continually change the orbit of a Jupiter orbiter in such a way that multiple near encounters with several satellites are achieved with a small expenditure of rocket fuel. This has been coined a game of ''Jovian billiards'' by G. Colombo, who

TABLE 3.3
Dynamical Coefficients of Planets

Planet	$J_2 \times 10^5$	$J_4 \times 10^5$	$\epsilon \times 10^5$	R (km)	Author
Mars	194.7 ± 0.1	——	521.0	3392	Woolard (1944)
	195.0 ± 0.2	——	521.5	3409	Wilkins (1966)
	196.6 ± 0.3	——	523.8	3393	Sinclair (1972)
	196.0 ± 1.8	−3.2 ± 0.7	521.0	3393	Born (1974)
Jupiter	1471 ± 22	−67 ± 56	6518	71432	de Sitter (1931)
	1472 ± 4	−65 ± 15	6521	71398	Null *et al.* (1974)
Saturn	1667 ± 3	−103 ± 7	9792	59670	Jeffreys (1954)
Neptune	490 ± 50	——	1710	22300	Eichelberger and Newton (1926)
	500 ± 50	——	1730	22300	Gill and Gault (1968)

suggested the ingenious multiple swing-by of Mariner 10 to Mercury. Thanks to spacecraft and artificial satellites, we have today very accurate information about the gravity fields of the Moon, the terrestrial planets, and Jupiter, as well as about the masses of the Galilean satellites.

In Tables 3.1, 3.2, and 3.3 are summarized the most important and most recent determinations of the above-discussed constants and their standard errors from analyses of the motions of satellites and spacecraft.

Constants From Ordinary and Mutual Phenomena

Visual timings and photometric observations of ordinary eclipses of the Galilean satellites have been an important source of positional data on the satellites. Thus, Sampson (1910, 1921) based his famous "Theory of the four great satellites of Jupiter" entirely on photometric eclipse observations made at Harvard and Durham Observatories (Pickering, 1907; Sampson, 1909). The lightcurves of the eclipse disappearances or reappearances, which are not instantaneous, yield better defined times than those obtainable from subjective visual observations (cf. Chapt. 5, Pascu). Sampson's analysis of the eclipse observations resulted not only in highly precise ephemerides for the Galilean satellites but also in a significant improvement of the then-accepted values for several associated constants, *viz.*, the satellite masses (Table 3.2), the equatorial radius of Jupiter (Table 3.3), and the orientation of the planet's equatorial plane, the light time for 1 AU, and the solar parallax. For these last two constants, he found 498.72 sec (499.012 sec) and 8.797 arcsec (8.794 arcsec), where the numbers in parentheses give the values now recommended by the IAU.

From the durations of visually observed eclipses of Saturn II–VI during 1905–08, Struve (1915) determined the values 60540 ± 80 km and 0.0980 ± 0.0030 (standard errors) for Saturn's equatorial radius and geometric flattening.

The occultations, transits, and shadow transits of the Galilean satellites are not nearly as useful as the eclipses, because of the difficulty of determining the times of contact with Jupiter's bright limbs. Such timings also are much more sensitive to disturbances in the atmospheres of Jupiter and the Earth.

It is a rather surprising fact that in spite of the availability of predictions, the mutual phenomena of Jupiter's satellites were not seriously exploited until 1973. Before then, only visual observations were made, mainly for reasons of curiosity. Accounts of such observations have been given by Peek (1958) and Fauth (1940).

Largely on the initiative of Brinkmann and Millis (1973), a worldwide campaign was organized to make photoelectric observations of the Jovian mutual satellite phenomena occurring in 1973–74. As already mentioned, several independent series of predictions had been published for the phenomena, which spanned an interval of more than a year from February 1973 to May 1974. Since Jupiter reached opposition in the middle of this interval, conditions were unusually favorable for observing the phenomena.

Lightcurves of nearly one hundred mutual occultations and eclipses have so far been communicated from many observatories around the world to the data bank established at the Lowell Observatory. Currently, only a fraction of these lightcurves have been analyzed. Here, we shall remark only briefly on the methods and the attainable results of such analyses and refer the reader to more detailed work.

During an occultation, the combined brightness of the two satellites, normalized to unity outside the event, will be given approximately by

$$I_{occ.} = 1 - \frac{\beta A}{1 + \gamma} \quad , \tag{6}$$

where γ is the ratio of the brightnesses of the occulting and occulted satellites just before or after the event, and A is the occulted area of average surface brightness β. The units are such that for a total occultation $\beta = A = 1$. If β varies considerably across the disk of the occulted satellite, it will be necessary to introduce a "surface-brightness map" by putting

$$\beta A = \sum_i \beta_i A_i \quad . \tag{7}$$

For an eclipse, the equation corresponding to eqn. (6) can be written

$$I_{ec.} = 1 - A_u + \int_0^{A_p} (1 - I_x) dA_p \quad , \tag{8}$$

where A_u and A_p denote the fractions of the satellite's disk covered by the umbra and the penumbra, respectively. At a given point in the penumbra, the light intensity I_x will be proportional to the unobstructed portion of the Sun's disk as seen from that point, corrected for its limb darkening. Since I_x must itself be obtained through an integration, a double integration is, in reality, implied in eqn. (8). The situation is further complicated by the probable need of a correction for light variations across the disk of the satellite due to albedo features or limb darkening (cf. Morrison and Morrison, Chapt. 16).

The areas A, A_u, and A_p in eqns. (6) and (8) can be calculated (Aksnes, 1974a) by means of heliocentric ephemerides of the Earth and of Jupiter and by Sampson's (1921) theory for the motions of the Galilean satellites. It is reasonable to assume that the areas so calculated will be in error mainly because of errors in the longitudes and in the adopted radii of the two satellites involved in a mutual event. Through a least-squares correction of these parameters, it should therefore be possible to bring the computed lightcurves into close agreement with the observed ones.

This has indeed already been demonstrated for several observed lightcurves of occultations of Europa by Io. Independent analyses by Duxbury et al. (1975) and by Aksnes and Franklin (1975a, b) have yielded consistent corrections to the longitudes of these two satellites, with probable errors of only a few seconds of time. (One second corresponds to path lengths of 17, 14, 11, and 8 km for Jupiter I–IV, respectively.) These accuracies are considerably better than those obtainable from observations of the ordinary eclipses, which are affected by uncertainties introduced by Jupiter's atmosphere. The values determined for Europa's radius (Io's radius is already quite accurately known from a stellar occultation; see Chapt. 13, O'Leary) are more discordant, ranging from 1480 to 1550 km, with a mean value of 1521 ± 27 km. One difficulty lies in the uncertainty in the observed brightness ratio γ, which through eqn. (6) is strongly correlated with Europa's radius.

The above-mentioned investigators, and Greene et al. (1975), have also found strong evidence for a bright polar cap around Europa's north pole, in agreement with an earlier, preliminary analysis by Murphy and Aksnes (1973). [Editor's note: Subsequent analysis (Aksnes and Franklin, 1975b), employing de Sitter's orbital theory, suggests that a latitude correction may eliminate the need for a polar cap.]

CALCULATION AND PUBLICATION
OF SATELLITE EPHEMERIDES

At this point, it is necessary to define precisely what we mean by the word *ephemeris*; it is a series of positions, or orbital elements from which positions can readily be calculated, tabulated at regular intervals of time. Sometimes first- or second-order differences of the tabulated quantities are also given for ease of interpolation. The purpose of such an ephemeris is to relieve the user of the often very lengthy calculations that application of a satellite theory requires. A satellite theory, in the sense used here, is a product of observations and a mathematical model (analytic or numerical) representing the motion of the satellite. The accuracies of existing satellite theories, and of the ephemerides computed from them, are probably limited more by the errors in the observations than by the inaccuracies in the orbital models.

Ephemerides of the satellites are published yearly in the A. E. and in *Connaissance des Temps*. Also, the *International Information Bureau on Astronomical Ephemerides* and the *Central Bureau for Astronomical Telegrams*, both operated by the IAU, will occasionally issue circulars containing ephemerides of satellites for limited intervals of time.

The A. E. contains approximate ephemerides for all the known satellites, except for Jupiter I–IV and VIII–XII, Saturn X (Janus), Uranus V (Miranda), and Neptune II (Nereid). The ephemerides are intended only for search and identification, not for the exact comparison of theory with observation. Tables are given for the times of geocentric eastern or western elongation, or for superior or inferior conjunction, and for the slowly changing values of the semimajor axis and the position angle of the satellite's apparent orbit on the celestial sphere. The satellite's position in the apparent orbit, approximated by an ellipse, can be read from another table with the time from the nearest elongation as argument. By means of these tables, the apparent distance and position angle of the satellite relative to the primary can readily be deduced. For Saturn I–VIII, elements are also given at 5-day intervals for the calculation of the radius vectors and longitudes in the true orbits. For the slow-moving satellites, Jupiter VI–VII and Saturn IX (Phoebe), only differential right ascensions and declinations relative to the primaries are tabulated, at intervals of 4 and 2 days, respectively. Herget (1968a, 1968b) has published similar tables for Jupiter VIII–XII for every 10th day between 1966 and 2000.

Since 1915, the Bureau des Longitudes has had the responsibility of publishing detailed ephemerides for the Galilean satellites in *Connaissance des Temps*. After 1915, the ephemerides have been computed, not from Sampson's Tables, but from analytic expressions (Andoyer, 1915).

We shall now quickly review the satellite theories currently used to calculate ephemerides and attempt to assess their accuracies. All of them were constructed several decades ago, mainly from visual micrometric observations of some antiquity (cf. Chapt. 5, Pascu). Although the satellites have received regrettably little attention during the last 40 years, by observers as well as by theoreticians, partial revisions have been made of most of the theories. These newer results, to which we shall refer, should be incorporated into the calculation of future ephemerides of the satellites (cf. Appendix by Seidelmann).

Mars' Satellites

The ephemerides of Phobos and Deimos are computed from mean orbital elements determined by Struve (1911) from micrometer measures of the satellites made between 1877 and 1909. Owing to their faintness and closeness to Mars, they can be observed only during a relatively short period of time around each opposition. The only perturbations included are the secular motions of the pericenters and of the nodes due to the action of Mars' oblateness and the Sun.

Revised elements have been published (but are not incorporated in the A. E.) by Burton (1929), Woolard (1944), Wilkins (1965, 1966), and Sinclair (1972a). Sinclair constructed a new theory that includes periodic perturbations, which were, however, found to have an insignificant effect on the mean residual and on the values of the constants. A total of 3,107 observations was used, covering a time span from 1877 to 1969. By omitting about 9% of the observations, a mean residual of 0.426 arcsec resulted. Born and Duxbury (1975) have determined extremely accurate ephemerides for Mars' satellites for the period November 1971 to October 1972 from 80 television photographs taken by the Mariner 9 spacecraft. Finally, Shor (1975) has made a new, and even more extensive, analysis of the motions of these satellites from 1877 to 1973.

Jupiter's Satellites

The calculation of the ephemerides and the phenomena of the Galilean satellites is based on Sampson's Tables (1910), which include all terms having coefficients of 1 arcsec or more. Since the data published are not intended for the comparison of observation with theory, such accuracy is not needed. A simplified method devised by Andoyer (1915) is therefore used. This simplification, which retains terms of amplitude 0.001 or more, introduces errors of at most a fraction of a minute in the calculated times of the phenomena. However, not surprisingly, now more than 50 years later, the predicted times are in error by up to several minutes. While most of this error can be removed simply by applying constant corrections to the computed longitudes, an urgent need now exists for a revised theory for the Galilean satellites. Marsden (1966) has extended the earlier work by the inclusion of neglected short-period terms. Lieske (1974, 1975) is revising Sampson's theory, and Ferraz-Mello (1966, 1975) and Sagnier (1975) are working on two entirely new theories; see also Vu and Sagnier (1974). Another interesting alternative would be to develop further de Sitter's (1931) elegant theory, which he unfortunately did not carry quite far enough for practical applications.

The ephemeris of Jupiter V is based on a theory due to van Woerkom (1950), in which only the secular perturbations due to Jupiter's oblateness are included. Sudbury (1969) revised this theory from old visual observations dating back as far as 1892 and new photographic observations made in 1954 and 1967. Because of the great difficulty in observing this satellite, its orbital longitude can be computed only to $0^\circ.1$ or $0^\circ.2$.

The differential right ascensions and declinations of Jupiter VI and VII given in the A. E. are calculated from the tables constructed by Bobone (1937a,b), which are based on an analytic theory that includes the main perturbations due to the Sun. The orbital elements were obtained from a 40-year arc of observations that gave geocentric residuals of 10 to 20 arcsec. For current times, though, the tables are likely to be in error by much more than that. Mulholland (1965) has applied a modification of Hansen's lunar theory to the motion of Jupiter VI.

The motions of Jupiter VIII–XII are very strongly perturbed, and at present their orbits can be represented to a sufficient accuracy only by means of numerical integration. By this technique, Herget (1968a, 1968b) has calculated ephemerides whose 1- to 4-arcsec accuracies are limited essentially by the errors in the available photographic observations of these very faint satellites.

Saturn's Satellites

The ephemerides of Saturn's six major inner satellites (SI–SVI) and of Iapetus are computed from the orbital elements derived by G. Struve (1924–33). These are mean elements, which include the most important perturbations arising from Saturn's oblateness and the mutual attractions, derived from a large number of micrometer measures made between 1789 and 1924 and a few photographic observations. The ephemeris of Hyperion is computed from the elements given by Woltjer (1928); that of Phoebe, from the theory by Ross (1905). Hyperion's motion is characterized by a libration caused by Titan (see Chapt. 8, Greenberg), while Phoebe's departs from an ellipse only slightly because of solar perturbations. More observations of Janus (Dollfus, 1967) are needed to define its orbital elements more accurately.

Modern observations indicate that the above-mentioned theories are now in error by from 0.1 to 0.7 arcsec for Saturn I–VIII, and by no more than 1 arcmin for Saturn IX, geocentrically. Although the orbits of most of these satellites were redetermined by Zadunaisky (1954), Kozai (1957), and Garcia (1972), these later results are in need of revision. Sinclair (1974b) has discussed the orbit of Iapetus. While Zadunaisky applied Delaunay's lunar theory to the motion of Phoebe, Elmabsout (1970) has devised a seminumerical theory for this satellite based on Hill's lunar theory. But Elmabsout did not attempt to fit his theory to observations.

Uranus' Satellites

No ephemeris is given for Miranda. The ephemerides of the outer four satellites are computed from elements derived by Newcomb (1875) and Struve (1913) on the basis of micrometric observations made between 1874 and 1911. These investigators found that to within the uncertainty of the observations (typically about 0.2 arcsec), the satellites move in circular orbits, all of which lie in the equatorial plane of Uranus. However, thanks mainly to a highly accurate series of photographic observations made with the 82-inch reflector at McDonald Observatory, Harris (1949) detected some definite eccentricities in the orbits of the four outer satellites, while he found a nearly circular orbit for Miranda. A later determination of the orbital elements of the Uranian satellites by Dunham (1971) essentially confirmed the values found by Harris, except for Umbriel's eccentricity and the motions of the pericenters of Ariel, Umbriel, and Oberon. A cogent investigation by Greenberg and Whitaker (1974) suggests that both the eccentricity and the inclination of Miranda's orbit are pronounced. The use of noncircular orbits for the five Uranian satellites does not significantly improve the residuals of the observations, which are of the order of a few tenths of an arcsecond. Greenberg (1975) summarizes the dynamics of the system.

Neptune's Satellites

The ephemeris for Triton is based on orbital elements determined by Eichelberger and Newton (1926) from observations in the interval 1848–1923. They derived a circular, retrograde orbit inclined at $159°\!.9 \pm 2°\!.3$ to Neptune's equator. Owing to the primary's oblateness, the line of nodes of Triton's orbit makes a complete revolution in 585 ± 66 years. This is in good agreement with an investigation by Gill and Gault (1968), who found the values $161°\!.14$ and 580.83 years for the same two parameters from combined visual and photographic observations in the interval 1887–1958.

Until the studies of Aksnes (1974b), no ephemeris has been available for Neptune's second satellite, Nereid, although it is an easy task to calculate an ephemeris from the purely elliptic elements published by van Biesbroeck (1957). From 44 observations between 1949 and 1969, Rose (1974) has determined a new set of elliptic elements for Nereid. He has confirmed van Biesbroeck's conclusion that the solar perturbations are smaller than the errors in the observations, whose mean residual is somewhat less than 1 arcsec.

Suggestions for Improving the Ephemerides

The need for improved ephemerides based on the most recent orbital theories for the satellites has been pointed out. The question also arises as to whether the existing *format* of these ephemerides is well adapted to modern needs. For instance, in addition to publishing printed tables, it would be useful to make the tables available in a computer-readable form on magnetic tape, paper tape, or IBM cards. Perhaps an even better solution would be to release certified and thoroughly documented computer programs, including test cases, for ephemeris computation, with the understanding that the user in his work would be expected to give due credit to the author of the program, as he would to the author of a regular publication. These are not new ideas. They have been discussed before by the members of IAU Commission 4 on ephemerides. Any new decisions or recommendations concerning ephemerides of satellites properly belong with this Commission.

ACKNOWLEDGMENTS

I am indebted to my colleague, B. G. Marsden, for several valuable suggestions and criticisms during this work. The support from the National Aeronautics and Space Administration under Grant NGR 09-015-213 is also gratefully acknowledged.

MOTIONS OF NATURAL SATELLITES

Jean Kovalevsky
Observatoire de Paris-Meudon
and
Jean-Louis Sagnier
Bureau des Longitudes, Paris

The equations of motion of planetary satellites are presented and the various types of disturbing forces are reviewed. From a consideration of forces, three classes of problems are defined and described. A general method to solve the equations is sketched in order to give the general form of their solution. Using Delaunay's algorithm, the equations are discussed and three types of solutions are obtained; they correspond more or less to the previously defined classes. Then the motions of all the natural satellites are reviewed and, for each group of them, the latest works on the theory of motion are presented.

EQUATIONS AND CLASSIFICATION OF PROBLEMS

A satellite is a celestial body, belonging to the solar system, whose distance from its primary planet is at all times much less than the distance between the planet and the Sun. Its motion essentially is ruled by the gravitational attraction of the planet. Although this planetocentric force is dominant, other forces—due to the Sun, to other planets, to other satellites, or to the nonsphericity of the primary—can be relatively important and cause large perturbations to the orbit. The geometrical character of the motion is shown by Aksnes (Chapt. 3, herein; see Figure 3.1). Sometimes the properties of the motion may be changed completely. We shall first consider the perturbations due to some other body that is considered to be a mass point, and then those caused by the nonsphericity of the primary.

In the first case, let us refer the motion to a system of rectangular axes, with an origin at the center of mass A of the central planet and parallel to fixed directions (Fig. 4.1).

G is the universal constant of gravitation; M, m, and m', respectively, are the masses of the planet A, of the satellite S, and of the disturbing mass-point B; additional notation is shown in Figure 4.1. Then, the absolute acceleration of the satellite S is

$$- GM \cdot AS/r^3 + Gm \cdot SB/\Delta^3,$$

while the acceleration of the system of axes is

$$Gm \cdot AS/r^3 + Gm' \cdot AB/r'^3,$$

[43]

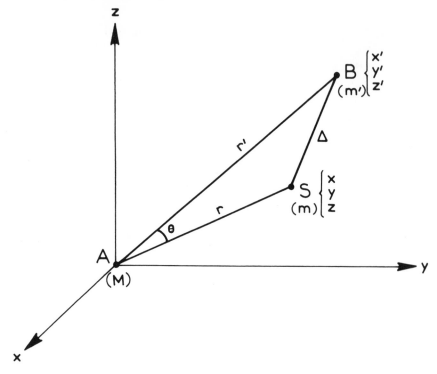

Fig. 4.1 Geometry of the problem of satellite motion.

so that the vectorial equation of motion is

$$d^2\mathbf{AS}/dt^2 = -G(M+m)\mathbf{AS}/r^3 + Gm' \cdot \mathbf{SB}/\Delta^3 - Gm' \cdot \mathbf{AB}/r'^3 , \qquad (1)$$

where bold face type indicates a vector directed from the first symbol to the second.

If one introduces the disturbing function R (see, for instance, Brower and Clemence, 1961a),

$$R = Gm'[\Delta^{-1} - (xx' + yy' + zz')r'^{-3}] , \qquad (2)$$

eqn. (1) may be written:

$$\begin{aligned}
d^2x/dt^2 &= -\mu x/r^3 + \partial R/\partial x \\
d^2y/dt^2 &= -\mu y/r^3 + \partial R/\partial y \\
d^2z/dt^2 &= -\mu z/r^3 + \partial R/\partial z
\end{aligned} \qquad (3)$$

where $\mu = G(M+m)$.

Three cases are to be considered:

Case 1: B is the Sun. It can be shown (Brouwer and Clemence, 1961a) that Δ^{-1} can be developed in a converging series of r/r', this quantity being, by definition, much smaller than unity. One has:

$$R = Gm'\left[r^2r'^{-3}(\frac{3}{2}\cos^2\theta - \frac{1}{2}) + r^3r'^{-4}(\frac{5}{2}\cos^3\theta - \frac{3}{2}\cos\theta) + \ldots\right] ,$$

so that R is of the order of $Gm'a^2a'^{-3}$, where a and a' are, respectively, the semimajor axes of the orbits of the satellite and of the planet. Let us call n and n' the corresponding mean motions. Since, by Kepler's third law, $Gm' = n'^2a'^3$, R is of the order of n'^2a^2. The coefficient n'^2 is small in comparison with the corresponding zero order quantity n^2, which is similarly present in the major terms of eqn. (3). Therefore, the disturbing force due to the Sun is characterized by the small quantity $(n'/n)^2$.

Case 2: B is a planet. The distances are of the same order of magnitude as in the case of the Sun, but the mass m_p of the disturbing planet is much smaller than the mass $m_\odot$ of the Sun. The characteristic small quantity is then $(m_p/m_\odot) \cdot (n'/n)^2$.

Case 3: B is another satellite. In this circumstance the distances r and r' are of the same order of magnitude. The smallness of the disturbing force is solely due to the smallness of m_s/M, where m_s is the mass of the disturbing satellite.

Let us now consider the perturbations due to the nonsphericity of the central planet. They are derived from the expression for the gravitational potential of the planet. In the general case of a planet with axial symmetry, using notation that is now classical for the Earth's potential (Hagihara, 1962), one has:

$$U = GM\,r^{-1}\left[1 - \sum_{n=2}^{\infty}(a_e/r)^nJ_n\cdot P_n(z/r)\right] = GM\,r^{-1} + U' ; \qquad (4)$$

J_n are the coefficients of zonal harmonics of order n and are dimensionless numbers; a_e is a scaling coefficient, usually taken equal to the equatorial radius of the planet; and P_n are Legendre polynomials of order n.

Actually, because of a lack of data and because of the smallness of their effect on satellite motions, it has been considered sufficient for planets other than the Earth to take into account only J_2 and J_4, and the following notation (de Sitter, 1924) is commonly used:

$$J = \frac{3}{2}J_2 \quad , \quad K = -\frac{15}{4}J_4 \quad .$$

In order to allow for the effects of this potential, one must add the partial derivatives of the disturbing term U' to the right-hand members of eqn. (3). Finally, in the planetocentric system of rectangular coordinates we have defined, eqn. (1) takes the form:

$$d^2x/dt^2 = -\mu x/r^3 + \partial R_\odot/\partial x + \sum_p \partial R_p/\partial x + \sum_s \partial R_s/\partial x + \partial U'/\partial x , \qquad (5)$$

with two similar equations in y and z. $R_\odot$, R_p, and R_s are, respectively, the disturbing functions due to the Sun, the planet p and the satellite s.

With $R = R_\odot + \sum R_p + \sum R_s + U'$, eqns. (3) are generally valid for the motion of any satellite. But many other forms of equations, using different variables, may be more useful or more efficient to solve or to discuss their solutions. Some of them will be used later.

Considering the relative order of magnitude of each term in R, one can classify problems of the motion of natural satellites.

Let us first remark that $\sum R_p$ is always much smaller than $R_\odot$, because of the presence of the factor $m_p/m_\odot$. Therefore the planetary perturbations are small in comparison with the solar perturbations and do not alter the general behavior of the motion. The same applies also to the so-called ''indirect planetary perturbations'' introduced by the non-Keplerian part of the apparent motion of the Sun relative to the central planet, due to the action of the perturbing planets.

Analyzing the other terms in R, one can distinguish three classes of problems.

Class 1: Close satellites. For the nearest satellites, especially for those which revolve around very oblate planets, the quantity U' is preponderant. The theory of motion is similar to those derived for artificial satellites of the Earth. This is the case for (i) Mars' satellites Phobos and Deimos, (ii) the fifth satellite of Jupiter, (iii) Janus, the most recently discovered satellite of Saturn, (iv) all five satellites of Uranus, and (v) Neptune's satellite Triton.

Class 2: Satellites mainly disturbed by the Sun. The motion is then essentially governed by $R_\odot$, even if direct or indirect planetary effects are not negligible. Two subclasses may be distinguished.

Class 2a. The solar disturbing force is not larger than 1% of the main central force. Perturbations are large, but the general behavior of the motion is elliptic. The typical case is the lunar theory. Other satellites in this class are (i) satellites VI, VII, X, XIII of Jupiter, and (ii) Saturn's satellites Titan, Rhea, Iapetus.

Class 2b. Solar perturbations are very strong. The orbit does not look like a Keplerian ellipse any more. Compare, for instance, the orbit of the Moon during one terrestrial year, shown in Figure 4.2, with that of the eighth satellite of Jupiter as given by Grosch (1948) in Figure 4.3. The fact that most of the orbits of the satellites of this class are retrograde probably has a cosmogonic reason, but that does not change the mathematical problem of their motion. These satellites are (i) Jovian satellites VIII, IX, XI, and XII, (ii) Phoebe (Saturn), and (iii) Nereid (Neptune).

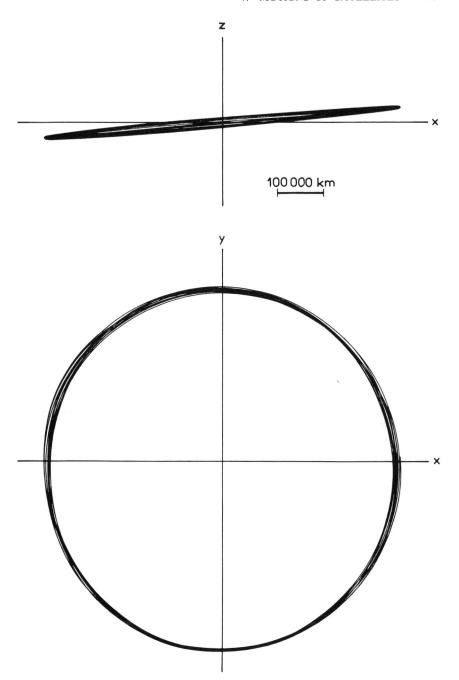

Fig. 4.2 Orbit of the Moon during one terrestrial year.

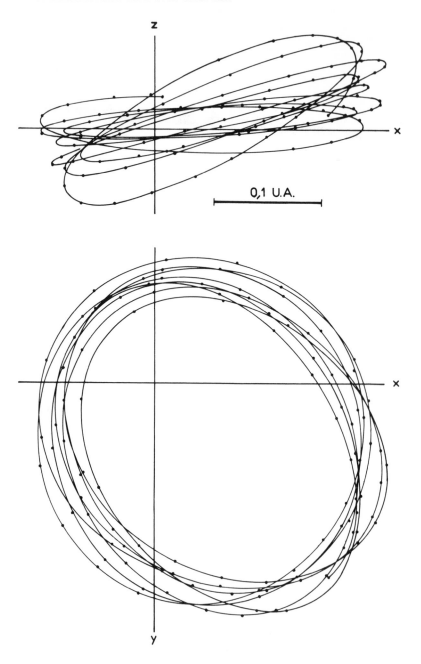

Fig. 4.3 Orbit of the eighth satellite of Jupiter
as given by Grosch (1948).

Class 3: Satellites disturbed by another satellite. When R_s is predominant in R, the situation is similar to that which exists among planets in the solar system. The theory is the same as that for the motion of a planet. But there is a new difficulty, since planetary observations span a maximum of a few hundred years, whereas one must describe quantitatively tens of thousands of revolutions in satellite theory (cf. Chapt. 5, Pascu).

Such "planetary problems" in satellite motions are met in the following cases: (i) the Galilean satellites of Jupiter (I, II, III, IV), (ii) the Saturnian pairs, Mimas-Enceladus and Tethys-Dione, and (iii) Hyperion as disturbed by Titan.

From a mechanical point of view, one must distinguish the case of interactions (several satellites with masses of the same order) from the case when one satellite has a much smaller mass than that of the disturbing body, and so does not noticeably perturb it.

It is obvious that a satellite does not belong purely to one class, and several types of perturbations may be simultaneously important. For instance, the Galilean satellites interact strongly with one another, and at the same time the first two Galilean satellites, Io and Europa, are strongly disturbed by the oblateness of the planet, while Callisto undergoes large disturbing effects due to the Sun.

In order to analyze in more detail the actual motions of satellites, it is now necessary to present some remarks on the methods that may be used to solve the equations of motion.

TYPES OF SOLUTIONS OF THE EQUATIONS OF MOTION

Eqn. (5) is the simplest for presentation, but it is not the most easily solved. It is a classical result of celestial mechanics that equivalent systems of equations, using different variables, may be more efficient. The use of osculating elements: semimajor axis a, eccentricity e, inclination i, longitude of node Ω, argument of pericenter ω, and mean anomaly M, lead to the well-known Lagrange equations. But it is preferable here to use a set of canonical equations, with the Delaunay variables:

$$L = \sqrt{\mu a} \quad , \quad G = \sqrt{\mu a}\sqrt{1-e^2} \quad , \quad H = \sqrt{\mu a}\sqrt{1-e^2} \cos i \ .$$
$$l = M \quad , \quad g = \omega \quad , \quad h = \Omega \ . \tag{6}$$

One can find proofs of these results in most treatises of celestial mechanics (e.g., Brouwer and Clemence, 1961a; Kovalevsky, 1967; Hagihara, 1972). The corresponding canonical equations are:

$$dL/dt = +\partial F/\partial l \ ; \ dG/dt = +\partial F/\partial g \ ; \ dH/dt = +\partial F/\partial h \ ;$$
$$dl/dt = -\partial F/\partial L \ ; \ dg/dt = -\partial F/\partial G \ ; \ dh/dt = -\partial F/\partial H \ ; \tag{7}$$

where the Hamiltonian $F = \mu^2/2L^2 + R$ is expressed in terms of Delaunay variables.

Let us take, as an example, the perturbations of a satellite by the Sun. Neglecting indirect perturbations, the coordinates of the Sun are periodic functions of the time with a period $P = 2\pi/\nu$. Since the coordinates x, y, z of the satellite are 2π − periodic functions of l, g, and h, the Hamiltonian F can therefore be expanded into a quadruple trigonometric series of the form:

$$F = \sum_{\alpha,\beta,\gamma,\delta} A_{\alpha\beta\gamma\delta}(L,G,H) \cos(\alpha l + \beta k + \gamma g + \delta h) , \qquad (8)$$

where $k = \nu(t - t_0)$ is the mean longitude of the Sun, and where other parameters (orbital parameters of the planet) are not explicitly written but are present in the coefficients A. The numbers α, β, γ, δ are positive or negative integers. This development is in general convergent when eccentricities are smaller than 0.6627 (Tisserand, 1896) as they are always for satellite problems.

A common and efficient method to get a formal solution to eqns. (7) and (8) is to apply a series of transformations of variables in such a way that the equations with the new variables are simpler. Many various methods have been proposed in this direction—for example, the methods of Delaunay, Lindstedt, Von Zeipel, and Hori (see Hagihara, 1972).

All these methods are based on the introduction of a new set of variables (l′, g′, h′, L′, G′, H′), which differ from the initial variables by quantities of the order of the small parameter characterizing the disturbing force:

$$L' = L + \Delta L \; ; \; G' = G + \Delta G \; ; \; H' = H + \Delta H \; ;$$
$$l' = l + \Delta l \; ; \; g' = g + \Delta g \; ; \; h' = h + \Delta h \; . \qquad (9)$$

The increments ΔL, etc., are chosen so that the new variables are canonical, and that, if F′ is the Hamiltonian expressed in terms of the new variables, the transformed equations are:

$$dL'/dt = +\partial F'/\partial l' \; ; \; dG'/dt = +\partial F'/\partial g' \; ; \; dH'/dt = +\partial F'/\partial h' \; ;$$
$$dl'/dt = -\partial F'/\partial L' \; ; \; dg'/dt = -\partial F'/\partial G' \; ; \; dh'/dt = -\partial F'/\partial H' \; . \qquad (10)$$

All the methods considered consist in choosing the transformation (9) in such a way that the new Hamiltonian is simpler, and the new eqns. (10) are easier to solve.

The most favorable case would be that all angular variables l′, g′, h′ do not appear in F′, so that the integration is straightforward, L′, G′, H′ being then constants, and l′, g′, h′ linear functions of time. As an example, the Delaunay transformation allows the elimination from F of any chosen term A cos(αl+ βg+ γh + δk), or, at least, the reduction of its order of magnitude with respect to the characteristic small parameter.

Such a transformation is precisely that which eliminates the specified term from the simplified Hamiltonian:

$$\Phi = \mu^2/2L^2 + P(L,G,H) + A(L,G,H)\cos(\alpha l + \beta g + \gamma h + \delta k) , \qquad (11)$$

the new Hamiltonian being then:

$$\Phi' = \mu^2/2L'^2 + P'(L',G',H') .$$

It is not necessary, in this review, to describe the various cases arising in the search for such a transformation. By a simple transformation of variables, putting

$$\theta = \alpha l + \beta g + \gamma h + \delta k ,$$

the solution of the canonical system (11) can be reduced to the solution of the following equations:

$$d\Theta/dt = \partial\overline{\Phi}/\partial\theta \; ; \; d\theta/dt = -\partial\overline{\Phi}/\partial\Theta , \qquad (12)$$

where $\overline{\Phi}$ is the Hamiltonian Φ expressed in new variables, and Θ is the new variable associated with θ. Other new metric variables are constant in the solution and can be considered as parameters of the problem. $\overline{\Phi}$ has the form $\overline{\Phi} = A + B\cos\theta = C$ (constant) since $\overline{\Phi} = C$ is an integral of the system.

Using this result, and eliminating θ between this integral and the first equation of (12), $d\Theta/dt = -B\sin\theta$, one gets $(d\Theta/dt)^2 = B^2 - (C - A)^2$, or:

$$t - t_0 = \pm \int d\Theta/\sqrt{(B+C-A)(B-C+A)} . \qquad (13)$$

According to the respective values of the parameters entering in A, B, and C, one obtains two limiting values Θ_1 and Θ_2 between which Θ oscillates. Two cases may arise:

Case a. There is no root between the two obvious roots $B-A = -C$ and $B+A = C$ of the denominator of (13), which respectively correspond to $\theta = 0$ and $\theta = \pi$. All values of θ can then be reached, and they repeat, as well as the values of Θ, with a period equal to:

$$p = 2\int_{\Theta_1}^{\Theta_2} d\Theta/\sqrt{B^2 - (C - A)^2} . \qquad (14)$$

After such a period, θ increases by 2π and one gets:

$$\theta = \frac{2\pi}{p} (t - t_0) + \text{Per}(t) \; ; \; \Theta = \text{Per}(t)$$

where Per(t) denotes periodic functions of time with period p. Owing to the behavior of θ, such a solution is called *circulatory*.

Case b. There exists another root, Θ_3, of $B^2 - (C - A)^2$, corresponding to a possible value θ_3 of θ, such that $A(\Theta_3) + B(\Theta_3) \cos \theta_3 = C$. Then θ cannot take any value and varies only between θ_3 and $-\theta_3$. This solution is a *libration-type solution*, where θ is a purely periodic function of time and does not circulate. Θ is also periodic and the period is given by eqn. (14), where one of the limits of integration must be replaced by Θ_3.

When such a transformation is applied to eqns. (7) and (8), the new Hamiltonian does not contain the term A$\cos\theta$. If similar transformations are performed for all significant terms of eqn. (9), one gets at last a system of variables (L*,G*,H*,l*,g*,h*) and a Hamiltonian F* such that the equations are:

$$dL^*/dt = \quad \partial F^*/\partial l^* \quad ; \quad dG^*/dt = \quad \partial F^*/\partial g^* \quad ; \quad dH^*/dt = \quad \partial F^*/\partial h^* \quad ;$$
$$dl^* /dt = \quad -\partial F^*/\partial L^* \quad ; \quad dg^* /dt = \quad -\partial F^*/\partial G^* \quad ; \quad dh^* /dt = \quad -\partial F^*/\partial H^* \quad , \tag{15}$$

with F* = F*(L*,G*,H*) independent of l*, g* and h*. Hence the first three equations yield: L* = L_0, G* = G_0, H* = H_0 (constants), and, therefore, the right-hand members of the last three equations are constant and the solution is:

$$l^* = n_l t + l_0 \quad ; \quad g^* = n_g t + g_0 \quad ; \quad h^* = n_h t + h_0 \ .$$

The three periods of the solution are derived from these expressions relative to mean angular arguments, and are functions of L_0, G_0, and H_0.

In order to return to the original physical elements, one then has to apply backward all the transformations performed. Again, two cases exist.

Case a. All the transformations are of circulatory type. Then all the equations have the same structure, and one finally gets formal solutions of the form:

$$L = L_0 \ + \sum_{\alpha\beta\gamma\delta} L_{\alpha\beta\gamma\delta}(L_0,G_0,H_0) \cos(\alpha l^* + \beta g^* + \gamma h^* + \delta k)$$
$$l = l^* + \sum_{\alpha\beta\gamma\delta} l_{\alpha\beta\gamma\delta}(L_0,G_0,H_0) \sin(\alpha l^* + \beta g^* + \gamma h^* + \delta k) \tag{16}$$

with similar expressions for the other unknowns. The metric variables L, G, and H, and hence a, e, and i, are multiperiodic (bounded) functions of time, the periods being the induced periods (here, the period of revolution of the planet, introduced by the argument k) and the proper periods of the problem, which are the mean periods of revolution of the satellite, of the pericenter and of the node (through the arguments l*, g*, and h*). The actual variations of the angular variables l, g, and h are the sum of the secular linear terms l*, g*, and h* and of similar multiperiodic functions of time. This type of motion will be called a *circulatory type* of motion.

Case b. When one of the transformations is of a libration-type, the solution itself has a formally different character, which we shall denote a *libration type* motion.

Let us call θ the linear combination of arguments which produces the libration-type transformation. Let us also call x and y two other angular variables out of l, g, and h, so that x, y and $\theta - \delta k$ are linearly independent.

Assigning x*, y*, and θ* the same meaning as l*, g*, and h* in the system of equations equivalent to eqn. (15), one obtains, after the elimination of all periodic terms in F, that the formal solutions are of the form:

$$L = L_0 + \sum_{\alpha\beta\gamma\delta} L_{\alpha\beta\gamma\delta}(L_0,G_0,H_0) \cos (\alpha\theta^* + \beta x^* + \gamma y^* + \delta k)$$

$$\theta = \theta_0 + \sum_{\alpha\beta\gamma\delta} \theta_{\alpha\beta\gamma\delta}(L_0,G_0,H_0) \sin (\alpha\theta^* + \beta x^* + \gamma y^* + \delta k) \qquad (17)$$

$$x = x^* + \sum_{\alpha\beta\gamma\delta} x_{\alpha\beta\gamma\delta}(L_0,G_0,H_0) \sin (\alpha\theta^* + \beta x^* + \gamma y^* + \delta k)$$

with similar expressions for G, H, and y.

The essential difference is that the libration argument does not circulate but oscillates around a value θ_0. The corresponding period is also a function of the constants of integration and of the parameters of the motion, but in an analytically quite different form, as discussed in many papers on resonance (see Chapt. 8, Greenberg) and by the main treatises in celestial mechanics. One of the main characteristics of such a motion is that the amplitude of the libration oscillation can be very large, independently of the order of magnitude of the disturbing force. (It is indeed an arbitrary constant of integration.)

It is to be remarked that the formal expressions obtained by Delaunay's method (or equivalent results obtained by other, generally more efficient, methods) are not convergent. But, in practice, if one takes a limited number of terms in F, and then in the solution, one gets expressions that represent the solution within a given precision, provided that one stays within a finite interval of time. (This restriction, anyhow, is necessary because of the errors in the constants of integration due to the finite accuracy of observations.)

This is discussed by Poincaré in Chapter 13 of his *Méthodes Nouvelles* (Poincaré, 1957), and this justifies the semianalytical methods. In such methods, the coefficients in the solutions are taken numerically, while the arguments of the trigonometric functions remain literal. The form of the solution (16) or (17) being known, the coefficients are computed numerically in such a way that expressions (16) or (17) verify the equations of motion and the conditions expressed by the results of the observations. Such numerical methods are generally used when the construction of formal literal solutions leads to unmanageable calculations.

Let us now see what kinds of solutions correspond to the classes of motion defined in the first part by a qualitative consideration of the forces present.

Class 1. Theories of this type of motion have been developed for artificial satellites, and the expressions obtained are more than sufficient when applied to natural satellites of this class. As an example, one will find developed expressions in Brouwer (1959) or Levallois and Kovalevsky (1971). They are of the form (16) for all cases except when the inclination of the orbit on the equator of the planet has the critical value ($\approx 63.4°$). In this case the argument of the pericenter is a librating quantity and the form of the solution is that of eqn. (17); see, for instance, Hori (1960). Other resonance conditions appear also when the revolution period of the satellite is commensurable with the rotation period of the planet, if the tesseral harmonics are nonnegligible. Then, again, libration-type solutions are valid. But such resonance situations do not appear for natural satellites.

Class 2a. This is typically the case of lunar theory. Then, the convergence of Delaunay-type theories of motion is ensured by the fact that the eccentricities of the satellite and planet orbits, as well as the inclinations, are small quantities of the same order as the ratio of the mean motions.

A literal solution, based on Delaunay's method, was given over one century ago up to the 7th order of the small quantities (Delaunay, 1860-67) and recently to much higher orders by Deprit, Henrard, and Rom (1971). Seminumerical methods, where the mean motions are taken numerically, were essentially developed by Brown (1896) and later improved by Eckert and Eckert (1967).

All these theories can be directly applied to all the satellites of this subclass, and the circulatory solution (16) characterizes their motion.

Class 2b. When the eccentricity or the inclination is large, one can generally eliminate without difficulty all terms, including l and k, but this is not the case for those terms that depend only on g and h.

A general discussion of this problem is shown in Figure 4.4 (Kovalevsky, 1966), as a function of $y = H_0/L_0$, and the integral defined by the reduced Hamiltonian $F = F_0 + A + B \cos 2g$, $d = F/L_0$. Two regions exist in the plane $(d - y^2)$, corresponding to circulatory orbits (A, including all satellites of Class 2a) and to orbits with a libration of the pericenter.

Natural satellites of Class 2b are all in region A. However, due to the magnitude of the eccentricity and inclination, perturbation methods are no longer applicable, and convergence is not insured for literal solutions. This is why the analytic solutions that have been constructed were seminumerical. Actually, in most cases, only numerical integrations of the equations of motion have been performed until now.

Class 3. The problem of the motion of a satellite disturbed by another satellite—or more generally, the problem of the motion of several satellites under the action of their mutual attractions—has a different character than in cases of Class 2, essentially because the ratio of the semimajor axes cannot be considered a small quantity, and, hence, the fast convergence of a series development in a/a' or n'/n does not exist. Actually, this is the situation in the

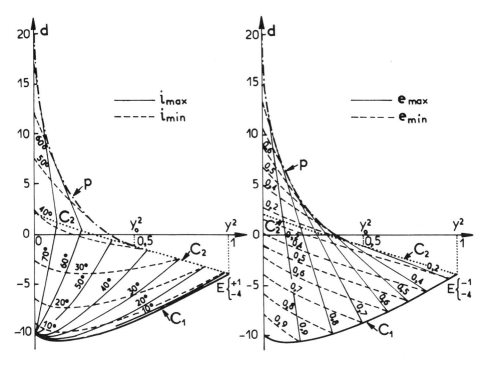

Fig. 4.4 Motion of satellites with large eccentricity and inclination
(from Kovalevsky, 1966). See discussion in text.

system of planets, and, comparing a satellite system to a small solar system, it is easy to understand that the problems concerning their motions are similar.

In fact, however, there is a fundamental difference. The motions of the planets have been observed for a maximum of a few hundreds of revolutions for the inner planets, or a few tens for the outer ones. In contrast, three hundred years of observations of Jupiter's inner satellites represent several tens of thousands of revolutions. Therefore, although the satellite theories may be relatively less precise than the planetary ones because of the large planetocentric errors of the observations (cf. Chapt. 5, Pascu), they must be valid for a much longer time as expressed in relative time units (number of revolutions). This implies, among other difficulties, that it is not usually possible to represent the long-period terms by power series in time as is classically done for the planets. A representation of the solution by purely trigonometric series is necessary for satellite systems.

Apart from the convergence problems and the length of some expressions, the form of the equations is similar to the one we have presented in the case of solar perturbations. This means that one can also get either circulatory (16) or libration type (17) solutions. The appearance of libration is generally due to the commensurability between mean motions of the satellites (cf. Chapt. 8, Greenberg).

In regard to the planetary-type satellite motions, resonances often are encountered. One may distinguish three types:

Type a. Resonance without interaction between satellites. This is the case of Hyperion perturbed by Titan, the mass of Hyperion being too small to noticeably disturb Titan. In this case, the quantity

$$\theta = 4l - 3l' - g - h \, ,$$

where l' is the mean anomaly of Titan, has a mean motion of about 1° per day, while the basic periods of the satellites are 21 and 16 days. In this case, the motion is a libration type motion that can be studied in a fashion similar to some asteroid orbits in resonance with Jupiter.

Type b. Interaction without resonance. This is the case of Callisto with any of the other Galilean satellites. The resulting motion is similar to planetary motion and is circulatory.

Type c. Interaction with resonance. This is the general case, since resonant pairs of satellites have often comparable masses. It brings great analytical difficulties, as discussed in Chapter 8 by Greenberg.

APPLICATION TO THE NATURAL SATELLITES

The different classes of motions we considered are widely represented among the natural satellites of the solar system. One must emphasize the fact that the physical parameters that can be best determined from the analysis of satellite motions are precisely those which are dominant in the corresponding equations. So, the classification of types of satellite motions is not only an indication of the kind of mathematical problems to be solved in the theoretical study but also of its possible physical result.

Specific results concerning particular satellites or groups of satellites will be listed later (see also Aksnes, Chapt. 3, herein). First, Table 4.1 summarizes approximate values, giving the main parameters of satellite orbits: period of revolution, semimajor axis a_R in planetary radii and a_0 in units of the planetary orbit's semimajor axis a_0 (the former being significant in considering the perturbation caused by the planetary oblateness and the latter in considering the solar perturbation), the value of n'/n (which indicates the rapidity of the convergence of the series representing the solar contribution to the actual motion), the eccentricity e, the inclination i_0 on the planetary orbit (when the solar perturbation is the main one) or i_e on the equator of the central planet (when the influence of its oblateness plays a major role). When significant, the order of magnitude of J_2 also is given for each planet. Asterisks following the name of some satellites mean that their mass is significant, indicating the possible importance of their gravitational attraction on other satellites.

TABLE 4.1
Main Features of Satellite Orbits

Satellite	Period of rev. (days)	Semimajor axis a_R	a_o	n'/n	e	i_o	i_e	J_2
Earth								
Moon	27.3	60	$3\cdot10^{-3}$	0.07	0.06	5°		10^{-3}
Mars								
I. Phobos	0.32	2.8	$4\cdot10^{-5}$	$5\cdot10^{-4}$	0.02	1°		$2\cdot10^{-3}$
II. Deimos	1.3	6.9	10^{-4}	$18\cdot10^{-4}$	0.003			
Jupiter								
I. Io*	1.8	5.9	$5\cdot10^{-4}$	$4\cdot10^{-4}$	0		0".1	
II. Europa*	3.6	9.3	$9\cdot10^{-4}$	$8\cdot10^{-4}$	0		1'	
III. Ganymede*	7.2	14.9	10^{-3}	$17\cdot10^{-4}$	0.002		5'	
IV. Callisto*	17	26.2	$2\cdot10^{-3}$	$39\cdot10^{-4}$	0.007		26'	
V. Amalthea	0.50	2.5	$2\cdot10^{-4}$	10^{-4}	0.003		0°.4	
VI. Himalia	251	160	10^{-2}	$0\cdot06$	0.16	28°		$1.5\cdot10^{-2}$
VII. Elara	260	163	10^{-2}	$0\cdot06$	0.21	28°		
X. Lysithea	260	163	10^{-2}	$0\cdot06$	0.11	29°		
VIII. Pasiphaë	744	327	$3\cdot10^{-2}$	$-0\cdot17$	0.41	148°		
IX. Sinope	763	330	$3\cdot10^{-2}$	$-0\cdot18$	0.32	153°		
XI. Carme	693	314	$3\cdot10^{-2}$	$-0\cdot16$	0.21	163°		
XII. Ananke	631	295	$3\cdot10^{-2}$	$-0\cdot14$	0.17	147°		
Saturn								
X. Janus	0.7	2.7	10^{-4}	$7\cdot10^{-5}$	0		0	
I. Mimas*	0.9	3.1	10^{-4}	$9\cdot10^{-5}$	0.02		1°.5	
II. Enceladus*	1.4	4.0	$2\cdot10^{-4}$	$13\cdot10^{-5}$	0.004		0	
III. Tethys*	1.9	4.9	$2\cdot10^{-4}$	$17\cdot10^{-5}$	0		1°.1	
IV. Dione*	2.7	6.3	$3\cdot10^{-4}$	$25\cdot10^{-5}$	0.002		0	
V. Rhea	4.5	8.7	$4\cdot10^{-4}$	$42\cdot10^{-5}$	0.001		0°.4	$1.7\cdot10^{-2}$
VI. Titan*	16	20.3	$9\cdot10^{-4}$	$15\cdot10^{-4}$	0.03		0°.3	
VII. Hyperion	21	24.6	10^{-3}	$20\cdot10^{-4}$	0.10		0°.4	
VIII. Iapetus	79	59	$2\cdot10^{-3}$	$74\cdot10^{-4}$	0.03	18°.4		
IX. Phoebe	551	215	$9\cdot10^{-3}$	0.05	0.17	175°		
Uranus								
I. Ariel	2.5	8.1	$7\cdot10^{-5}$	$8\cdot10^{-5}$	0.003		0	
II. Umbriel	4.1	11.4	$9\cdot10^{-5}$	$14\cdot10^{-5}$	0.004		0	
III. Titania	8.7	18.6	$2\cdot10^{-4}$	$28\cdot10^{-5}$	0.002		0	
IV. Oberon	13.4	24.9	$2\cdot10^{-4}$	$44\cdot10^{-5}$	0.001		0	
V. Miranda	1.4	5.5	$5\cdot10^{-5}$	$5\cdot10^{-5}$	0		0	
Neptune								
I. Triton*	5.9	16	$8\cdot10^{-5}$	$-9\cdot10^{-5}$	0		160°	$5\cdot10^{-3}$
II. Nereid	360	250	10^{-3}	$60\cdot10^{-4}$	0.07	28°		

* indicates satellites of mass sufficiently great to possibly assert noticeable gravitational attraction on other satellites.

The values discussed below are listed in Tables 1.2 and 1.4 of Morrison *et al*. (Chapt. 1) and Aksnes (Chapt. 3). Seidelmann (Appendix), Pascu (Chapt. 5), and Aksnes give references for theories and observations.

The Moon

The Earth's natural satellite must be considered somewhat separately here, due to the specific conditions of its observation, the major requirements in accuracy for the theory of its motion, and its relative mass, which is much greater than that of any other satellite.

The mass of the Moon (relative to the mass of the Earth) was formerly obtained by the analysis of the lunar inequality (due to the motion of the Earth around the barycenter of the Earth-Moon system), since one could not study its action on any other body. Two methods are now used, the lunar inequality being derived from radar measurements of the inner planets or Doppler observations of space probes (Mariner 2, Mariner 4), while the direct dynamical analysis of lunar space probes (Ranger) also gives good results. This revitalization of classical methods now makes the mass of the Moon one of the best known astronomical constants: $m_{\mathbb{C}}/m_{\oplus} = 1/81.300$.

The significant relative mass of the Moon does not give rise to specific problems in the theoretical study of its motion, the main difficulty of which is the great accuracy needed in the solution of this Class 2a problem. Any good literal lunar theory would then be more than sufficient to be applied to any other satellite mainly perturbed by solar actions. Delaunay's theory may be used in this manner, but one must remember its principal drawback, the slow convergence with respect to n'/n, which limits its application to satellites not too far from the planet.

The motion of the Moon is characterized by a feature one does not have to consider for other natural satellites: the reaction of the tidal bulge of the Earth, which is taken into account by means of empirical terms in the longitude of the Moon. (We must, however, note that tidal bulge is the probable cause of secular acceleration of Phobos; see also Chapt. 7, Burns, and Chapt. 14, by Pollack.)

The Satellites of Mars

Phobos is a typical Class 1 satellite, while Deimos belongs to Classes 1 and 2a. Both have negligible masses, which do not induce interactions and therefore cannot be evaluated by gravitational methods.

An interesting feature of Phobos' motion is the alleged acceleration of the mean longitude, similar to the acceleration of artificial satellites of the Earth at low altitude. Sharpless (1945) made the suggestion, but the analysis of observations made by Wilkins (1967) shows that the poor quality of the observations does not enable us to demonstrate such a tiny effect. Sinclair (1972a) and Shor (1975), in comparing their theories to all available observations, reached the same conclusion but suggested that a small secular acceleration may exist, a conclusion in agreement with Mariner 9 observations (Born and Duxbury, 1975; see Chapt. 14 by Pollack).

One must mention here the difficulty of observing such small bodies so near to the central planet (Chapt. 5, Pascu). This explains why the only physical parameters to be safely determined from the analysis of their motions are the mass of Mars and the main coefficient J_2 of the expansion of its potential, although J_4 and the direction of the pole could, in principle, be derived from a good knowledge of the orbit of Phobos. Mariner 9 observations have provided a much better definition of the Martian gravity field (Born, 1974; Jordan and Lorell, 1975).

The best numerical result now obtained is for the reciprocal mass of Mars. It has been derived from the solar perturbations of Deimos as well as from those of both satellites (Wilkins, 1967) and, with a much smaller uncertainty, from Mariner 4 observations (Null, 1969). J_2 has also been computed from classical observations (Woolard, 1944; Wilkins, 1967; cf. Burns, 1972; Born, 1974).

A special effort to solve the observational problem, such as Pascu's experiment (1967, Chapt. 5 herein, 1975), is desirable in order to obtain a better description of the orbits, but the classical methods can no longer compete with the analysis of the motion of space probes to derive dynamical parameters: for instance, the motion of Mars' orbiters gave all the coefficients of the potential up to the eighth-order harmonics (Born, 1974; Jordan and Lorell, 1975).

The Inner Satellites of Jupiter (I to V)

These satellites belong simultaneously to several classes. One observes a progressive transfer from Class 1 to Class 2a when following the sequence V-I-II-III-IV. At the same time, the four Galilean satellites (I to IV) have nonnegligible masses, leading to strong interactions. Moreover, these interactions are enhanced by the existence of two approximate resonances between the mean longitudes (Chapt. 8, Greenberg). The relations between the mean motions are:

$$n_1 - 2n_2 = n_2 - 2n_3 \approx 0 \ ,$$

implying the exact resonance:

$$n_1 - 3n_2 + 2n_3 = 0 \ .$$

As a consequence, these satellites are good indicators of many dynamical parameters. Let us leave the mass of the planet to be derived from the solar perturbations of more distant satellites or the perturbation of spacecraft. The fifth satellite is in principle the best to obtain the coefficients of the potential and the direction of the pole, but it is better to include the Galilean satellites in the analysis, since the innermost satellite presents the same observational problems as those described for the satellites of Mars. On the other hand, several satellites are needed in order to separate the effects of J_2 and J_4. This has been done by de Sitter (1931) and van Woerkom (1950). The direction of the pole of Jupiter and its precession were determined by Sampson (1921). H. Struve (1906) derived also the direction of the pole from the motion of the fifth satellite.

A most interesting feature of the Galilean satellites is the possibility of determining their relative masses, since the interactions and the resonances are strong. Sampson (1921) introduced one set of values in his theory of motion which is still in use.

Better values were produced by de Sitter (1931), but these results have now to be compared to the determination of Null *et al.* (1974) from the analysis of the motion of Pioneer 10, which gives also J_2, J_4, the direction of the pole and the mass of the planet.

We already quoted the difficulty of observation of the fifth satellite, analogous to that of Mars' companions. The problem is different for the Galilean satellites, the main distances of which are larger and the magnitudes smaller. Here, one has to seek greater accuracy, and there is a serious need of modern long-focus photographic observations (see Chapt. 5, Pascu).

The Outer Satellites of Jupiter (VI-XIII)

These bodies cluster in two groups within which the kinematical properties are very similar. The first group (VI, VII, X, and XIII) includes orbits with notable eccentricities and inclinations, at a distance of about 160 Jovian radii. The satellites of the second group (VIII, IX, XI, and XII) have retrograde motions, large eccentricities, and a mean distance of about 300 planetary radii (see Table 1.2).

For all these bodies, the main perturbation is the solar one, so that they allow good determinations of the mass of Jupiter. Goldreich (1965b) noticed near-commensurabilities of the mean motions with the mean motion of the Sun:

$$n_6 - 17n_\odot \approx n_7 - 17n_\odot \approx n_8 - 6n_\odot$$
$$\approx n_9 - 6n_\odot \approx n_{10} - 17n_\odot$$
$$\approx n_{11} - 6n_\odot \approx n_{12} - 7n_\odot \approx 0$$

Analytical theories are here difficult to derive, since the eccentricities and n'/n are large. Let us mention the study of the motion of satellite X by Lemechova (1961). Seminumerical theories are easier to obtain—for instance, see Kovalevsky's work on satellite VIII (1959)—but it represents the actual motion to only one or two minutes of arc. This is why the most current and efficient tool here is numerical integration, which gives an accuracy of a few arcsec. The method was used by Herget (1967) and Bec (1969), who obtained the reciprocal mass of Jupiter, from the comparison of the motions of satellites VIII and IX, respectively, with numerical integration.

The problem of the accuracy of the observations is not as critical as it is for the Galilean satellites, so that long focus astrometry is not needed. The main problem is the lack of observations: some satellites have not been observed at all for ten years. Thus a greater number of observations even of medium quality are required first in order to improve our knowledge of the constants of integration. One has to use large astrographs, since the magnitudes of these small bodies range between 15 and 19.

The Inner Satellites of Saturn (I-V and X)

These satellites are disturbed by the effect of the oblateness of the planet and by the ring, while solar perturbations are noticeable only for Rhea. The masses are small, but four of them may be determined since resonances enhance the interactions (Chapt. 8, Greenberg). Enceladus and Dione admit the critical argument $2l_4 - l_2 - \bar{\omega}_2$, so that $2n_4 - n_2$ is very small, and Mimas and Tethys are subject to the resonance of argument $4l_3 - 2l_1 - \Omega_1 - \Omega_3$, so that $2n_3 - n_1$ is also small. Classical determinations of masses have been accomplished for Mimas (G. Struve, 1930, and Jeffreys, 1953), Enceladus (G. Struve, 1930, and Jeffreys, 1953), Tethys (G. Struve, 1930), and Dione (G. Struve, 1930, and Jeffreys, 1953).

The mass of Rhea is poorly known, since there is no resonance. G. Struve (1930) estimates it between 20 and 50×10^{-7}, while Jeffreys (1953) obtains $(32 \pm 38) \times 10^{-7}$.

Jeffreys (1954) derived also values of J_2 and J_4 for Saturn from the motions of the inner satellites. The direction of the pole of Saturn was determined by G. Struve (1930), and its precession by H. Struve (1898).

The orbit of the recently discovered satellite X (Janus) is known with a very low accuracy and cannot at present give reliable information on the gravitational field of the planet.

The Outer Satellites of Saturn (VI-IX)

Titan has the largest mass in Saturn's satellite system. It can be rather well determined, since its action on the motion of Hyperion is a strong resonant one, the critical argument being $4l_7 - 3l_6 - \bar{\omega}_7$ (Eichelberger, 1911; Woltjer, 1928; Jeffreys, 1954; Greenberg, Chapt. 8, herein). The resonance acts only on the motion of Hyperion, whose mass is too small to induce sizeable perturbations of Titan and hence cannot be determined in this way.

Iapetus' orbit is characterized by a high inclination, and is perturbed by the oblateness of the planet and the solar action (both being the same order of magnitude), as well as by the action of Titan. Its motion has been studied by G.

Struve (1933) and more recently by Grebenikov (1958, 1959) and Sinclair (1974b). Delaunay's method has been unsuccessfully applied to Phoebe (Zadunaisky, 1954), while Elmabsout (1970) adapted Hill's lunar theory to this case. Jeffreys (1954) derived the value of the reciprocal mass of Saturn from the analysis of the seven largest satellites, but a better determination is given by the study of the motion of Jupiter.

Although these satellites (except for Phoebe) are brighter, the problem of their observation is the same as for Jupiter's outer satellites, except for Titan, which is more like the Galilean satellites.

The Satellites of Uranus

These are Class 1 satellites. It seems that the important action of the J_2 perturbation ensures the stability of these orbits, which are nearly perpendicular to the orbit of the planet. Note that $n_5 - 3n_1 + 2n_2$ is very small (compare with Jupiter's three first satellites, for which this quantity is exactly zero).

The study of these satellites allowed van den Bosch (1927), Harris (1949) and Dunham (1971) to obtain values for the reciprocal mass of the planet. The accuracy of the observations is not good enough to give reliable values of J_2. Nevertheless, H. Struve (1913) deduced the direction of the pole without any sensible precession. See Greenberg (1975) for a more recent discussion. Favorable times to determine the mutual inclinations occur every 42 years (1966, 2008, . . .), when Uranus' equatorial plane contains the radius vector.

The Satellites of Neptune

Triton is a satellite of Class 1 and appears as a good indicator of the direction of the pole of the planet. The coefficient J_2 for Neptune has been obtained but is poorly determined since the orbit is nearly circular. The analysis of this orbit also gives the value of the reciprocal mass of Neptune. All these results were obtained by Gill and Gault (1968).

The mass of Triton may be determined in the same way as the Moon's, by analyzing the ''lunar inequality'' it induces on Neptune's orbit when it moves around the common center of masses (Alden, 1943).

Nereid has no sensible mass. Van Biesbroeck (1957) and, more recently, Rose (1974) used it to obtain the value of the reciprocal mass of the planet.

ASTROMETRIC TECHNIQUES FOR THE
OBSERVATION OF PLANETARY SATELLITES

Dan Pascu
U.S. Naval Observatory

The construction of the great refractors in the late 19th century made it possible to obtain precise positions for all known planetary satellites. Prior to that time, precise positions from phenomena and heliometer observations were limited to the Galilean moons and Titan. The mean error of the visual micrometer observations made with the great refractors was about ±0.4 arcsecs with the major portion due to systematic errors in the measurement of the planetary disk.

In 1885 H. Struve introduced a scheme in which only positions of one satellite relative to another are used in the orbital adjustment. The use of these intersatellite positions resulted in a significant improvement in the satellite orbits due to the elimination of the systematic error in the measurements on the planet. The mean error of one intersatellite observation obtained visually is about 0.2 arcsecs. Unfortunately, Struve's method is neither generally applicable nor always advantageous and it has thus been necessary to continue making observations of satellites relative to the planet.

At the turn of the century, the introduction of the photographic technique was successful largely due to Struve's method. At present the photographic technique has completely replaced the visual micrometer technique for satellite observation. In the modern photographic technique, as in the classical visual technique, the main difficulty lies in the determination of the position of the primary with an accuracy comparable to that of the satellite. There have been some modest successes in this, using observational and statistical techniques. The mean error for one exposure is less than ±0.1 arcsec for a planet-satellite position as well as for an intersatellite position. The smaller errors of the photographic observations are due to the virtual elimination of the personal errors.

For the future, spacecraft imaging techniques and Earth-based radar techniques hold considerable promise. The expected precision of these observations is on the order of 10 km. For the satellites of the outer planets, these observations would be from one to two orders of magnitude more precise than conventional photographic observations. Unfortunately, the expense and irregularity of these observations make it unlikely that they will replace the conventional photographic ones in the foreseeable future.

The heyday of the observational and orbital study of planetary satellites was the period from 1874 to about 1928. The increased activity during this period was directly related to the progress made in the mid-nineteenth century on the theories of the motions of the major planets. It was clear to Newcomb that any improvement on Leverrier's Tables would involve a new discussion of the fundamental constants, including the planetary masses. At that time the mass of a planet was determined from the size of the orbits of its satellites.

The phenomenal surge in observations, however, did not come until the construction of the great refractors late in the nineteenth century. The great refractors were instruments "par excellence" for the astrometric observation of satellites. Not only was it possible to observe easily the faintest moons, but the long focal lengths of these instruments made it possible to increase the accuracy of the orbital scales. In the introduction to his work on the Uranian and Neptunian systems, Newcomb (1875) wrote

...when the 26 inch Equatorial with an object glass nearly perfect in figure, was mounted at the Naval Observatory, the observation of the satellites of the outer planets, with a view of determining not only the elements of their orbits, but more especially the masses of the planets, was made the first great work of the instrument....

Most of the dozen large refractors constructed in the three decades following the construction of the Naval Observatory's 26-inch were employed in the astrometric observation of satellites. Certainly the reputation gained by Asaph Hall for his discovery of the Martian moons must have been an important personal motivation for other observers. It is not surprising, thus, that the most diligent observer of that time, E. E. Barnard, discovered the fifth satellite of Jupiter with the largest refractor of that day.

Before the close of the nineteenth century the problem of the masses of the outer planets (Mars through Neptune) was solved to Newcomb's (1895) satisfaction. Research on the satellites did not diminish but rather increased. The remarkable discoveries made on the complex motions in the Jovian and Saturnian systems caused astronomers to consider these systems as small-scale replicas of the solar system. In his Darwin Lecture, de Sitter (1931), in speaking about the Galilean system, pointed out that:

The interval of 321 years since the discovery of the satellites thus is, in the number of revolutions, equivalent to nearly 18,000 years of the four inner planets, and to more than 1,100,000 of the outer planets. During all this time the general aspect of the system has not changed, and especially the stability appears to be unimpaired. This enormous magnification of the time scale makes the system of the satellites of special interest for the study of secular and long periodic perturbations....

It is clear that by the beginning of the twentieth century the emphasis in satellite research had changed from obtaining the dynamical parameters of the primary to the determination of the motions and dynamical parameters of the satellites themselves.

These fifty years of activity increased the number of known satellites by 50% (from 17 to 25), largely due to the construction of the great refractors and the introduction of photography. The massive observational and theoretical effort of this period produced dynamical constants of the planets and satellites still valuable, and the orbital theories of this era are used by the national almanac offices for the computation of present ephemerides (cf. Appendix, by Seidelmann, and Aksnes, Chapt. 3, herein).

The dynamical study of planetary satellites prospered and then languished along with the rest of celestial mechanics. As in the latter field, satellite studies attracted fewer young astronomers; when the outstanding astronomers who had devoted their time to this work died, the field went into a steep decline. Though observations were continued at a much reduced pace until the early 1950s, interest in the dynamical studies of the planetary satellites was revived more recently with the aim of improving the satellite ephemerides for the purpose of the spacecraft reconnaissance missions to the outer planets (Aksnes, Chapt. 3, herein, and Appendix by Seidelmann).

In view of this renewed interest in the positional obervation of the satellites, it is important to describe the classical techniques of observation as well as those being used today. The modern techniques were designed to remedy the shortcomings of the classical methods and, though they have been largely successful in this, some of the problems remain to this day.

CLASSICAL OBSERVATIONAL TECHNIQUES:
THE VISUAL METHODS

Observations made for the purpose of orbital adjustment were, until very recently, of two kinds: (a) the timing of phenomena and (b) the measurement of positions in the tangent plane. Recently the optical spacecraft technique has added a new dimension to the positional observation of planetary satellites; this technique, while a modification of type (b), should be especially valuable for the outer planet satellite systems because of its potential for precision. Soon it will also be possible to obtain ground-based radar observations of the Galilean moons and a few other large satellites. This is an entirely new type of observation, which, if the radar observations of the planets themselves are any indication of its accuracy, will make a valuable, though limited, contribution.

Phenomena

The phenomena include such planet/satellite events as eclipses, occultations, and shadow transits (cf. Aksnes, Chapt. 3). They also include the mutual phenomena of the satellites. The major advantage in the use of such observations is that they are essentially independent of instrumental characteristics. Unfortunately, they can be used successfully only for orbital adjustment for the Galilean moons. Phenomena for the other satellites are too rare, or the satellites are too faint and too close to a bright primary, for reliable observations. The principal phenomena observed for the Galilean moons are eclipses due to the primary. The fact that eclipses are frequent and can be observed with small instruments made them of great practical value in the determination of geographic longitude. It is for this reason, and the belief that phenomena observations were superior to astrometric observations, that eclipse timings were made to the exclusion of astrometric observations up until 1890.

Eclipse observations do have their shortcomings: notwithstanding that the in-plane parameters such as the motion in longitude are accurately determined from them, the scale and orientation of the orbits are not. It was because of this weakness in the eclipse observations that J. C. Adams persuaded Gill (1913) to undertake his famous series of heliometer observations at the Cape in 1891. Another objection to the use of satellite phenomena is the apparent limitation in the accuracy attainable due to systematic errors. Sampson (1910) showed that the Harvard eclipse series was affected by systematic effects which he attributed to an oversimplification of the eclipse model. The timings thus are also dependent on such nonorbital parameters as the refraction and absorption of light in the Jovian atmosphere, and the shape and surface brightness distribution of the satellites (see Cruikshank, 1974). After reviewing the results from several series of eclipse observations, de Sitter (1931) concluded that

> observations of eclipses, however carefully made, cannot determine the time with a greater accuracy than $\pm 10^s$, and this limit cannot be lowered by combining a great number of observations.

A modern series of photoelectric eclipse observations made by Kuiper and Harris in the early 1950s has been reduced using a more sophisticated eclipse model (Harris, 1961). A discussion of these observations, as well as those of Cruikshank and Murphy (1973) and Greene et al. (1971), is given by Peters (1973). The large residuals after solution tend to confirm de Sitter's conclusions.

Recent activity in this area has focused on the mutual phenomena of the satellites [see Aksnes and Franklin (1975) and Duxbury et al. (1975)]. Although the solutions for the longitude corrections appear to be quite accurate, the out-of-plane parameters cannot be determined well, and the observations are much too rare.

Observations of Position

Observations of the second type—positions in the tangent plane—include visual observations made with a heliometer or filar micrometer and photographic observations (including spacecraft observations). Hypothetically, observations of this type may be obtained for all satellites with few restrictions due to config- urations. Astrometric observations also may be used to determine a complete set of orbital parameters. Thus, this type of observation is, and has been, the most widely used for the majority of the satellites.

THE HELIOMETER

The heliometer is a form of double-image micrometer developed to measure large angles—in particular the solar diameter (from which its name is derived). Its application to satellite observation is due to this property and to its additional capacity for reducing personal systematic errors.

In the heliometer the object glass is divided in half, forming two images in the focal plane whose separation is a function of the separation of the two halves of the objective along their common side. By turning the objective such that the direction of the line of section is in the position angle of the line joining the two objects to be measured, the separation of the two objects is obtained by first superposing the two images of the same object and then superposing the image of the first object on the image of the second. The position angle is then the angle that the line of section makes with the hour circle.

Since the separation and rotation of the lens sections must be accurately measured, the dimensions of the heliometer objective are necessarily small. Thus, the application of this instrument to the positional measurement of satellites has been limited to the Galilean moons and Titan. This is the main drawback of the heliometer for such work. Although the heliometer was not successful in completely eliminating the systematic personal error in the measurements taken with respect to the primary, the errors appear to be smaller than those of other methods. In fact, the heliometer mass determinations for Jupiter and Saturn were among the most accurate before 1870.

The most successful program of observation carried out with the heliometer was the series of intersatellite measurements of the Galilean moons made with the Cape heliometer by Gill and Finlay in 1891 and by Cookson in 1901 and 1902. (See *Annals of the Cape Observatory* 12, 1915). The probable error of the separations was estimated by de Sitter (1931) as being less than $\pm0.''1$. After a comparison of the relative accuracy of the photographic and heliometer observations, de Sitter (1931) concluded that, for equal focal length, the heliometer observations were superior, and he made an appeal for the construction of a heliometer of focal length from 5 to 7 meters. To the writer's knowledge no such instrument has ever been built.

THE FILAR MICROMETER

While the two principal advances introduced by the great refractors for satellite observation are increased light-gathering power and a long focal length, significant improvements were also made in the stability of the mounting and the accuracy of the clock drive—both necessary for precise measurement. In addition to these advances in the telescope, the filar micrometer reached a high level of development at this time. Principal improvements were the introduction of thin wires such as spider webs, the movable reticule box, finely threaded screws, and various forms of achromatic eyepieces. For a detailed description of the history and construction of the filar micrometer, see the articles by Sir David Gill in the 9th and 11th editions of the Encyclopaedia Britannica.

By far the greatest number of astrometric observations of satellites have been made with the filar micrometer. The reason is simply that, before the photographic technique was fully developed, this was the only form of observation for most satellites and by the time the photographic technique had become competitive, observational work had declined.

The basic micrometer for satellite observations around the turn of the century included a reticule box with a fixed wire parallel to the micrometer screw and two wires perpendicular to it. One of these two wires was fixed while the other was moved by the micrometer screw. The whole reticule box was made to be moved by a coarse screw in the direction parallel to the fine micrometer screw. Since a large field was needed, the eyepiece was usually an achromatic wide-field type which was mounted on a sliding plate so that a large portion of the focal plane could be inspected. The wires or the field could be illuminated with various intensities and colors, and a red illumination was commonly used for contrast. Various forms of occulting bars, smoked mica, or red glass filters were commonly fixed in the eyepiece in order to reduce the brightness of the primary.

The measurements usually were made in separation and position angle, although a higher degree of accuracy was attained in rectangular coordinates for large separations. On occasion, the position wire was oriented in the equator of the planet, and "longitude" and "latitude" measurements were made. Barnard made these types of measurements for the fifth satellite of Jupiter. For position angle, either the fixed (position) wire or the movable (micrometer) wire was used. Once the celestial equator was set by trailing an equatorial star along the wire, both the satellite and planet were bisected by the wire. If the movable wire was used for the position angle measurement, the micrometer head was turned 90° for the measurement of separation. The method of double distances was used for the separations; a bisection (or limb setting) was made on the planet with the fixed wire, while the satellite was bisected with the movable wire. The bisections were then repeated with the wires interchanged—giving twice the separation. This technique was used to avoid the necessity of obtaining the coincidence of the wires. Measurements on the planet were made in three ways. First, the planet was simply bisected. This was always done for Neptune, usually done for Uranus, but seldom done for the remainder of the planets because of their large diameters. Asaph Hall made an exception to this—he almost always bisected the planet. A second classical method was to make tangential settings on the limbs of the planet, make a correction for phase, and obtain the center by dividing the difference between the limb settings by two. For Saturn it was preferable to set the micrometer wire on the edge of the rings because of their better definition— particularly when the rings obstructed the limbs. It was generally assumed, in the reduction to center, that the centers of figure of the planet and of the rings were the same. The third method is similar to the last one; a tangential setting was made only on one limb because of difficulties in making a setting on the other limb. The diameter of the planet was then measured with the micrometer, and this measured semidiameter was used to reduce the limb measurement to the center of the planet's disk.

The accidental error of a separation with respect to the primary usually varied widely—from $0.15''$ to $0.5''$ (m.e.)—depending on a number of factors such as the number of bisections made, the observer, the brightness of the planet and satellite, the size and proximity of the primary, the zenith distance, the atmospheric seeing, and so forth. In H. Struve's (1898b) discussion of the nineteenth century observations of the Martian moons, it can be seen that the errors vary in this range but for no apparent reason. The accidental error of a position angle ($=$sdp with the separation s and the differential of position angle dp) is generally greater than that for separation, contrary to what is found for double stars. Sometimes the reason for this is simply that more measurements were made for the separations than for the position angles. This was often the case when the mass of the planet was the main objective of the work. For large separations, however, the larger errors of the position angles were due to the limitations of the position circle. Although the circles could generally be read to one hundredth of a degree, their accuracy was somewhat less. Hall (1885), in his work on Iapetus, found that of 30 residuals greater than $1.0''$, 26 were in position angle. He attributed this to the inferiority of the position circle—pointing out that an error of $0.1°$ at greatest elongation ($550''$) would result in an error of $1.0''$ in sdp. Better results for large separations were obtained by the measurement of $\Delta\alpha\cos\delta$ by the timing of transits of the planet and satellite across the position wire. The difference in declination was then determined by the use of the micrometer screw. Other sources of error in the measurement of large separations are due to incoherent seeing oscillations and the inability of the observer to see both limb and satellite simultaneously.

There are three sources of systematic errors in the micrometer observations. Some of these are peculiar only to the micrometer observations, and a few of them still plague us today.

Instrumental Errors. These include primarily the screw value (scale) and orientation parameters, but there also are indications of residual effects due to the colors of the objects and the micrometer illumination. The major systematic error in the orientation due to the instrument would arise from a misalignment of the polar axis. An error would result if an equatorial star was used to align the position wire. This error could be minimized by using the planet, or a star close to it, for the alignment of the position wire.

Because of its importance to the determination of the planetary mass from the scale of a satellite orbit, the instrumental error of main concern is the value of the revolution of the screw. An error in the screw value propagates into the mass with a factor of 3. A carefully determined screw value (including possible periodic or progressive errors) is accurate to one part in 10^4. Thus an error of 3 parts in 10^4 could at best be obtained in the mass of the primary. This accuracy was never realized from those satellites which have an orbital semimajor axis less than $100''$ because the accidental error on the semimajor axis was generally greater than $\pm 0.01''$.

For Saturn this precision was obtained on occasion from Titan and Iapetus and for Jupiter from Ganymede and Callisto. In these cases the effort spent in determining an accurate value for the revolution of the screw was justified. In the determination of the screw value systematic differences occur between the values obtained from different techniques. These differences are usually attributed to instrumental limitations; however, there were also sizeable personal errors involved—though the techniques were designed to minimize them. Asaph Hall (1893) indicated that the value of the revolution of the screw (for the micrometer of the 26-inch telescope) to be used with his measures was different from that determined by Holden (1881) by one part in 10^3.

Personal Errors. These include the systematic errors due to the observer, such as the magnitude error in bisecting a stellar image and the position angle error due to the orientation of the line joining the objects to the horizontal. To minimize this position angle error one should observe so that the line joining his eyes is parallel to (or normal to) the line between the primary and satellite. Occasionally the satellite system was observed east and west of the meridian in order to change the angle of the line between the planet and satellite by 90° with respect to the horizon.

The magnitude error arises when the observer does not estimate the true center of an image in bisecting it but makes his bisections consistently to one side of center. This error is proportional to the size and/or brightness of the image. In photographic work only the size would be important, but in visual work both are a factor. In the visual micrometer observations most of the error had to be in the bisections of the planet, and since means were used to diminish the brightness of the planet without minimizing the systematic error, the apparent size of the planet must have been the main factor in the origin of this error. Attempts to eliminate the magnitude error were usually accomplished by use of a reversing prism eyepiece. Van den Bos was one of a few who used it for double stars, but it has almost never been applied to satellites. The main reason given is that for those systems in which the primaries were bisected, the satellites were too faint for the use of the prism. For Saturn and Jupiter, where limb measures were usually made rather than bisections of the whole disk, the applicability of the reversing prism is doubtful. In addition, the reversing prism eyepiece is difficult to use; thus it has rarely been employed.

This magnitude/size error is just one facet or explanation of a larger problem in the measurement of the planetary disk. This problem was most thoroughly discussed by Hermann Struve (1888) following his many analyses of satellite observations. Struve studied what he referred to as the difference between the optical center of the planet and the center of gravity. This is more correctly expressed as the difference between the observed center (however that may be obtained) and the center of figure of the planet. This difference will be referred to as the center of figure error (CFE). Explanations for the phenomenon indicated that it was of "astronomical" origin and these will be discussed below. That there was a personal component to the CFE was shown in a relatively late paper by Asaph Hall Jr. *et al.* (1926). Observations of the Martian moons were made in the same manner by two observers, Hall and Bower. Their (O−C)'s indicated, first, that "the observed distances and position angles are too small when the satellite is observed in eastern elongation and too large when observed in western elongation." Secondly, the variation in the apparent magnitude of the CFE between the two observers indicated that personality was a sizeable factor. The fact that the sign of the (O−C) varied with the side of the planet is quite consistent with the personal magnitude/size error.

One might expect that measurements made on the limb of the planet would be free from personal errors. This is not so. In the measurement of photographic images of planets by tangential settings on the limbs, it was found that a "direct minus reverse" difference of about 20 μm or more was consistently obtained. This was found to be due, in large part, to the weakness of a tangential setting technique using a screw-type (backlash) measuring device. The criteria used for making a judgment of tangency on the right or lower limbs were different from those on the left or upper limbs. For a screw-type measuring device in which measurements are made from right to left and from bottom to top, the instant of tangency (on the right and bottom limb) is taken to be when the bright space separating the approaching black planetary limb and the black wire disappears as first contact is made. This presents no difficulty; the problem arises on the opposite limb. When the wires are brought across the large planetary disk, the wires superposed on the disk become difficult to see and one must judge the approach to tangency by the visible parts of the wires. The criterion for the instant of tangency for these limbs is the instant of last contact—just before the wire separates from the limb and the first particle of light is seen. In fact, one cannot make this judgment with confidence until he has passed tangency and first sees the light. By doing this one overestimates the opposite limb. If one tries to judge tangency before last contact is broken, one usually underestimates the limb by a larger margin. In the photographic technique, the reversal of the plate will eliminate the error. In the visual micrometer technique, however, a reversing prism was not used and, in any case, it would *not* have detected or removed this error because the direction of the motion of the wires is also reversed. Thus the wire makes first and last contacts on the same limbs. This error could have been eliminated simply by reversing the micrometer head 180° and repeating the measurements—no prism was needed.

Astronomical errors. In this chapter, astronomical errors do not mean spherical corrections, such as differential refraction of parallax, but rather errors arising from some physical phenomenon not directly associated with the instrument or personality. As might be expected, these have to do primarily with the center of figure error. Two classical explanations for the CFE were offered by H. Struve (1898b, 1903). The first was due to atmospheric dispersion. Not only would this distort (to some extent) the planetary disk, but in the presence of a significant color difference between the planet and satellite, their separation in zenith distance would also be altered. The second source might be explained as differential irradiation along the limb of the planet due to intensity variations such as the bright polar cap of Mars, the dark polar regions of Saturn, or the region of the terminator. This latter effect, along with the personal errors, must account for the major portion of the CFE.

Intersatellite Measurements

Hermann Struve was the giant of observational and orbital work on the natural satellites. In the 30 years from 1885 to 1915 he analyzed the observations made with the great refractors and made a definitive study of the motions of almost all of the known satellites of the solar system. In his observational work, he took great pains to eliminate sources of accidental and systematic error.

In his first major work on the Saturnian system, Struve (1888) introduced into general use a new technique for the observation and orbital study of satellites using only the position of one satellite with respect to another. These "intersatellite positions" or "intersatellite connections" are obtained by observing the satellites in pairs, in position angle and separation or in rectangular coordinates x, y, in the same manner as the "planet-satellite connections." For the orbital adjustment with these observations one begins with the same conditional equations in X and Y as for the planet-satellite connections and forms new conditional equations by subtracting the corresponding equations for the two satellites. The $(O - C)$'s then refer to the intersatellite positions and the corrections to the orbital parameters of both satellites occur in the equations. With this technique Struve was able to reduce the observational accidental errors and eliminate the systematic error due to measurements made on the planet. There is however a serious drawback to this scheme, which was known to Struve, and that is that the eccentricity, e, and the longitude of the apse, π, are not well determined. Laves (1938), in his treatise on the Saturnian system, derives the error of edπ (where dπ is the differential in longitude), based on unpublished lectures of H. Struve. He finds that the error of edπ from intersatellite coordinates is 30 times greater than that derived from planet-satellite connections. For those satellites with eccentric orbits this has serious consequences since, if the eccentricity and apse are not well determined, then neither is the semimajor axis, nor correspondingly the mass of the primary. In subsequent work on the Saturnian system, Struve (1898a) made planet-satellite connections of the four outer moons as well as intersatellite connections for all of them. He discovered a systematic error in the y coordinates for the planet-satellite connections and treated it by introducing a term, Δy, into the conditional equations in y. This technique was also used on other satellites which were referenced to the primary such as JV (Struve, 1906). The introduction of these observational parameters into the conditional equations, however, is counterproductive when they are highly correlated with one or more of the orbital parameters (such as the eccentricity).

In the ensuing years, Struve recommended intersatellite observations or mixed intersatellite and planet-satellite positions for a particular system depending on the problems and configurations in it (Struve, 1903).

MODERN OBSERVATIONAL TECHNIQUES:
THE PHOTOGRAPHIC TECHNIQUE

The application of photography to satellite astrometry was made first for the Galilean satellites in 1891 at several observatories. Plates were taken at Pulkovo by Kostinsky (Renz, 1898), at Helsingfors by Donner, and at the Cape—all with Carte du Ciel astrographs. Only the small scale of the instruments prevented these observations from being more competitive with the visual methods. The application of photographic techniques during the early part of the century was directed more to the discovery and orbital work of the faint outer satellites of Jupiter and Saturn using short focus astrographs. The early application of the long-focus telescope to the photographic technique was deterred because the inferior photographic emulsions of that time required long exposures and permitted the bright planetary image to grow too rapidly—obliterating the inner, fainter satellites. Exceptions to this situation are the Galilean system and the Neptune-Triton system. For the latter system photographic observations with a long focus instrument were first made at Greenwich from 1901 to 1910 (see Greenwich Observations for 1904). These observations did not result in the photographic technique taking the place of the visual technique for Triton nor in its establishment as a competitive alternative. The observations were stopped, probably because a systematic difference was found in the scale of the orbit of Triton between the photographic and visual observations; and for some reason, as often occurs nowadays, systematic errors were attributed to the photographic observations rather than the visual. Thus, it was to de Sitter's credit that for 30 years he pursued the photographic technique for the Galilean moons, first with short-focus astrographs and finally with the long-focus refractors, and proved the superiority of the photographic technique for the Galilean satellites. His success influenced others to apply photographic methods and was directly responsible for the long and successful series of photographic observations of the Saturnian system begun by Alden and O'Connell (1928) in 1926. De Sitter's discussion of Alden's 1927 and 1928 observations of the Galilean satellites at the Yale Southern Station indicated a precision of $\pm 0\,.''06$ (p.e.) for an intersatellite position obtained from one exposure. Similar results were found by Garcia (1970, 1972) in his discussion of the photographic series of Saturn's satellites begun by Alden and O'Connell. G. Struve (1928) compared the relative precision of the 1926 Alden photographic observations of Saturn's satellites with his own micrometer observations made at the same time. He concluded that the photographic observations are generally twice as precise as the visual ones. In addition, the photographic technique saves both telescope time and time in the reductions. The photographic plate also becomes a permanent record which may be referred to in the future.

One by one, successful series of photographic observations were obtained for each planetary system or satellite. In some cases, good photographic observations were not obtained until the late 1960s (e.g., Jupiter V by Sudbury, 1969). The photographic technique has replaced the visual technique for all satellites (excluding the Moon), and there no longer are plans for visual observations (cf. Appendix, by Seidelmann).

The conversion from the visual technique to the photographic technique was made possible largely due to the orbital adjustment procedure for intersatellite observations introduced by H. Struve. Thus, the overexposed primary was of little concern. One serious objection to the use of intersatellite observations has already been mentioned, but there are other problems: Observationally the method is not feasible for Neptune's satellites or for Jupiter V because of a large magnitude difference between the satellites; nor is the method feasible for Jupiter VI-XIII and Phoebe because of the great extent of their systems. The method also does not work well for the Martian satellites, as will be explained. Further problems are encountered in the comparison of theory with observation: There are as many as twice the unknowns to determine from the intersatellite observations as for the planet-satellite observations. This alone will degrade the precision of the corrections; but, in addition to this, statistical correlations, which often arise between the corrections to the parameters of both satellites, will further confound the results.

Another problem is encountered when the observations for one satellite are more accurate than those for the other, or when the theory for one of the satellites is more complete than that for the other. Both the observational and theoretical problems for Phobos, for example, are greater than those for Deimos. An orbital adjustment in this case, from intersatellite positions, would be at the expense of Deimos. For the Galilean system also (Arlot, 1975), the failure to include an important term for one of the satellites (probably J III) in the orbital adjustment using the intersatellite observations has resulted in a degradation of results for the other satellites. It is clear that the use of planet-satellite connections in the comparison of theory with observation is superior to the use of intersatellite connections provided that the position of the primary can be determined with a precision comparable to that of the satellites.

In the modern photographic technique there are basically two problems which are dealt with. The first has to do with the procedures used for reducing the photographic plates. The second is concerned with the means, whether by observation or ephemeris, of determining the center of figure for the planet at the time of the exposure.

Plate Reduction Procedures

Two procedures are used in satellite work for reducing photographic plates. The first, the trail/scale method, is derived from the visual technique. A trail of a star or satellite image is exposed on the plate along with the satellite system. This trail (if it is a stellar trail) represents the equator of date and is used to orient the plate in the measuring engine. The measured X and Y coordinates are converted to angular measure by applying the scale value for the instrument given in arcsec/mm. Corrections must be applied to the orientation for the motion of the system, precession, and nutation, while the scale value is adjusted for atmospheric compression and dispersion (for details of this technique see van de Kamp, 1967). This technique is generally sufficient for small scale orbits, but the large scale orbits require greater precision. A well determined semimajor axis will have an error of about $\pm 0.005''$. Since nightly scale variations of 1 part in 10^4 are common even for astrometric refractors, the mean scale value is insufficient for orbits with semimajor axes greater than one arcmin. This problem is partially remedied by taking nightly scale plates (when possible). The precision of the scale is then limited by the accuracy of the atmospheric corrections to it. Unfortunately, this will not improve the orientation which can, in general, be determined by a star trail to only $\pm 0.01°$ for a well-adjusted telescope. The accuracy of the orientation can be improved somewhat if a star bright enough to be trailed is located a little east of the field. After the last exposure the drive is turned off and, when the star reaches the field of the plate, the shutter is opened and a trail impressed on the plate. This eliminates the errors due to clamping, unclamping, re-engagement of the drive gears and moving the telescope. This requires a somewhat extraordinary circumstance, however; and a more realistic approach would be to take a fainter star east of the field, allow it to drift into the field and expose it, for the same duration, at equal distances from the center of the field (allowing it to drift between the exposures). This requires only two re-engagements of the drive gears. Further improvement of the orientation from a trail is obtained by making measurements on the trail, fitting them with a quadratic, and reducing the measures of the planet and satellites to it. This minimizes errors due to the deviation of the optical axis from the center of the field as well as errors from oscillations due to seeing. If more than one exposure of the satellite system is taken on a plate, reduction to the trail exposure must be accomplished by use of two well-spaced field stars of similar magnitude. In the absence of these, a trail should be made for each exposure unless the mechanism for separating the exposures contributes a negligible error to the orientations.

An alternative method, the method of plate constants, can improve both the orientation and the scale, but not without a significant increase in labor, and not before certain requirements are fulfilled. In this method one must have a good configuration of background reference stars on the plates with known positions. The differences between the tangent plane coordinates of the reference stars and their measured coordinates are expressed as a power series in the measured coordinates and occasionally the magnitudes. The coefficients of this series then represent corrections to the field, and in particular, to the scale and orientation of the plates. From these expressions, the positions of the satellites and planet are found from their measured coordinates. The errors of the scale and orientation are a function of the positional and measuring errors of the reference stars, the number of reference stars, the shape and size of their configuration, and the validity of the model.

A few practical problems are encountered in the use of this technique for satellite observation using long-focus telescopes. First, even if the limiting magnitude of the satellite plates is 12, there will be a substantial fraction of fields which will have too few stars or too poor a configuration for an accurate determination of the orientation and scale parameters. In these cases the trail/scale method has to be used. Second, even in the event that there is a satisfactory field of background stars, either their positions will not be known at all or with an accuracy insufficient to improve on the trail/scale method. The only reasonable way to solve this problem is to take several field plates with a short-focus astrograph. Reduction of these plates with one of the modern reference catalogues (AGK3R, AGK3, SRS, or Yale) will result in a precision of $\pm 0.04''$ in the relative positions of the fainter stars. With these procedures it is expected that, for well exposed images, the mean error for one exposure of the planet-satellite or intersatellite positions will be as small as $\pm 0.05''$, with a scale uncertainty of 1 part in 10^5 and an orientation error smaller than $0.01°$ (Pascu, 1972).

A difficult problem to contend with in the use of plate constants is that of bridging the magnitude gap from the catalogue reference stars (9-10 vis) to the fainter reference stars on the satellite plates. For stars to about 16th or 17th magnitude, a coarse objective grating technique may be used with a short-focus astrograph (Eichhorn, 1971).

A novel technique is planned for obtaining positions of the faint outer satellites of Jupiter in a cooperative program between the Naval Observatory and Hale Observatories (Appendix, Seidelmann). A neutral density filter placed near the focal plane of the 48-inch Schmidt will reduce, by about 8 magnitudes, the light of all stars (especially the catalogue reference stars) except those in a clear oblique band (about 5 inches wide) in which the satellite system is located. This procedure eliminates the need for an intermediate faint reference star system. Both the objective grating technique and the filter technique remove only those magnitude errors which affect the image shape such as coma or guiding.

The Problem of Coordinate Origin

We have mentioned here two observation types (or coordinate origins) for satellite observation: the planet-satellite connection and the intersatellite connection. For the latter, as mentioned before, there was (and is) no problem in obtaining photographic observations, provided that at least two satellites are visible. In the former case there is a problem and it is solved along two lines— one observational, and the other by use of the planetary ephemeris.

EPHEMERIS METHOD

In this method the absolute position (α, δ) is obtained for a satellite from a plate using the method of plate constants. The position of the satellite with respect to the planet is then obtained from the ephemeris position of the planet at the instant of observation. This type of observation was very uncommon in visual work but was made on occasion for J VI. In photographic work it is quite common and, for lack of a better technique, is used almost exclusively for the outer satellites of Jupiter, Phoebe, and Nereid (Roemer and Lloyd, 1966). Although the positions are determined from several reference stars, the catalogue errors, instrumental magnitude errors, and ephemeris errors of the planet limit the precision of these observations to between 0.3 arcsec to 4 arcsec. Refinements to this technique have been made by Garcia (1970, 1972), Dunham (1971), and Sinclair (private communication, 1974; 1974b). Garcia attempted to obtain planet-satellite positions by applying corrections to the ephemeris of Saturn, using meridian circle observations of the planet made concurrently with the photographic observations of the satellites. The planet-satellite residuals remained large and systematic, probably because the catalogue errors were the major contributor and not the ephemeris of Saturn. In addition, the meridian circle observations of Saturn are not more accurate than $\pm 0.4''$. Garcia abandoned this approach and used only intersatellite positions. Dunham introduced a novel, though questionable, solution for the Uranian system. He obtained nightly differences to the ephemeris position of Uranus relative to the Astrographic Catalogue. This was done by assuming that the differences in α and δ for a particular night were equal to the average of the $(O-C)$'s of the satellites (with respect to the ephemeris position of Uranus taken over all of the satellites and all plates taken on that night (about 5 plates/night). New $(O-C)$'s for the satellites were obtained by subtracting these values from the normal positions for the night, with the result that the sum of the *(O−C)*'s for the normal positions of the satellites in each coordinate is zero for each night. One might expect that the average accidental error due to the observations would be close to zero, but there is no reason to expect the average error due to the computed postions to likewise be zero. In taking out the systematic observational error in the positions of the satellites, Dunham has also taken out the average error due to their ephemerides. Dunham points out that a better solution to the problem would be to correct local catalogue errors—obtaining accurate relative positions of the reference stars. This would enable one to model the systematic difference between the ephemeris and catalogue positions of the planet in the conditional equations of the satellites.

Just such an approach was taken by Sinclair (1974b) for Saturn's satellites: Iapetus, Titan, Rhea, and Dione. Positions of these satellites were obtained on the AGK3 when Saturn was near a stationary point. The accuracy of the relative positions of the reference stars was improved in the process by use of overlapping plates. The difference between the ephemeris position of Saturn and its position on the AGK3 was modeled by a constant term plus a term linear with time and included in the orbital adjustment equations for the four satellites.

This method appears to work best when there are several satellites and when they are observed well throughout their orbits. For one or two satellites whose observations are restricted around the elongations (J V and Martian satellites) it is difficult to separate the corrections to the position of the planet and to the eccentricity and apse of the satellite.

OBSERVATIONAL METHOD

In the observational method, the point is to obtain a measurable image of the primary in the same exposure with the satellites. It is important to equalize the intensity of the planet and satellite(s) in order to minimize magnitude errors— especially that due to guiding. There are three basic techniques available which are astrometrically sound and are widely used in stellar astrometry for magnitude reduction. These are the objective grating technique, interruption techniques, and filter techniques.

The Coarse Objective Grating Technique. In this technique a coarse objective grating is constructed which will produce first order images of the planet that are equal in intensity to the satellite. The separation of these spectral images from the planet should be made similar to that of the satellite. This technique has been applied successfully only to the Neptune-Triton system (Alden, 1943). The technique works for Neptune apparently because its diameter is very small. An attempt to apply it to Uranus by van Biesbroeck (Dunham, 1971) produced elongated images of the planet which could not be reliably used.

Interruption Techniques. Interruption consists of occulting the light from the planet during part of the exposure for the satellites with some form of occulting device. The image of the planet is obtained by taking several shorter exposures during the exposure for the satellites. As might be expected, one of the earliest applications of an interruption technique was to the Neptune-Triton system. The reason for this is simply that, of the two satellite systems to which early photography could be applied (Jupiter and Neptune), intersatellite positions could not be obtained for Triton. The technique is described in the Greenwich Observations for 1904. A small opaque disk was mounted on a moveable rod and operated by an electromagnet. During the 15-20 minute exposure for Triton, Neptune was given a 0.1 second exposure every 20 seconds.

Perhaps the most significant application of an interruption technique was that of Petrescu (1935, 1938, 1939) to Jupiter and its Galilean system in 1934. A double occulting assembly was used for the magnitude reduction of both Jupiter and the satellites to the background stars, thus making it possible to improve the scale and orientation parameters over and above those derived from de Sitter's technique. These plates are the first in the modern epoch for the Galilean system, and several attempts are being made to have these plates remeasured and re-reduced using modern instruments and catalogues.

Not all applications of the interruption technique have been successful. One notable failure was B. P. Sharpless' series of photographic observations of the Martian satellites in the years 1939, 1941 and 1943 with the Naval Observatory's 40-inch Ritchey-Chrétién. Sharpless' (1939) technique was similar to that used successfully on Neptune at Greenwich, but the light diffracted around the occulting disk edges from the much brighter primary distorted the image of Mars. Thus, Sharpless never used the positions of the satellites relative to Mars but adjusted the orbits of both satellites using intersatellite positions. Recently, Kanaev (1970) has had some success on Mars using a vibrating slit assembly.

Filter Techniques. In the filter method a neutral density filter is used to attenuate the light of the planet by an amount necessary to equalize the intensity in the image of the planet to that of the satellites. Filter techniques were developed only in the middle and late 1960s and have proven to be much more successful than interruption techniques. A fairly extensive application of the filter technique to several planets and their brighter satellites was made by Soulié *et al.* (1968) using a short-focus Carte du Ciel instrument. No details of the observations are given, and the small scale of this instrument indicates that the observations of the satellites are not valuable themselves but may serve to obtain accurate positions for the planets only.

In 1967 two independent studies were made to test the feasibility of the filter technique. Sudbury (1969) obtained a series of photographic observations of Jupiter V with the Helwan reflector, while Pascu (1967, 1972) used the Naval Observatory's 61-inch astrometric reflector for the Martian satellites. Sudbury's filter consisted of a small triangular piece of filter (neutral) glass, 2 mm thick, which was polished optically flat and had an optical density of 3. This small filter was attached to the surface of the color filter placed immediately in front of the photographic plate. Special precautions were taken to insure that the filter combination was perpendicular to the optical axis since the light from the planet had to pass through 2 mm of glass more than the light from the satellite and reference stars. Sudbury appears to have had a problem in obtaining the center of figure for Jupiter. Although the solution for his complete set of observations shows no significant deviation of the center, Sudbury finds large night-to-night variations (up to 0.4 arcsec) in his photographic observations, which are larger than those he finds for the earlier visual observations. These nightly deviations of the center are attributed by him to systematic variations in the settings on the limbs. One might agree with Sudbury about the visual observations, but it is difficult to see what systematic setting errors (which are personal errors) on the photographic plates have to do with the observing conditions on the night they were taken. On the other hand, the observations for J V have never been satisfactorily represented by the theoretical model. It is quite likely that adequate account has not been taken of the perturbations due to Io.

For Pascu's observations of the Martian satellites, a metallic film filter suggested by L. W. Fredrick (personal communication, 1966) was used. The filter consists of an optically flat GG14 filter with a small, partially transparent metallic film ''spot'' deposited on its center by evaporation. The spot filter is composed of chrome or nichrome, is neutral in transmission, and has an optical density of 3.0. In the observing procedure, the spot is placed over the planet and attenuates its light by an amount necessary to equalize the intensity with that of the satellites. Kodak 103aJ plates were used in order to reduce the exposure time and reduce the extent of the planetary halo. Figure 5.1 shows a 20-second exposure taken with the 61-inch reflector at Flagstaff in 1969. The results of the orbital analysis for the 1967 observations gave no indication of systematic errors in the observations (Pascu, 1975). An analysis of the variation of the $(O-C)$'s from consecutive exposures gave an expected mean error of ± 0.09 arcsec for one exposure, and the orbital adjustment tends to confirm this, giving residuals after solution of ± 0.1 arcsec.

An inexpensive alternative to the metallic film filter is to make the filters from Kodak HRP or type 649 plates. These emulsions are low in granularity, they are thin, and the unexposed portions can be made perfectly clear on developing. Such a filter has been used by Pascu (1973b) to obtain plates of the Galilean satellites with the McCormick refractor. Figure 5.2 shows a portion of a plate. The filter is a composite of three densities in order that magnitude compensation can be made between the planet and satellites, between the satellites themselves, and between the whole system and 10th mag background stars. A preliminary discussion by Arlot (1975) of some preliminary observations (trail/scale reduction) indicates that the precision of these observations is comparable to that for the Martian satellites. A sizeable center of figure error was discovered and found to be correlated with the phase of the planet. This residual phase effect results from the fact that the geometric terminator is not seen because of limb darkening. As concerns the use of the planet-satellite observations in the orbital adjustment, the problem appears adequately solved by the inclusion of terms in the conditional equations which have the same signature as the phase.

Observations of Saturn's system have been started at the Naval Observatory (26-inch telescope) by Pascu and Fiala using a filter technique. Figure 5.3 shows that all satellites brighter than Mimas can be obtained in a one-minute exposure with this telescope. The feasibility of using measurements with respect to the planet or rings has not yet been tested.

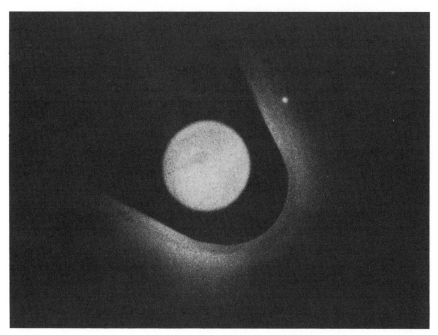

Fig. 5.1. The satellites of Mars taken with the 61-inch astrometric reflector at Flagstaff, Arizona, on May 28, 1969. The 20 second exposure was taken at 8^h38^m U.T. The plate is a 103aJ in combination with a GG14 filter. The shadow around the planet is that due to the partially transparent metallic chrome filter. Phobos is the brighter inner satellite. (North is at bottom, east at right.)

Astrometrically, the use of metallic film filters is sound. They are microscopically thin, do not affect image quality, and can be made uniformly and neutrally dense. The density of the filter in a particular passband is obtained by equating the light intensity in the planetary and satellite images. At opposition, the density in magnitudes is given by

$$D = \left[m_s(1,0) - m_p(1,0) \right] - 5 \log d_p$$

where $m_s(1,0)$ and $m_p(1,0)$ are the opposition magnitudes of the satellite and planet reduced to unit distance from the Earth and the Sun, and d_p is the planetary diameter in units of the satellite diameter (usually the seeing diameter). This formula will work reasonably well for a uniformly illuminated disk such as that of Mars; but for a heavily limb darkened planet such as Jupiter, the limbs will be underexposed. In these cases it is advisable that the filters be made at least 1 magnitude lighter, and that correct definition of the limbs be achieved by adjusting the exposure time. For Jupiter it will be found that when the limbs are well exposed, the planet, in general, will show no features. For Saturn, two

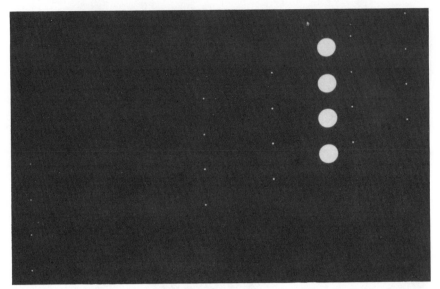

Fig. 5.2. Four exposures of the Galilean system taken with the McCormick refractor on June 7, 1968. The exposures, ranging from 30 to 40 seconds, were taken between 1^h58^m U.T. and 2^h01^m U.T. The filter used here was made from type 649F plates. The plate is a 103aJ in combination with a GG14 filter. Trails of the satellites are used for the orientation of the plates. The fifth image is that of a star. (North is at bottom, east at right.)

filters will be needed over the long run: one for use on the disk of the planet and one for use on the rings when they obscure a large portion of the planet (in particular the N and S limbs). The cases of Uranus and Neptune are relatively straightforward, except that the effect of seeing on the size of the planetary disk is not negligible and adjustment of the density of the filter must be made. In the use of the filter, the metallic film must be kept close to the photographic plate (2 mm) in order to avoid undue diffraction (see Figs. 5.1 and 5.3).

The introduction of photography into satellite observation has been responsible for a considerable reduction in the accidental and systematic errors of observation. This is due to the virtual elimination of the personal components of these errors. Although there is still a CFE, it can be accounted for in the orbital analysis in some cases since it is not as subject to personal variations. There are, however, three photographic effects which have caused some misgivings about photographic observations. These effects are the gelatin effect caused by an overexposed primary, turbidity caused by the halo light, and guiding errors caused by magnitude differences. The guiding errors can be minimized by equalizing the light from the primary to that of the satellites and, in addition, by guiding on the planet and minimizing the exposures. Flash sensitization of the

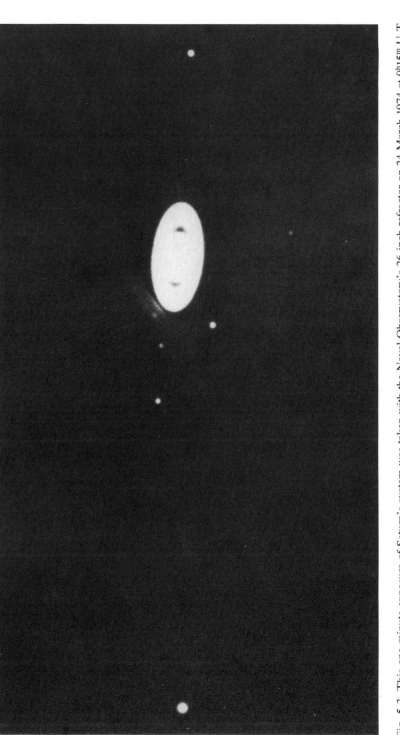

Fig. 5.3. This one-minute exposure of Saturn's system was taken with the Naval Observatory's 26-inch refractor on 24 March 1974 at 0^h15^m U.T. The plate is 103aG in combination with a GG14. The shadow around Saturn is due to a partially transparent inconel coating filter. From left to right the satellites are Titan, Dione, Enceladus, Tethys, Mimas, and Rhea. The faint image below the planet is a star. (North is at bottom, east at right.)

plates would be counterproductive because it would cause the planetary halo to grow too rapidly. In the case of the gelatin effect, no such effect has been shown for satellite work using modern emulsions. In any case, preventing overexposure is not a difficult problem. Turbidity is not strictly a photographic effect but arises also in visual observations. It has not been shown to be a problem in satellite photography. Minimizing turbidity must be done by minimizing the halo light. This is usually accomplished by using a long focal length instrument at a high altitude site. If there is a sizeable color difference between planet and satellite, the observations may be made in that passband in which the planet is fainter. Sinton's (1972) methane filter technique is not yet suitable for astrometric use.

Spacecraft Techniques

Observational techniques from a spacecraft have been applied only to the Martian moons (Born and Duxbury, 1975). In observing the planetary satellite, the pointing direction of the camera is known with respect to the planet, the Sun, and Canopus. This pointing accuracy for the Martian moons is a few hundredths of a degree. Thus, in the close encounter photographs of the satellites, the position of the imprinted reseau is known with this precision. In this sense, the spacecraft technique is similar to the photographic technique—the position of the satellite is found relative to fiducial points. The great advantage in this technique over the ground-based photographic technique is in the spacecraft's proximity to the satellite. A pointing error of $0.02°$ at 50,000 km from the satellite corresponds to an error of $0.072''$ as seen from the Earth (50×10^6 km). At 50,000 km the spacecraft technique is just competitive with the best ground-based observations of the Martian satellites at the most favorable opposition. For smaller distances of the spacecraft from the satellite, and for unfavorable oppositions, the spacecraft technique is more accurate. The major advantage of the spacecraft technique is that this accuracy is the same (in theory) for the distant planets. Thus, spacecraft observations of the satellites of Neptune could be two orders of magnitude more precise than the best Earth-based observations.

Radar Observations

Starting in 1975, range and range rate observations of the Galilean moons have been made at Arecibo (F. Drake, personal communication). The purpose of these observations is to obtain positions of Jupiter itself which, because of its deep atmosphere, is not a good radar target. In order to obtain these positions, the orbits of the satellites must be adjusted to satisfy the radar observations.

The precision of these observations is expected to be about 1 km (Pettengill, personal communication, 1974), which makes them more precise than spacecraft observations. The fact that they are ground-based observations makes it quite likely that they will be continuous, which gives them an added edge over spacecraft observations. However, unlike the tangent plane observations (which include spacecraft observations), radar observations cannot be used to determine a complete set of orbital parameters. The orientation parameters of the orbital planes, for example, cannot be determined from radar observations. Nor is it likely that radar observations can be obtained for more than about a half-dozen satellites in the foreseeable future.

LIMITS OF COMPLETENESS

It was a common practice to make informal searches for new satellites whenever a new large telescope was installed, although few negative results were ever published. The question of completeness in recent times was most thoroughly investigated by Kuiper (1961a). In the middle 1950s, he investigated the region from the planet out to the stable satellite boundary for all the planets but Mercury and Jupiter. In the former case no search was made, while for Jupiter the search was limited to the region within 20 arcmin of the planet. A complete search of the outer regions of Jupiter had already been made by Nicholson (1939). The limits of completeness vary widely for Kuiper's survey; from 14–17 mag (photographic) for the inner regions to 18–22 mag for the outer regions. Thus, close satellites as large as 500 km in diameter would not have been detected for the outer planets.

Morrison *et al.* (Chapt. 1) report on other additional satellite searches since that of Kuiper's: Dollfus' (1967) discovery of Janus, Sinton's (1972) investigation of the Uranian system with a methane absorption filter, and the Mariner 9 search for additional satellites of Mars (Pollack *et al.*, 1973a). The two latter investigations had negative results, while all three of them were very restrictive in their scope and did not substantially improve on Kuiper's completeness limits. A search of Mercury made by Mariner 10 has been reported by Murray *et al.* (1974). The completeness limit established in this study is that no satellite larger than 5 km (of the same albedo as Mercury) exists within 30 Mercury radii of the planet. This corresponds to a completeness limit of about 17 mag visual.

A new search of the outer region of Jupiter is being carried out by C. Kowal (private communication, 1974) with the Palomar 48-inch Schmidt and sensitized IIIaJ plates. A thirteenth satellite has been reported by him (Kowal *et al.*, 1975) of the 20th magnitude visual. This survey is expected to extend the completeness limit in this region to 21 mag visual.

The discovery of Janus by Dollfus (1967) is not a serious challenge to the limits set by Kuiper since the magnitude of Janus is approximately the same as the completeness limit (14 mag visual). It is surprising, however, that no observations of Janus have been made since its discovery. This is due to the belief that it can be observed only when the Earth or Sun passes through the plane of the rings. This view is somewhat pessimistic. When Saturn is at mean opposition distance, Janus will be 4 arcsec from the outer edge of the rings when it is at maximum elongation. Figure 5.3 shows a measurable image of Mimas at 4 arcsec from the edge of the ring on a plate taken with the USNO 26-inch refractor in the heart of Washington, D.C. The rings were fully opened at the time this plate was taken. It is the author's·opinion that similar results for Janus could be obtained with a reflector (provided that the secondary supports are not in the East-West direction) which has a focal length about 3 times that of the USNO 26-inch refractor and which is located at a good high altitude site.

ACKNOWLEDGMENTS

I wish to express my gratitude to Charles Worley for the many informative discussions concerning the micrometer observations, and to Alan Fiala for translating portions of the German references. I have also had fruitful discussions with R. L. Duncombe, K. Aa. Strand, and P. K. Seidelmann.

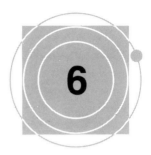

ROTATION HISTORIES OF THE NATURAL SATELLITES

6

Stanton J. Peale
University of California, Santa Barbara

Several aspects of the rotational history and current rotation states of the natural satellites are considered. Only the regular satellites as well as Iapetus, Hyperion, Triton, and the Moon are tidally evolved. The remaining irregular satellites essentially retain their primordial spins except for a relaxation to principal axis rotation shared by all the satellites. None of those tidally evolved satellites whose rotation periods remain unobserved are expected to be in non-synchronous spin states stabilized by spin-orbit coupling. Tidal evolution of the obliquities of the satellites, most of which are in inclined orbits which precess about a nearly invariant plane, is evaluated in the framework of the generalization of Cassini's laws for the Moon. Either of two final obliquities, separated by an angle which is greater than the orbital inclination to the invariable plane, can be selected by the tides. The selection is determined for most of the tidally evolved satellites under the assumption of hydrostatic equilibrium. Exceptions to this last assumption are the Moon, Phobos, Deimos, and Iapetus, where the latter's obliquity shifts by more than 8° for hydrostatic versus moon-like gravitational asymmetries. This large shift is proposed as an observational test for the composition of Iapetus. The observed shapes of Phobos and Deimos are used to calculate nearly resonant, forced librations in longitude of 4.8° and 0.5°, respectively; the former has been observed.

The study of the rotation of the natural satellites in the solar system really is not separable from the study of the rotation of all other objects in the solar system, since similar processes affect them all. However, the 33 known satellites provide many examples and tests of our ideas of rotational evolution under a fairly wide range of conditions. This chapter represents a first attempt to apply all the recent advances in the theory of rotation to the entire group of natural satellites as well as to include applications of older ideas.

The observed rotation of the various objects in the solar system has always led to questions of the origin of the particular rotation states. These rotation states often place constraints on the process of origin of the bodies themselves, such as the widely accepted accretion mechanism for forming the planets, satellites, and other solar system bodies from smaller objects (Safronov, 1969; Chapt. 26, Safronov and Ruskol). Of course, we see the rotational states only as they exist today, and primordial states often are completely masked by the dissipation of energy in a wobble motion or by the exchange of angular momentum between spin and orbital motion through tides. It is important to understand the details of these evolutionary processes if we hope to understand what constraints, if any,

[87]

current spin states place on primordial conditions of origin. The current unusual spin of Uranus with its spin axis nearly in its orbit plane is almost certainly that which it had immediately after its formation was completed, since tides from the Sun and the Uranian satellites cause little change during the entire history of the solar system (cf. Greenberg, 1975; Singer, 1975). But the equally unusual retrograde spin of Venus must be substantially altered from its primordial state. If a core-mantle interaction stabilizes the retrograde Venusian spin against the gravitational tides, which would otherwise stand it upright, we could constrain the primordial spin to being at least slightly retrograde, as is that of Uranus (Goldreich and Peale, 1970). However, we are frustrated even in this case from inferring primordial conditions with confidence from the contemporary highly evolved spin state, since an accelerating atmospheric thermal tide could turn Venus upside down from a prograde rotation.

This frustration in deducing the past spin states from current highly evolved states is the rule in most investigations to date. By evaluating the magnitude of dissipative effects, we can, however, determine the extent to which an initial spin has been altered. For a relatively isolated body, the only change in the spin will be the assumption of a minimum energy configuration consistent with a conserved angular momentum. This wobble decay has been calculated by Burns and Safronov (1973) for the asteriods, where it is shown to be too rapid for the wobble to persist between collisions. We apply a similar calculation to each of the natural satellites in the solar system to obtain the characteristic times for the decay of any wobble motion to the smooth rotation about the principal axis of maximum moment of inertia. These extremely small time constants allow us to assume principal axis rotation in all subsequent discussions of the satellites, but at the same time they erase any information a wobble might imply about the satellite's rotational history.

The natural satellites also lose spin angular momentum to their orbits. The differential gravitational field of the planet distorts a satellite into a slightly cigar shape. This distortion on the Earth, due to the Moon and Sun, results in the well-known ocean and solid body tides. Only the latter will result on the satellites, since oceans do not exist there; however, Titan (and perhaps others) does have an atmosphere. Ideally the tidal bulge would be aligned with the tide-raising body, but dissipation of tidal energy results in a phase lag in the response of the satellite to the disturbing potential. This means that high tide at a specific point occurs after the tide-raising body has passed the meridian through that point. The tidal bulge on the satellite is no longer aligned with the planet, and the same differential gravitational field which generates the tide also exerts a torque on the tidal bulge, thereby changing the spin rate (cf. Chapt. 7, Burns, particularly Fig. 7.1).

If the satellite is in a circular orbit, and its spin and orbital angular momentum remain parallel, the angular momentum transfer from the satellite spin to its orbit continues until the satellite keeps one face toward its planet by rotating synchronously with its orbital motion. In this rotation state, the tidal bulge remains aligned with the planet, and further angular momentum transfer between the satellite's spin and orbit occurs only on the much longer time scale of the orbital evolution. The Moon is the obvious example of a satellite that has reached this end-point of tidal evolution. The reader is referred to Munk and MacDonald (1960) for the elementary theory of tidal interaction.

The time scale for such simple tidal evolution as determined for each satellite is representative of all the processes involving a transfer of angular momentum between spin and orbit. The rotational damping time scale is also determined for Saturn's ring particles as a function of particle size and position in the ring, since two sets of observations are consistent with synchronous rotation of these particles.

Several complications in the evolution of a satellite's spin angular momentum result if we relax the above conditions of orbit circularity and alignment of orbital and spin angular momenta. Some orbits maintain a substantial eccentricity, and the end point of tidal evolution may not be the synchronous rotation just described. For a certain class of tidal models, the tidal torque averaged over an orbital period vanishes at a rotation rate somewhat greater than the synchronous value, and the spin magnitude at the end point of tidal evolution would be above the synchronous one (Peale and Gold, 1965).

For satellites with a permanent asymmetry about the spin axis, a resonant torque on this asymmetric mass distribution can counteract the tidal torque and stabilize the spin angular velocity at a non-synchronous value which is a half-integer multiple of the orbital angular velocity. The planet Mercury, which is in the 3/2 spin resonance, is the only known example of this end point of tidal evolution (Pettengill and Dyce, 1965). Theoretical discussions of this type of spin-orbit coupling are given by Colombo (1965), Goldreich and Peale (1966), Colombo and Shapiro (1966), and in a review by Goldreich and Peale (1968).

We have calculated stability criteria and capture probabilities for the 3/2 spin resonance, which is the most likely non-synchronous, commensurate spin. Nine of the ten satellites, which are determined to be tidally evolved but which have not yet been observed to be synchronously rotating, were included in this analysis. None of the satellites is likely to be in the 3/2 spin resonance primarily because of their small orbital eccentricities.

The libration in longitude of Phobos and Deimos about the observed synchronous spin is discussed because of the large amplitude of such a libration observed for Phobos (Duxbury, 1973, and Chapt. 15 herein). This libration would have important implications for the recent collisional history of these satellites if it were a free libration, but a near-resonant forced libration of the proper amplitude accounts for the observation.

Up to this point we have been concerned with the magnitude of the spin vector, which is appropriate if the spin stays perpendicular to the orbit plane. If, on the other hand, the spin vector is not perpendicular to the orbit plane, the tidal bulge is carried out of the plane of the orbit and a torque results which changes the direction of the spin vector as well as its magnitude. The angle between the orbital and spin angular momenta is called the obliquity, and this angle as well as the spin magnitude has an evolutionary history due to tidal interaction. An orbit *fixed in space* leads to an obliquity history which depends on initial conditions, but in every case the obliquity is zero at the end point of tidal evolution (Goldreich and Peale, 1970).

Satellite orbits are not fixed in space but precess with nearly constant inclination about some nearly invariant plane—usually the planet's equatorial plane. The final obliquity is no longer zero but may assume one of two stable values. The choice between these two values of the final obliquity and the magnitude of each depends on the values of several fixed parameters and on the initial conditions (Peale, 1974). The dependence of the positions of these final states on the various parameters and the criteria for selection of one or another of the final states by the tides will be discussed. The approximate final obliquity of each satellite for which sufficient data exists is determined. Differences in the principal moments of inertia necessary for these calculations are determined from hydrostatic equilibrium for most of the satellites, but consequences of non-isostasy are pointed out. In particular, Iapetus would have measurably different obliquities for isostasy compared with moment of inertia differences similar to those of the Moon. This provides an observational test for determining the internal strength and hence some estimate of the composition of Iapetus.

The Iapetus example points out one of the more important motivations for the study of rotation in the solar system—the determination of bounds on the internal structure of the various objects. Certain minimum internal strengths are required to maintain given deviations of a body from hydrostatic equilibrium (cf. Johnson and McGetchin, 1973). Going from a hydrostatic to a mildly non-hydrostatic configuration for Iapetus leads to a measurable shift in the final obliquity. Mercury and likewise any satellite must support a permanent non-axially symmetric distribution of mass to remain stable in a non-synchronous spin resonance. Venus would need a liquid core to be captured in a spin resonance with the Earth's orbital motion (Goldreich and Peale, 1967), and possibly to stabilize the 180° obliquity (Goldreich and Peale, 1970). Very refined measurements of rotational variations reveal much about internal properties as well as about internal and surface dynamics for the Earth (Munk and MacDonald, 1960), although the necessary precision of measurement will probably never be obtained on any natural satellite except for the Moon (Williams *et al.*, 1973). Nevertheless, we will learn a great deal about the natural satellites as measurements of their rotational states improve. Perhaps they hold some clues or answers to some of the larger puzzles concerning the origin and history of the solar system.

The results of all the calculations about the rotation of the natural satellites are given later in Table 6.1. Also included in this table are the supporting data for the calculations. These data were taken collectively from the reviews by Morrison and Cruikshank (1974), Newburn and Gulkis (1973), and Brouwer and Clemence (1961b); the data are summarized in Chapter 1, Tables 1.2 and 1.4, by Morrison *et al.* Whitaker and Greenberg (1973) are the source of data for Miranda. Wherever the radius of the satellite was uncertain by a large amount, either the preferred value given by Morrison and Cruikshank or the average of the extremes was used. A satellite density of 2 g-cm^{-3} was assumed when no value was available, except where known densities of nearby satellites differed substantially. In the latter case a density close to that of the nearby satellite was adopted. Only those numbers necessary for the calculations are included in Table 6.1.

WOBBLE

Since no body in the solar system is either perfectly rigid or perfectly elastic, an arbitrary rotation of a semi-rigid satellite is expected to decay toward the minimum energy configuration corresponding to the given angular momentum (Lamy and Burns, 1972). For nearly spherical bodies, the rotational energy is approximately

$$E = \frac{L^2}{2I} \tag{1}$$

where L is the angular momentum and I is the moment of inertia about the instantaneous spin axis. Minimum energy for given L thus corresponds to rotation about the principal axis of maximum moment of inertia. The dissipation of the rotational energy results from the motion of the spin axis relative to the principal axes fixed in the body whenever the spin axis does not coincide with the axis of minimum or of maximum moment of inertia. The periodic distortions of the body due to the changing position of the equatorial bulge are not conservative and lead to the principal axis rotation corresponding to minimum energy.

The time constant for the exponential decay of the wobble amplitude is given by Peale (1973)

$$\tau = \frac{3GIQ}{k_2 a_e^5 \dot{\psi}^3} \tag{2}$$

for isolated, nearly spherical bodies. In eqn. (2) G is the gravitational constant, a_e is the mean equatorial radius and $\dot{\psi}$ is the spin angular velocity. The Love number k_2 is discussed by Munk and MacDonald (1960, p. 23) and is defined by

$$k_2 = \frac{1.5}{1 + \frac{19\mu}{2\rho ga_e}} \qquad (3)$$

for a homogeneous sphere with μ, ρ and g being, respectively, the coefficient of rigidity, the mean density and the surface acceleration of gravity. The specific dissipation function Q is defined by (Munk and MacDonald, 1960; MacDonald, 1964; cf. Burns, Chapt. 7, herein for further discussion)

$$\frac{1}{Q} = \frac{1}{2\pi E^*} \oint \frac{dE}{dt} \, dt \, , \qquad (4)$$

where E* is the peak energy stored in an oscillating system and the integral represents the energy dissipated over a cycle of oscillation. If the dissipation is small,

$$\frac{1}{Q} = \epsilon \, , \qquad (5)$$

where ϵ is the phase lag of the response of the body to a forced periodic oscillation. This relation was used to derive eqn. (2). Finally, I is the (average) moment of inertia about the center of mass of the satellite. We shall adopt $I = 0.4 \, M_s a_e^2$, taking a homogeneous sphere for all the satellites, where M_s is the satellite mass. Central condensation is unlikely to reduce the coefficient of $M_s a_e^2$ below 0.3, and the error introduced is considerably less than the uncertainty in Q.

For all the satellites in the solar system, the second term in the denominator of eqn. (3) is much larger than 1, and we can use the approximation

$$k_2 \approx \frac{3\rho ga_e}{19\mu} = \frac{4\pi\rho^2 Ga_e^2}{19\mu} \, . \qquad (6)$$

This approximation is not appropriate if a satellite has a relatively thin crust over a liquid interior like some of the models proposed by Lewis (1971a). The Love number would then approach the fluid value of 1.5 and relaxation to principal axis rotation would be essentially immediate. Otherwise we can substitute eqn. (6) into eqn. (2), write the other parameters in terms of a_e and ρ, and obtain

$$\tau = \frac{38\mu Q}{5\rho a_e^2 \psi^3} \, . \qquad (7)$$

Eqn. (7) is identical to that obtained by Burns and Safronov (1973) and McAdoo and Burns (1974) except the numerical coefficient is as much as an order of magnitude smaller in eqn. (7). However, we shall find the time constants for all the satellites to be small in any case so that this difference in the coefficient has little practical significance.

The satellites, of course, are not isolated bodies but suffer significant periodic conservative torques and dissipative tidal torques due to the gravitational fields of their primaries. We shall consider the consequences of these torques in more detail subsequently, but let us note here that the tides decrease the time for relaxation to principal axis rotation such that eqn. (7) is an upper bound for given values of the parameters. Tides actually dominate in the lunar case where τ is reduced by about a factor of four from its tide-free value (Peale, 1976).

In Table 6.1, we have listed the wobble relaxation time $\tau \times Q^{-1}$ for each of the natural satellites in column 7. An average density $\rho = 2$ g-cm^{-3} and rigidity $\mu = 5 \times 10^{11}$ dynes-cm^{-2} was assumed for all satellites, since uncertainties in Q may be much greater than the probable deviation of the true values of μ and ρ from these choices. The rotation period P (Table 6.1, column 4) was assumed synchronous with the mean orbital motion for those satellites whose tidal evolution times are short compared to the age of the solar system, whereas a period of 1/2 day was chosen for those satellites whose rotation states have not been substantially affected by the tides. This latter period is slightly longer than the mean rotation period of the asteroids (McAdoo and Burns, 1973; Burns, 1975).

For values of Q between 100 and 1,000, only Hyperion and Iapetus have wobble decay times comparable to the age of the solar system. But the inverse cube dependence of τ on $\dot{\psi}$ means that the wobble decay for these satellites was much shorter in the past when they were presumably rotating at higher angular velocities. The wobble can be excited by collisions with asteroidal-sized meteorites, but the large mass of Saturn means that most of these objects should be eliminated from orbits which intersect Saturn's orbit relatively early in the history of the solar system. Hyperion and Iapetus would then rotate unmolested by large impacts throughout most of history, and we would expect principal axis rotation for these satellites as well.

From the preceding arguments concerning Hyperion and Iapetus and the short time constants for wobble decay of the remaining satellites, we can assume that the principal axis rotation has obtained throughout most of history for all the natural satellites. Burns and Safronov (1973) concluded that the collision frequency between asteroids is too low to maintain a significant wobble amplitude for observable asteroids. The sizes of the smallest known satellites are comparable to those of the asteroids discussed by Burns and Safronov, but the collision frequency with exterior bodies must be much less because of rapid elimination of such objects by the primary. Hence, our assumption of principal axis rotation for the satellites throughout most of history seems relatively secure. But when the opportunity arises, perhaps a close look at Hyperion and Iapetus would be warranted, in case unusual circumstances coupled with their relatively long decay times have prevailed.

TIDAL DECAY OF ROTATION

The time scale for reaching a final obliquity or a non-synchronous but stable rotation are both of the same order of magnitude as the time scale for reaching synchronous rotation in a circular orbit. We can therefore use this latter time-scale, which is easy to calculate, to determine which satellites are likely to have reached the end of the tidal evolution of their spin angular momenta. The time rate of change of the spin angular velocity for a circularly orbiting satellite with zero obliquity is given by (cf., MacDonald, 1964; Peale, 1974; Burns, Chapter 7 herein)

$$\frac{d\dot{\psi}}{dt} = \frac{-3k_2 GM_p^2 a_e^5}{Ca^6Q} \quad , \tag{8}$$

where M_p is the planetary mass, a is the orbital semimajor axis (radius), and C is the maximum moment of inertia about the center of mass. We shall assume a constant value for Q, use the approximation of eqn. (6) for k_2 with $\rho = 2$ g-cm^{-3} and $\mu = 5 \times 10^{11}$ dynes-cm^{-2}, and let $C = 0.4 \, M_s a_e^2$ to arrive at a time for changing $\dot{\psi}$ by $\triangle\dot{\psi}$ of

$$T = 2.4 \times 10^{10} \frac{P_0^4 \, \triangle\dot{\psi}Q}{a_e^2} \text{ years} \quad , \tag{9}$$

where P_0 is the orbital period in days, $\triangle\dot{\psi}$ is measured in radians/sec and a_e is in km. The change $\triangle\dot{\psi}$ and hence T is maximized by choosing the initial value of $\dot{\psi}$ such that $a_e \dot{\psi}^2 = GM_s/a_e^2 = 4\pi G\rho a_e/3$, which is the limiting velocity for rotational stability for a spherical body of zero strength. This initial value of $\dot{\psi} = 7.5 \times 10^{-4}$ radians/sec (period = 2.3 hrs). The final value of $\dot{\psi}$ is the synchronous value for each satellite.

The time T (xQ^{-1}) required for complete tidal retardation of the rotation of each satellite is given in the eighth column of Table 6.1. With $100 \lesssim Q \lesssim 1000$, the satellites are seen to fall rather conclusively into two categories—those which are certainly completely tidally evolved and those which have essentially retained their primordial (or since the last collision) spin angular momenta. Possible exceptions to the absoluteness of this organization are Hyperion, Iapetus, and Oberon, the latter only if its radius is near the lower limit. However, we have indicated by asterisks in column 8 of Table 6.1 positive observations of synchronous rotation, and Iapetus, whose decay time from eqn. (9) is most marginal, is known to be synchronous with essentially absolute certainty (Widorn, 1950; Morrison and Cruikshank, 1974; Morrison et al., 1975). This illustrates that the rotational decay times in Table 6.1 tend to be upper bounds because of the extreme value chosen for the initial spin rate. The most recent confirmation of the predictions of eqn. (9) (Burns, 1972) are the observed synchronous rota-

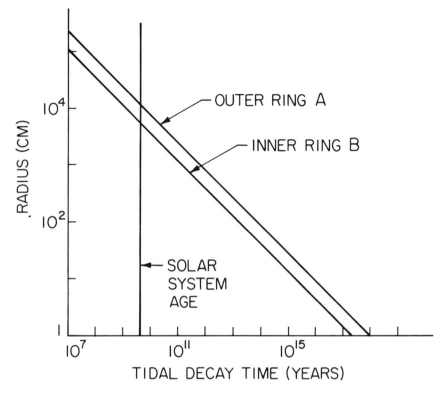

Fig. 6.1. An upper bound on the tidal damping for particles in Saturn's rings with $Q = 100$.

tion of the satellites of Mars (Pollack *et al.*, 1972; Chapt. 14, Pollack; Chapt. 15, Duxbury).

Since Hyperion is inside the orbit of Iapetus and has a smaller decay time, and Oberon is much less marginal, we shall assume that all three satellites have reached an end point of tidal evolution. The satellites J 6 through J 13, S 9 (Phoebe), and N 2 (Nereid) have such long tidal decay times that essentially the only significant changes in their rotation following collisions has been the relaxation to principal axis rotation. We expect to find the spin vectors of these satellites to have random orientations and magnitudes. The latter may be typical of asteroids. Brightness variations of Phoebe are consistent with either a 9- or 13-hour period (Andersson, 1972; 1974). The grouping of tidally evolved and unchanged spin angular momenta is essentially the same as that of "regular" and "irregular" satellites except for Iapetus, the Moon, Triton, and perhaps Hyperion.

Thirteen of these 23 tidally evolved satellites have confirmed synchronous rotations (Morrison and Cruikshank, 1974; Franz and Millis, 1975; Chapt. 1, Morrison *et al.*). We shall discuss in the next section the possibility of finding any of the remaining tidally evolved satellites stabilized in non-synchronous but resonant spin states by spin-orbit coupling.

Using the same assumption for density and rigidity as for the satellites, we show in Figure 6.1 the rotational decay times for the particles in Saturn's rings (cf. Chapt. 19, Cook and Franklin) in the range of orbit radii from the inside of the B ring to the outside of the A ring as a function of the particle size. Only ring particles larger than several hundred meters could have had their rotations damped in the age of the solar system if Q = 100. Since collisions surely have occurred much more recently, we might expect these and even larger ring "rocks" to be tumbling unless some mechanism other than the tides can be invoked to retard the rotation. Perhaps we should look seriously for such a mechanism, since at least two sets of observations are consistent with synchronous rotation for a significant fraction of the ring particles (Ferrin, 1974; Morrison, 1974b).

SPIN-ORBIT COUPLING

The likelihood of finding a satellite in a nonsynchronous resonant spin state can be inferred from the criterion for stability of the resonance against disruption by tides and from the probability of capture into the resonance as the satellite passes through a commensurate spin. The stability criterion and an estimate of a capture probability are obtained by long published procedures (Goldreich and Peale, 1966; Colombo and Shapiro, 1966).

For an angular velocity which is nearly commensurate with the orbital mean motion we can write

$$\psi = p'M + \psi_0 , \qquad (10)$$

where ψ is the Euler angle locating the axis of minimum moment of inertia from the ascending node of the equatorial plane of the satellite on the orbit plane, p' is a half-integer, M is the mean anomaly and ψ_0 is the value of ψ when the satellite is at the pericenter of its orbit. If $\dot{\psi} \approx p'n$, where n is the orbital mean motion, then ψ_0 is a slowly varying quantity whose variation is governed by the equation (eqn. 63, Peale, 1973)

$$C\ddot{\psi}_0 = \frac{2GM_pM_sa_e^2}{a^3} \sum_q F_{220}(Z = 1) G_{20q}(e) \times$$
$$C_{22} \sin\left[2\gamma + (2 + q)M - 2(p'M + \psi_0)\right] + \frac{\partial V_T}{\partial \psi} , \qquad (11)$$

where F_{2mp} (Z) and G_{2pq} (e) are functions of the obliquity $i = \cos^{-1}Z$ and expansions of the orbital eccentricity e, which have been tabulated by Kaula (1966, pp. 34 and 38). The indices mpq are integers; $0 \leqslant m \leqslant 2$, $0 \leqslant p \leqslant 2$, $-\infty < q < \infty$. By choosing $Z = 1$, we assume that the spin angular momentum is perpendicular to the orbit plane with the result that only $p = 0$ terms occur in the sum. The angle γ locates pericenter in the orbit plane from an axis fixed in that plane, which we may consider inertial for our purposes here. The harmonic coefficient C_{22} of the satellite's gravitational field is defined by

$$C_{22} = \frac{B - A}{4M_s a_e^2} , \qquad (12)$$

where $A \leqslant B \leqslant C$ are the principal moments of inertia of the satellite. The last term in eqn. (11) is the tidal torque about the spin axis, where V_T is the gravitational potential from the tidal distribution of mass.

As short period terms will be small amplitude, we average eqn. (11) over an orbit period holding all parameters constant except M. All the periodic terms vanish in the averaging process except that for which

$$2 + q - 2p' = 0$$

and eqn. (11) becomes the pendulum equation (Goldreich and Peale, 1966),

$$C\ddot{\delta} = -\frac{3}{2} n^2 G_{20(2p'-2)}(B - A) \sin 2\delta + \langle \mathscr{T} \rangle, \qquad (13)$$

where $\langle \mathscr{T} \rangle$ is the average tidal torque and $\delta = \psi_0 - \gamma$.

The criterion for stability of a nonsynchronous spin resonance is that the maximum restoring torque of the pendulum (the coefficient of sin 2δ) should exceed the averaged tidal torque. If we use eqn. (6) for k_2, eqn. (8) for $\langle \mathscr{T} \rangle/C = \langle \mathscr{T} \rangle/0.4 \ M_s a_e^2$ and rearrange the terms, the stability criterion for the lowest order and most stable nonsynchronous spin state ($p' = 3/2$) becomes

$$\frac{B - A}{C} \gtrsim \frac{15}{19} \frac{n^2 \rho a_e^2}{\mu Q G_{201}} = 4.8 \times 10^{11} \frac{a_e^2}{Q P_o^2 e} , \qquad (14)$$

where $G_{201} = 7e/2$, $\rho = 2$ g-cm^{-3}, $\mu = 5 \times 10^{11}$ dynes-cm^{-2} have been used. The numerical coefficient corresponds to a_e being expressed in kilometers and the orbital period P_o in days.

The limiting values of $\left[(B-A)/C\right] \times Q$ for stability of the 3/2 spin resonance are given in column 9 of Table 6.1 for 9 of the 10 tidally evolved satellites. The eccentricity of Janus' orbit is unknown, although it is likely to be very small. For $Q = 100$, the lower bounds on $(B-A)/C$ range from 2.2×10^{-9} for S 7 (Hyperion) to 7.8×10^{-5} for U1 (Ariel). Even the largest value is less than the

2×10^{-4} value for the Moon, and one might conclude that any of these 9 satellites might be stable in the 3/2 spin resonance.

However, the lower bounds on $(B - A)/C$ given by eqn. (14) cannot be taken too seriously because of the dependence on the rigidity μ. The value of 5×10^{11} dyne-cm^{-2} used here is "earth-like" and may not be appropriate if the satellites have relatively thin ice crusts over liquid interiors as in the models of Lewis (1971a). In this latter case there would be no hope of stabilizing the satellite at a commensurate but nonsynchronous spin.

The probability of capture into the 3/2 spin resonance is perhaps a better criterion for judging whether or not to expect some satellite to be in a nonsynchronous commensurate spin state. This probability is independent of both the rigidity and the magnitude of the tidal torque, although it does depend on the tidal model. The expressions for the probability P' of capture into a stable resonance as a satellite passes through a commensurate spin angular velocity are given by Goldreich and Peale [1966, eqns. (18), (20)] and Peale [1974, eqn. (18)].

$$P' = \cfrac{2}{1 + \cfrac{\pi \left[p' - (1 + 15e^2/2)/(1 + 3e^2) \right]}{2 \left[3 \dfrac{B - A}{C} G_{20(2p' - 2)} \right]^{\frac{1}{2}}}} \; ; \tag{15a}$$

$$\text{if} \quad \frac{1}{Q} \propto \text{frequency}$$

$$P' = \frac{\langle \mathscr{T} \rangle (\dot{\psi} > p'n) - \langle \mathscr{T} \rangle (\dot{\psi} < p'n)}{\langle \mathscr{T} \rangle (\dot{\psi} > p'n)} \; ; \tag{15b}$$

$$\text{if} \quad Q = \text{constant}$$

for two rather extreme tidal models, where $\langle \mathscr{T} \rangle$ is again the average tidal torque and p' is a half integer.

As before, we choose to consider the lowest order resonance above synchronous rotation as being the most likely. In this case $p' = 3/2$ and $G_{20\,(2p' - 2)} = G_{201} = 7e/2$. As P' in eqn. (15a) increases with $(B - A)/C$, we choose the rather large value of 10^{-4} to maximize P'. Also terms second order in e^2 can be neglected in eqn. (15a) with little error. For $Q = $ constant, we can write (Goldreich and Peale, 1966, eqn. 22)

$$\langle \mathscr{T} \rangle \propto \sum_{p' = -\infty}^{\infty} \left[G_{20(2p' - 2)} \right]^2 \operatorname{sgn}(\dot{\psi} - p'n) \; . \tag{16}$$

For the 9 natural satellites being considered here, e is sufficiently small that we can neglect terms higher than second order in e in eqn. (15b) and write

$$P' \approx 2 \left[G_{201}(e) \right]^2 , \tag{17}$$

except perhaps for Hyperion whose eccentricity is 0.104.

The above assumptions are used in eqns. (15) and (16) to calculate the capture probabilities given in columns 10 and 11 of Table 6.1. Fourth-order terms in e are included in eqn. (16) for Hyperion. The capture probabilities for the 3/2 resonance are also given for the Moon, Phobos and Deimos for comparison. However, instead of 10^{-4} the value of $(B-A)/C = 2 \times 10^{-4}$ for the Moon and values of 0.228 and 0.204 were calculated respectively for Phobos and Deimos from the triaxial ellipsoids given by Pollack et al. (1973a; Chapt. 15, Duxbury), where a uniform density was assumed. (Of course, these latter three satellites are known to have escaped permanent capture into the 3/2 resonance.) It is interesting that the large asymmetry of the Martian satellites leads to such a large capture probability for a frequency dependent Q.

For the 9 satellites (excluding Janus) which are possible candidates for a nonsynchronous, commensurate angular velocity, only Hyperion, with its large orbital eccentricity has a non-negligible capture probability and that only for Q = constant. We would then expect to find these satellites synchronously rotating (or nearly so if they have insufficient rigidity to maintain a permanent axial asymmetry), except possibly Hyperion. However, the orbital mean motion of Hyperion is commensurate with that of Titan in the ratio of 3:4, and the eccentricity of Hyperion is known to increase secularly with time in this resonance (Greenberg, 1973a, Chapt. 8 herein; Yoder, 1973; Colombo et al., 1974). This means that the probability of capturing Hyperion into the 3/2 spin resonance was much smaller in the past. This reasoning leads us to expect to find Hyperion also (nearly?) synchronously rotating, which leaves the planet Mercury as probably the only example of nonsynchronous spin-orbit coupling in the solar system.

However, the photometric observations of Uranus' satellites by Steavenson (1948, 1964) are not consistent with this conclusion. The satellites U3 (Titania) and U4 (Oberon) were found to vary in magnitude with a period which was not necessarily that of the orbit. In addition, this variability persisted when the satellite orbits were nearly in the plane of the sky. The latter observation implies large obliquities for these satellites. The short tidal damping times imply that both satellites should be in synchronous or nearly synchronous rotation or in nonsynchronous rotation commensurate with the orbital mean motion. But the very small orbital eccentricities would appear to disallow the latter. It now appears likely that the photometry of the reference stars used by Steavenson was incorrect (Andersson, 1974).

If the obliquities are indeed large, spin resonances involving functions of the obliquity [F_{2mp} (i)] in the restoring torque of the libration equation (11) are possible. However, use of the general form of the Hamiltonian in the variational equation for the spin angular momentum (Peale, 1973) shows that the only such resonances possible involving the harmonic coefficient C_{22} correspond to $p' = 1$, 0, -1. The first corresponds to synchronous rotation, the second to no rotation at all, which is not consistent with the observed brightness variations, and the third corresponds to a retrograde synchronous rotation which is really synchronous for a redefined obliquity. A higher order obliquity-type spin resonance necessarily involves higher order harmonic coefficients of the gravitational field. The coefficient of the restoring torque in the pendulum equation is thus much smaller due both to the smaller values of higher order gravitational harmonics and to additional factors of a_e/a. Both stability and capture become much less likely.

We shall see that the tides eventually reduce the obliquity to zero for fixed orbits, which would reduce the restoring torque to zero in an obliquity-type spin resonance. The obliquity can be stabilized at a large value if the orbit plane precesses about the equatorial plane while maintaining a constant inclination. Even this is a possibility of exceedingly low probability, which is reduced to zero as the orbital inclination goes to zero (Peale, 1974). Non-zero orbital inclinations for the outer satellites of Uranus have not been observed.

Since the above observations are not compatible with the expected tidal evolution of the spin angular momenta of the Uranian satellites, we hope for new precision photometry of these satellites which can either show that no problem exists or more accurately define any unusual rotation states to motivate a theoretical attack. Recent photoelectric observations of Oberon (U 4) have been made (Andersson, 1974) and they strongly suggest its synchronous rotation.

LIBRATION OF PHOBOS AND DEIMOS

An observed example of a relatively large amplitude libration about the synchronous rotation state is that of Mars' satellite Phobos, which appears to librate with about a 5° amplitude (Duxbury, 1973; Chapt. 15, herein). A moderate amplitude, forced libration of Phobos was predicted by Burns (1972) and an amplitude of approximately 5° was calculated by Colombo (private communication, 1973) and separately by Burns (private communication to T. Duxbury, 1972) by utilizing the observed shape of Phobos (Chapt. 14, Pollack; Chapt. 15, Duxbury). The large amplitude libration results from the nearness of the free libration period,

$$\frac{P_0}{\left[3(B - A)/C \right]^{1/2}} \approx 0.39 \text{ days} \tag{18}$$

for $(B-A)/C = 0.228$, to the orbital period of $P_0 = 0.31891$ days.

The situation is easily analyzed from the Hamiltonian development of Peale (1973). Substituting eqn. (28) into the first of eqn. (19) of this latter paper yields

$$C\ddot{\psi} = \frac{2GM_sM_pa_e^2}{a^3} \sum_{p=0}^{2} \sum_{q} G_{2pq}(e) F_{2mp}(i) \times$$

$$C_{22} \sin\left[(2-2p)(\gamma-\Omega) + (2-2p+q)M - 2\psi\right] \,, \tag{19}$$

where we have kept only the lowest order terms in a^{-1} in the expansion and where Ω is the longitude of the ascending node of the equator plane on the orbital plane. If we write $\psi = M + \psi_0$, where ψ_0 is the value of ψ at pericenter, assume the obliquity i = 0, set the now arbitrary value of $\Omega = 0$, substitute eqn. (12) for C_{22} and keep only the first order terms in e, eqn. (19) becomes

$$\ddot{\delta} + \frac{3n^2(B-A)}{2C}\sin 2\delta = \frac{3}{2}n^2\frac{(B-A)}{C} \times$$

$$\left[\frac{e}{2}\sin(\delta-M) - \frac{7e}{2}\sin(\delta+M)\right] \,, \tag{20}$$

where $\delta = \psi_0 - \gamma$ measures the libration angle, which is the angle between the axis of minimum moment of inertia and the direction to the orbit pericenter when Phobos is at the pericenter. If δ remains near zero, eqn. (20) becomes the ordinary forced harmonic oscillator equation

$$\ddot{\delta} + \omega_0^2\delta = 2\omega_0^2 e \sin nt \,, \tag{21}$$

where $\omega_0 = n\sqrt{3(B-A)/C}$ is the free libration frequency and n is the orbital mean motion. The forced libration is then

$$\delta = \frac{2\omega_0^2 e}{n^2 - \omega_0^2}\sin nt \,. \tag{22}$$

With $2\pi/\omega_0 = 0.39$ days, n = 0.31891 days, e = 0.021, the amplitude of the libration is given by

$$\delta_{max} = 8.46 \times 10^{-2} \text{ radians} = 4.86° \tag{23}$$

which is very close to that observed. (See Chapt. 15, Duxbury.)

A similar calculation for Deimos yields a forced libration amplitude of 0.5°, which may not be discernible in the Mariner 9 data.

Since the observed libration of Phobos is most likely forced rather than an as-yet-undamped free libration, it cannot be used to infer any recent collisional history for this satellite, but it may be valuable in estimating the satellite's value of $(B-A)/C$.

OBLIQUITIES

If the orbit plane of a satellite is nearly fixed in inertial space, the spin axis of a satellite with a nonzero obliquity will precess about the normal to the orbit plane while keeping the obliquity constant. This well-known effect is due to the gravitational torque which the primary exerts on the equatorial bulge of the satellite. A unit vector along the spin angular momentum thus describes a trajectory which is a line of latitude on a unit sphere centered on the satellite center of mass. The spin vector remains stationary in the orbit frame of reference only when it is either parallel or antiparallel to the orbital angular momentum. This situation is depicted in Figure 6.2, which is a projection of the unit sphere onto a plane containing the orbit normal. The dotted lines are projections of possible trajectories of the spin axis on the unit sphere, and S_1 and S_3 indicate the two possible invariant positions of the spin vector which are normal to the orbit plane.

The tidal torques slowly change the obliquity and thus change the positions of the spin-axis trajectories on the unit sphere. For large spin angular velocities the tides drive the spin axis toward a trajectory on the unit sphere with an obliquity between 0 and 90° (Darwin, 1879b; Goldreich and Peale, 1970). The obliquity corresponding to this trajectory, where the tidal change in the obliquity vanishes, depends on the spin angular velocity—being close to 90° for very large spin rates and shrinking toward 0° as the spin ψ approaches 2n. The obliquity is reduced to 0° for $\psi \lesssim 2n$, where the exact value of ψ near 2n for which this is true depends on the eccentricity and the tidal model (Peale, 1974).

The trend of the obliquity toward an angle slightly less than 90° for high spin rates can be understood in terms of the different rates of tidal reduction of the components of the spin angular velocity perpendicular to, and parallel to, the orbit plane. There is no tidal reduction in the spin angular velocity of a satellite (or planet) if the spin axis is aligned with the direction toward the primary. Averaged around the orbit, the tidal reduction of the parallel component of the spin is only about one-half the reduction of the perpendicular component for equal values of the two components. The trajectory for which di/dt vanishes corresponds to equal fractional rates of reduction of the two components. This also shows directly why the time scale for changing the obliquity is comparable to that for reducing the spin magnitude. The shrinkage of the trajectory on which di/dt vanishes toward zero obliquity means that tides, if acting alone, always ultimately bring a satellite or planet in a *fixed orbit* plane to zero obliquity, provided the endpoint of tidal evolution has $\psi \lesssim 2n$. (See under Spin-Orbit Coupling.) A core-mantle interaction or an accelerating atmospheric thermal tide was invoked by Goldreich and Peale (1970) to stabilize the Venusian obliquity near 180°. Titan is the only satellite known to have a substantial atmosphere (Chapt. 20, Hunten; Cruikshank, 1974), and Titan is sufficiently far from the Sun that thermal effects probably are weak. The core-mantle interaction is

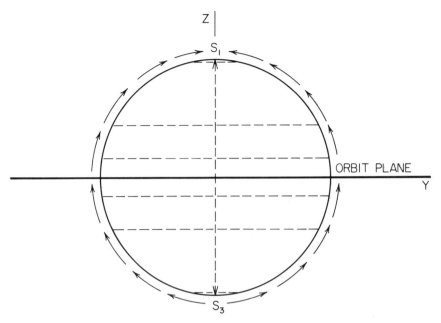

Fig. 6.2. Precession trajectories of the satellite spin vector on a unit sphere showing stable, fixed positions of the spin vector for a fixed orbit plane. Arrows indicate tidal drift of the obliquity.

important only if the initial obliquity is greater than 90°. We will ignore both these effects in discussing the evolution of satellite spins but keep them in reserve for future surprises.

The simple tidal evolution of the obliquity for fixed orbits becomes considerably more complicated if we allow the orbit to vary in a regular way. The simplest such variation is a constant precession about the normal to some nearly fixed plane while the orbit plane maintains a constant inclination to this plane. This type of motion is exhibited in an average sense by the lunar orbit relative to the ecliptic plane, by Mercury's orbit relative to the plane of the solar system, by the regular natural satellite orbits relative to the equatorial planes of their primaries, and by Iapetus' orbit relative to the Laplacian plane. Essentially all the satellites which are tidally evolved are thus in such precessing orbits, although the inclinations may be small, and the tidal evolution of their obliquities is more complicated than that outlined above for the fixed orbit. One immediately suspects that the final obliquity is also not 0°, since the orbit normal is no longer fixed.

A generalization of Cassini's laws describing the rotation of the Moon (Colombo, 1966; Peale, 1969; Beletskii, 1972; Peale, 1974) provides the proper framework for the discussion of the obliquities of satellites in precessing orbits. For the Moon these laws are (Kaula, 1968, p. 183):

1. The rotation of the Moon is synchronous with its orbital motion.
2. The spin axis of the Moon maintains a constant inclination to the ecliptic plane.
3. The spin axis, the orbit normal, and the ecliptic normal remain coplanar.

The generalization of these laws is best understood by reference to Figure 6.3. Like Figure 6.2, the circle is the projection of the unit sphere on the plane containing the orbit normal, but with the added restriction that the plane also be perpendicular to the node of the orbit plane on the (nearly) invariable plane. The precession trajectories of the spin vector on the unit sphere are now determined by the intersection of the parabolic cylinder

$$(Z - K_1)^2 = K_2(Y - K_3) , \qquad (24)$$

where X, Y, Z are direction cosines of the spin vector relative to the XYZ axes, respectively, with the Z-axis along the orbit normal, the X-axis along the ascending node of the orbital plane on the invariable plane; two of the three constants are given by (Peale, 1974):

$$K_1 = \frac{\cos \iota - 2S}{2(R + S)} \qquad (25)$$

$$K_2 = \frac{\sin \iota}{R + S} \qquad (26)$$

with

$$R = \frac{3n^2 (C - A/2 - B/2) G_{210}(e)}{4\dot{\psi}\mu'C} \qquad (27)$$

$$S = \frac{3}{16} \frac{n^2}{\dot{\psi}\mu'} \frac{(B - A)}{C} G_{201}(e) ; \qquad \frac{\dot{\psi}}{n} = \frac{3}{2}$$

$$S = \frac{3}{16} \frac{n^2}{\dot{\psi}\mu'} \frac{(B - A)}{C} G_{200}(e) ; \qquad \frac{\dot{\psi}}{n} = 1 \qquad (28)$$

$$S = 0 ; \qquad \frac{\dot{\psi}}{n} \neq p' ,$$

where ι is the inclination of the orbit plane relative to the invariable plane and μ' is the magnitude of the precessional angular velocity of the orbit plane. The third constant K_3 contains the Hamiltonian of the system in the XYZ frame of the reference which rotates with the orbit.

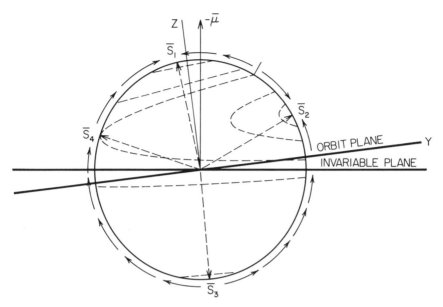

Fig. 6.3. Precession trajectories of the satellite spin vector on a unit sphere showing the location of Cassini spin states for a uniform precession of the orbital plane. Arrows indicate tidal drift of the obliquity.

Since the probability of a stable, resonant, but nonsynchronous spin is negligibly small for all the tidally evolved satellites, we shall assume $\dot{\psi}/n = 1$ in all cases. This choice means that K_1 and K_2 are determined once and for all when the other constant parameters in the problem are known. The shape of the parabolic cylinder and its plane of symmetry are therefore fixed. The Hamiltonian, and hence K_3, can have a limited range of values consistent with the given values of the parameters. The limited arbitrariness in the value of K_3 results from the change in the gravitational potential energy as a function of the satellite's orientation in the field of the primary.

All the possible trajectories of the spin vector on the unit sphere are then represented by the intersection of the sphere with the parabolic cylinder of fixed shape as the latter is moved parallel to its plane of symmetry. These trajectories are indicated by the dotted lines in Figure 6.3. The spin axis remains stationary in the orbit frame of reference for extreme values of the Hamiltonian (and hence K_3) which correspond to the points where the parabolic cylinder is tangent to the unit sphere (Peale 1969, 1974). These fixed positions are indicated by S_1, S_2, S_3, S_4 in Figure 6.3, of which only the first three are stable as indicated by the trajectories which pass close to each position. Since the orbit normal, the invariable plane normal and the spin vector remain coplanar for these positions, we call the positions Cassini spin states or Cassini positions. Mercury and the Moon occupy states 1 and 2, respectively.

Comparing Figures 6.2 and 6.3, we see that the orbit precession generally introduces two additional positions of the spin vector in which this vector is fixed relative to the orbit. In addition, the precession trajectories of the spin vector no longer correspond to a constant obliquity but can be much more complicated, especially about state 2.

The Cassini states correspond to extremes in the system Hamiltonian. We would then expect any dissipative process such as tidal friction to drive the obliquity to one of these extremes just as the spin is ultimately driven to zero obliquity for a fixed orbit. But there is now a choice of final states—none corresponding to zero obliquity—and this choice depends on the values of the various parameters in the problem and on the initial conditions or the conditions at the time of capture into a spin resonance (Peale, 1974). The tides always remove a satellite from state 3 but can select either of the other two stable states (S_1, S_2) as an endpoint of evolution. As in the fixed orbit case there are inter-mediate precession trajectories toward which the spin is driven as the satellite slows down. These intermediate trajectories shrink toward either state 1 or state 2 depending on the particular values of the parameters and vanish at one of these states when $\dot{\psi}$ approaches 2n.

Extreme values of K_1 essentially determine the evolution of the spin and the ultimate choice between states 1 and 2. If $0.3 \lesssim K_1 \lesssim 0.9$, initial conditions (or conditions at the time of capture into a spin resonance) determine the ultimate choice. If $-0.3 \lesssim K_1 \ll 1$, the axis of the parabolic cylinder eqn. (24) passes close to the center of the unit sphere, state 2 lies nearly in the orbit plane and state 1 is very close to the orbit normal (see Figure 6.3). In these circumstances K_2 is also small and the parabolic cylinder is very thin. If the spin is within the trajectory through the unstable state 4, the tides tend to drive the spin toward state 2. K_1 changes substantially as the spin is captured from a non-resonant to a resonant value, since S assumes a nonzero value. The critical trajectory through state 4 thus changes its position at this time and the spin would have to be within the new critical trajectory at the time of capture into the resonance in order to eventually be trapped in state 2. The smallness of this region means this is an event of low probability. For small values of K_1, we therefore expect to find the satellite in state 1, which is very close to the orbit normal. The planet Mercury appears to satisfy these criteria, and its obliquity is known to be very small (B. A. Smith, private communication, 1973). In addition, the obliquity is here an adiabatic invariant, and the spin will follow the Cassini state through the relatively slow orbit variations (Peale, 1974).

In the opposite extreme of $K_1 > 1$, the axis of the parabolic cylinder does not pass through the unit sphere at all. States 1 and 4 cease to exist, and state 2 has now moved to a relatively small obliquity. The only choice of tidal evolution is now state 2 which is displaced from the orbit normal by an angle which is greater than the orbital inclination relative to the invariable plane. It is impossible for the spin vector to remain between the orbit normal and the normal to the invariable plane (Peale, 1969). Here, the fluctuations in the orbit parameters will be relatively rapid and the spin will occupy the average position of the Cassini state. The Moon satisfies these criteria and is known to be in state 2.

For intermediate values of K_1, either state 1 or state 2 may be selected with comparable probabilites. The choice depends on whether the spin happens to be inside or outside the trajectory through state 4 at the time of capture into the here-assumed synchronous spin resonance. Fluctuations of the spin from both the instantaneous and average positions of the Cassini state of the same order as the state position fluctuations themselves will also occur.

Except for the Moon, Phobos, and Deimos, we know nothing about the moment differences of the natural satellites which determine the final Cassini states. However, we do know μ' and ι for many of the satellites and we can determine the value of K_1 and thus infer the final state for values of the moment differences corresponding to hydrostatic equilibrium. This may be the most appropriate choice if a satellite is composed primarily of ice-like materials (Johnson and McGetchin, 1973). This exercise has been completed with the hydrostatic moment differences for both tidal and rotational deformation being determined from expressions given by Peale (1973) where a fluid Love number $k_f = 3/2$ was assumed. The values of $(C - A/2 - B/2)/C$, $(B-A)/C$ and K_1 are given for all the tidally evolved satellites in Columns 12 through 14 in Table 6.1. The values for Phobos and Deimos are not hydrostatic but are based on the known ellipsoidal shapes, and the known lunar values of the moment differences are used. Since Iapetus is of special interest because of its large (8°) orbit inclination to the Laplacian plane (Brouwer and Clemence, 1961b), we have also given the value of K_1 for the larger non-hydrostatic value of the moment differences of $(C - A/2 - B/2)/C = (B - A)/C = 10^{-4}$ to illustrate the dependence of the conclusions drawn on the assumptions made. The precession period of the orbit node and the value of μ'/n used in the calculations are also included in Table 6.1 for reference.

The large value of K_1 in Table 6.1 for the Moon reinforces the statement above pointing out the observed occupancy of state 2 by the Moon. The very large moment differences for Phobos and Deimos make their values of K_1 very "Mercury-like" so the Martian satellites should both have obliquities which correspond to the instantaneous positions of the Cassini states 1, which are close to the orbit normals. Measurements of the actual obliquities would yield bounds on the moment differences (Peale, 1969), but the magnitudes of the angles are considerably below current observational errors (Peale, 1972).

The hydrostatic values of the moment differences for the remaining satellites which are tidally evolved yield values of K_1 which are all very "Mercury-like" except for Iapetus, where $K_1 = 7.67$ is even larger than the lunar value. This large value of K_1 would place Iapetus definitely in state 2 with an obliquity

TABLE 6.1
Rotational Data for the Natural Satellites

Satellite		1 a_e (Km)	2 e	3 M_p/M_s ($\times 10^4$)	4 P (Days)	5 $2\pi/\mu'$ (Yrs)	6 μ'/n ($\times 10^{-4}$)
E1	Moon	1738			27.322	18.61	40.195
M1	Phobos	12.1			0.31891	2.262	3.850
M2	Deimos	6.8			1.2624	57.329	0.603
J1	Io	1820		2.37	1.7691	7.4589	6.493
J2	Europa	1550		3.96	3.5512	30.2349	3.216
J3	Ganymede	2635		1.23	7.1546	136.59	1.434
J4	Callisto	2500		2.09	16.689	562.25	0.813
J5	Amalthea	70	0.003	41300	0.49818	0.39361	34.652
J6	Himalia	80			0.5		
J7	Elara	30			0.5		
J8	Pasiphaë	8			0.5		
J9	Sinope	10			0.5		
J10	Lysithea	9			0.5		
J11	Carme	11			0.5		
J12	Ananke	8			0.5		
S1	Mimas	200	0.0201	1540	0.94242	0.98468	26.204
S2	Enceladus	250		669	1.3702	2.3603	15.894
S3	Tethys	500		90.9	1.8878	4.9841	10.370
S4	Dione	575		49.1	2.7369	11.1719	6.394
S5	Rhea	800		17.6	4.5175	35.8209	3.453
S6	Titan	2500	0.0290	0.41	15.945	720.865	0.606
S7	Hyperion	80-460	0.104	14.1	21.277	1300	0.448
S8	Iapetus	900		9.31	79.331	3000	0.724
S9	Phoebe	30-160			0.5		
S10	Janus	110		10200	0.74896	0.62092	33.0
U1	Ariel	300-1700	0.003	1.38	2.520		
U2	Umbriel	200-1100	0.004	5.02	4.144		
U3	Titania	360-2000	0.0024	0.72	8.706		
U4	Oberon	330-1900	0.0007	0.87	13.46		
U5	Miranda	110-650	0.017	12.4	1.4135	15.8	2.449
N1	Triton	600-3500		0.03	5.877	585	0.275
N2	Nereid	100-600			0.5		

Column	=	explanation

1 = satellite radius
2 = orbital eccentricity
3 = primary/satellite mass ratio
4 = period of rotation
5 = period of orbit node

6 = ratio of orbit precession to orbital mean motion
7 = time constant for wobble decay
8 = time to reach synchronous rotation; asterisks indicate those satellites for which synchronous rotation has been observed.

7	8	9	10	11	12	13	14
τQ^{-1} (10^3 Yrs)	TQ^{-1} (10^4 Yrs)	$(B-A)Q/C$ ($\times 10^{-4}$)	P' Q = Const	P' $Q \propto 1/f$	$\dfrac{C-A/2-B/2}{C}$	$\dfrac{B-A}{C}$	K_1
110	*320		0.074	0.027	5.13×10^{-4}	2.3×10^{-4}	4.60
3.5	* 0.088		0.011	0.44	0.196	0.228	−.224
680	* 91		1.9×10^{-4}	0.18	0.176	0.204	−.225
0.026	* 0.005				5.96×10^{-3}	7.15×10^{-3}	−.175
0.29	* 0.12				1.52×10^{-3}	1.82×10^{-3}	−.122
0.83	* 0.66				5.74×10^{-4}	6.89×10^{-4}	−.103
12	* 22				1.54×10^{-4}	1.84×10^{-4}	.0409
0.40	0.018	3.2	2.2×10^{-4}	4.5×10^{-3}	7.43×10^{-2}	8.91×10^{-2}	−.207
0.31	1.1×10^{9}						
2.2	9.1×10^{9}						
31.0	5.3×10^{12}						
20	9.3×10^{12}						
24	1.6×10^{11}						
16	6.4×10^{12}						
31	8.0×10^{12}						
0.33	0.014	0.98	9.9×10^{-3}	0.012	5.98×10^{-2}	7.18×10^{-2}	−.208
0.65	* 0.060				2.42×10^{-2}	2.90×10^{-2}	−.196
0.42	* 0.086				1.38×10^{-2}	1.65×10^{-2}	−.192
0.97	* 0.38				5.44×10^{-3}	6.53×10^{-3}	−.171
2.3	* 1.5				1.92×10^{-3}	2.30×10^{-3}	−.138
10	19	0.41	0.021	0.014	1.09×10^{-4}	1.30×10^{-4}	.0556
100	9.2×10^{3}	0.002	0.236	0.026	6.53×10^{-5}	7.82×10^{-5}	.122
700	$*8.7 \times 10^{4}$				4.70×10^{-6}	5.64×10^{-6}	7.67
					(1.0×10^{-4})	(1.0×10^{-4})	(.186)
0.77	1.7×10^{10}						
0.55	0.041				0.104	0.124	−.213
0.25	0.27	78	2.2×10^{-4}	4.5×10^{-3}	6.09×10^{-3}	7.31×10^{-3}	
2.6	4.0	9.7	3.9×10^{-4}	5.2×10^{-3}	2.26×10^{-3}	2.72×10^{-3}	
6.7	30	11	1.4×10^{-4}	4.0×10^{-3}	5.13×10^{-4}	6.15×10^{-3}	
28	200	14	1.2×10^{-5}	2.2×10^{-3}	2.14×10^{-4}	2.56×10^{-4}	
0.19	0.30	6.0	7.1×10^{-3}	0.011	1.95×10^{-2}	2.34×10^{-2}	−.224
0.76	* 2.0				1.83×10^{-4}	2.20×10^{-4}	−.154
0.016	1.0×10^{9}						

Column	=	explanation
9	=	lower bound on $(B-A)/C$ for stability of 3/2 spin resonance
10	=	probability of capture into 3/2 spin resonance for Q = constant
11	=	probability of capture into 3/2 spin resonance for $Q \propto$ (frequency)$^{-1}$
12	=	hydrostatic values of $(C - A/2 - B/2)/C$ except for E1, M1, M2, S8 (lower)
13	=	hydrostatic values of $(B - A)/C$ except for E1, M1, M2, S8 (lower)
14	=	parameter determining selection of Cassini spin state

exceeding 8°. On the other hand, if we calculate K_1 for the larger moment differences of 10^{-4}, Iapetus, too, would be Mercury-like and would be in state 1 with a smaller obliquity. These two extremes provide an observational test for determining whether Iapetus is composed mainly of ice or rock, since the former would favor the hydrostatic moment differences and hence occupancy of state 2, whereas the latter implies more internal strength, moment differences of lunar magnitude and occupancy of state 1. The only reason this difference may some-day be observable from spacecraft is that these two states for Iapetus must be separated by at least 8°. See also Ward (1975b).

One might also argue for nonhydrostatic values for the moment differences of some other tidally evolved satellite and hence possible occupancy of state 2. However, the differentiation between states 1 and 2 for these satellites would be difficult, since the minimum separation between the two states is only the rela-tively small orbital inclination. Only for Iapetus does a successful observation seem possible. For this reason we have not determined K_1 for non-hydrostatic values of the moment differences for the remaining satellites.

SUMMARY

If we adopt Q = 100 (comparable to that of the Earth's mantle) in column 7 of Table 6.1, the upper bounds on the time constants for an exponential decay of satellite wobble motion range from 1,600 years for Nereid (N2) to 6.8×10^7 years for Deimos for the assumed or measured rotation periods given in column 4. Iapetus and Hyperion have the exceptionally large time constants of 8.1×10^8 and 9.7×10^8 years, respectively, under these same assumptions. However, tides will tend to shorten these decay times, and the faster rotation periods of the past lead to much shorter times. The infrequency of large meteoroid impacts means that the satellites have thus maintained principal axis rotation throughout most of history and should be so rotating today. We should look closely at Iapetus and Hyperion, however.

If the same value of Q = 100 is adopted in column 8 of Table 6.1, the times for the regular satellites to reach synchronous rotation from an initial rotation period of 2.3 hours range from 5,000 years for Io (J1) to 2×10^8 years for Oberon (U4). All of these regular satellites have almost certainly reached an endpoint of the tidal evolution of their spin angular momenta in the 4.6×10^9 years available. We include the Moon, Triton, Iapetus, and Hyperion in the group of tidally evolved satellites—the Moon and Iapetus from observation, Triton from its short time for tidal decay and Hyperion because it is interior to Iapetus' orbit. The tidal decay times for Iapetus and Hyperion exceed the age of the solar system for moderate values of Q, but the times in Table 6.1 are only upper bounds because of the extreme value of initial spin angular velocity and the assumption of earth-like rigidity. The remaining irregular satellites have such long decay times for any reasonable value of Q that they essentially retain their primordial spin angular momenta. Of the 23 satellites which are tidally evolved by these criteria, 13 have confirmed synchronous rotation periods, while the others have not yet been measured.

The long tidal decay times for Saturn's ring particles suggest that they should be tumbling as the result of relatively frequent interparticle collisions. However, the observations of Ferrin (1974) and Morrison (1974b) are consistent with synchronous rotation for some major fraction of the particles. Is there some other dissipative mechanism keeping the particles in synchronous rotation?

None of the ten tidally evolved satellites whose rotation periods have not yet been determined should be captured in nonsynchronous, commensurate spin states as indicated by the capture probabilities into the 3/2 resonance in columns 10 and 11 of Table 6.1. These low probabilities result primarily from the small orbital eccentricities. The relatively large capture probability for Hyperion (S7) is sharply reduced in the past by a smaller orbital eccentricity. These conclusions, when applied to the Uranian satellites, mean that Steavenson's photometric observations (1948, 1964) of these satellites which had implied a nonsynchronous rotation and large obliquities are not consistent with the expected tidal evolution (nor are they with the recent observations of Andersson, 1974).

The values of the constant K_1 in the last column of Table 6.1 for the values of the moment differences given in columns 12 and 13 (mostly hydrostatic) are consistent with nearly all the tidally evolved satellites most probably occupying Cassini spin state 1 with an obliquity near 0°. The exceptions are the Moon, which is known to occupy Cassini state 2 with an obliquity of about 6.5°, and Iapetus, which should be in state 2 with an obliquity greater than 8° if it is near hydrostatic equilibrium. Iapetus would most probably be in state 1 with an obliquity near 0° if it possesses nonhydrostatic moment differences comparable to those of the Moon. The large difference in internal strength implied by these two extremes can serve as a practical observational differentiation between an ice-like and a rock-like interior for Iapetus. The changes in obliquity for nonhydrostatic moment differences for the remaining satellites are probably too small to be observable because of the small orbital inclinations relative to the respective (nearly) invariable planes.

Finally the observed 5° libration of Phobos is not an indication of a recent collision since it is fully accounted for as a nearly resonant forced libration.

ACKNOWLEDGMENTS

It is a pleasure to thank the members of the Laboratory of Atmospheric and Space Physics at the University of Colorado for their kind hospitality and support during this writing. This research was supported by the Planetology Program, Office of Space Science, NASA, under grant NGR 05-010-062.

DISCUSSION

DALE CRUIKSHANK: What is the effect on particle alignment (toward sychronous rotation) of inter-particle collisions in Saturn's rings?

S. J. PEALE: One would expect collisions to maintain some kind of equilibrium distribution of random rotation states, which would begin relaxing toward the synchronous state only after the collisions had become sufficiently rare.

IGNACIO FERRIN: I have made a calculation showing that the synchronous rotation is maintained under collisions. At most, the particle oscillates after the collision with an amplitude of the order of 10° around the radial direction. The main reason for this is that the gradient of the gravitational field is very strong at the small distances considered (~2 Saturn radii): it maintains the orientation of the particle toward the planet despite perturbations.

S. J. PEALE: I suspect that the results of these calculations may be spurious due to over-restrictive assumptions about the nature of the collisions and/or the behavior of the particles after collision. The effect of the gravity gradient depends on the assumed particle mass distribution. Also, the particles have been assumed to have reached synchronous rotation without libration before collisions were considered.

ORBITAL EVOLUTION

Joseph A. Burns
Cornell University

The orbital evolution of a large satellite is governed primarily by tidal interactions between the satellite and the planet it orbits. Tides raised on a planet by a satellite transfer energy and angular momentum to the satellite orbit; this changes the semimajor axes of satellite orbits, increasing the size of those orbits where the satellite mean motion n is smaller than the planetary angular velocity ω, and decreasing those where n > ω. Such tides have caused substantial changes for satellites of the terrestrial planets, which may explain why there are no satellites about Mercury and Venus. For Jovian and Saturnian satellites, such tides probably are only important in bringing about some of the observed orbital resonances. Tides raised on satellites generally cause decreasing orbital eccentricities, indicating why close satellites always have nearly circular orbits. Tidal effects on orbital inclination are relatively smaller.

Different processes of orbital evolution dominate for small bodies; their effects probably are critical in positioning material in the primordial dust cloud so that satellite coagulation may occur. A qualitative description is given of the orbital results of gas drag, radiation pressure, Poynting-Robertson drag and electromagnetic forces. Some of these processes are important today in determining the orbital evolution of small particles in circumplanetary space.

This review is concerned with the orbital evolution of particles in circumplanetary space. The discussion is largely qualitative, emphasizing simple physical principles. It attempts to give an understanding of the important phenomena and to provide a guide to the current literature. Processes which govern the orbital evolution of large bodies, that is, the bodies seen today, will be distinguished from those which principally affect small bodies—dust particles.

The mechanisms which are important for large bodies involve tides in the planet and/or the satellite itself, as well as third body gravitational interactions with other satellites or the Sun. Tidal processes, in particular, will be discussed in some detail since they are crucial especially in the histories of satellites of the terrestrial planets and most likely in setting up many of the orbital resonances seen in the satellite systems of the outer planets (cf. Chapt. 8, Greenberg). Tides also determine the rotation histories of the natural satellites, as explained in Chapter 6, by Peale. Gravitational interactions will be mentioned only insofar as they may secularly affect satellite configurations over the age of the solar system. This will include short discussions about the long-term stability of satellite systems and about the possibility of capture of the irregular satellites of Jupiter and

[113]

Saturn. Other classical effects of gravity fields on satellite orbits are described in more detail in Chapter 4 by Kovalevsky and Sagnier and in Chapter 3 by Aksnes.

The orbital evolution of small particles also is considered, since they undoubtedly were prevalent at the time of the origin of the satellite systems and played a fundamental role in the birth of the satellites (cf. Chapt. 23, Cameron; Chapt. 24, De et al.; Chapt. 26, Safronov and Ruskol). Such particles continue to exist today in the neighborhoods of the planets, either as debris captured from interplanetary space or as pieces of ejecta, originating in collisions with satellites. Any forces (such as radiation pressure, gas drag, or Poynting-Robertson drag) which are proportionate to a particle's cross-sectional area produce accelerations that vary as the inverse of the particle radius. These forces, therefore, normally have significant effects only for small particles. Collective motions also can be important in the initial circumplanetary nebula. This review merely sketches the effects of such processes on the orbital evolution of small particles.

The orbits of particles of an intermediate size range (meters to kilometers) are measurably affected only by classical celestial mechanics effects since such particles are too small to produce significant tides but too large to be influenced by the non-classical effects discussed above.

GENERAL CONSIDERATIONS

Long-time orbital evolution calculations are used in trying to provide information on where satellites originated. In essence one attempts to answer the questions: Did the satellites begin their lives at roughly the same distances fron the planets as they are now? Are the regularities that we see today in satellite orbits—in particular, low eccentricities, small inclinations, and orbital resonances—conditions of origin or results of later processes? How probable are captures and escapes from satellite orbits? Of course, rather than answering these questions, one would prefer to be able to say that at time t, satellite x was precisely at position r or had exactly orbital elements a_i. Unhappily, such cannot be done: equations are too approximate, processes too sporadic, knowledge too meager, and the past too distant to permit extrapolating with confidence far back in time. The best that can be accomplished is to construct probable scenarios or to place bounds, generally based on either conservation laws or on the form of the dynamical equations which give the perturbations of the satellite orbital elements.

Since our discussion usually will be restricted to qualitative arguments we often will make the simplifying assumption that the satellite mass m is negligible in comparison to the planetary mass M. This is equivalent to assuming that the satellite is orbiting a fixed mass center. With the possible exceptions of the Earth-Moon, where $m/M \sim 10^{-2}$, and Neptune-Triton, where $m/M \sim 10^{-3}$, this is a good approximation (cf. Table 1.4).

One of the most useful tools available for estimating past orbital positions is the work-energy theorem. Following MacDonald (1964), the total energy E contained in a satellite's orbit of semimajor axis, a, moving about fixed mass M is

$$E = - GMm/2a, \tag{1}$$

where G is the universal gravitational constant. This result is also true in the two body problem when m/M is not very small; then $M/(M+m)$ of the energy is stored in the smaller, faster moving mass.

The change in the energy (1) of the satellite orbit, that is, the work, governs the manner in which the orbit size varies:

$$\triangle a = (2a^2/GM)\triangle E/m. \tag{2}$$

So, knowing the work done by the various forces over all time will give the initial orbit size after a simple integration. However, to arrive at a fairly accurate answer, $\triangle E$, which often is a function of a or m, must be well known, and normally it is not. Frequently, the most that is known is the sign of $\triangle E$ which only gives the direction of change of a: work done *on* an orbit expands that orbit while work done *by* it collapses the orbit.

One can also write a simple expression for the angular momentum H of a satellite orbiting a fixed planet (MacDonald, 1964),

$$H = m \left[GMa(1-e^2) \right]^{\frac{1}{2}} \tag{3}$$

or, inverting (3) with (1), the orbital eccentricity e is

$$e = \left[1 + 2H^2E/(G^2M^2m^3) \right]^{\frac{1}{2}}. \tag{4}$$

In the two body problem most of the angular momentum is contained in the satellite orbital motion. Note that considerations of conservation of angular momentum for planet-satellite systems must include the angular momentum stored in the bodies' rotations, particularly that of the more massive planet, as well as their orbital motions. In fact, the only satellite which contains much relative angular momentum in its orbit is the Moon (Burns, 1975). All other satellite orbital angular momenta are negligible in comparison to those of their primaries' rotations, saying that most satellites have not noticeably influenced their primary's rotation. It is curious that in the Sun-planet system virtually all angular momentum is in the planetary orbital motions: apparently other processes are important there.

The orbital eccentricity is affected by changes in both orbital energy and angular momentum. Differentiating eqn. (4),

$$\Delta e = e^{-1}[e^2 - 1]\left[(1/2)(\Delta E/E) + (\Delta H/H)\right]. \tag{5}$$

Since ΔE and ΔH can have either sign, the two terms may compete with one another to determine whether the orbit is circularized with time or not. Obviously the eccentricity does not change if, and only if, $\Delta E/E = -2\,\Delta H/H$ over each complete orbit. This condition will be further discussed following eqns. (8).

The orbital inclination i and its variation can be similarly discussed. The inclination of the orbit plane relative to some "fixed" plane—customarily either the planetary equatorial plane or the planet's orbit plane—is defined by

$$\cos i = H_n/H, \tag{6}$$

where H_n is the component of angular momentum normal to the fixed plane. Differentiating this,

$$\Delta i = (H^2/H_n^2 - 1)^{-\frac{1}{2}}\left[(\Delta H/H) - (\Delta H_n/H_n)\right]. \tag{7}$$

Another, complementary approach that tells how satellite orbits have evolved is the use of the classical perturbation equations of celestial mechanics. The perturbation equations written in terms of force components (called Gauss's equations) are more indicative of the effects of particular dynamical processes than the Lagrange planetary equations, those written with a perturbation potential or disturbing function (Chapt. 4, Kovalevsky and Sagnier). Each can be found in any celestial mechanics text (e.g., Brouwer and Clemence, 1961a, or Roy, 1965). A heuristic derivation of the six perturbation equations, based on elementary dynamical principles, is contained in Burns (1976b). The perturbation equations in terms of force components are

$$\frac{da}{dt} = \frac{2}{n(1 - e^2)^{\frac{1}{2}}}\left[S\,e\,\sin v + (1 + e\cos v)T\right] \tag{8a}$$

$$\frac{de}{dt} = \frac{(1 - e^2)^{\frac{1}{2}}}{na}\left[S\sin v + T(\cos \mathscr{E} + \cos v)\right] \tag{8b}$$

$$\frac{di}{dt} = (na^2)^{-1}(1 - e^2)^{-\frac{1}{2}}\quad W\,r\cos(\Omega + v) \tag{8c}$$

plus three similar equations for orbital elements which are not of particular interest for long-time orbital evolution calculations. (S,T,W) are the components of perturbing force per unit mass in directions radially outwards, transverse to S in the orbit plane in the direction of motion, and normal to the orbit plane so as to form a right hand triad. The true anomaly is v and the eccentric anomaly $\mathscr{E}$, while r is the radial distance, n the mean motion, and Ω the argument of pericenter (cf. Fig. 3.1). Of course, eqns. (8a) and (8b) can be derived directly from (2) and (5) by accounting for those forces which change the angular momentum and those which do work. Similarly, eqns. (2) and (5) can be developed immediately from (8a) and (8b). To find eqn. (8c) from (7), only the angular momentum and torque in the equatorial plane should be considered. As can be found from eqn. (8b), the eccentricity will be unchanged after one orbit if, and only if,

$$\int_0^{2\pi} \left[S \sin v + T(2 \cos v + e \sin^2 v) \right] (1 - 2 e \cos v) dv = 0 \qquad (9)$$

is satisfied to first order in e.

PROCESSES AFFECTING LARGE BODIES

In this section we will consider phenomena that are particularly effective on the satellites of today, that is, bodies that are kilometers or more in size. Most of the discussion will concern tides, which are qualitatively well understood, and for which quantitative theories have been developed. After describing the physical principles governing tidal evolution, we list the appropriate equations, and then application is made to the Earth-Moon problem, the motion of the Martian satellites, and the orbital collapse of hypothetical satellites of the other terrestrial planets; brief mention is given of the way in which tides may account for the orbital resonances seen in the satellite systems of the outer planets. We also discuss the long-term evolution of satellite orbits under gravitation (capture and stability problems) and the effect of collisions with debris. A summary of all these processes is given in Table 7.2.

Tidal Evolution

PHYSICAL ARGUMENTS

The manner in which tides affect orbits is easily illustrated (cf. Kaula, 1964; MacDonald, 1964; Goldreich and Soter, 1966; Goldreich, 1972). Consider two bodies near one another, calling one the satellite and the other the planet. This designation is arbitrary, and the effects to be discussed are obviously symmetric although they generally proceed faster for the satellite because of its smaller mass. The presence of the satellite distorts the gravitational field of the planet and vice versa. The bodies, since they are not rigid, respond to the total gravitational field—their own plus that of the disturbing body. Looking now at the planet only, its response would be symmetric about the line of centers as shown in Figure 7.1a if it were made of perfect material. The double tidal bulge arises because only at the planet's center does the gravitational pull of the satellite exactly counterbalance the centrifugal "force" caused by the planet's accelerated circular motion about the system's center of mass. Elsewhere, the material nearer the satellite than the planetary center has a larger gravitational attraction, but the same orbital angular velocity (and therefore a smaller centrifugal "force") as the planet's center, while the material on the opposite side has a lesser gravitational attraction and a greater centrifugal "force": it is this imbalance of forces that elongates the planet. Since the planet's material is neither purely elastic nor a perfect fluid, energy is lost in straining it. This energy loss delays the material's response as it does in the simple mass-spring-dashpot system and, if the planetary rotational angular velocity ω differs from the satellite's orbital angular velocity n, the bulge rotates from under the satellite that caused it. The misaligned bulge will lead the satellite position when $\omega > $ n, as shown in Figure 7.1b, and will lag for $\omega<$n as seen in Figure 7.1c. The circumstance in which $\omega>$n is most common in the solar system, excepting Phobos and retrograde satellites, and will be the one which is discussed below unless otherwise noted.

Since the bulge is misaligned in the (assumed) symmetric gravity field of the satellite, a torque is produced on the planet which attempts to align the bulge. When $\omega >$n, this slows the planet's rotation. The kinetic energy taken out of the rotation is partly put into the satellite orbit and partly dissipated as heat in the planetary interior. This process will continue until $\omega=$n. As described by Peale (Chapt. 6), the same has happened to the satellite's rotation except it has proceeded to its conclusion for close satellites, accounting for the synchroneity between the rotation and revolution periods of all inner satellites.

The expressions given by Peale (Chapt. 6) for the characteristic slowing time of satellite rotations are of course applicable here as well for the planetary slowing. Tides raised by the Sun have similarly influenced the rotations of the inner terrestrial planets in a significant way (Goldreich and Peale, 1968).

The angular momentum lost from the planet's rotation is transferred to the satellite's orbital motion to conserve the total angular momentum. In the case illustrated in Figure 7.1b, the satellite is pulled forward along its orbit and moves away from the planet as indicated by eqns. (2) and (8a); the reverse occurs for $\omega<$n.

The orbital inclination is also affected since the planet's rotation carries the delayed tidal bulge out of the satellite's orbital plane, unless i=0; this bulge is then able to produce a torque on the satellite orbit plane. From the standpoint of an angular momentum conservation principle, evolution of the orbital angular momentum vector implies that tides will cause the planetary obliquity to change unless there is a fortuitous alignment between angular momentum vectors.

(a)

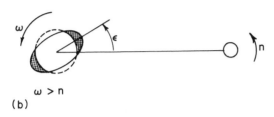

$\omega > n$

(b)

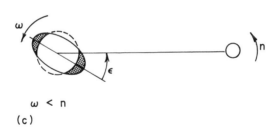

$\omega < n$

(c)

Fig. 7.1. (*a*) Planetary tides. The gravitational attraction of the satellite distorts the planet, which responds immediately, assuming perfect materials. No angular momentum nor energy is transferred to the satellite's orbit since the gravitational field of the planet is symmetric about the line of centers. (*b*): Planetary tides for the usual case in the solar system, $\omega > n$. The presence of friction delays the tidal bulge, which rotates ahead of the perturbation that caused it by the lag angle ϵ. Angular momentum and energy are given from the rotation of the planet to the orbit of the satellite: the planet's rotation slows and the orbit grows. (*c*): Planetary tides for $\omega < n$, including friction. This situation holds for Phobos and retrograde satellites. The time-delayed tidal bulge lags behind the satellite position by lag angle ϵ. Angular momentum and energy are transferred by tidal forces from the satellite's orbit to the planet's rotation. The orbit shrinks while the planet's rotation increases.

EVOLUTIONARY DIAGRAMS

The transfer of angular momentum between the planetary spin and the satellite orbit occurs in a manner such that the total mechanical energy (gravitational potential energy plus the rotational and orbital kinetic energy) decreases, since some energy is lost to heating the planetary interior. The total mechanical energy and angular momentum of the system uniquely specify the planetary rotational velocity ω and the orbital semimajor axis for a circular satellite orbit. To correctly evaluate a and ω, we must include in expressions (1) and (3) the energy, $(1/2)\ \omega \cdot \underline{I} \cdot \omega$, and angular momentum, $\underline{I} \cdot \omega$, of the planet's rotation. Counselman (1973), considering a low mass satellite in a circular equatorial orbit about a homogeneous planet, plotted constant angular momentum and constant energy contours on diagrams whose axes correspond to orbit size and planetary rotation rate. A constant angular momentum line is followed in the direction of decreasing total system energy throughout the evolutionary track of the system for given initial conditions regardless of how the angular momentum is transferred. It is found that the evolutionary track can end in any of three possible states: (i) satellite escape, (ii) orbital collapse onto the planet's surface, or (iii) a synchronous state in which the rates of satellite rotation, planet rotation and orbital revolution are all the same; the last can be an unstable configuration. The speed at which the evolutionary track is traversed is not given by this method and the approach to the end states mentioned above may be so slow that the designation "end state" is meaningless.

Greenberg (1974) generalized Counselman's tidal orbital evolution diagram to permit a non-negligible satellite mass and a non-homogeneous planet; more importantly he included a non-zero orbital inclination. The system angular momentum remains conserved but is now a vector: the directions of the orbital **H** and the spin **H** are not the same nor constant. Several sections are taken through the resulting three-dimensional diagram to illustrate characteristic properties of evolutionary tracks. The same end states exist in Greenberg's diagrams as Counselman's. Four-dimensional evolutionary diagrams, including the orbital eccentricity, might be instructive and should be developed. Greenberg (1974) applies his constructions to a study of possible histories of the Uranian system (cf. Singer, 1975). He (Greenberg, 1975) elsewhere considers the classical dynamics of the Uranian system as well.

SEMIMAJOR AXIS

Now we wish to be more quantitative. The tidal torque felt by a planet of radius R and density ρ, and caused by tides raised on it by a satellite of radius r at a distance a, is approximately

$$N = (8\pi/5)(GmR^4/a^3)(\rho h \sin 2\epsilon)_p, \qquad (10)$$

where ϵ, as illustrated in Figure 7.1b and c, is the tidal lag angle, that is, the angle between the maximum tidal bulge and the planet-satellite line of centers (H. Jeffreys, 1970). This assumes that only the principal semidiurnal tide (sometimes called the M_2 tide) produces a torque and that only solid body tides occur. Usually the semidiurnal tide is most significant, although if other tides had large lag angles, they too could produce large torques (cf. Lambeck *et al.*, 1974). The maximum height h of the semidiurnal tide is

$$h = \left[3/4\,(m/M)\,(R/a)^3 R\right] (5k_2/3). \tag{11}$$

The term in brackets is the height of the equilibrium tide (i.e., the tidal potential divided by surface gravity), and the second factor, through the Love number k_2, is a correction that accounts for the additional disturbance caused by the self-potential of the tide and includes the effect of the planet's rigidity μ in reducing the tide. The theoretical Love number for a homogeneous elastic body is

$$k_2 = \frac{3/2}{1 + 19\mu/(2g\rho R)}\,, \tag{12}$$

g being the planet's surface gravity. The first term in the denominator is ignorable for bodies smaller than about the size of Mars.

Knowing the torque permits one to calculate the time to tidally slow the planet's rotation. Since $N = I\dot{\omega}$, where I is moment of inertia and the dot denotes time differentiation, the characteristic slowing time for a homogeneous sphere is

$$T = 16\rho\omega a^6/(45GM^2 k_2 \sin 2\epsilon) \tag{13}$$

in which it is tacitly assumed that a does not change over T (cf. Goldreich and Soter, 1966; or Chapt. 6, Peale). The slowing of the Earth's rotation was first recognized by Halley in 1695, and some sixty years later Kant invoked tides as an explanation.

By action-reaction, the torque slowing the planet acts as well on the satellite orbit. The work done on the orbit is Nn. Substituting this with eqn. (10) into eqn. (2) gives

$$\dot{a} = 3(G/M)^{1/2}\,mk_2\,a^{-11/2}\,R^5\sin 2\epsilon. \tag{14}$$

Note the strong dependence on a and that large satellites move faster than small. Eqn. (14) may be integrated by assuming k_2 and ϵ to be independent of the time t:

$$a^{13/2} = a_0^{13/2} + (39/2)\,(G/M)^{1/2}\,mk_2 R^5 \sin 2\epsilon\,(t-t_0) \tag{15a}$$

$$(a/a_0)^{13/2} = 1 - (13/3)\,(\dot{n}_0/n_0)\,(t-t_0)\,, \tag{15b}$$

where a_0 is the semimajor axis at time t_0. Only the magnitude of sin 2ϵ is not yet specified; it will be discussed below. As shown in Figure 7.1, the sign of ϵ is usually taken to be the sign of $\omega - n$. Singer (1968, 1972) has emphasized, however, that for a satellite on an eccentric orbit which passes within the synchronous orbit position, probably a more proper choice would be for sign ϵ to be given by sign $(\omega - dv/dt)$, where dv/dt is the instantaneous orbital angular velocity. In this circumstance, over the part of the orbit when the satellite is beyond the synchronous position, the situation in Figure 7.1b occurs while over the rest of the orbit Figure 7.1c depicts the process. Since the sign of ϵ changes along the orbit path, the evolution of a is slowed because à is positive for part of the orbit but negative over the rest. The situation is temporary, however, because changes in the eccentricity, as described below, rapidly move the satellite either entirely within or beyond the synchronous orbit.

ECCENTRICITY AND RADIAL TIDES

The tidal forces that change the orbit's size, of course, also affect the other orbital elements (cf. eqns. 8), as was first discussed by Darwin (1879a, 1880). More elaborate and general developments of the governing equations are those by Kaula (1964) and MacDonald (1964). Kaula (1964) extended and systematized the original approach of Darwin (1879a, 1880), in which the tidal potential is expanded in a Fourier time series, to include all orders and degrees. In this treatment the phase lag of each Fourier component appears separately to be specified by the particular tidal model chosen. Kaula's expressions are derived from the Lagrange planetary equations written in terms of the perturbation potential; they are complete but cumbersome because of their use of the mean anomaly which is awkward for high-eccentricity orbits. MacDonald (1964) on the other hand developed simpler expressions based on eqns. (8) in terms of the true anomaly v. This more easily permits averaging over the orbit; his expressions assume however only a second degree tidal distortion is important, similar to our treatment above, and that the phase angle lag of the tide is small and constant over the orbit. Singer (1968) has used an approach similar to MacDonald's but one which allows for a frequency dependent dissipation function. Lambeck (1975) (cf. Lambeck et al., 1974) has tabulated expressions for the secular variations of the lunar orbit due to the principal frequencies of the solid body and ocean tides; he first estimates the amplitudes of the major components in the ocean tides from tidal models (see, e.g., Hendershott, 1972) and then uses these strengths and phase lags to find the importance of ocean tides.

Darwin's results have been applied by Jeffreys (1961) to find how the eccentricity and inclination of satellite orbits are influenced by various models of tidal friction. Lambeck (1975) claims that the N_2 ocean tidal response determines the rate of change of e. Jeffreys (1961) showed that usually, although not always, the eccentricity will grow under the action of planetary tides. Referring to eqn. (5), we see that this occurs despite the fact that the satellite's already positive angular momentum is increased by the tidal torque because the energy term dominates. Similar arguments show that satellites passing within the synchronous orbit will have their orbits circularized, as calculated by Singer (1968, 1972), since the torque at pericenter decreases the apocenter height while that at apocenter increases pericenter. This circularization occurs relatively rapidly and leaves the satellite orbit either entirely within or beyond the synchronous position. Hence, the evolution of the semimajor axis only pauses around the synchronous position.

One of the remarkable regularities in satellite orbits is the near zero values of eccentricities for all close satellites (cf. Table 1.2). Yet we saw above that planetary tides generally cause increasing eccentricities (Jeffreys, 1961). The explanation for this contradiction has been provided by Goldreich (1963) following a suggestion of Urey and others. Consider a satellite whose rotation is synchronously locked to its orbital motion as shown in Figure 7.2. If the satellite is on an elliptic orbit, it experiences an oscillating tidal strain, that is, the distortion in Figure 7.2 changes with the orbital period. This oscillation dissipates energy for anelastic materials. The satellite tide is a radial, or "push-pull," tide. Because of the synchronous lock, angular momentum cannot be transferred to the satellite orbit by such a tide. Therefore, considering only energy loss in the satellite, eqn. (5) for orbits of low e reads

$$e \, \Delta e \approx -\Delta E/(2E), \qquad (16)$$

and, since E is negative, energy loss in the satellite decreases eccentricities. Because e is generally small, small energy losses are effective; of course for larger e, $\Delta E/E$ will increase so that decay from large e can also be efficient. In physical terms, the work that is done on the orbit arises because the maximum tidal strain takes place after pericenter due to the time delay generated by energy loss; this means that more work is done on the passage outbound from pericenter than during the inbound passage. Goldreich (1963) calculates that this effect of decreasing e overwhelms that of the planetary tides for all cases where tidal effects might be significant, except for Phobos, Deimos, and possibly the Moon and Amalthea. The conclusion depends on the tidal model chosen and on the satellite composition; it should be checked for the satellites of the outer planets using the subsequently developed water and ice interior models of Lewis (1971a, 1973). New information on the structure of the lunar interior should replace the simple model used by Goldreich to see if $\dot{e}_p$ in the lunar case continues to remain nearly equal to $\dot{e}_s$.

INCLINATION

The inclination will also be changed by tides, as can be seen in eqn. (8c); physically this occurs because the planetary rotation carries the tidal bulge out of

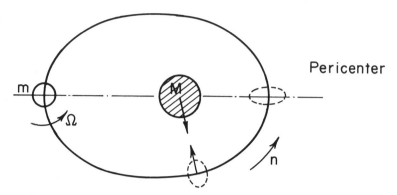

Fig. 7.2. Radial tides in satellites. The satellite is assumed to have its rotation synchronously locked with its revolution (n = Ω). As the satellite approaches pericenter, its distortion increases. Maximum distortion occurs slightly after pericenter, because of an anelastically delayed response. Angular momentum is not transmitted by radial tides but energy is lost from the satellite orbit.

the satellite orbit plane, producing a torque normal to the orbit plane [cf. eqn. (7) or Kaula, 1971]. Darwin (1880) first calculated these effects, which were later confirmed by Jeffreys (1961) for a specific tidal model. They showed that the orbital inclinations of a satellite with respect to its proper plane should change with time in the opposite sense to the change of a. This agrees with conservation of orbital angular momentum, ignoring the angular momentum taken from the primary. Considering only the dominant term in the tidal torque, Kaula (1964) finds that the rate of change of i is

$$\frac{di}{dt} = \frac{\cos i - 1}{2a(1 - e^2)^{1/2}} \frac{1}{\sin i} \frac{da}{dt} \approx -\frac{i}{4a}\left(\frac{da}{dt}\right), \qquad (17)$$

where the approximation holds for small i and e. Since, as will be seen below, (ȧ/a) is small for all extant satellites except the Moon, i similarly has not changed much. The apparent complication introduced by the precession of the primary will later be shown to be unimportant. This discussion includes only the M_2 solid body tide, whereas Lambeck (1975) claims that the contributions of K_1 and O_1 tides are of similar magnitude.

TIDAL LAG ANGLE AND Q

One might tend to believe from the above that, after selecting a tidal model, one need only to integrate the evolution equations back 4.6×10^9 years to find the original state (a_0, e_0, i_0) of any satellite system. This is false not only because processes other than tides act but also because, even for the well studied Earth, the tidal lag angle ϵ is not precisely known today and is yet more obscure in the past. Not knowing ϵ means that the time scales for all the tidal effects are uncertain, and thus one does not accurately know where a satellite joined the evolutionary trail along which it moves today.

Indeed on the Earth where the tidal delay might be thought to be at least measurable, effects such as local geology and nearby ocean loading make tidal observations difficult and inaccurate (cf. the discussion in Lambeck *et al.*, 1974). Furthermore, for a complete tidal understanding a Fourier analysis of the various frequency tidal components must be carried out to find each delay individually since theory indicates that tidal response might be a function of frequency (see below, however). In the Earth's case the distinction between the large amplitude ocean tides, which are strongly dependent on specific boundary shapes and sizes, and the smaller amplitude solid body tides further complicates matters. There is still considerable controversy over the relative importance of solid body and ocean tides in dissipating energy (see below). Even if accurate measurements could be properly made, one still could not confidently extrapolate into the past. For then, the Earth's interior temperature distribution and structure were perhaps different, the ocean tidal amplitudes and energy losses were certainly modified by the unknown continental distribution and coastal morphology, and the solid body tidal amplitudes were larger which, according to some theoretical models, influences the relative energy dissipation. All these complications have been generally overlooked simply because there is as yet no manageable way to deal with them.

Until these difficulties are resolved for Earth, extrapolation to the other terrestrial planets is premature. And, of course, many of these complications are not unique to Earth so that even if the various effects can be separated in the terrestrial case the situation is not clear elsewhere. It is probable that careful analysis of the motions of the Martian satellites (Chapt. 14, Pollack, and below) will permit the magnitude of solid body tides to be determined on a terrestrial planet and in this way hint at how important solid body tides are for the Earth. Work is also progressing on separating the effects of ocean tides from solid body tides. However, to extend these results directly to the outer planets requires a very large act of faith since the important energy dissipation mechanisms for the solid terrestrial planets are undoubtedly different than those operating in the outer solar system.

The usual manner of treating the lag angle depends on a simple calculation given in Munk and MacDonald (1960) and MacDonald (1964). Discussing an oscillating body with energy dissipation, it was shown for $\epsilon << 1$ and slow tidal motions that

$$\sin 2\epsilon \approx 1/Q \doteq \frac{1}{2\pi E^*} \oint \frac{dE}{dt}\, dt \ , \tag{18}$$

where E* is the peak energy stored in the body during a complete cycle and the line integral gives the energy dissipated over a cycle; hence the energy lost is a simple function of the lag angle and is a given fraction of the energy stored. The dimensionless material parameter Q is called the specific dissipation function, or the quality factor in other applications (e.g., circuit theory and acoustics). A Q should be evaluated for each oscillation frequency; however, it is usual in orbital evolution calculations to ascribe all energy loss to the second degree semidiurnal tide and quote a single Q value as the tidal effective Q for that planet.

Although intuition as well as theory suggests that Q should be a function of the tidal amplitude and oscillation frequency, experiments and observations show it to be fairly constant (Knopoff, 1964; Kaula, 1968; Stacey, 1969). Note, however, that this approximate constancy is not precise enough to distinguish between various tidal theories nor to completely unravel the puzzles of satellite orbital evolution. In the solar system the range of relevant frequencies for the assumed dominant second order tidal pulse goes from 4×10^{-5} hz for Phobos to 5×10^{-8} hz for Venus; tidal amplitude strains range from a maximum of about 10^{-1} for the nearest satellites, such as Phobos or Amalthea (Goldreich and Soter, 1966; Soter and Harris, 1976; Morrison and Burns, 1976), to essentially zero.

Observations show that Q varies by somewhat more than an order of magnitude over the range of geophysically interesting frequencies from 10^{-7} hz to 10^{-1} hz. A low value of $Q \approx 10$ seems to pertain to the terrestrial tidal problem (see below) whereas the damping of the Earth's latitude variation corresponds to a number nearer 30. Seismic waves and free oscillations damp with Q's running from a few hundred to a few thousand depending on their frequency (Kaula, 1968; Jeffreys, 1970). Laboratory studies of seismic waves with frequencies up to 10^6 hz passing through various types of rock have typical values for Q of a few hundred. Since free oscillations of different periods sample different parts of Earth's interior, one can attempt to find dependences on temperatures, pressures or material composition. Increased temperature should tend to increase energy absorption while pressure will suppress it (Lagus and Anderson, 1968). The preliminary gross picture (Stacey, 1969) is one of a Q of several hundred in the uppermost surface layers dropping rapidly to a low of about 80 in the asthenosphere and then rising to about 1,000 at depths greater than 1,000 km, below which it seems to be constant. The ocean tides which also dissipate considerable energy—perhaps much more than the solid tides—complicate the picture considerably (see below). When their effect is subtracted, it appears that the Q of the solid Earth is about 100 with a low estimate of 60 (Lagus and Anderson, 1968; Lambeck, 1975). Further experimental and theoretical work on dissipation mechanisms for real materials is sorely needed if we are to ever trace the history of the natural satellites.

Estimates of Q elsewhere in the solar system have been presented in a classic paper by Goldreich and Soter (1966). Their discussion will be updated in sections directly following on the tidal evolution of specific satellite systems.

EARTH-MOON SYSTEM

For obvious reasons, the techniques and insights sketched above have most often and most systematically been applied to the orbital evolution of Earth's Moon and only rarely to the long term dynamics of other satellites. Although the primary motivation for tidal evolution calculations is to determine the Moon's origin, possible lunar origin schemes will not be emphasized here since they are described in Chapter 27 by Wood, as well as in many journal articles (e.g., Kaula, 1971, and Kaula and Harris, 1975, and the references therein) and in books (e.g., Marsden and Cameron, 1966; Kaula, 1968; or Jeffreys, 1970). Many thorough attacks—some analytical and some numerical—have been made on the evolution problem (cf. Darwin, 1880; Gerstenkorn, 1955, 1968, 1969; MacDonald, 1964, 1966; Kaula, 1964; Goldreich, 1966a; Singer, 1968, 1970a, 1972). Excellent papers (Kaula, 1971; Kaula and Harris, 1973, 1975; Öpik, 1972) have critically reviewed the dynamical aspects of lunar origin, while others (Goldreich, 1972; Urey and MacDonald, 1971) have given more heuristic treatments. Because of all this work only a few phases of the dynamical evolution of the Earth-Moon system will be dealt with here; we will not give the problem the depth of treatment it clearly deserves.

The tidal evolution of the Earth-Moon system is more complex than for any other planet-satellite system. This arises for several reasons: (i) the Moon's orbit is strongly perturbed by the Sun; (ii) solar tides significantly slow the primary's rotation which they do not elsewhere; (iii) the large relative mass ratio m/M means that much angular momentum has been added to the lunar orbit; and (iv) both ocean tides and solid body tides affect the evolution.

As an example of the type of result one can get from lunar orbital evolution calculations, and also because they are the most comprehensive, we present plots from Goldreich's principally numerical integrations (1966a) as Figure 7.3. Goldreich neglected eccentricity effects, using his previous result (Goldreich, 1963) to justify considering the lunar eccentricity in the past comparable to the present value (0.055) or smaller. He accounted for the Earth's forced precession and the lunar orbital precession due to the Sun plus the terrestrial oblateness, and analytically averaged his equations over the appropriate precession periods. He included the slowing of the Earth's rotation by solar and lunar tides but neglected tides in the Moon and did not explicitly consider ocean tides.

Figure 7.3 uses the semimajor axis as the ordinate so that questions of time scale (see section on Q and below) are deferred until later. The displayed results apply for a phase lag that is proportional to rate but the numerical conclusions are only slightly modified for a rate-independent phase lag. The two tracks shown in Figures 7.3 represent maximum and minimum values. Variations on the left sides of Figures 7.3a and b occur because of the precession of the Earth and the lunar orbit under the solar torque whereas that on the right side of Figure 7.3c depicts the fact that at large distances the Moon's orbit maintains an essentially fixed inclination to the ecliptic while the Earth precesses.

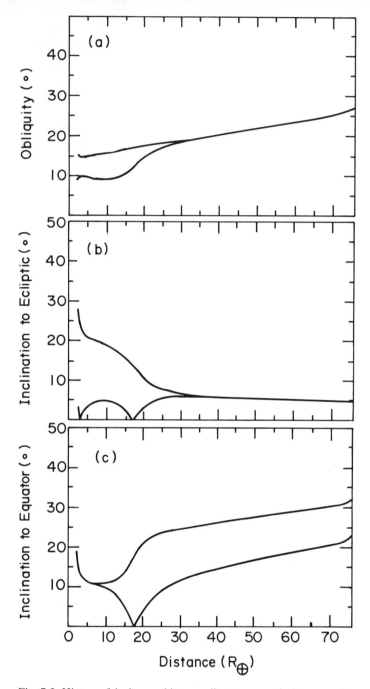

Fig. 7.3. History of the lunar orbit, according to a numerical computation by Goldreich (1966a). Phase lag is taken to be proportional to rate, and solar tides are included. Tides in the satellite and orbital eccentricity are neglected. The two branches show maximum and minimum values. (a): The obliquity of the Earth's equator to the ecliptic. (b): The inclination of the Moon's orbit to the ecliptic. (c): The inclination of the Moon's orbit to the Earth's equator.

One principal result with profound implications for fission theories of lunar origin is seen in Figure 7.3c: the inclination of the lunar orbit relative to Earth's equator is significantly non-zero for all close radii. This conclusion is unaffected even when arbitrary changes are made in the Earth's obliquity and also does not seem to be significantly influenced by the amplitude and frequency dependence of the planet's Q (Goldreich, 1966a). Until this finding can be discredited, fission theories—all of which require origin near the equator—have serious problems. Rubincam (1975; cf. Kaula and Harris, 1975) assiduously tried to find ways around this difficulty but still ended up with some inclination, even for quite artificial tidal models.

The inclination can also be affected by a push-pull type of tide. Ward (1975a) has investigated the past orientations of the lunar rotation axis and found that, when moving between about 30 R_E and 40 R_E, the Moon had an obliquity lying between ~ 25° and ~ 50°. Hence, even though the lunar rotation was synchronously locked at this time, a large tidal bulge moved back and forth over the lunar surface and so energy was lost. This energy heats the lunar interior and has a total deposited energy that is comparable to, or larger than, the energy lost in tidally slowing the initial lunar rotation. It is withdrawn from the energy stored in the inclination motion [Ward, private communication, 1975; cf. eqn. (7)], and thereby makes the lunar inclination problem seen in Figure 7.3c worse.

As previously stated, results such as those displayed in Figure 7.3 do not say *where* the Moon joined its evolutionary path. Nevertheless, proponents of lunar capture might be tantalized to see in Figure 7.3 that large inclinations are required near the Earth; we caution that Goldreich's neglect of e becomes important at small a and so his diagrams should not be taken literally near the Earth. Gerstenkorn (1955) had earlier found that integration back in time eventually led to the Moon's orbit flipping over the Earth's pole and finally receding from the Earth on a hyperbolic, high-inclination path. He concluded that the Moon was captured at a large inclination near the Earth. These calculations have been revised (Gerstenkorn, 1968, 1969) but the conclusion remains essentially the same. Capture in various modes has been seriously considered by many other distinguished scientists (Alfvén, 1963; Alfvén and Arrhenius, 1969; Lyttleton, 1967; Singer, 1968, 1970a, 1972; Urey and MacDonald, 1971). In all cases, capture must occur relatively close to the primary; otherwise tidal processes cannot account for the energy loss needed to turn a hyperbolic orbit into a captured elliptical one (Kaula and Harris, 1973). This means the process is highly improbable since the target cross section is small, particularly if the Moon is taken to be on an originally solar orbit considerably different than the Earth's so as to explain their disparate compositions. An origin by capture also suffers from other inherent difficulties as outlined by Wood in Chapter 27, Kaula (1971), Öpik (1972), and Kaula and Harris (1975).

For the accretion model of lunar origin (Chapt. 26, Safronov and Ruskol; cf. Chapt. 24, De *et al.*), the evolution calculations of Goldreich (1966a) suggest that coagulation must occur between 10 and 30 Earth radii (that is, near a_{crit} as given by eqn. 22) in order to avoid having the Moon presently be in the ecliptic, unless it has since been struck by a sufficiently massive planetesimal or has had subsequent out-of-plane tidal torques (Darwin, 1879a, 1880; Rubincam, 1975).

Only the tidal lag angle ϵ is needed to assign a time scale to the changes illustrated in Figure 7.3. As previously mentioned, it is impossible to make actual measurements of ϵ that have much real meaning for any calculation of the time scale of the past history of the Earth-Moon system. Although crude estimates of the Q distribution in the Earth's interior are available, we are uncertain as to precisely where the tidal energy is dissipated (cf. Kaula, 1964), to what extent it is somewhere in the solid Earth or in its oceans. Because of these difficulties, a more direct tack is frequently taken: that of observing the current orbital evolution and then inferring a gross tidal Q without asking how or why this Q is generated. The total Q is calculated by rashly assigning all the energy loss to the semidiurnal solid body tide as we have done above.

To see the contemporary lunar orbital evolution, the deceleration of the lunar longitude is observed. Any change in the semimajor axis directly involves the orbital mean motion through Kepler's third law: for a fixed mass center

$$\dot{n}/n = -(3/2)\dot{a}/a = -3\left[a/(GM)\right](\dot{E}/m). \tag{19}$$

In the case of the Earth-Moon system, solar tides must be included as well as the energy lost in the Earth's rotation and the fact that the longitude measurements are made from the Earth's surface, which itself is rotating and decelerating (Lambeck, 1975).

The value for $\dot{n}$ usually employed (e.g., MacDonald, 1964; Kaula, 1968) is based on listings of lunar and solar eclipses by Spencer-Jones and Fotheringham more than a third of a century ago; it is $-22''/\text{century}^2$, corresponding to a Q of about 13 or a tidal lag angle of about 2.2°. The results of many recent measurements of $\dot{n}$ are tabulated by Kaula and Harris (1975); the new techniques used range from employing historical records of eclipses or lunar occultations to counting the growth lines indicative of tidal rhythms on fossil corals and seashells. These more modern measurements produce values of $\dot{n}$ lying between $-18''/\text{century}^2$ and $-65''/\text{century}^2$; most, however, are clustered near $-40''/\text{century}^2$, and Lambeck (1975) gives a mean value for a subset of the measurements of $-(1.9\pm0.2)\times10^{-23} \text{ rad/sec}^2$ or $-(40\pm4)''/\text{century}^2$. If one considers that the tidal energy is contained only in the semidiurnal tides, the Q associated with the mean deceleration is only about 7 or $\epsilon\approx4°$ (Lambeck, 1975), saying that virtually all the stored energy is lost during the oscillation. Note that such a small Q is an order of magnitude smaller than the smallest estimated Q in the solid Earth (Stacey, 1969; Lagus and Anderson, 1968).

The aforementioned "time scale problem" comes down to this: using $-40''/$

century[2] in eqn. (19) and then eqn. (15) [or employing eqn. (20) below] brings the Moon back to the vicinity of the Earth 0.9×10^9 yr ago, at a time well after it must have been there according to lunar and terrestrial data (MacDonald, 1964; Kaula, 1971; Kaula and Harris, 1975). Even use of the classical deceleration value of $\dot{n}$ returns the Moon much too fast (1.6×10^9 yr ago). So the following situation arises: we have no completely adequate way to calculate Q, particularly for ocean tides, and yet if we extrapolate into the past with the contemporary effects of tidal dissipation we arrive at an answer which superficially appears absurd.

A resolution of this quandary is suggested in a recent important calculation (Lambeck *et al.*, 1974; Lambeck, 1975) of the energy lost in ocean tides. The result is in qualitative agreement with several other recent attempts (see Hendershott, 1972, and other references in Lambeck, 1975) but seems more complete. Lambeck finds that only the second degree wavelength components in the *ocean* tide cause significant secular changes in the lunar orbit. This is done by calculating the secular perturbations of the Moon's orbit using available tidal models for the principal semidiurnal and diurnal tides. Lambeck's calculated estimate of the $\dot{n}$ caused by ocean tides alone is $-35 \pm 4''$/century[2], in near agreement with the astronomical estimates of the total deceleration. These are difficult calculations which, because of their profound implications, should be repeated and improved by other workers. Accepting the results at face value, they indicate that much— if not virtually all—of the secular change in the Moon's longitude can be accounted for by dissipation of tidal energy in the oceans; it is not necessary to invoke significant energy sinks in the Earth's mantle or core.

This important conclusion is in qualitative agreement with previous suggestions that ocean tides account for most of the Earth's tidal energy dissipation. These suggestions were based on calculations—which are now being questioned since they seem incorrect and since other oceanic dissipation mechanisms are available (T. Gold, private communication, 1975)—showing that the energy flux across the entrances of shallow seas is a major fraction of the total energy lost per unit time (cf. Jeffreys, 1970). Lambeck's work is independent of these calculations in its basis of computation and does not identify where dissipation takes place in the ocean since he computes the mean rate of work per unit area done by the Sun and Moon on the entire ocean surface.

The total rate at which energy must be dissipated in the Earth, according to the astronomical estimates of its rotational deceleration rate, is 5.7×10^{19} erg/sec; the oceans dissipate 5.0×10^{19} erg/sec, according to Lambeck (1975). If one accepts the two numbers as accurate, only the difference of 0.7×10^{19} erg/sec must be lost through solid body tides. Bounds can be placed on the solid body Q of both the Earth and Moon by ascribing the $\dot{n}$ that remains, after accounting for ocean tides, to solid body tides: assuming the Q's to be the same, Lambeck (1975) estimates that Q equals 100 with a lower bound of about 60. These values are not in contradiction to previous bounds for the Earth (Stacey, 1969; Lagus and Anderson, 1968); however, Q for the Moon is much higher (Toksöz et al., 1974b).

The dominance of the oceans in dissipating energy today means that the global tidal Q which is presently observed may be abnormally low because of an unusual ocean/continent configuration. In the past when the continents were clustered into a single Pangaea and ocean levels were lower, energy loss may have been substantially less and orbital evolution correspondingly slower (cf. Kaula and Harris, 1975; Lambeck, 1975). Thus to integrate the orbit backwards with contemporary values of Q is foolhardy. Even with this possible resolution, the time scale problem remains one of the more nagging frustrations of solar system cosmogony: we are so close to the solution of a crucial problem but we cannot attain the complete answer because of one missing number.

THE MARTIAN SATELLITES

The dynamical characteristics of the Martian satellites as known prior to Mariner 9's visit are summarized by Burns (1972) while a tabulation of the spacecraft results pertaining to the Martian system including Phobos and Deimos is presented by Jordan and Lorell (1975).

From the standpoint of orbital evolution, the evaluation of the secular acceleration of the mean motion $\dot{n}$ of Phobos is needed to set the time scale, similar to the lunar case described above. A surprisingly large value ($18.8°$/century^2) for $\dot{n}$ was once thought to exist (Sharpless, 1945) but was later discredited (Wilkins, 1965, 1966, 1967, 1968, 1969; Burns, 1972). More recent evaluations (Sinclair, 1972a; Shor, 1975), developed from much larger data sets, indicate smaller, more explainable values for $\dot{n}$. Sinclair (1972a) found $\dot{n}=9.6°$/century^2 but that his value strongly depended on the data subset processed. Shor (1975), by including a large number of previously unused Russian observations, reported a value of $14.4°$/century^2; this result is not strongly dependent on the particular data subset used, and therefore appears reliable. The results for Deimos are inconclusive but it seems to have a secular deceleration, as would be expected for a tidally produced motion. These secular accelerations have lately been shown to be consistent with the Mariner 9 television observations of the satellite locations (Born and Duxbury, 1975).

The numerous clever explanations that were advanced to account for the secular acceleration of Phobos—particularly Sharpless's abnormally large one—are summarized by Shklovskii and Sagan (1966) and by Burns (1972). The latter author concludes that tidal torques alone are of the correct magnitude to explain the entire effect and are most plausible. Pollack (Chapt. 14 herein) and Smith and Born (1976), after estimating the mass of Phobos, have computed the work done by the orbit [see eqns. (10)–(12) above] to evaluate the Martian lag angle. They find that $\epsilon \sim 0.3°$ or that $50 < Q < 150$. Note the similarity of these values to the result of Lambeck (1975) for the terrestrial solid body tides. The requisite Q suggests partial melting in the Martian interior (J. Pollack, Chapt. 14 herein).

A point worth emphasis is that $\omega < n$ for Phobos, a unique case in the solar system for prograde satellites; this means that energy is withdrawn from this satellite's orbit. The loss of energy will eventually cause the orbit to collapse onto the Martian surface (cf. Fig. 7.1c). The future orbital lifetime t* can be predicted by setting a=0 in eqn. (15b):

$$t^* = (3/13)(n_0/\dot{n}_0) , \tag{20}$$

where the terms on the right are the current values of the mean motion and the secular acceleration. Accepting Shor's value (1975) for the acceleration, Phobos will crash into Mars a mere 50 million years from now. Integrating backward in time (Burns, 1972; Pollack, Chapt. 14, herein) places Phobos near the synchronous orbit position 4.5×10^9 yr. ago. Deimos, because of the tidal torque's strong dependence on a [cf. eqns. (14) and (15)] as well as perhaps the fact that $\omega - n$ is small, currently has a secular deceleration from the influence of tides which is at least three orders of magnitude less than that of Phobos. Hence Deimos' orbit has not been appreciably affected by tides; this is emphasized by the fact that it lies near the stationary position, away from which it is driven by tidal torques. Interestingly, the two satellites originated close to one another, with Phobos just inside and Deimos just outside the stationary orbit position (cf. Pollack, Chapt. 14 herein). The use of contemporary value of ϵ in both of the above calculations is certainly more correct for the Martian satellites than it was for the Moon since the Martian tides are smaller and ocean tides are not a worry.

The orbits of both Martian satellites have decreasing eccentricities owing to tides in the planet (Jeffreys, 1961), in contrast to most other satellites. Furthermore, push-pull tides in the Martian satellites themselves cause decreases in e (Goldreich, 1963) with the effects of tidal friction in the planet being far more important, a unique case in the solar system. Goldreich (1963) concludes, by summing the two effects, that Phobos's eccentricity, which is now fairly small (e=0.015), could have been decreased considerably by tides on Mars. In contrast, Deimos's e, which is currently very small (e=0.001), has been much less strongly altered, and thus Deimos particularly must have originated in a nearly circular orbit.

The orbital inclinations of the satellites are little affected by tides. As can be seen from eqn. (17) and the results on a described above, for Phobos di/dt>0 but is small while for Deimos di/dt<0 and definitely negligible. The smallness of each has been found analytically (Jeffreys, 1961) and numerically (see Wilkins, 1968). So the Martian satellites, if only acted on by tides, can be said to have originated in equatorial orbits. The precession of Mars can be ignored, as explained later.

Singer (1968), employing a frequency-dependent tidal theory, has developed a flow diagram in (a,e) space which illustrates the evolution of Martian system. While this plot shows that the eccentricity of Phobos may have been substantial at one time and suggests that the satellites may be captured bodies, more recently Singer (1970a) has rejected capture because the evolutionary time scale is inadequate by at least an order of magnitude.

Phobos and Deimos, therefore, each appear to have originated in nearly circular orbits of small inclination, assuming that only classical tides determine their subsequent orbital evolution. Deimos started its journey through time near its current orbit, whereas Phobos was considerably closer to the stationary orbit and may have been in a more elliptic orbit than it is today. Origins which will produce such primordial orbits are qualitatively described by Goldreich (1966a), Burns (1972) and Pollack (Chapt. 14, herein). Hartmann *et al.* (1975) also consider the origin of the Martian satellites.

HYPOTHETICAL SATELLITES

As described above, the Moon is being driven away from the Earth, slowing its orbital motion as well as the Earth's rotation. Eventually (many billions of years in the future) the month and day will be the same: $\omega = n$ (Goldreich, 1966a). Because of solar tides, the Earth's rotation will continue to be retarded, yielding $\omega < n$ (Fig. 7.1c), and the Moon will then begin to gradually reapproach the Earth (cf. Counselman, 1973). Solar tides presently only remove 2% of the terrestrial angular momentum; however, they will be relatively more important in the future.

Processes similar to the projected future development of the lunar orbit may have occurred for the inner terrestrial planets, which are seen not to have satellites. The rate at which these processes would proceed is much faster with Mercury and Venus for, as eqn. (13) implies, slowing by solar tides is very effective in the inner solar system. This is attested to by the long rotation periods of Mercury (58 days) and Venus (-243 days) (Goldreich and Peale, 1968), both of which have likely been decelerated by solar tides (Goldreich and Soter, 1966). To illustrate the phenomenon: If the condition $\omega = n$ were to occur precisely at one-half the age of the solar system, a hypothetical satellite would be seen today at the orbital distance at which it had been born, having spent half its life moving away from the planet and the last half moving toward its primary. If the turn-around occurred at less than 0.5 of the solar system age, the satellite would be closer to the planet today than any other time, perhaps even on its surface.

Burns (1973) first pointed out that in most cases hypothetical satellites that formed around Mercury and Venus would subsequently have crashed onto their primaries' surfaces by the present time through the mechanism described above. He noted that Kepler's third law written in terms of the relative distance $\alpha = a/R$ gives $n \sim \rho^{1/2} \alpha^{-3/2}$. Thus, hypothetical satellites of the inner planets, located at relative distances similar to those of the outer planets, orbit only somewhat faster than satellites of the outer planets; characteristic periods are a few Earth days. Hence, as soon as the planetary rotation periods lengthen to more than a few Earth days, planetary tides will start to draw in any satellites since then $\omega < n$. Since Pluto's rotation period is near six days (Allen, 1963), it too could be sweeping up any satellites that happen to orbit it.

Taking $k_2 = 0.05$, 0.25, and 0.28, and $Q = 100$, 100 and 13 for Mercury, Venus, and Earth, respectively, and assuming conservatively that turnaround occurred 2.3×10^9 yr ago, Burns plots from eqn. (15)

$$(\alpha/\alpha_{EM})^{13/2}(\mu_{EM}/\mu) \; < 0.174 \quad \text{Venus}$$
$$< 0.035 \quad \text{Mercury} \tag{21}$$

as the collapse criterion; μ is the relative mass ratio and EM refers to the Earth-Moon values. Only very small satellites ($R \leqslant 10$km) formed at $\alpha_0 > 10$ can survive the pull of tidal friction.

Ward and Reid (1973) later independently reached conclusions similar to Burns (1973) by numerically integrating the orbital evolution equation, including both satellite and solar tides, for various initial conditions. Their results differ from Burns's only for the case when Q is just sufficient for the phenomenon to work. Reid (1973, 1974) extended these ideas to the loss of bodies originally orbiting satellites; Gold (1975a) instead feels that three and four body interactions will determine the stability of satellite-orbiters rather than tides (cf. Reid, 1975).

Ward and Reid (1973) also explicitly considered the probability that the satellite will break up under tidal stresses as it moves within the Roche limit. Harris (1975) showed that subsequent self-collisions among the debris resulting from the Roche fracture will break the particles down until they are no more than a few kilometers in diameter. These smaller particles will tidally evolve more slowly than mentioned above because of their smaller mass (cf. eqn. 14), perhaps preventing total collapse from ever being completed (Öpik, 1972; Harris, 1975). However, tidal breakup likely occurs considerably closer to the surface than believed by Harris: Aggarwal and Oberbeck (1974) have found that the inclusion of the structural strength of the satellite moves the break-up distance into ~ 1.2R for reasonable material properties; the classical Roche limit for a satellite with no strength and the same density as the planet is ~ 2.5R.

Satellites lost by tides will strike the planetary surface on grazing trajectories; the resulting craters should be elongated in the orbit plane—probably near the equator—and have asymmetrical crater blankets. Such surficial scars have not been seen by Mariner 10 on Mercury (Burns, 1976a). Substantial amounts of energy will be released in such collisions. McCord (1968) and Singer (1970b) have postulated that this energy release may be responsible for the present dense atmosphere of Venus and have even proposed that Venus's anomalous, slow, retrograde spin might be caused by the collapse of a retrograde satellite orbit.

McCord (1968) showed that most retrograde satellite orbits have quite limited lifetimes under tidal action; McCord (1966) had earlier considered the eventual demise of Neptune's Triton.

A satellite could be lost as well by escape to heliocentric orbit if its eccentricity became very large (Harris and Kaula, 1975).

TIDAL FRICTION IN THE OUTER SOLAR SYSTEM

All of the major satellites of the outer planets are relatively much closer to their primaries than is the Moon (cf. Table 1.2). Since these satellites are beyond the synchronous orbit position (because of the rapid planetary spins), they are being driven away from their primaries by tides (Fig. 7.1b), except for the retrograde Triton. A lower bound can be placed on the Q of the primary by considering the orbital evolution of the satellites under tidal forces. Assuming that the satellite originated 4.5×10^9 yr ago at the surface of the planet (more correctly, at the Roche limit) and was driven by tides to its current orbital radius, Goldreich (1965b; Goldreich and Soter, 1966) used eqn. (15a) with (13) to give a bound on sin2ϵ or Q. (A modification of this bound is described below.) This assumes that the formalism developed above for the orbital evolution under solid body tides is applicable to the gaseous outer planets. There is no reason besides ignorance to accept this (see T. Gold's comment at the end of this chapter).

As can be seen from eqn. (14), the rate of retreat is proportionate to the satellite mass and $(\dot{a}/a)\sim a^{-13/2}$ so that inner satellites move much more rapidly away from their primaries. This means that, as a satellite orbit evolves, its mean motion may come into a low order commensurability with that of another (assumed independently-evolving) satellite. Such resonances are seen in the solar system, many more than would be expected by chance (Roy and Ovenden, 1954, 1955; Goldreich, 1965b; cf. Chapt. 8, Greenberg). Goldreich (1965b) has demonstrated that for many of the resonances present in the solar system the mutual gravitational interaction between the two satellites is strong enough to maintain the commensurability regardless of the upsetting influence of tides. The mechanism for capture into a resonance is qualitatively discussed by Greenberg *et al.* (1972; cf. Chapt. 8, Greenberg) and analytically described by Greenberg (1973a). Other aspects of resonances are considered by Allan (1969), Greenberg (1973b), and Sinclair (1972b, 1974a, 1975a, 1975b). Sinclair (1975a) has shown that tidal friction is an unlikely cause of the resonances of the Galilean satellites; some other mechanism, perhaps an early gas or electromagnetic drag, is needed to set up their precise Laplace relation (cf. Mogro-Campero, 1975; Morrison and Burns, 1976).

The presence of these commensurabilities means that, as the innermost satellite is driven away from its primary by tides, it pushes ahead the satellite with which it is in resonance. This is accomplished through mutual gravitational perturbations which distribute the angular momentum received from the primary in such a manner as to preserve the commensurability. These interactions slow the orbital evolution of the inner satellite and thereby considerably lower the bound on Q described above (Goldreich, 1965b; Goldreich and Soter, 1966).

Upper bounds on Q may be estimated if one accepts a tidal origin for the commensurabilities (cf. above, and Chapt. 8, Greenberg). That is, the fact that commensurabilities occur indicates that significant tidal evolution must have taken place. The bounds placed on the Q's of the outer planets by these consider-ations are for Jupiter: 10^5–10^6; for Saturn: (6 to 7) $\times 10^4$; and Uranus: $> 7 \times 10^4$ (Goldreich and Soter, 1966). The Q of Neptune cannot be found by this technique since its small satellite Nereid is not appreciably affected by tides while Triton, due to its retrograde orbit, is approaching Neptune.

Note that the Q's for the outer planets are two to three orders of magnitude larger than those found to be acting in the inner solar system as well as those of most terrestrial materials. Such a difference is not too surprising: the disparity in composition, and internal temperature and pressure may well account for it. Discussions of the Q's of the outer planets have been provided by Goldreich and Soter (1966), Hubbard (1974), and D. J. Stevenson (private communication, 1974), but none is generally accepted. Neither turbulence nor molecular viscos-ity seems capable of producing Q's of the correct magnitude. Work should continue since understanding Q for the outer planets is one way to learn about their interior structure and also to test our ideas about Q in the inner solar system.

An estimate of Q for Neptune has been proposed by Trafton (1974c). Noting that Neptune has a higher brightness temperature than Uranus in spite of Nep-tune's greater distance from the Sun, Trafton proposed that the extra internal heating is caused by frictional dissipation of the tides raised in Neptune by the massive Triton. The added heat requires $k_2/Q = 2.4 \; (+3.3, \; -1.7) \times 10^{-3}$ for Neptune, yielding $Q = 170 \; (+435, \; -100)$ for a fluid planet. Since this Q is considerably lower than elsewhere in the outer solar system, and since other energy sources are available to heat Neptune, the result must be viewed skepti-cally.

McCord (1966) had earlier studied the orbital evolution of the retrograde Triton. He found that its orbit is unstable (cf. Figure 7.1c) and will collapse until the satellite crashes into Neptune or breaks up under strong tidal forces near Neptune (cf. Harris, 1975). McCord (1968) generalized these results to find the conditions (i.e., limiting values of satellite masses and initial orbital radii) under which hypothetical retrograde satellites will be lost in the age of the solar system, showing that such losses would be common.

Gravitational Effects

INCLINATION ABOUT A PRECESSING PLANET

At first glance, one of the real puzzles of solar system mechanics is the fact that satellites—at least those near their primaries—have very small inclinations. This is surprising because the solar torque causes the planets to precess and thus one might think that, even if a satellite is in a low inclination orbit now, it will not remain so if the planet precesses from beneath it.

Goldreich (1965a) has derived the equations which govern the rate of change of the inclination of a satellite orbit to the equator of an oblate precessing planet, neglecting mutual and solar perturbations of the satellite, and ignoring drift under the action of the tides. He found (see also Goldreich, 1966a) that if the motion of the satellite's ascending node on the equatorial plane has a period (cf. Chapt. 3, Aksnes) which is short compared with the planet's precession period, then the satellite's inclination to the planet's equator will remain essentially constant as the planet precesses. This criterion holds for most major satellites, excepting the Moon, and is similar to the breakdown made in Chapter 4 by Kovalevsky and Sagnier into inner satellites whose orbits are dominated by planetary oblateness and outer satellites whose orbits are determined by solar perturbations (cf. Brouwer and Clemence, pp. 66-69, 1961b). Goldreich (1965a) further showed that the inclinations of the outer satellites remain constant with respect to their planet's orbital plane. He also demonstrated that slow changes of a planet's obliquity will not affect the inclinations of nearby satellites.

The division between "inner" and "outer" satellites (i.e., those with fixed inclinations relative to their primaries and those with constant inclinations relative to the planet's orbital plane), according to Goldreich (1966a), is located at

$$a_{crit} \approx \left[2(C - A) \, a_\odot^3 / M_\odot \right]^{1/5} , \qquad (22)$$

where $a_\odot$, $M_\odot$ are the planet's orbital semimajor axis and the solar mass, and $(C - A)$ is the difference between the planet's moments of inertia. For outer satellites the "oblateness" corresponding to inner satellites (which gravitationally act like an equatorial bulge) should be included. We list the critical position a_{crit}/R in Table 7.1. This shows that all satellites—except for Triton which is approaching Neptune (see above)—lying within their respective critical distances have nearly equatorial orbits ($i \lesssim 1.5°$; cf. Table 1.2). Nonequatorial orbits are held by all which lie beyond a_{crit} as would be expected even if these satellites had originated in the equatorial plane.

LONG-TIME STABILITY FOR THE REGULAR SATELLITES

Stability has many different definitions and criteria. Here, because more precise work has not been done for satellite systems, we will use a layman's understanding of stability: Have satellite configurations changed "much" under purely gravitational influences?

The important question, "Are the inner satellites forever bound to their primaries?", can be answered positively. In the framework of a circular restricted three body problem with its usual assumptions of a circular planetary orbit and an infinitesimal satellite mass, all the regular satellites lie well within

TABLE 7.1
Location of Equatorial and Nonequatorial Satellites

Planet	a_{crit}	Equatorial Satellites	a/R	Nonequatorial Satellites	a/R
Earth	10*	Moon	60.2		
Mars	13	Phobos	2.8		
		Deimos	6.9		
Jupiter	32†	Amalthea	2.6	Inner Cluster	~160
		Io	6.0		
		Europa	9.5	Outer Cluster (R)	~310
		Ganymede	15.1		
		Callisto	26.6		
Saturn	43‡	Janus	2.7	Iapetus	59.3
		Mimas	3.1	Phoebe (R)	216
		Enceladus	4.0		
		Tethys	4.9		
		Dione	6.3		
		Rhea	8.8		
		Titan	20.4		
		Hyperion	24.7		
Uranus	84§	Miranda	5.1		
		Ariel	7.5		
		Umbriel	10.5		
		Titania	17.2		
		Oberon	23.0		
Neptune	70			Triton (R)	14.6
				Nereid	227

Note: a_{crit}, the critical semimajor axis, is given by eqn. (22). For orbits with semimajor axes much larger than a_{crit}, the orbit plane precesses about the pole of the planet's orbit, whereas for those much less than a_{crit} the precession takes place about the planet's rotation pole. After Goldreich (1966a).

* 17 with the terrestrial oblateness when the Moon is at $10R_E$
† 38 including the "oblateness" due to Ganymede
‡ 57 including the "oblateness" due to Titan
§ assumes $J_2 = 0.017$
(R) indicates a retrograde orbit

the zero velocity surface surrounding the planet (Hagihara, 1961). The zero velocity surface is sometimes called the Hill curve and is defined by the value of the Jacobi constant, that is, the relative energy in the coordinate system rotating with the mean motion of the primaries. Perturbations from other satellites or planets can be shown to not disturb the satellite across the Hill curve.

Answers to most other questions on satellite stability can be suggested only by analogy with the planetary system. A quite important distinction between the satellite and planet problems, however, should be noted and that is the matter of time scale (cf. Chapt. 4, Kovalevsky and Sagnier; Chapt. 5, Pascu). Typical satellite periods are days, while planetary orbits have periods of years. Hence satellites have taken 10^{11} or 10^{12} revolutions about their primaries, whereas planets have gone around "only" 10^9 times and so stability criteria must be that much more severe for satellites. Furthermore, direct comparison with the planetary case may be misleading because the criteria depend on the relative masses of the objects, which differ in the two cases, and because mutual perturbations are strong on the satellites while solar perturbations act only on the satellites (see Chapt. 5, Kovalevsky and Sagnier). The presence of satellite resonances means that one member of the system cannot be perturbed without a reaction on the rest of the system; this usually increases the stability (Hagihara, 1961).

Brouwer and van Woerkom (1950) have used the Laplace-Lagrange theory of secular perturbations to first order in the disturbing masses and second degree in e and i. They develop analytical expressions for the characteristic frequencies and amplitudes of the planetary orbital elements of Mercury through Neptune. The results for e and i show them to be composed of several periodic functions and to have small amplitudes. Contemporary numbers ordinarily give reasonable estimates of their values at any time. Plots have been developed from these expressions by Brouwer and Clemence (1961b), Murray et al. (1973), and Cohen et al. (1973). The first presented only a few results to illustrate their nature, while Murray et al. merely gave a plot of the periodic variation of the Martian eccentricity for the last 10^7 years. Cohen et al. provided a graphical representation of the orbital elements for all the planets over a time span of 10^7 years centered at the present; they have tested the results based on the secular theory of Brouwer and van Woerkom (1950) versus a numerical integration valid over 10^6 years, and the agreement is quite good. Ward (1974b) has plotted the Martian obliquity and inclination.

Similar calculations should be carried out for satellite systems; these will be difficult to do because of the strong mutual interactions (cf. Chapt. 8, Greenberg) and because of the large number of revolutions that must be considered. By analogy with the planetary system we anticipate, however, that the orbital elements (a,e,i) of the inner regular satellites will not change much over the solar system age.

LONG-TIME STABILITY FOR THE IRREGULAR SATELLITES: ORIGIN BY CAPTURE?

Hagihara (1961) has also considered the outer satellites in the context of the circular restricted three body problem. He states that, within the approximations valid for the three body problem, the prograde outer satellites will remain in orbit about their primary forevermore since the zero velocity surface surrounding the planet encloses them. For the outer Jovian group (J8, J9, J11, J12), which are on retrograde orbits, the criterion does not disclose stability, but it does show that the orbits of Saturn's Phoebe and Neptune's Triton, which are also retrograde but relatively closer to their primaries and further from the Sun, are stable. Retrograde orbits have been shown to be more stable than direct orbits semi-analytically by Moulton (1914b) and others employing the usual Laplace theory of secular perturbations. Numerical computations by Chebotarev (1968) and coworkers, and Hunter (1967), who investigated primarily circular satellite orbits about Jupiter, confirm Moulton's result. In a more general circumstance Hénon (1970) numerically found that there is apparently no upper bound on the dimensions of some retrograde quasi-periodic orbits and that retrograde orbits are stable for a wider range of Jacobi constants than are the corresponding direct orbits. More realistic and complete computer models could be useful in better understanding the origin of the irregular satellites.

Since Newton's equations are time-reversible, the fact that satellite escape is not possible means that capture of the present satellites has not occurred—at least not in a straightforward manner. Capture (and escape) in the present system might not be seen in current models because of ignored terms (rare close approaches between satellites, ellipticity of the Jovian orbit, or planetary perturbations) or because of energy dissipation effects which cannot be easily modelled (satellite collisions or early orbital drag by a gaseous disk). The latter point needs emphasis: capture is a reversible process unless energy dissipation intervenes to lower the satellite's Jacobi constant. This has been illustrated in the computer calculations of Everhart (1973) which show that capture in the Sun-Jupiter system is a temporary event, as well as being extremely rare (cf. Byl and Ovenden, 1975).

Capture, nevertheless, is the most commonly accepted origin for the outer Jovian satellites and has been suggested even for Phoebe (Kuiper, 1951). The comparatively small sizes of the outer satellites, their irregular orbits and the great distances from their primaries all speak to an origin quite different than that of the inner regular satellites (cf. Chapt. 23, Cameron; Chapt. 24, De et al.). The literature on the orbital aspects of capture in the Jovian system has been reviewed by Greenberg (1976a) whereas Morrison and Burns (1976) describe the observed physical properties of the Jovian satellites while considering orbital questions of an origin by capture.

The outer Jovian satellites are often thought to be captured from the nearby asteroid belt (Kuiper, 1951), perhaps shortly following the solar system's origin when the space density of asteroids in the vicinity of Jupiter was considerably higher than it is today (Kaula and Bigeleisen, 1975; Weidenschilling, 1975). The triangular Lagrangian points in the Sun-Jupiter circular restricted three body problem are frequently taken to be convenient storage places to permit later capture, although the Jacobi integral at L_4 or L_5 is considerably smaller than that needed to pass through the inner Lagrangian point L_2.

The similarity of the orbital elements of both clusters of the irregular Jovian satellites has indicated to some that the members of each group originated from the same event. Kuiper (1956) proposed that the two groups result from two recaptured, previously escaped moons which broke during passage through the gaseous envelope of proto-Jupiter. Colombo and Franklin (1971) suggested that a single collision occurred between objects (either two satellites, two asteroids, or a satellite and an asteroid) inside Jupiter's sphere of influence, which eliminated excess orbital energy relative to Jupiter, permitting capture (see Greenberg, 1976a). Bronshten (1968) hypothesized two separate collisions.

Bailey (1971, 1972) has treated capture in a Sun-Jupiter elliptic restricted three body problem and proposed that capture may occur solely through the inner Lagrangian point, and only at either Jupiter's perihelion or aphelion where Jupiter's radial velocity is zero. He claims that perihelion captures in the model lead to direct orbits which lie very near the actual inner irregular satellite cluster, whereas captures at aphelion produce retrograde orbits with semimajor axes approximately those of the outer satellite cluster. Bailey's analytical treatment is confusing but his results are so striking as to make one wish to believe them. However, these results are now in considerable dispute: numerical calculations (Hunter, 1967; Everhart, 1973) have demonstrated many captures and escapes at points other than the inner Lagrangian point; Davis (1974) has called attention to the fact that the Jovian orbital elements which are necessary to produce Bailey's remarkable numerical agreement occur only occasionally because Jupiter's orbit varies under the action of the other planets (Brouwer and van Woerkom, 1950; Cohen et al., 1973); and Heppenheimer (1975), while developing his own capture criterion and overcoming some of Davis's objections, has noted several analytical errors in the original calculation. A more detailed critique of Bailey's proposal (1971, 1972) is given by Heppenheimer (1975) and Greenberg (1976a).

Morrison and Burns (1976) point out that one should not too facilely assume capture as a proven scenario for the origin of the outer Jovian satellites. The apparent lack of large brightness variations for any outer satellite and the unusual color of J6 (cf. Andersson, 1974; Morrison and Burns, 1976) argue that these satellites may not be simply captured asteroids but could represent a class of objects with a special origin or which has been acted upon by processes not influencing the Trojan asteroids. Lastly, the size distribution of the outer Jovian satellites does not appear to be correct for either asteroids or the collision debris as suggested by Colombo and Franklin (1971): each of these categories of fragments would have many more small ones than are seen. Further observations are needed to confidently classify the outer Jovian satellites as asteroidal.

Collisions With Debris

During all stages of their lifetimes, but particularly during their early development, satellites suffer collisions with smaller particles: debris. The debris may either be in orbit about the same primary or come from a heliocentric path. The former case presently may occur by impacts with particles in a planetary dust belt (Soter, 1971) or an extended planetary exosphere (Roy, 1965); primevally such collisions would be with yet-unaccumulated circumplanetary debris. Heliocentric particles (detritus from asteroidal collisions or cometary dust) still impact satellites today (Bhatnagar and Whipple, 1954; Kerr and Whipple, 1954) whereas in the past heliocentric (interplanetary) material brought to the satellites through collisions accounts for the primary growth of a protosatellite in some satellite origin theories (cf. Chapt. 26, Safronov and Ruskol).

The momentum transferred to the satellite by these impacts naturally affects the satellite orbit. The specific effect that occurs depends on whether mass is added to —or subtracted from—the satellite during a collision and *that* is an unknown function of the relative impact velocity, and the masses and compositions of the two bodies. In the work that follows, we will assume that the satellite accretes mass.

Following Kaula (1971), we write the total angular momentum stored in both the satellite and planet orbits about the system's center of mass as

$$H = Mm\left[Ga(1-e^2)/(M+m)\right]^{\frac{1}{2}} ; \qquad (23)$$

eqn. (3) is (23) in the limit $M \gg m$. Throughout the growth phase the mass of the planet as well as the mass of the satellite changes and so all the terms in eqn. (23), except G, will be considered variable here. It has even been proposed that G may be decreasing (Dicke, 1966; van Flandern, 1975), according to some cosmologies. One can investigate the change in the orbital size due to the infalling of matter by neglecting the eccentricity and taking the time derivative of eqn. (23):

$$\dot{H}/H \approx \dot{m}/m + 1/2\,(\dot{M}/M + \dot{a}/a) \qquad (24)$$

for $m/M \ll 1$. If the matter coming from heliocentric orbit has a random distribution of angular momentum, then

$$\dot{a}/a = -(\dot{M}/M + 2\dot{m}/m) , \qquad (25)$$

showing that added matter causes the orbit to collapse. Lyttleton (1967) and Clark *et al.* (1975) have postulated that the rapid growth of proto-Earth may have permitted the Moon to be captured from heliocentric orbit.

So, besides being moved out by tidal friction (cf. eqn. 14), the satellite can be brought in with mass accreted by the primary. Mass accreted by the satellite itself is relatively more efficient in changing its orbital semimajor axis. Harris and Kaula (1975) have considered the details of the satellite accumulation process for the Earth-Moon case (see also Chapt. 26, Safronov and Ruskol). They have evaluated a drag force corresponding to the m term in eqn. (25), improving an approximate result of Kerr and Whipple (1954), and given diagrams based on the numerical integration of eqn. (25) with a tidal friction term added. Harris and Kaula (1975) conclude from their model that the Moon could have grown to its present size if the Moon embryo were introduced when the Earth was about 1/10 its final mass. During this growth the Moon remained at about 10 Earth radii.

Accretion drag may be important in forming some satellite resonances, notably the Titan-Hyperion one (see Harris's comment following Chapt. 8, Greenberg; Harris and Kaula, 1975). During the final stages of accumulation, we may ignore M/M and expect that $\Delta m/m$ for neighboring satellites will be inversely proportional to the satellite radius, assuming that the satellite mass does not significantly attract the intersecting particle orbits. The current position of the Saturnian satellites hints that Rhea and Hyperion, both small, have moved in toward Saturn by this mechanism to produce the Titan-Hyperion resonance and Rhea's somewhat unusual orbital spacing.

Accretion drag will tend to lower the satellite's orbital eccentricity but this seems to be a smaller effect than that of tidal friction (Harris and Kaula, 1975).

The small particles that make up the debris currently orbiting the planets have motions slightly different from the satellites themselves because the forces per unit mass acting on them are not quite the same (see below). Comparison with atmospheric drag effects on the orbits of artificial satellites (cf. Roy, 1965; Shapiro, 1963) indicate that the orbital elements a and e, and usually i, will decrease because of collisions. This can also be seen from simple physical arguments based on eqns. (2), (5), and (7). Consider a limiting case in which the satellite moves through a stationary cloud of particles, which are symmetric about the equatorial plane of the primary. Collisions dissipate energy, causing the orbit to shrink, according to eqn. (2). Angular momentum also is lost from the satellite to the cloud, circularizing the orbit (cf. eqn. 5). The satellite will attempt to align its angular momentum with that of the cloud and, given enough time, would ultimately succeed in doing so; this process means that the orbital inclination slowly decreases. Effects like atmospheric drag, however, are not likely to be too important today for natural satellites because of the smallness of the impacting mass.

PROCESSES AFFECTING SMALL PARTICLES

For use in later chapters concerning the origin of satellites and for the sake of completeness, we now give a short, qualitative description of how an initial circumplanetary cloud would evolve and then discuss the motion of an individual small particle. In the latter case we are discussing a different class of "satellite." For these bodies, forces which are proportional to cross-sectional area will be important because they produce accelerations which are inversely proportional to the particle's radius. Such forces will include radiation pressure, Poynting-Robertson drag, and frictional drag. The first two effects have been discussed by Burns *et al.* (1976a, b). Electromagnetic forces, while more difficult to compute, may also most affect small bodies; our presentation on them will summarize Shapiro (1963) and Peale (1966). Tides, since their effects are proportionate to mass (cf. eqn. 14), will have little influence on small particles. The acceleration due to gravity is independent of satellite mass, and therefore the discussion given earlier in this chapter and in more detail in Chapter 3 by Aksnes and in Chapter 4 by Kovalevsky and Sagnier need not be repeated. Table 7.2 summarizes the overall dynamical effects of these processes.

Gross Evolution of a Circumplanetary Cloud

Imagine a cloud of particles orbiting a planet. Ignoring the mass of the cloud, the precession of an individual particle's orbit plane will occur either about the planet's rotation pole or about the pole of the planet's orbit plane, depending upon whether the body is well within or considerably beyond a_{crit} (see eqn. 22, or Chapter 3 by Aksnes). This precession (which depends on the particle's a and e) spreads the satellite cloud symmetrically about either plane shortly after the cloud is placed about the planet regardless of its original structure (cf. Goldreich, 1965a; 1966a). Collisions amongst the particles making up the cloud cause energy to be lost from the orbital motion but conserve the total angular momentum of the cloud about the planet. This causes the particles comprising the cloud to collapse into a single plane so that they may maximize their orbital moments of inertia about the central body. The same process has occurred in the solar nebula, as put forth in Chapter 23 by Cameron. This flattening has been numerically demonstrated by Brahic (1975). It produces a large increase in the space density of material since the three-dimensional distribution of matter reduces to an essentially two-dimensional disk; this higher density may permit the gravitational instability of Goldreich and Ward (1973) to be initiated (cf. Chapt. 23, by Cameron).

Even after the cloud has collapsed to a thin sheet, collisions will continue to occur unless all orbits are non-interesting, that is, perfectly circular. To conserve angular momentum while losing energy, the disk starts to spread—some particles are lost to the planetary surface, others escape. These processes may be occurring today in Saturn's rings (cf. Brahic, 1975) but probably do not have any long time effects on the current distribution of ring material since they are counteracted by motions resulting from satellite perturbations (Franklin *et al.*, 1971).

Radiation Pressure

The photons in a radiation field carry not only energy but also linear momentum. When a particle absorbs or reflects this momentum, it feels a force, called the radiation pressure, according to the impulse-momentum theorem and action-reaction. It will be shown below that this effect is important only for a limited range of particle sizes.

The Sun has a luminosity $L_\odot$ ~3.9 × 10^{33} erg/sec (Allen, 1963); this radiant energy streams away from the Sun at the speed of light c. The solar energy flux $\mathscr{S}$ at distance r (or R measured in AU) from the Sun is

$$\mathscr{S} = L_\odot/(4\pi r^2) = 1.388 \times 10^6 R^{-2} \text{ erg/cm}^2\text{-sec.} \tag{26}$$

The momentum flux u carried by this energy flux is u = $\mathscr{S}$/c, radially outward. As a body intercepts the momentum, redirecting some of it while the rest is absorbed (to eventually be isotropically re-emitted as thermal radiation in order to maintain thermal equilibrium), the body feels a pressure

$$P = f(\mathscr{S}, \lambda) \mathbf{u} \cdot \hat{n}/c \tag{27}$$

for a flat surface with unit normal $\hat{n}$. The quantity $f(\mathscr{S}, \lambda)$ is the efficiency factor which is discussed below; it is an idealization to summarize the way a surface responds to a radiation field. In the geometrical optics limit, when the particle radius $\mathscr{s}$ is much larger than λ (some characteristic wavelength of the radiation field), the efficiency factor is constant. It may then be written

$$f = 1 + \kappa \,, \tag{28}$$

where the reflectivity κ is zero in the limiting case of a perfect absorber and is one for a perfectly reflecting surface; the value of κ can be negative for dielectric particles.

To appreciate the magnitude of the radiation pressure force, it is usually compared to the gravitational force of the Sun on the particle:

$$\beta \equiv F_{rad}/F_{grav} \tag{29a}$$

$$= K/(\rho \mathscr{s}) \,, \tag{29b}$$

where ρ is the particle density and K is constant in the geometrical optics limit:

$$K = 3fL_\odot /(8\pi cGM_\odot) = 1.2 \times 10^{-4} f \quad g-cm^{-2} \tag{30}$$

from eqns. (26) and (27). Eqn. (29b) illustrates two important facts: in interplanetary space β is independent of distance to the Sun (assuming f to be independent of temperature) and the radiation pressure becomes important solely for small particles. We will consider the radiation pressure force to be an important perturbation for circumplanetary particles only if it is comparable to the solar gravitational force.

Using eqns. (29) and (30), $\beta\approx1$ when s is about $1/2$ μm. Since the solar spectrum peaks at about $1/2$ μm, the radiation pressure force (unhappily) is significant just when the approximation used to calculate it is no longer valid. To actually compute f(s, λ), one needs to know precisely how a particle of radius s scatters and absorbs light of wavelength λ given the particle's optical properties as a function of λ. Then one can integrate over the energy distribution contained at different wavelengths in the solar spectrum to find β, which now will be a complicated function of s. In general, $f \equiv f_{abs} + f_{sca} (1 - < \cos \theta>)$, where f_{abs} and f_{sca} are, respectively, the absorption and scattering efficiency factors, each calculated from Mie scattering theory (van de Hulst, 1957); $<\cos \theta>$ accounts for the asymmetry of the scattered radiation.

These Mie scattering calculations have recently been carried out by Gindilis *et al.* (1969), Lamy (1975), and Mukai and Mukai (1973) (see Mukai *et al.*, 1974). Figure 7.4 shows the results of Lamy (1975) for various materials; the conclusions of the other authors are roughly the same and generally agree with the earlier but cruder results of Shapiro *et al.* (1966). At long wavelengths $\beta\sim(\rho s)^{-1}$, accurately reflecting the classical result (29b) for the geometrical optics limit. Near the peak in the solar spectrum the radiation pressure force is about equal to the gravitational force while for smaller particles f (s, λ) rapidly drops to zero and stays there. The latter effect can be easily understood: the particles are so small that they no longer disturb the radiation field. As Figure 7.4 depicts, radiation pressure on typical materials in the solar system is important only for particles 10^{-1} μm to 1 μm in radius.

Dynamical Effects of Radiation Pressure

In interplanetary space, the radiation pressure—since it is proportional to r^{-2}, just as is the gravitational force—does nothing to change the character of a particle's motion. The only effect is that the particle moves as though it were orbiting a less massive Sun. Of course, when $\beta>1$, an elliptic orbit becomes hyperbolic since the force is then repulsive. Such effects are discussed by Burns *et al.* (1976a, b).

Because the size of every planet's sphere of influence is a negligible fraction of

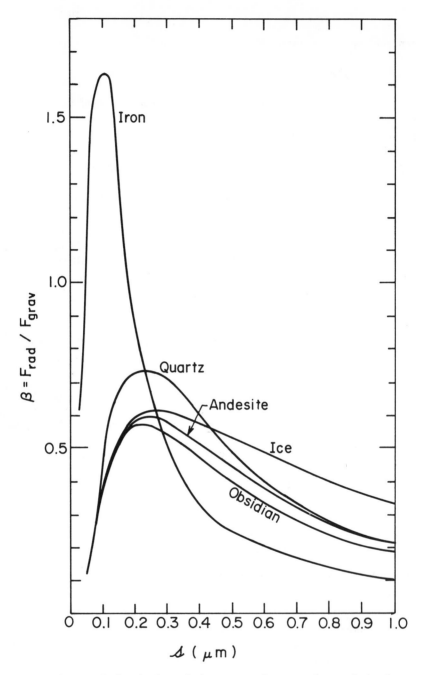

Fig. 7.4. Ratio of solar radiation pressure force to solar gravitational attraction as a function of particle radius s for spheres of astrophysically important materials. Taken from Mie calculations by Lamy (1975).

the distance to the Sun, any particle in orbit about a planet does not appreciably change its distance to the Sun. Thus to first order we can assume that the solar radiation force on a planet-orbiting particle is a constant force radially away from the Sun. We do not consider higher order terms nor the radiation pressure due to light reflected from the planet.

To change the orbit size, eqn. (2) shows that work must be done on the orbit. In a constant force field, work is only accomplished by absolute displacements in the field. Since after one orbit the particle returns to approximately where it was, no work is done by the radiation pressure force and therefore the orbit size does not vary. This ignores the likelihood that the orbit passes through the planetary shadow. Since the particle enters and leaves the shadow at different solar distances (see Fig. 7.5), work will be done over one orbit of the particle, and minor, temporary changes in a will occur. However, as will be mentioned directly below, under radiation pressure the orbit will precess in its orbit plane so that when the longitude of pericenter has moved through π radians all possible shadow orientations will be sampled. Thus the integrated effect of the radiation pressure on the semimajor axis, even including shadowing, will add to zero (see Peale, 1966, 1968).

The constant force of the radiation pressure will produce a total torque relative to the planet on the orbit when integrated over a complete particle revolution which, after multiplication by time, changes the orbital angular momentum. This change in the orbital angular momentum results because the particle spends more time at apocenter, at which position it has a larger moment arm relative to the planet's center (cf. Fig. 7.5). The resultant torque has two effects: First, a torque lying in the orbit plane is produced as can be easily seen by considering the limiting case in which the particle orbit plane is normal to the planetary orbit plane. Any noncircular, inclined orbit will feel a total torque which will rotate the orbit plane, keeping the inclination constant. Second, consider a particle whose orbit lies in the planet's orbit plane. A torque normal to the orbit will result for the reasons given above; this torque, according to eqn. (5), produces changes in the orbital eccentricity e and also causes the orbit to precess in the orbit plane, that is, the location of pericenter revolves relative to the Sun-planet line. Peale (1966) calculated that major changes in e are produced which are periodic because the precession of the orbit means that a torque which increases e will soon become a torque which decreases e. His results (which are based on Shapiro (1963) and which neglect shadowing, planetary oblateness, second order terms and orbital inclination) can be written, including the Mie modification, as:

$$\triangle a = 0 \tag{31a}$$
$$e^2 - e_0^2 = (9/2)\beta^2 (a/a_p)(M_\odot /M_p)(1 - e_0^2)(1 - \cos n_p t) , \tag{31b}$$

where e_0 is the initial value of eccentricity and the subscript p indicates values for the planet's orbit. Since the right hand side of eqn. (31b) is periodic, no secular

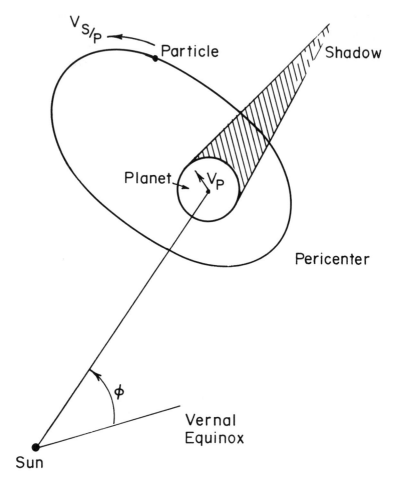

Fig. 7.5. Passage of particle through planetary shadow. The particle generally enters into, and emerges from, the solar shadow at different solar distances. After Shapiro (1963).

effects take place unless the perturbations in e are large enough to cause the particle to strike the planet's surface (or that of an inner satellite) during one of its excursions. Peale (1966) found that in the geometrical optics limit such perturbations occur when the particle size is less than 1 or 2 μm. However, now that Mie calculations have shown that radiation pressure itself is only significant for a limited size range, these dynamical processes must be considered only for particles between about 0.05 to 2 μm in radius.

Poynting-Robertson Effect

The Poynting-Robertson effect is a first-order drag force which arises from the radiation pressure because of the transverse motion of the particle. The most correct historical derivation is due to Robertson (1937). His derivation is, however, unnecessarily abstruse because it is based on a general relativity treatment, even though the relevant effect is a classical one; it also does not provide a general result because only perfectly absorbing particles are considered, whereas small particles are rarely good absorbers.

A more careful derivation by Burns *et al.* (1976a, b) separates the effects of the radiation energy scattered by the particle, that transmitted by it, and that which it absorbs to instantaneously re-emit isotropically. Using the transformation laws of special relativity, the Cornell authors evaluate the energy and momentum in the incoming beam and the outgoing radiation as seen in both the solar reference frame and the particle reference frame. The difference in the momentum of the incoming and outgoing radiation must be taken from the particle's momentum in order to conserve total momentum, electromagnetic plus mechanical. Since momentum per unit time is force, this momentum difference can be placed on the right hand side of the equation of motion. Burns *et al.* (1976a, b) conclude that Newton's law for a particle moving through a radiation field is:

$$m\dot{\mathbf{v}} = (f \times \mathscr{A}/c)\left[(1-\dot{r}/c)\hat{\mathbf{r}} - \dot{\mathbf{v}}/c\right] , \qquad (32a)$$

where $\dot{r}$ is the particle's radial velocity and $\hat{\mathbf{r}}$ is the radial unit vector. The first term in the brackets of eqn. (32a) is the radiation pressure force (cf. eqn. 27) as slightly modified because of the Doppler-shifted radiation; the second term is what will be called the Poynting-Robertson drag, since it opposes the velocity. Equation (32a) can be rearranged to read instead:

$$\ddot{\mathbf{r}} = (\beta GM_{\odot}r^{-2})\left[(1-2\dot{r}c^{-1})\hat{\mathbf{r}} - r\dot{\theta}c^{-1}\hat{\theta}\right] , \qquad (32b)$$

where the last term in the brackets is the transverse velocity.

Other effects due to the relative velocity of the particle and the solar limb have been investigated by T. McDonough (private communication, 1975) and have been shown to produce forces which are counterparts to the Poynting-Robertson effect.

Dynamical Effects of Poynting-Robertson Drag

Since we are interested in particles which move in the solar radiation field but which are orbiting a *planet,* the **v** portion of the force term in eqn. (32a) must include both the planetary orbital velocity $\mathbf{v}_p$ and the "satellite" orbital velocity $\mathbf{v}_{s/p}$ relative to the planet. The work done on the orbit per unit time is the scalar product of the force with the satellite orbital velocity $v_{s/p}$. Integrated over a complete revolution, $v_p \cdot v_{s/p} = 0$ and thus only the satellite orbital velocity is important in determining the work done on the orbit (Shapiro, 1963). The drag, according to eqn. (2), causes the particle orbit to collapse. The change in a over one particle orbit, ignoring shadow effects, is

$$\triangle a = -2\pi a \, \beta^* \left[v_{s/p} c^{-1} \right] Z(\phi) \, , \tag{33}$$

where β^* is the ratio of the radiation pressure to the *planetary* gravitational force (cf. eqn. 29) while $Z(\phi)$, which is of order 1, is a positive function of the angular position ϕ of the planet (Shapiro, 1963). Eqn. (33) is accurate to zeroth order in e and first order in v/c. One can attempt to use eqn. (33) to place bounds on the size of particles in Saturn's rings by assuming that Poynting-Robertson drag should not cause orbital collapse over the age of the solar system. Calculations give a lower bound of a few centimeters (Pollack *et al.*, 1973b) but they are probably improperly applied since perturbations by the satellites, particularly Mimas, form a dynamical barrier, which is only semi-permeable, in the vicinity of the gaps in the rings. This barrier prevents major secular effects from Poynting-Robertson drag on ring particles (Franklin *et al.*, 1971). However, because of collisional momentum-sharing, it is possible that a few large satellites can slow this process. Harris (private communication, 1975; cf. Harris, 1975) computes that a ring mass greater than 10^{-10} Saturnian masses is necessary to resist inward spiraling under Poynting-Robertson drag.

The time scale for orbital collapse is given by $\tau = (a/\triangle a) \, P$, where P is the particle's orbital period. From eqn. (33) this is

$$\tau = c\beta^{-1}(a_p^2/GM_\odot) \, , \tag{34}$$

which in the case of a strongly perturbed dust grain ($\beta \approx 1$) orbiting the Earth is $\sim 10^3$ years (cf. Peale, 1966).

The total Poynting-Robertson force is proportional to $v_p + v_{s/p}$. Since the second term merely adds and subtracts from the much larger v_p, the average force—and therefore the average torque—is proportional only to v_p. The force associated with this term can be broken into two parts, a tangential component and a radial perturbation (cf. eqn. 9); each is a trigonometric function of the orbital angle. These forces produce only short periodic effects on the eccentricity. The precession of the longitude of pericenter gives a periodic variation in eccentricity, which Shapiro (1963) expresses as

$$\triangle e = 3\pi\beta^* \left[v_p/c \right] Y(\phi + \pi/2) \, , \tag{35}$$

where Y is a periodic function of ϕ which is of order 1. The eccentricity therefore returns to its previous value after one planetary orbit.

Electromagnetic Effects

Small particles in space can develop substantial electric charges on them through photoionization by the solar radiation field. The exact electric potential Φ developed on a particle's surface results from a competition between photoionization and collisions with the charges in the surrounding ambient plasma (Peale, 1966). A potential Φ of $\sim +10$ volts is often taken to be present on interplanetary particles but this is based on the subtraction of two large numbers and so is questionable. For a spherical particle, an electric charge $e \sim \Phi/s$ is associated with the potential. Thus electromagnetic forces per unit mass are most effective for small particles, going like s^{-4}.

The dynamics resulting from electromagnetic forces are probably important since the forces can be relatively large but, unfortunately, they are very model-dependent. One simple conclusion is that magnetic forces do not change a since they do no work. The most recent and most elaborate calculation is that of Mendis and Axford (1974). They investigated the motion of charged particles in model planetary magnetospheres in an attempt to explain the photometric regularities observed for the satellites of the outer planets as functions of their orbital phase angles (cf. Chapt. 9, Veverka). Mendis and Axford (1974) found intriguing similarities between the distribution of impact sites on the satellite surfaces from charged particle collisions and the photometric properties seen. Other dynamical studies are those of Shapiro (1963), Peale (1966, 1968) and Gold (1975b).

An interesting result is that one of the electromagnetic forces, the Coulomb drag or charge drag, changes sign at the synchronous orbit position. A charged particle experiences a drag as it moves through a plasma because it interacts through electromagnetic forces with the electric charges in the plasma that lie near its path. The plasma in circumplanetary space will be primarily tied (ignoring drift motions) to the planetary magnetic field lines, which are assumed to rotate with the planet. Hence, when inside the synchronous orbit position, the particle will be moving more rapidly than the ambient plasma and will experience a drag to slow it. Beyond the synchronous orbit the situation reverses and the particle is pushed forward by the plasma. Hence Coulomb drag tends to move particles away from the synchronous orbit position unless it is counteracted by other processes. The competition between charge drag and Poynting-Robertson drag, for example, may produce high particle concentrations; so can Poynting-Robertson drag and orbital resonances (Gold, 1975b).

The changes in the satellite eccentricity due to charge drag also depend on whether or not the particle is inside the synchronous orbit. For a particle inside the synchronous position, the effects are the same as for atmospheric drag: the larger force acts at pericenter, decreasing the eccentricity. Beyond synchronous orbit, the larger Coulomb force occurs at apocenter and the eccentricity can increase.

Discussion

The manner in which the processes described above change the orbital energy and angular momentum are summarized in Table 7.2. As intimated, the effects on the orbital evolution of a small particle for any single process are usually calculable, although frequently complex. However, when all processes are included, the problem is no longer tractable. Since many of the effects compete with coefficients that are often model-dependent, one cannot yet describe with any conviction the orbital evolution of a circumplanetary cloud. Further study of these processes is necessary if we are to perceive the earliest stages of satellite development.

CONCLUDING REMARKS

Substantial progress in understanding the orbital evolution of both satellites and small particles has been brought about since the mid-1960s. It seems likely that a complete grasp of the consequences of tides will soon be forthcoming; this most important feature of the Earth-Moon evolution will require a further appreciation of the distinction between ocean tides and solid body tides which can most convincingly be accomplished by observing the action of solid body tides on other satellites, particularly Phobos. Long-time orbital evolution calculations with purely gravitational terms should be carried out for satellite systems; it may be that, before such calculations can be significant, a new approach for dealing with stochastic, dynamical systems will be needed. The dynamics of small particles in orbit about a planet have not received the detailed treatment that they merit.

Studies into the origin of satellite systems have a major advantage over similar investigations for the solar system as a whole since more examples of the species are available. Even with this advantage, it is unlikely that the orbital evolution calculations will ever be able to say with absolute certainty that a particular satellite originated in a specific manner at a particular point in space-time. We can even now, however, eliminate certain origin scenarios as dynamically implausible. Perhaps that is the most we can hope for.

ACKNOWLEDGMENTS

I thank Steven Soter and Alan Harris for speedily reviewing and helping to improve this chapter after it was turned in very late by a shamefaced author/editor. The research was supported in part by NASA.

TABLE 7.2
Dynamical Effects for Different Physical Processes

Process	Orbital Energy	Orbital Angular Momentum	Comments
Processes Important for Small Bodies			
1. Radiation Pressure			Significant for $0.1-1\mu m$ particles. See Figure 7.4.
Short-Term	$(\downarrow\uparrow)$	$\downarrow\uparrow$	
Long-Term	no secular change	no secular change	No secular effects on a and e; orbit and orbit plane precess; elimination possible by collision with planet or satellite.
2. Poynting-Robertson Drag	$\downarrow$	$(\uparrow)\downarrow$	Shrinking orbits; only short period effects on e; inclination damped.
3. Charge Drag			Results strongly dependent
Inside synchronous	$\downarrow$	$\downarrow$	on magnetospheric model
Outside synchronous	$\uparrow$	$\uparrow$	and actual photoionization process.
4. Gas Cloud	$\downarrow$	no change	Collapse to disk which then
Self-collisions		in total H	spreads.
5. Solar Gravity	Orbit plane precesses around pole of planet's orbit. No secular change in a or e.		
6. Planetary Oblateness	Orbit plane precesses around planet's rotation pole. No change in a or e. Collisions cause collapse to disk in equatorial plane.		
Processes Important for Large Satellites			
7. Tides in Planets			See Figure 7.1.
Inside synchronous	$\downarrow$	$\downarrow$	Orbit collapses to planetary surface.
Outside synchronous	$\uparrow$	$\uparrow$	Orbit grows; e increases.
8. Tides in Satellites	$\downarrow$	no change	See Figure 7.2. Circularizes orbit.
9. Gas Drag	$\downarrow$	$\downarrow$	Smaller, circular orbits.
Interplanetary Dust or Atmosphere			Depends on whether accreting mass or not.

Note: This table gives only gross features: see the text for qualifying remarks. The manner in which the change in orbital energy and angular momentum affects the orbital elements (a,e,i) is given by eqns. (2), (5) and (7). ( ) indicates shadowing must be included. Processes 5), 6) and 9) are classical effects that can be important for both large and small bodies.

DISCUSSION

T. GOLD: In the case of the major planets, atmospheric tides of a very different nature from the solid body tides considered by Burns may be at work. In particular, when the velocity of the tidal wave on the planet exceeds the velocity of a gravity wave, a non-linear regime will be set up. A small scale height in the atmosphere may imply a breaking wave travelling around the planet. The importance of such a type of tidal action is that it may drive out satellites not only at rates that are complex functions of magnitude and speed but also to distances where the physical regime at the planet changes in character.

ORBIT-ORBIT RESONANCES
AMONG NATURAL SATELLITES

Richard Greenberg
University of Arizona

A qualitative, physical description of the orbital resonance phenomenon is stressed. Various examples observed in the satellite systems are discussed, and recent theoretical work on the formation of resonances is summarized.

A qualitative physical description of the basic resonance mechanism will be presented in order to develop for the novice an understanding of orbital resonances. Perhaps this qualitative viewpoint will reinforce the insight of the specialists as well. Examples of orbital resonances found among the satellites will be briefly described; inferences about the satellites and their evolution history from the existence of resonances will be discussed; and, where appropriate, the direction that future work may take will be suggested.

Because of these limited objectives, a great deal of important theoretical work will not be discussed. However, several of the referenced works (*e.g.,* Tisserand, 1896, and Hagihara, 1972) provide detailed reviews of the more technical aspects of this subject. See also Kovalevsky and Sagnier (Chapt. 4, herein). A complete list of primary sources can be obtained from the combined bibliographies of the cited references.

QUALITATIVE DESCRIPTION OF THE
RESONANCE MECHANISM

In general, a resonance occurs when the periodic behavior of a dynamical system is matched by some periodic driving force. If a satellite system has a configurational periodicity, the mutual perturbations will have the same period, thus enhancing the perturbations and yielding an orbital (or "orbit-orbit") resonance. The periodicity occurs if, and only if, "commensurabilities" (small-integer ratios) exist among the orbital periods. The resonance is said to be stable (or "locked") if the enhanced mutual perturbations maintain the commensurability against other, disruptive influences.

In order to see how such a mechanism can work, consider the following simple model based on the resonance between Titan and Hyperion (Fig. 8.1). The inner satellite of the pair, Titan, is taken to have a circular orbit coplanar with that of the outer one, Hyperion, whose orbit has significant eccentricity (e ≈ 0.1). The

[157]

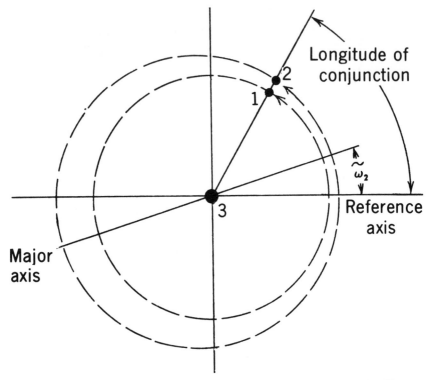

Fig. 8.1. A simplified model of the orbits of Titan (1) and Hyperion (2). The two satellites are shown in conjunction relative to Saturn (3). The dashed lines depict the orbits and $\tilde{\omega}_2$ represents the outer satellite's longitude of pericenter. The orbit of the inner satellite is taken to be circular.

satellites' orbital periods are near a ratio of ¾. Between conjunctions of the satellites relative to the planet, Titan makes four complete revolutions about Saturn, while Hyperion makes three. This commensurability implies that the longitude of conjunction varies slowly.

Hyperion's mass is negligible, so its effect on Titan can be ignored. The effect of Titan on Hyperion need be considered only near conjunction, where the satellites are close to one another and where the attraction is relatively strong. Suppose conjunction occurs after Hyperion's pericenter and before apocenter, as in Figure 8.1. At conjunction, Titan (satellite 1 in Figure 8.1) exerts a force on Hyperion (satellite 2) which is directed radially in toward Saturn, while Hyperion is moving outward as it moves from pericenter to apocenter. Thus energy is removed from Hyperion's motion. Moreover, because the two orbits are diverging at this conjunction, the satellites would actually be closest to one another shortly before conjunction. Titan, having a greater angular velocity, would be behind Hyperion at closest approach and would therefore remove energy from Hyperion's orbit. The loss of energy decreases Hyperion's period (cf. Chapt. 7, Burns). Although this effect is small at each conjunction, it is enhanced by the repetition of this configuration. As Hyperion's period decreases, the ratio of the orbital periods increases above ¾ so that subsequent conjunctions occur closer to Hyperion's apocenter. Similarly, if conjunction occurs after apocenter, it is driven back toward apocenter. The gravitational interaction tends to maintain conjunction at a certain longitude; that is, it tends to maintain the commensurability.

The occurrence of conjunction at the longitude of Hyperion's apocenter is, therefore, a stable configuration. The behavior is closely analogous to that of a pendulum. A pendulum can oscillate about the stable equilibrium position, or, given enough kinetic energy at that position, it can circulate through 360°. Likewise, in the orbital resonance model, conjunction can oscillate (or "librate") about Hyperion's apocenter, or, if at the stable configuration the ratio of orbital periods is far enough from commensurability, conjunction can circulate through 360°. The observed Titan-Hyperion case is, in fact, librating with an amplitude of 36° and period of 1.75 yr. Thus, on the average, the ratio of orbital periods relative to the longitude of apocenter is maintained at ¾.

In this model the variation of the orientation of Hyperion's major axis has not yet been considered. In fact, the repeated radial perturbative forces exerted on Hyperion near apocenter cause the longitudes of the apsides to regress. In the last century, before resonances were understood, the observed regression surprised astronomers who expected the opposite effect due to Saturn's oblateness (Newcomb, 1891). For conjunction to continue to oscillate about apocenter requires that, on the average, the ratio of the sidereal periods must be slightly less than ¾, as is observed in the Titan-Hyperion case.

Variation in the rate of apsidal regression can be important for cases with very small orbital eccentricity. As will be discussed, Hyperion may have had such an eccentricity in the past. Suppose that in our model we take Hyperion's $e \sim 0.01$. The stability mechanism discussed above is weakened because it depended on Hyperion's significant eccentricity. The principal effect of Titan on Hyperion is a radial force exerted at conjunction. The effect of such a radial force varies sinusoidally with longitude (Fig. 8.2). If exerted within 90° of apocenter, the line of apsides regresses; if exerted within 90° of pericenter, the line of apsides advances. If exerted after apocenter and before pericenter, e increases; if exerted after pericenter and before apocenter, e decreases (cf. Burns, Chapt. 7). Suppose conjunction repeatedly occurs at point B, shortly before apocenter. Apocenter will regress and, because e decreases, apocenter will be accelerated toward conjunction. Similar arguments show that if conjunction occurs at any longitude within 90° of apocenter, apocenter is accelerated toward conjunction. And if

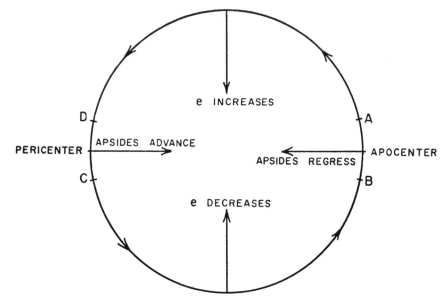

Fig. 8.2. The orbit of a satellite with very small eccentricity. Radial forces exerted on the satellite at various points in its orbit are shown with arrows labeled according to their effects on the orbit. Such forces might be due to another satellite at inferior conjunction. If conjunction occurs at A or B, the resonance mechanism tends to accelerate apocenter toward the longitude of conjunction; if conjunction occurs at C or D, the resonance mechanism tends to accelerate pericenter toward the longitude of conjunction.

conjunction occurs within 90° of pericenter, pericenter is accelerated toward conjunction. Thus, for cases where this low e mechanism is dominant, conjunction can be stable at apocenter or at pericenter.

In order to analyze the resonance problem properly, a mathematical description of the phenomenon is required. Although this review attempts to maintain a qualitative viewpoint, a brief mention of the analytic approach may be a useful introduction to other chapters. Clearly, an expression for the longitude of conjunction is crucial.

The longitude of conjunction of two satellites is a "stroboscopic" function of time; it is only meaningful at the instants of conjunction. But a continuous function connecting the stroboscopic points can be defined. For example, in the Titan-Hyperion case, the following definition is possible:

$$\text{Longitude of conjunction} \equiv 4 \times (\text{longitude of Hyperion}) - 3 \times (\text{longitude of Titan}).$$

When the satellites' longitudes have the same value, this function also takes that value, so this continuous function matches the stroboscopic points. For satellites in resonance, the longitude of conjunction varies slowly compared to the mean motions. The coefficients 4 and -3 are selected so that the continuous function varies slowly near the ¾ commensurability of periods.

The Titan-Hyperion resonance is characterized by libration of the conjunction longitude about Hyperion's apocenter. The mean longitude of Hyperion equals the true longitude at this point. This is also nearly true for Titan, since its orbit has a low eccentricity. Thus, the resonance can be described by the statement that the "resonance variable," ϕ, defined as

$$\phi \equiv 4\lambda_2 - 3\lambda_1 - \tilde{\omega}_2 ,$$

librates about the value 180°. Here λ_1 is a mean longitude and $\tilde{\omega}$ is the longitude of pericenter. In any pair of satellites, subscripts 1 and 2 refer to the inner and outer one, respectively.

The analysis of any resonance reduces to a study of the behavior of its resonance variable through application of Lagrange's equations for the variation of orbital elements (cf. Kovalevsky and Sagnier, Chapt. 4, and Burns, Chapt. 7). The expressions for the behavior contain the "disturbing function" which describes the mutual perturbing forces of the two satellites. The disturbing function is customarily expanded into a Fourier series. Most of the terms in the series contain sines and cosines whose arguments circulate within a few orbital periods. The effects of such short-period terms can usually be ignored. However, "critical terms," which contain the resonance variable as their argument, cannot be ignored. See Chapter 4, by Kovalevsky and Sagnier, for details.

Eliminating most of the terms of the disturbing function is equivalent to ignoring the effect of Titan on Hyperion except near conjunction. Thus the physical description of the resonance given in this section closely parallels the mathematical analysis.

The model that we have discussed has been simplified to facilitate description of the basic mechanism. It does retain the essential features of orbit-orbit resonance. It must be emphasized that the qualitative description of its behavior is based on a mathematical analysis (Greenberg *et al.*, 1972; Greenberg, 1973a).

CHARACTERISTICS OF ACTUAL RESONANCES

The Titan-Hyperion case is a useful example to illustrate the fundamental principles of the resonance mechanism. The other satellite resonances are not so simple, as will now be seen.

In the Saturn satellite system there are two other resonant pairs. In the Enceladus-Dione case the periods have a ratio of ½ and conjunction librates about pericenter of the inner one with an amplitude $<1°$ and period of about 12 yr, that is, the resonance variable $2\lambda_2 - \lambda_1 - \tilde{\omega}_1$ librates about zero. The inner satellite has the smaller mass and larger e. The mechanism can be interpreted qualitatively in terms of a simplified model analogous to the Titan-Hyperion model, but a rigorous analysis is greatly complicated because the masses and eccentricities are comparable. Thus, the mutual interactions must be considered.

In the Mimas-Tethys resonance the periods have a ratio of ½ and conjunction librates about the midpoint of the ascending nodes of the two satellites on Saturn's equatorial plane, that is, the resonance variable $2\lambda_2 - \lambda_1 - (\Omega_2 + \Omega_1)/2$ librates about zero, where Ω_i is the longitude of an ascending node. The amplitude is about 48° and the period is 71 yr. This resonance is of the "inclination type" because the longitude of conjunction is locked to the orientation of nodes rather than of apsides. Both qualitative and rigorous investigations of inclination-type resonances show that mutual perturbations tend to maintain conjunction 90° from the mutual nodes of the satellites. This effect combined with the precession of the orbital planes due to Saturn's oblateness yields the observed nature of the resonance (Greenberg, 1973b).

The structure of Saturn's rings (see Table 1.5 and Cook and Franklin, Chapt. 19) can be interpreted in terms of resonances. It has long been observed that the gaps in the rings correspond to orbits which would have near-commensurabilities with the inner satellites of Saturn. Most notably, particles near the Cassini division would have ½ the period of Mimas. Presumably, particles experiencing resonant perturbations by the satellites were unable to maintain nearly circular orbits. In such regions collisions were common and cleared the gaps. Franklin and Colombo (1970; cf. Greenberg, 1976b) have modeled the structure of the rings by assuming that the particles now move in non-colliding orbits. Further, it is assumed that particles near commensurabilities are in stable resonances. They note that, as has been demonstrated by our qualitative model, the closer such particles are to exact commensurabilities, the larger the orbital eccentricity must be. And the larger the eccentricity, the less densely the orbits can be packed. Franklin *et al.* (1971) account for the shift of the gaps from exactly commensurable positions by invoking the oblateness of the planet and the mass of the rings which tend to advance the apsides. The advance is compensated by a slight adjustment in the ratio of orbital periods, analogous to the adjustment required to maintain the resonance in our qualitative discussion. Further references are given in Chapter 19, by Cook and Franklin.

The inner three Galilean satellites of Jupiter are involved in a resonance with the resonance variable $\lambda_1 - 3\lambda_2 + 2\lambda_3$ locked to a value of 180°, where the subscripts 1, 2, and 3 refer to Io, Europa and Ganymede, respectively (cf. Chapt. 3, Aksnes). This expression implies that whenever Europa and Ganymede are in conjunction with respect to Jupiter (i.e., whenever $\lambda_2 = \lambda_3$), Io is 180° away. The three satellites are prevented from lining up on the same side of Jupiter. The commensurability relation between mean motions is given by differentiating the resonance variable with respect to time, yielding $n_1 - 3n_2 + 2n_3 = 0$. Taken by adjacent pairs the mean motions have ratios of 2:1. Thus, expressing the resonance variable as $(2\lambda_3 - \lambda_2) - (2\lambda_2 - \lambda_1)$ demonstrates that the longitudes of conjunction of pairs are separated by $180°$.

Laplace (1839) first demonstrated the stability of this resonance, and the relation has been named after him. Refinement of the theory is a complicated problem and a current one, as indicated by the number of papers presented on this subject at IAU Colloquium No. 28 (Arlot, 1975; Ferraz-Mello, 1975; Lieske, 1975; Sagnier, 1975; and Sinclair, 1975b).

The inner three satellites of Uranus (Miranda, Ariel, and Umbriel or Uranus 5, 1, and 2, respectively) have a commensurability relation nearly identical to that of the Galilean satellites (cf. Greenberg, 1975). In this system $n_5 - 3n_1 + 2n_2 = -0°.08$/day. Thus $\lambda_5 - 3\lambda_1 + 2\lambda_2$ circulates through $360°$ in 12.5 yr. (Harris, 1949). Despite the near-commensurability there is no evidence that these satellites are locked into a stable resonance. This case also differs from the Galilean satellites in that, taken by pairs, the mean motions are not near ratios of small whole numbers.

The four outer satellites of Uranus have a more exact commensurability relation:

$$n_1 - n_2 - 2n_3 + n_4 = 0°.0034/\text{day}$$

(Harris, 1949). However, this is not the type of relation associated with stable resonances. For a resonant interaction the sum of the integer coefficients in the commensurability relation must be zero. Otherwise the interaction would depend on the choice of reference longitude (Ovenden *et al.*, 1974). The sum here is not zero, so this near commensurability simply represents a special distribution of orbital radii, not a resonance relation *per se*.

Despite the diversity of these resonances, they have two interesting common characteristics. First, in each case the stable configuration at conjunction is a "mirror configuration." In such a configuration all the bodies lie on a single plane and all velocity vectors are normal to that plane. Subsequent behavior is a mirror image of previous behavior (Roy and Ovenden, 1955). Second, stable configurations tend to keep satellites apart. For example, conjunction is kept away from the apocenter of an inner satellite, from the pericenter of an outer satellite, or from a third satellite. This latter property demonstrates that resonances help to maintain the stability of satellite systems by preventing near collisions.

APPLICATIONS

Until the 1950s resonances were studied in order to construct accurate ephemerides of the satellites, besides being studied simply because they are interesting phenomena. Moreover, the enhanced mutual perturbations allowed determination of resonant satellites' masses (Chapt. 1, Morrison *et al.*; Chapt. 3,

Aksnes). Excellent reviews of older works are given by Tisserand (1896), Brouwer and Clemence (1961b) and Hagihara (1972). The details of resonance mechanisms are still under study. Shelus *et al.* (1975) describe current work on refining the theory of other resonances.

A great deal of recent research has been directed toward using the existence and properties of resonances as clues to the origin and evolution of the solar system. In 1954 Roy and Ovenden, leaving questions of stability aside, pointed out that there are more nearly-commensurable pairs of mean motions in the solar system than one might expect if the orbits were randomly distributed. Of course it is important to avoid any tendency to attach too much significance to numerological coincidences (Molchanov, 1968; Dermott, 1969). The details of Roy and Ovenden's statistical argument were modified by Goldreich (1965b) and Dermott (1968), but the conclusion is still accepted: the large number of near-commensurabilities needs to be explained.

Goldreich (1965b) suggested that resonances among satellites might have evolved due to tidal dissipation. He pointed out that tides raised on a planet by a satellite tend to vary the satellite's orbital period at a rate dependent on the satellite's mass and distance (see Chapt. 7, Burns). Moreover, he showed that mutual interaction is strong enough to maintain the commensurability of resonant satellites' periods even against the upsetting influence of the tides. Thus it was reasonable to assume that, unless the resonances were formed before their environment became essentially as it is today, the satellites were originally nonresonant and their periods evolved independently until stable resonances were reached and maintained.

On the basis of this assumption, Goldreich and Soter (1966) placed limits on the tidal energy dissipation parameter, Q, for Jupiter and Saturn. They reasoned that the existence of close-in satellites places an upper limit on tidal evolution (a lower limit on Q) while the resonances imply that substantial tidal evolution did take place.

Indeed, introducing substantial tidal evolution into the model of the Titan-Hyperion resonance shows that capture into libration from circulation can occur (Greenberg *et al.*, 1972; Greenberg, 1973a). Suppose that initially Titan's orbital radius was smaller than it is today. Its orbit would expand due to tidal evolution. Hyperion's orbit would not expand since it is too small and too far from Saturn to raise appreciable tides. Suppose further that Hyperion's e was small (~0.01) when the viscous medium of the early solar system cleared. As Titan moved out, the ratio of orbital periods would approach ¾ from below so that conjunction would circulate in a retrograde direction relative to Hyperion's line of apsides. As the ratio approached ¾, the circulation would slow. Thus, as conjunction passed through the regions of increasing and decreasing e (Fig. 8.2), e would achieve increasingly higher maxima and lower minima. Eventually, the conjunction longitude would circulate so slowly that the minimum value of e could nearly reach zero allowing apocenter to regress rapidly enough to overtake the conjunction longitude. After apocenter regressed past the conjunction longitude, its rate of regression would decrease, allowing conjunction to overtake apocenter. The low e stability mechanism mentioned previously would maintain the resonance thereafter. After stable resonance was achieved, Titan would continue to spiral out from the planet. It would push Hyperion out so as to maintain the resonance. During this phase Hyperion's e would be increased to its current value.

Although this capture mechanism is described here in a qualitative way, it has been analyzed rigorously (Greenberg, 1973a). The results have been confirmed by Yoder (1973) and by Colombo et al. (1974).

One appealing aspect of this capture mechanism is that Hyperion's e increases only after resonance is achieved, thus avoiding the possibility of catastrophic near-collision of the two satellites near Hyperion's pericenter (Sinclair, 1972b). There is, however, a problem with this mechanism. If Goldreich and Soter's value of Q for tides raised on Saturn by the inner satellites is applied to tides raised by Titan, the evolution process requires at least 50 aeons. Thus, in order to permit tidal capture, Q for tides raised by Titan must have been smaller than for tides raised by Mimas. The implication is that Q is a function of tidal amplitude and frequency. Alternatively, this particular resonance may have formed by a chance distribution of orbital elements in the forming solar system (Goldreich, 1965b). Or perhaps some other dissipative mechanism was responsible for capture into libration (cf. Harris' discussion at end of this chapter).

That a dissipative mechanism is required is demonstrated by analogy with a pendulum. A pendulum cannot change from a state of circulation to oscillation unless energy is removed by friction or some other means. Similarly, in the satellite resonance problem passage from circulation into libration requires a change in the energy of the system. One must be careful not to extend the pendulum analogy too far. There is a tendency to assume on such grounds that small amplitude librations correspond to long-established resonances. However, after libration is achieved in the Titan-Hyperion model, the amplitude remains nearly constant even as tidal evolution continues (Greenberg, 1973a).

Sinclair (1972b, 1974a) has pursued the difficult problem of tidal locking of the Enceladus-Dione and Mimas-Tethys resonances by using an analytic and, where necessary, a numerical approach. In each case Sinclair finds that capture into the present resonance is only possible if independent tidal evolution brought the commensurability of periods to ½ from a previously lower value. This criterion is barely met by assuming Q to be the same for tides raised by each satellite. Any model of an amplitude- or frequency-dependent Q ought to meet this requirement.

In the Enceladus-Dione case, Enceladus' e is small enough that once the present resonance is reached, capture is assured just as in the Titan-Hyperion case. However, in approaching the present resonance, the satellite must pass through several inclination-type resonances and an e-type resonance involving Dione's apsidal line. Whether or not the system is captured in these resonances depends on the precise initial conditions of the system. Since the initial conditions are not known, Sinclair adopts a probabilistic approach. He finds that the probability of avoiding capture in these other resonances was about 0.25.

In the Mimas-Tethys case Sinclair finds the probability of capture into the present resonance to have been about 0.04. Although this value is low, we must bear in mind that even improbable events can occur.

The evolution of the Laplace resonance among the Galilean satellites has not yet been successfully studied. Sinclair (1975a, 1975b) has a formulation for a simplified model of this resonance which may permit a study of tidal evolution.

A different approach to the problem of the tidal evolution of resonances was taken by Allan (1969). Allan set aside questions about capture from circulation into libration. Instead, he investigated the tidal evolution of the Mimas-Tethys resonance backward in time from its present state. He found that, going back in time, the amplitude of libration increased until it reached its upper limit on the verge of circulation. At that time, Mimas' semimajor axis was about 0.992 times its present value. If Goldreich and Soter's value of Q is correct, this state was reached about 2×10^8 yr ago. Thus Allan obtains an estimate of the age of this resonance lock. Unfortunately, Allan's analysis depends on certain specific properties of the Mimas-Tethys resonance, so it cannot be applied to other cases. However, some progress in this area has been reported by Yoder (1975).

Goldreich (1965b) has proposed a method by which the present rate of tidal evolution might be measured. He noted that tidal evolution causes resonance variables to oscillate about slightly different stable values than the nominal values given here. These shifts may become detectable as improved theories (Chapt. 3, Aksnes; Chapt. 4, Kovalevsky and Sagnier) and observational techniques (e.g., radar observations; cf. Chapt. 5, Pascu) become available.

The tidal dissipation model may not be the only possible explanation of resonance capture. It is simply the most widely and successfully analyzed mechanism. An alternative that should be investigated is a model including viscous drag during the early solar system (cf. Harris' discussion at end of this chapter). A problem that might be encountered is the tendency for drag to lower orbital eccentricities, and thus one would have to find a way to explain Hyperion's relatively large eccentricity.

While these dissipative mechanisms may be responsible for locked resonances, they do not explain the large number of circulating near-commensurabilities among planets and satellites. Perhaps the preference for near-commensurability is a result of some aspect of the formation of the solar system (Dermott, 1968); e.g., resonance-induced collisions, such as those that apparently cleared the gaps in Saturn's rings, might have generated gravitational instabilities in the nebulae around planets. Thus satellites might have tended to form in orbits nearly commensurable with those of older satellites. Such possibilities have been discussed by Gold (1975b).

Ovenden *et al.* (1974) suggest that a prevalence of near-commensurabilities may represent the state at which mutual influence on orbital periods and radii is minimal (the "principle of least interaction action"). The rate of evolution in this state might be so slow that the probability of observing near-commensurabilities at any epoch is high. The principle appears successful in the Jovian and Uranian systems. Ovenden *et al.* find that the present distributions are long-lived. But serious difficulties arise when the principle is applied to the system of planets (Ovenden, 1972). Ovenden's results have been greeted with some skepticism (*e.g.,* Dermott, 1973); wider acceptance of the principle will require a detailed scrutiny of the computations.

CONCLUSION

Orbital resonances can provide a great deal of information about the natural satellites. Traditional lines of research reveal the masses and motions of resonant satellites. Recent considerations of resonances are helping to provide clues to the origin and history of the satellites. The existence of so many commensurabilities must be explained by any theory of satellite formation and evolution.

ACKNOWLEDGMENTS

Preparation of this review was supported by NASA Grant NSG 7045. The author's participation in I.A.U. Colloquium No. 28 was made possible by a travel grant from the Organizing Committee and by funds from the Lunar and Planetary Laboratory.

DISCUSSION

ALAN HARRIS: I wish to comment further on the possible formation of Titan-Hyperion resonance by a drag mechanism as mentioned at the end of the article.

Collisions with interplanetary matter, which can be very important during the final stage of planetary accretion, will change the satellite's semimajor axis (see also Chapt. 7 by Burns). The incoming matter would bring with it only a negligible fraction of the appropriate orbital angular momentum, and hence the orbit would decay from an increase in satellite mass without the corresponding increase in angular momentum:

$$\frac{\Delta a}{a} \approx -2\frac{\Delta m}{m} .$$

It is unimportant whether the interplanetary matter Δm sticks or is lost again as ejecta.

For a given flux of interplanetary matter, $\Delta m/m$ for neighboring satellites will be proportional to the area-to-mass ratio of each satellite, that is, inversely proportional to the satellite radius. So smaller satellites will suffer a greater loss of orbit radius than larger satellites. The effect is thus ideal for producing resonances between satellites of greatly differing size, notably Titan-Hyperion. A value of $\Delta a/a \gtrsim 0.04$ is needed to produce the Titan-Hyperion resonance, which requires a value of $\Delta m/m$ of only 0.02 for Hyperion, and only 0.002 for Titan (Yoder, 1973).

This hypothesis is further strengthened by the unusual spacing of Titan's neighbors, Rhea and Hyperion. These small satellites appear to have been displaced inward with respect to Titan from the positions one would expect from a satellite "Bode's Law," exactly as predicted by this effect. It should be noted that capture into resonance can only occur when orbits approach each other, not when they are moving apart (Yoder, 1973). Hence Hyperion can become captured into resonance with Titan while Rhea cannot.

R. GREENBERG: The drag mechanism for resonance capture ought to be investigated further. In particular, would a be the only orbital element to experience a secular variation? How would drag affect other elements, such as e and $\bar{\omega}$? Would their variation affect the capture possibility?

If a drag mechanism can be shown to be viable, it would eliminate the need for an uncomfortably small Q for tides raised on Saturn by Titan. In that sense it might be more plausible than the tidal origin hypothesis.

PART III
PHYSICAL PROPERTIES

PHOTOMETRY OF SATELLITE SURFACES

Joseph Veverka
Cornell University

Current information on the phase coefficients, opposition effects, phase integrals, and orbital lightcurves of satellites is reviewed, and areas where more information is needed are identified. There is an especially urgent need for detailed photometric work on the fainter satellites, such as Amalthea (J5), Himalia (J6), Hyperion (S7), and Phoebe (S9).

This chapter reviews some basic photometric concepts as they apply to satellites; then the photometric data for each satellite are summarized. A few topics of special interest are discussed in some detail, and a number of important observations are listed which should be carried out in the immediate future.

The notation adopted and the photometric definitions used are consistent with those of Harris (1961). The longitude of a satellite in its orbit is denoted by θ and is measured so that $\theta = 0°$ at superior geocentric conjunction, and $90°$ at eastern elongation. The solar phase angle is denoted by α. This is the angle ($\leq 180°$) at the center of the satellite between the directions of the Earth and the Sun. In what follows, r denotes the Sun-satellite distance and $\triangle$ the Earth-satellite distance.

It should be kept in mind that while a few satellites are easy to observe, photometric work on many others is notoriously difficult. For inner satellites (for example, Phobos, Deimos, Mimas, Enceladus), scattered light from the primary is a major problem. For some of the moderately faint outer satellites (for example, Himalia, Phoebe, Hyperion), simply finding them in a star field can be a challenge. In fact, a few satellites are so faint that they have only been detected photographically.

All Earth-based observations of satellites are limited in several important respects: First, only certain spectral regions are accessible, and within any of these the attainable spectral resolution is limited by the faintness of the satellites. Second, only disk-integrated parameters can be measured, since the apparent size of all satellites is small. Finally, the range of observable phase angle is very limited ($47°$ at the distance of Mars, $12°$ for Jupiter, $6°$ for Saturn, $3°$ for Uranus, and only $2°$ for Neptune).

[171]

PHOTOMETRIC PARAMETERS

Brightness Variations in General

When satellite observations are corrected for changes in Earth-satellite and Sun-satellite distances, brightness variations remain which can be separated into three types: (i) brightness variations related to the longitude θ of the satellite in its orbit about the primary—the usual case of synchronous rotation being assumed (cf. Chapt. 6, Peale); (ii) brightness variations due to changes in solar phase angle α; and (iii) secular or transient brightness changes not related to either of the above.

All satellites are subject to changes of type (ii), and most also to those of type (i). There is no evidence that (iii) is important for the majority of satellites, but temporary brightenings of Io following some eclipse reappearances have been reported (see Morrison and Cruikshank, 1974, for a review), while a gradual brightening of Titan seems to have occurred in recent years (Chapt. 22, Andersson). Variations of type (iii) are discussed later; for now, we concentrate on the first two. Initially it will be assumed that a separation of the two effects has been achieved.

Brightness Variations Due to Changes in Solar Phase Angle

PHASE COEFFICIENTS

We begin by assuming that brightness variations of type (i) are negligible or that they have been removed from the data by one of the procedures described later.

It is usual to plot the apparent brightness, expressed in magnitudes as a function of solar phase angle, and reduced to either $\Delta = r = 1$ AU, or to the mean opposition. Neglecting any possible "opposition surge," the rate of change of magnitude with phase angle is usually linear for $5° \lesssim \alpha \lesssim 50°$, and the slope of this linear part of the phase curve is called the *phase coefficient,* β (mag/deg). It is a good idea to indicate over what range of phase angles a given β was determined by writing, for example, β ($10°$–$20°$). If it is important to indicate the wavelength one can use a subscript and write, for example, β_V ($10°$–$20°$). Phase coefficients can also be written as $dV/d\alpha$; this notation has the advantage of indicating the wavelength involved, but it does not indicate the range of phase angles to which the phase coefficient applies.

If the satellite's reduced brightness were simply proportional to the illuminated area seen from Earth, the brightness variation with phase angle would be given by:

$$B(\alpha) = \frac{(1 + \cos\alpha)}{2}$$

and the resulting phase coefficient, near opposition, would amount to 0.002 mag/deg. Most solar system objects have much larger phase coefficients—often as much as 15 times larger for bodies with negligible atmospheres. Since only one satellite (Titan) has an appreciable atmosphere, the following discussion is restricted to phase coefficients of ''airless'' bodies.

The problem of what information about the surface layer is contained in an observed phase coefficient has been discussed by Hapke (1963; 1966), Irvine (1966), and Veverka (1971c), among others. Satellite surfaces consist of individual scattering elements, possibly individual particles or crystals. The observed phase curve will depend on three factors: (i) The effective scattering phase function and albedo for single particles making up the surface. These two parameters will be determined by the particle size, the particle shape, and the optical constants of the materials making up the surface. (ii) The shadowing function of a surface element (Irvine, 1966) determined largely by small-scale texture. Small-scale means surface elements large enough to contain a statistically significant sample of scattering particles, but small compared to the scale of the topography or relief. The shadowing function depends mostly on how the particles making up the surface are packed together, that is, on the porosity of the surface. (iii) The macroscopic shadowing function due to surface topography (e.g., craters) (Hapke, 1966; Hämeen-Antilla et al., 1965).

The last of these effects can be dominant. Two planets may have identical surfaces on a small scale, but if one of the two has a much rougher topography, its observed phase coefficient will be much larger, even though returned ''soil'' samples from each would show identical values when measured in the laboratory. Veverka (1971c) gives examples of calculations which indicate that the phase coefficient of a model planet which has a lunar-like surface layer can be doubled by increasing the large-scale roughness. These calculations assume that multiple scattering is negligible, which is certainly not true for many satellite surfaces. However, it is still probably true that large-scale surface roughness will have a dominant effect on the observed phase coefficients even for very bright surfaces.

At the present time it is impossible to derive surface parameters uniquely from observed phase coefficients, but a few generalizations can be made: (i) For a given surface microstructure and single-particle albedo, satellites with the roughest topography will have the largest phase coefficients. (ii) Other things being equal, a surface in which multiple scattering is important will have a lower phase coefficient than one in which single scattering dominates. This is obvious since multiple scattering tends to destroy shadows.

Only for the satellites of Mars and Jupiter is the range of phase angles observable from Earth sufficient to reach the linear portion of the phase curves ($\alpha \gtrsim 5°$) which is not affected by the opposition surge. For Saturn's satellites and beyond, *only* the region that contains the opposition surge can be observed.

OPPOSITION EFFECT

At small phase angles ($\alpha \lesssim 5°$), most surfaces show a surge in brightness called the "opposition effect." From the details of this surge, it should be possible to derive information about the physical properties of the surface layer—especially its texture. The term "opposition effect" first appears in a paper by Gehrels (1956), who discovered the phenomenon in his observations of asteroid 20 Massalia. At that time it was generally believed that common terrestrial powders do not show opposition effects, but Oetking (1966) has since pointed out that the effect had been masked in previous measurements by the use of excessively large detector apertures. Using a carefully designed small-aperture photometer, he observed the effect in numerous laboratory samples. Measurements by Coffeen (1965) suggest that the opposition effect is largest at wavelengths for which the sample reflectance is lowest. A similar inverse relationship between the magnitude of the opposition effect and surface reflectance was reported by O'Leary and Rea (1968) for Mars and Mars-like laboratory samples. This behavior is not unexpected since shadows, which are needed to produce an opposition surge, are destroyed by multiple scattering. Qualitatively, no strong wavelength dependence of the opposition effect is expected as long as multiple scattering is negligible (that is, as long as the reflectance of the sample is less than ~ 0.10).

A surprising result of Oetking's experiments is that even very bright powders can have appreciable opposition effects. Unfortunately, no laboratory studies of the opposition effects of frosts exist, but Oetking found that aluminum oxide surfaces having reflectances of 70% may show opposition effects as large as 10%—although the average is closer to 3–5%. These pioneering laboratory studies of opposition effects need to be continued and extended.

An approximate but useful model of the opposition effect for dark surfaces was given by Hapke (1963). A more exact treatment by Irvine (1966) followed. The connection between these two models and their applicability to real surfaces is discussed by Veverka (1970). Irvine's model considers a surface made up of large, dark particles whose radius $r \gg \lambda$, and whose albedo is so low that multiple scattering may be neglected. All particles are assumed to be spherical and of identical radius. The opposition effect of such a layer turns out to be determined largely by the compaction parameter,

$$ D = \frac{3}{4\pi} \rho/\rho_0 \ , $$

where ρ = density of a macroscopic volume element and ρ_0 = density of a single particle. It also depends on the phase function of a single particle $\Phi(\alpha)$, but not directly on the single-particle scattering albedo $\tilde{\omega}_0$, since it is assumed that multiple scattering is negligible (i.e., $\tilde{\omega}_0$ is small). Typical curves taken from Veverka (1970) are shown in Figure 9.1.

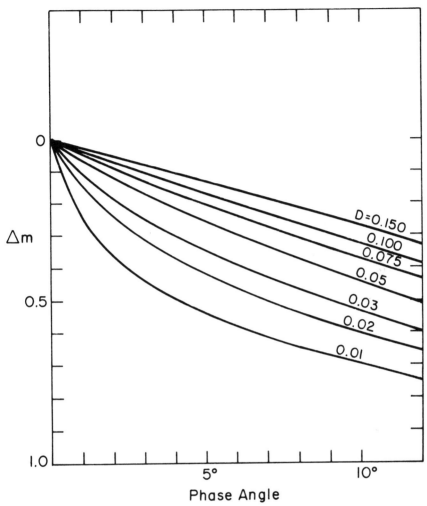

Fig. 9.1. Opposition surges, calculated using the model of Irvine (1966), for surfaces of different compaction. D is the compaction parameter (see text); as it decreases, the surface becomes more porous. (From Veverka, 1970)

The total magnitude of the opposition effect according to this model is independent of particle shape, although the particle shape does affect the shape of the phase curve (W. Irvine, private communication, 1974). Also, a distribution of particle sizes could be included (see Kawata and Irvine, 1974).

Hapke's treatment of the problem leads to considerably simpler equations and gives somewhat similar answers, provided that his parameter h is related to D by $h = 2\pi D$. A comparison of the two models shows that Hapke's model becomes inaccurate for $D \ll 0.1$.

One cannot expect that the assumptions on which Irvine's model is based hold exactly for real surface layers, in which multiple scattering is not entirely negligible and the individual particles cannot have identical phase functions. Surprisingly, however, the model does seem to fit available observations of dark surfaces and appears useful in establishing relative values of D (Veverka, 1970).

Judging from Irvine's model, the opposition surge for dark surfaces depends largely on the shadowing function of a surface element (determined by the surface texture and porosity). To a lesser degree, it depends on the particle phase function; it is largely independent of large scale roughness since shadows due to topography do not come into play effectively near opposition.

For a planet whose surface is covered by a material which scatters according to the Hapke-Irvine law, the disk integrated opposition surge is almost identical with that of a small region near the sub-Earth point. In other words, the opposition surge is largely independent of position on the disk. This is probably not true for layers in which multiple scattering is important, and it may therefore be deceptive to compare directly the measured disk integrated opposition surges of bright satellites with laboratory values. Work, both experimental and theoretical, on opposition effects of bright surfaces is urgently needed.

Opposition surges have definitely been observed for each of the Galilean satellites (cf. Chapt. 16 by Morrison and Morrison and later in this chapter) and possibly also in the case of Deimos (cf. Chapt. 14 by Pollack and later in this chapter).

For Saturn's satellites only the opposition effect part of the phase curve is observable, and the "phase coefficients" usually quoted are linear approximations to the observed trends between $\alpha = 0°$ and $6°$. For Iapetus, there is definite evidence that the phase curve is non-linear in this range (Franklin and Cook, 1974; Morrison et al., 1975b). For Rhea and possibly Dione the curve may also be non-linear, but for Titan the data indicate that a linear fit is preferable (Noland et al., 1974b).

The Irvine model of the opposition surge can be applied only to relatively dark surfaces in which multiple scattering can be neglected. Table 9.1 gives values of D, the compaction parameter, required to explain the opposition surges of Callisto and of the dark side of Iapetus. The quantity Δm defined later under *Photometry of Satellites* is a measure of the opposition surge. Values of Δm are taken from that section, and those of D are based on the graphs in Figure 9.1. Very underdense surfaces are indicated. Note that D is really a model parameter, and that it is difficult to connect it precisely with the compaction of the real surface layer, which has particles that are nonspherical and of different sizes, etc. For comparison, in the lunar surface D ~ 0.02 (Veverka, 1970).

Irvine's theory cannot be applied to Rhea, Dione, Tethys, or to the bright side of Iapetus, since the albedos involved are too high for multiple scattering to be negligible. It is an observed fact, however, that even such bright surfaces can

TABLE 9.1
Opposition Effects of Dark Satellites
(For the lunar surface D ~ 0.02;
for a surface of "normal" compaction D ~ 0.13.)

Satellite	ΔV (mag), Observed Opposition Surge	D, Approximate Surface Compaction Parameter (from Fig. 9.1)
Callisto (leading side)	> 0.25	< 0.025
Callisto (trailing side)	> 0.13	0.04
Iapetus (leading side)	> 0.2	< 0.03

have appreciable opposition effects, suggesting that many satellite surface layers tend to be very porous and underdense.

Observations of the satellites of Uranus sufficient to show opposition effects have not been achieved yet (Andersson, 1974).

Brightness Variations Due to Changes in Orbital Phase

Given limited observations, the problem of separating brightness variations related to orbital longitude from those related to changes in solar phase angle is non-trivial. In the past, this separation was usually attempted in one of the following ways: (i) by assuming that solar phase effects are negligible compared to orbital brightness variations (Harris, 1961; Blanco and Catalano, 1971, in the case of Rhea); (ii) by assuming particular values of the phase coefficient and fitting for the orbital lightcurve by least squares (Johnson, 1969); or (iii) by solving simultaneously for sinusoidal orbital brightness variations and linear solar phase angle effects (Noland *et al.*, 1974b).

These procedures have various obvious defects. Evidently, the first is inaccurate for satellites whose orbital lightcurve amplitude is small or whose phase coefficient is large. The second cannot give good results for satellites whose orbital lightcurve amplitude is large, since a large amplitude implies a heterogeneous distribution of surface materials of different albedos, and probably different phase coefficients. The assumption of a sinusoidal orbital lightcurve, while adequate for some satellites (Rhea and Iapetus, for instance) is definitely inappropriate for others (*e.g.*, Callisto and Ganymede).

It is also undesirable to assume that the shape of the orbital lightcurve is independent of either wavelength or phase angle. In other words, the reduction procedure should not assume that there is one particular orbital lightcurve which applies at all wavelengths and at all phase angles. The shape of Io's orbital lightcurve depends very strongly on wavelength (Johnson, 1969; Morrison *et al.*, 1974; Chapt. 16, Morrison and Morrison), and Callisto provides an example of a satellite whose orbital lightcurve depends significantly on solar phase angle. According to Harris (1961) and Johnson (1971), near $\alpha = 8°$ the amplitude of Callisto's lightcurve is about 0.3 mag in V but becomes imperceptible for $\alpha < 1°5$—the larger opposition effect of the darker side compensating for the albedo differences at $\alpha = 8°$. More observations are needed to confirm this unusual behavior.

Simplifying assumptions will still be necessary in future exploratory work, especially for the fainter satellites. For bright satellites, however, phase curves and phase coefficients should be measured as functions of orbital longitude θ and wavelength λ—a relatively easy observational task (Morrison *et al.*, 1974; Franklin and Cook, 1974).

Determining orbital lightcurves uniquely is more difficult, but this can be achieved at certain phase angles. For example, for Jupiter and Saturn there are periods of about two months during each apparition in which the solar phase angle varies by less than $1°0$, and periods of about two weeks in which it changes by only $0°1$. Thus, the orbital lightcurves of the Galilean satellites and of Saturn's inner satellites could be obtained at almost constant solar phase angle (about 12° for the Jupiter satellites and 6° for those of Saturn). These lightcurves need not necessarily apply to other solar phase angles, as the case of Callisto suggests.

Finally, it should be noted that as some of the smaller, fainter satellites are studied in detail, two standard prejudices may have to be abandoned in analyzing their light variations, viz. synchronous rotation and spherical shape (cf. Chapt. 6, Peale). Already there is some evidence that J6 and Phoebe may not rotate synchronously (Andersson, 1974), while some of the satellites to be studied intensively in the near future (J5 and Mimas, for example) are sufficiently small that they could be irregular in shape as Phobos and Deimos are known to be (Chapt. 15, Duxbury). A close irregular satellite would probably be in synchronous rotation and hence show two brightness maxima and minima per revolution; it may also be tidally distorted (Soter and Harris, 1976).

Phase Integrals

By definition (Russell, 1916), the phase integral of a satellite or planet is given by:

$$q = 2\int_0^\pi \phi(\alpha) \sin \alpha \, d\alpha \ ,$$

where α is the phase angle and $\phi(\alpha)$ is the disk-integrated brightness of the planet at a phase angle α relative to its brightness at opposition. The parameter depends on how the scattered light is distributed over the entire sky; therefore its value cannot be determined from observation of $\phi(\alpha)$ over a limited range of phase angles near opposition. For the satellites of the outer planets only such a small range of phase angles can be observed from Earth, and therefore the phase integrals of these objects are uncertain.

Note that q is a function of wavelength: $q(\lambda)$. Generally, one would expect q to be significantly wavelength dependent for materials whose spectral curves are steep; for grey materials the wavelength dependence of q should be weak.

For some satellites the equivalent bolometric value of q can be derived from radiometry (cf. Chapt. 12, Morrison). If we know the radius of a satellite, its effective bolometric Bond albedo can be derived from radiometry under certain assumptions. Also, since the radius is known, the effective geometric albedo can be calculated and the effective value of q found (Morrison, 1973b; Chapt. 12, herein):

$$\bar{q} = \bar{A}_B/\bar{p} \ .$$

Note that these are effective bolometric quantities.

Using this procedure, Hansen (1972) finds $\bar{q} = 0.8$ for both Io and Europa; Morrison (1973b) derives $\bar{q} = 1.0 \pm 0.3$ for Io, and $\bar{q} = 1.0 \pm 0.1$ for Europa; and Morrison and Cruikshank (1974) adopted values of $\bar{q} = 0.8$ for Ganymede, and $\bar{q} = 0.6$ for Callisto. Applying similar arguments, Morrison *et al.* (1975) find $\bar{q} \simeq 1.3$ for the bright side of Iapetus. These results are discussed in Chapter 12 by Morrison.

Two approximate methods can be used to estimate upper limits for phase integrals. The first—useful for very bright objects such as Europa—applies the fact that at any wavelength $A_B(\lambda) \leqslant 1$, and therefore

$$q_{max}(\lambda) \leqslant \frac{1}{p(\lambda)} \ .$$

For example, for Europa, $p(0.55\mu m) \sim 0.7$ and $q_{max} \leqslant 1.4$.

The second method yields a useful upper limit on q for texturally complex, dark surfaces. For such surfaces, a magnitude plot of $\phi(\mu)$ against $\mu = \cos\alpha$ is convex downward and has a straight line segment at small phase angles ($10° \leqslant \alpha \leqslant 50°$). Stumpff (1948) noted that if the tangent to this curve near opposition is used as an upper limit on $\phi(\mu)$, then:

$$\frac{d\phi(\alpha)}{d\alpha} \equiv -\beta\alpha \ \text{(in magnitudes)}$$

$$\text{and } q_{max} = 2\int_0^\pi e^{-\left\{\frac{0.4}{\ln\beta\alpha}\right\}} \sin\alpha \ d\alpha = \frac{2(1 + e^{-a\pi})}{(1 + a^2)} \ ,$$

where a $= 52.77\beta$ (mag/deg).

When Stumpff's approximation is valid, this equation gives useful values of q_{max} for $\beta > 0.02$ (see Fig. 9.2). Laboratory studies are needed to see to what

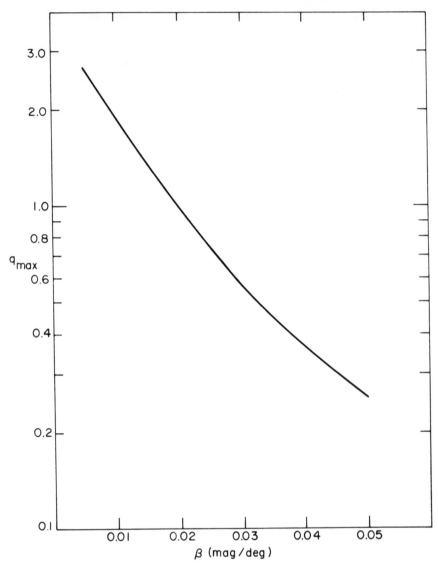

Fig. 9.2. Upper limit on the phase integral, using Stumpff's approximation, as a function of the phase coefficient β.

extent Stumpff's approximation holds for various materials. Present evidence is that it holds, at least, for dark materials. Even though Stumpff's approximation is valid for a Lambert surface, it does not lead to a useful upper limit on q in this case due to the low value of $\beta(\sim 0.002$ mag/deg) (see Fig. 9.2).

Stumpff's approximation could be applied to the dark side of Iapetus if we knew the value of the phase coefficient with the opposition surge removed. Guessing from the data that in the visible, $\beta \gtrsim 0.03$ mag/deg would mean that $q_{max} \lesssim 0.6$.

Pollack *et al.* (1973a) have derived approximate phase functions for Phobos and Deimos up to phase angles of 80°. Over this range, the phase functions are similar to the lunar one. Hence, for Phobos and Deimos, $q \sim 0.6$—the lunar value.

The general futility of deriving phase integrals from observed phase coefficients for satellites of the outer planets can be emphasized by approximating the phase integral using a two point Gaussian quadrature (Veverka, 1971d).

$$q = 2 \int_{-1}^{+1} \phi(\mu) d\mu \simeq 2\{\phi(54°) + \phi(128°)\} \ .$$

Even if $\phi(128°) << \phi(54°)$, one must still estimate the satellite's brightness at 54°. Even in the case of the Galilean satellites the observations extend to only 12°!

Whenever $\phi(128°)$ can be neglected in comparison with $\phi(54°)$:

$$q \sim 2\phi(54°)$$

which is to be compared with an empirical approximation suggested by Russell (1916):

$$q \sim 2.17 \ \phi(50°) \ ,$$

commonly referred to as Russell's phase rule (Harris, 1961).

Geometric Albedos

Geometric albedos for many satellites are listed in several previous reviews (for example, Harris, 1961; Morrison and Cruikshank, 1974; Chapt. 1, Table 1.4, Morrison *et al.*). Here, we wish to stress some of the uncertainties in these determinations. Several crucial steps are involved in calculating a satellite's geometric albedo: (i) The absolute size of the satellite must be known. Unfortunately, many satellite diameters are still uncertain or unknown. (ii) The satellite's observed brightness must be extrapolated to zero phase. Different investigators do this in different ways, often on the basis of insufficient data. (iii) The Sun's brightness over the spectral range in question must be known. Absolute solar fluxes may still be uncertain by several percent.

It is instructive to remember that the adopted geometric albedo of Io, near 0.6μm, dropped by 30% in the time between the reviews of Harris (1961) and Morrison and Cruikshank (1974) mostly due to readjustments in the accepted values of the satellite's radius and in the Sun's brightness.

At present, the geometric albedos of only six satellites (Phobos, Deimos, Io, Europa, Ganymede, and Callisto) are well known; those of two more (Titan and Rhea) are tolerably well determined, and for three others (Iapetus, Dione and Tethys), meaningful approximate values exist.

The principal uncertainty is usually the satellite's absolute size. For Phobos and Deimos, the size is known directly from Mariner 9 imagery (Pollack *et al.*, 1973a; Veverka *et al.*, 1974; Chapt. 14, Pollack; Chapt. 15, Duxbury), while for the Galilean satellites and the larger satellites of Saturn this information has been pieced together from a variety of Earth-based techniques, including micrometric measurements, radiometry, and stellar and lunar occultations (Dollfus, 1970; Morrison and Cruikshank, 1974; Elliot *et al.*, 1975; cf. Chapt. 1 by Morrison *et al.*). In V, the geometric albedos of Phobos and Deimos are about 0.06 (Veverka *et al.*, 1974; Zellner and Capen, 1974); in accordance with Table 16.1 they range from about 0.21 for Callisto to about 0.70 for Europa and Io; for Titan, Rhea, Dione, and Iapetus, the values given by Morrison and Cruikshank (1974) are 0.21, 0.6, 0.6, and about 0.1, respectively. Using the recently determined lunar occultation diameters of Saturn's satellites (Elliot *et al.*, 1975), the geometric albedo of Titan would be about 30% lower and that of Iapetus about 40% higher. Estimates of all geometric albedos are tabulated in Table 1.4.

The very useful information about the composition of a satellite's surface that can be derived from spectral reflectance measurements is reviewed in Chapter 11, by Johnson and Pilcher. In that context, it is also valuable to have an idea of the absolute surface reflectance at a given wavelength. The relationship between the normal reflectance, $r_n(\lambda)$, of a satellite's surface and the satellite's geometric albedo, $p(\lambda)$, will depend on the degree of limb darkening. For a limb-darkened planet, the geometric albedo, as defined by Russell (1916) or Harris (1961), cannot be equal to the normal reflectance of the surface point at the center of the visible disk.

The term "geometric albedo" applies only to a planet and essentially is the ratio of the backscattered energy from the planet compared to that from a *flat* standard disk under the same illumination and viewing conditions. The standard disk consists of a perfect Lambert surface which has a projected area equal to that of the planet and its normal coincides with the backscattering direction.

If the above comparison could be repeated, using—instead of the planet—a flat disk made of the planet's surface material and having the same projected area as the planet, the ratio of the backscattered energy from this disk compared to that from the standard disk is the normal reflectance of the planet's surface. It is this quantity that can be measured in the laboratory.

The definition of r_n is perfectly logical; that of p may not appear to be so at first sight, but its main advantage is that the quantity so defined is simply related to the Bond albedo A_B and to the phase integral q: $A_B = pq$.

For an object like the Moon, which at zero phase shows no limb darkening,

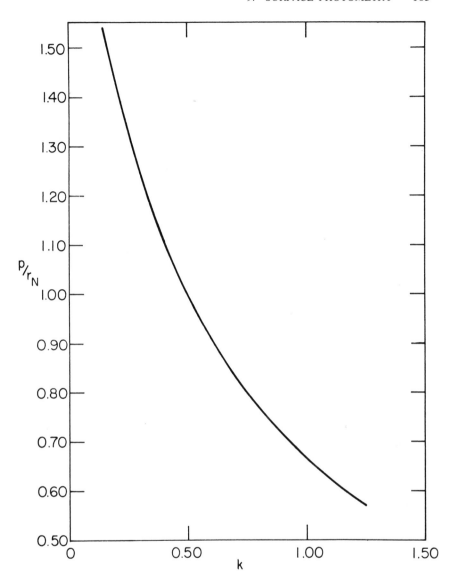

Fig. 9.3. Ratio of the geometric albedo (p) to the normal reflectance (r_n) as a function of the Minnaert exponent k at opposition.

$p = r_n$. For a planet which at zero phase shows Lambert limb darkening, $r_n = 1$ (if the Lambert surface is perfect and absorbs no photons), while $p = \frac{2}{3} r_n$.

If the limb darkening is described by a Minnaert exponent k (i.e., $B = B_0 \cos^k i$, where i is the incidence angle, and $\alpha = 0$), p is related to r_n by

$p = r_n(k + 0.5)^{-1}$; this is shown in Figure 9.3. Since in general $k = k(\lambda)$, even relative spectral reflectance measurements will be affected by this conversion, as has already been noted by McCord and Westphal (1971).

PHOTOMETRY OF SATELLITES

Satellites of Mars

New, recently published measurements of the integrated brightness of Phobos and Deimos by Pascu (1973a) and by Zellner and Capen (1974) are compared with Kuiper's original values in Table 9.2. Since both satellites are irregular (Pollack et al., 1973a; Chapt. 15, Duxbury), they probably show orbital brightness variations, but so far these have not been detected by Earth-based observers. From his photographic measurements, Pascu finds linear phase coefficients of 0.04 mag/deg for Phobos and 0.03 mag/deg for Deimos. The photoelectric observations of Zellner and Capen give a phase coefficient of 0.036 mag/deg for Deimos and an opposition effect of about 0.2 mag. Thus, the phase coefficients of both satellites exceed the lunar value, probably due to both a lower albedo and a rougher surface texture (see also Pollack et al., 1973a).

The data of Zellner and Capen consist of observations on three separate nights. On two nights ($\alpha = 10°.8$ and $18°.5$), Deimos was observed at western elongation. On the third night ($\alpha = 1°.3$), it was observed at eastern elongation. The phase coefficient derived by the authors is based on the assumption that orbital brightness variations are small. Whether or not this is a reasonable assumption cannot be answered with the data in hand.

Satellites of Jupiter

The photometry of the Galilean satellites has been expanded and synthesized in two recent papers (Morrison et al., 1974; Chapt. 16 by Morrison and Morrison). In addition to obtaining an extended series of important new observations, these authors reduced all previous measurements (Stebbins, 1927; Stebbins and Jacobsen, 1928; Harris, 1961; Johnson, 1971; Blanco and Catalano, 1974b) to a common scale. Their results are adopted here as the best available (Tables 9.3, 9.4, 16.1 to 16.4). More recent results by Millis and Thompson (1975) are not discussed.

Photometric information on Jupiter's other satellites is totally inadequate. This situation is especially shocking when one considers that two of these objects, J5 and J6, are moderately bright. The only photoelectric observations of J6 are those of Andersson (1974). He finds V = 14.82 ± 0.04 at mean opposition distance and $\alpha = 2°.5$ The colors are B–V = 0.68, U–B = 0.46, while the phase coefficient between 2° and 8° is about 0.04 mag/deg. For J5 a visual brightness estimate ($+13$) by Barnard exists (van Biesbroeck, 1946); for the remaining satellites only approximate photographic magnitudes are available (Harris, 1961; Morrison and Cruikshank, 1974; cf. Table 1.4).

TABLE 9.2
Satellites of Mars

Satellite	V_0	(B − V)	(U − B)	Observer	Method†	Reference
I Phobos	*11.6±0.1	0.6	—	Kuiper	pe	Harris (1961)
	11.2±0.1	—	—	Pascu	pg	Pascu (1973a)
	11.4±0.2	—	—	Zellner and Capen	pe	Zellner and Capen (1974)
II Deimos	*12.8±0.1	0.6	—	Kuiper	pe	Harris (1961)
	12.34±0.05			Pascu	pg	Pascu (1973a)
	12.45±0.05	0.65±0.03	0.18±0.03	Zellner and Capen	pe	Zellner and Capen (1974)

* Not corrected for phase
† pg = photographic; pe = photoelectric

The orbital lightcurves of the Galilean satellites are intriguing. They are distinctly non-sinusoidal, and their amplitudes and shapes are color dependent. This color dependence is especially striking for Io (Figure 9.4), and probably contains important information about the relative spectral reflectance of the "bright" and "dark" areas on this satellite. [Fig. 9.4 is later considered as Figs. 11.5b and 16.1.]

Generally speaking, the leading sides of the inner three Galilean satellites—Io, Europa, and Ganymede—are brighter than the trailing sides; for the outer satellite, Callisto, the reverse is true. There is also an interesting trend for the color variation with orbital phase to be most pronounced for the satellites closest to Jupiter. Both of these trends are discussed more fully later under "Special Topics."

In view of the non-sinusoidal nature of the lightcurves and of their wavelength dependence, the separation of brightness variations due to changes in orbital phase from those due to changes in solar phase angle should be done separately for each orbital longitude and each wavelength. This can be accomplished by plotting the observed brightness (for given θ and λ) against solar phase angle and finding $\beta(\theta, \lambda)$ as previously discussed. This is essentially the procedure adopted by Morrison and Morrison in Chapter 16.

For Jupiter's satellites the excursion in phase angle is only 12°, but this is sufficient to permit a separation of the "opposition surge" from the "linear part" of the phase curve (Figs. 9.5, 16.5). The results of Morrison and Morrison (Tables 9.3, 9.4, 16.1 and 16.3) are based on the following procedure: Beyond $\alpha > 6°$ a linear phase law is assumed; the opposition effect is modelled by fitting a parabola through the points at $\alpha < 6°$, so that its first derivative is continuous with the straight line segment at $\alpha = 6°$. Three parameters describe the complete

TABLE 9.3

Magnitude Phase Coefficients and Opposition Surges for the Galilean Satellites. Comparison of data found by Morrison and Morrison (Table 16.2) (top), with those inferred by Veverka (1970) from the data of Stebbins (1927) and Stebbins and Jacobsen (1928) (bottom).

	Io	Europa	Ganymede	Callisto (L)	Callisto (T)
$\dfrac{dV}{d\alpha}$ $(\alpha > 6°)$	0.022±0.003	0.006±0.003	0.018±0.002	0.030±0.003	0.030±0.004
$\triangle V$	0.17 ±0.03	0.09 ±0.03	0.07 ±0.03	0.25 ±0.04	0.13 ±0.05
$\dfrac{dM}{d\alpha}$ $(\alpha < 6°)$	0.029±0.005	0.012±0.005	0.023±0.005	0.032±0.005	0.032±0.005
$\triangle M$	0.07 ±0.05	0.07 ±0.05	0.04 ±0.05	>0.32 ±0.05	>0.12 ±0.05

TABLE 9.4

Color Amplitudes and Color Phase Coefficients for the Galilean Satellites (from Morrison et al., 1974)

	Io	Europa	Ganymede	Callisto
Color amplitudes of lightcurves at $\alpha = 6°$				
$\triangle(b-y)$	0.12±0.03	0.04±0.01	0.00±0.02	0.02±0.02
$\triangle(v-y)$	0.32±0.03	0.10±0.02	0.03±0.02	0.06±0.03
$\triangle(u-y)$	0.66±0.05	0.43±0.04	0.23±0.04	0.14±0.05
Color phase coefficients				
$\dfrac{d(b-y)}{d\alpha}$ $(\alpha > 3°)$	0.003	0.006	0.000	0.002
$\dfrac{d(v-y)}{d\alpha}$ $(\alpha > 3°)$	0.005	0.003	0.003	0.004
$\dfrac{d(u-y)}{d\alpha}$ $(\alpha > 3°)$	0.000	0.010	0.007	0.004

phase curve: V at $\alpha = 6°$, dV/dα for $\alpha > 6°$, and V at $\alpha = 0°$. Thus V(α) = V(0) +Aα + Bα^2, and a measure of the opposition surge can be defined:

$$\triangle V = V(6) - V(0) - 6\frac{dV}{d\alpha}$$

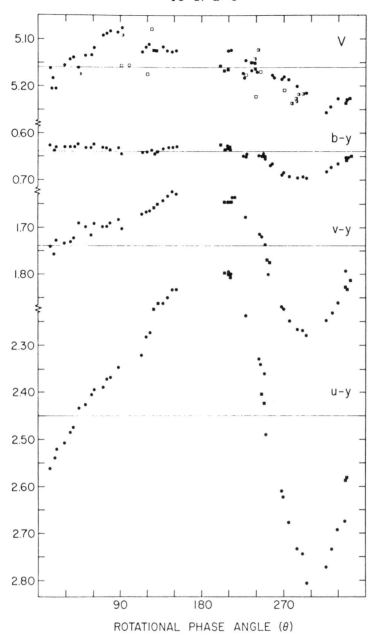

Fig. 9.4. Brightness and color variation of Io with orbital phase (from Morrison *et al.*, 1974). For the V magnitudes, filled symbols are Mauna Kea points, half-filled symbols are from Blanco and Catalano (1974b), and open symbols are from Johnson (1969).

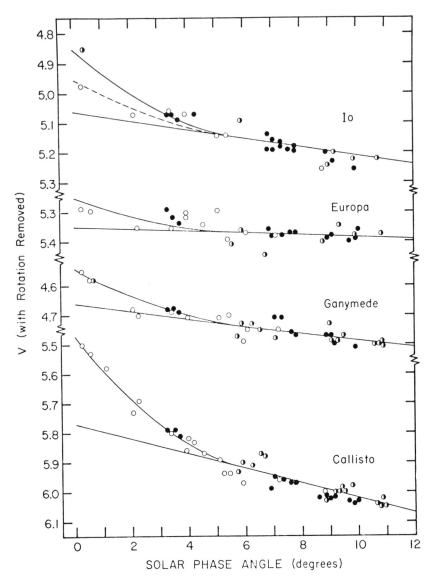

Fig. 9.5. Dependence of mean V magnitude on solar phase angle (α), "corrected" for rotational variations. The symbols are the same as those in Figure 9.4. Another possible fit to the Io data is indicated by the dashed curve.

which is equivalent to linearly extrapolating from $\alpha = 6°$ to $\alpha = 0°$, obtaining $V'(0°)$, and letting

$$\Delta V = V(0°) - V'(0°).$$

[Note that other definitions of opposition surges exist in the literature, cf. Irvine and Lane (1973)]. The bottom two rows in Table 9.3 give values for dM/dα and △M obtained by Veverka (1970) directly from the data of Stebbins (1927) and Stebbins and Jacobsen (1928). Here M is used to denote the satellite magnitude in the photometric system used by Stebbins and Jacobsen, which according to Harris (1961) is closer to the standard B filter than to V. With one exception, the agreement between the values of Morrison and Morrison (Chapt. 16 herein) and those of Veverka (1970) is satisfactory when it is realized that different effective wavelengths are involved. The only significant disparity occurs for Io: the opposition surge derived by Morrison and Morrison is significantly larger than that found by Veverka. The cause of the disagreement is evident in Figure 9.5. The large opposition surge found by Morrison and Morrison results from a single observation by Blanco and Catalano (1974b) which makes Io considerably brighter near α ~ 0° than the measurement of Stebbins and Jacobsen (1928) would suggest. Clearly, more observations of Io at very small phase angles are needed.

Several authors have commented on the phase coefficients of these satellites (Dollfus, 1962; Johnson, 1969, 1971; Veverka, 1970; Chapt. 16, Morrison and Morrison). All Galilean satellites show definite opposition surges. For Ganymede and Europa this surge is comparable to that expected for a frost-covered planet (Veverka, 1973c). For Callisto (and perhaps for Io) the effect is significantly larger.

Stebbins and Jacobsen (1928) first noted that the two sides of Callisto have identical phase coefficients beyond α ≳ 6° but have strikingly different opposition surges—indicating that the texture of the two sides is markedly different, as discussed previously. Away from opposition the trailing side of Callisto is brighter than the leading side; however, near opposition there is almost no difference in the brightness of the two sides since the opposition surge of the dark side is much larger than that of the bright side. This difference in the opposition surges may in part be due to the different albedos of the two sides (the surge being decreased by more multiple scattering on the bright side), but it is more likely that the difference results from a more porous microstructure on the dark side of Callisto.

Except for Europa, the phase coefficients of the Galilean satellites are considerably larger than those expected for smooth frost-covered planets, which would be about 0.002 mag/deg according to Veverka (1973c). For Europa, large scale surface roughness could easily account for the slightly higher value, but the much larger phase coefficients of the other satellites are not compatible with pure frost surfaces.

For satellites with steep spectral curves, such as Io, the phase coefficients and opposition surges must be color dependent. Morrison et al. (1974) have determined color phase coefficients for the Galilean satellites, but the implications of these data are not understood at present. No data on the color dependence of the opposition surges have been published.

It is important to realize that the above discussion is oversimplified in a fundamental way. The Galilean satellites show orbital brightness variations because they have "spotted" surfaces. The surface materials are therefore photometrically inhomogeneous, and the color phase coefficients for different faces of a satellite must differ. Thus, the mean phase coefficients used above correspond to values averaged over all longitudes. It is probable that the dark faces of satellites will have higher phase coefficients than these mean values, while the bright sides will have lower ones.

Since the lightcurve amplitudes in V are small for all four Galilean satellites, the mean values of the phase coefficients are probably close to the actual values for various areas on the satellite surfaces. This will no longer be true at wavelengths for which the lightcurve amplitudes are large.

In general, it is important to remember that the "dark" and "bright" areas on a satellite's surface will have different phase coefficients, and that the mean value usually published is some average of these. Bearing this in mind, the relatively large phase coefficients of Europa and Ganymede (0.006 and 0.018 mag/deg; Table 16.3) can be reconciled with the spectrophotometric fact that the surfaces of these two satellites are covered only partially by water frost (Pilcher et al., 1972; Chapt. 11, Johnson and Pilcher). The larger mean coefficients can be obtained by postulating that the "dark" regions on these satellites, which presumably are not covered by pure frost, have quite sizeable phase coefficients. It should be recalled that according to Dollfus and Murray (1974) the relative contrast of the "dark" spots on Ganymede is much larger than on Europa, while Pilcher et al. (1972) conclude that the water frost cover on Ganymede is less extensive than that on Europa (cf. Gehrels, Chapt. 18). Both facts lead one to expect that the dark area contribution to the mean phase coefficient is larger for Ganymede than for Europa, thus possibly accounting for Ganymede's large phase coefficient.

Satellites of Saturn: Inner Satellites

Due to their proximity to the rings, photometry of the inner satellites of Saturn is difficult. No photoelectric measurements of Janus or Mimas exist. The most recent estimate of the brightness of Mimas is V ~ 12.9 (Koutchmy and Lamy, 1975; cf. also Franz, 1975), while the magnitude of Janus is said to be 13.5 to 14 (Morrison and Cruikshank, 1974).

Typical scans across Enceladus, Tethys and Dione, obtained with a V filter by Franz and Millis (1973, 1975), are shown in Figure 9.6. Scattered light from the rings and planet contributes significantly to the observed signal, and this contamination varies non-linearly with distance from the rings. Conventional correction methods, in which sky measurements are made just "inside" and "outside" the satellite (or just "below" and "above") yield reasonable results only for Dione; for Tethys such photometry is marginal, while for Enceladus it is of almost no value, except near elongation (Noland *et al.*, 1974b). According to Franz and Millis (1973), area scanning yields magnitudes correct to ± 0.05 mag even for Enceladus. The errors in conventional photometry are at least this large for Dione, and larger, in general, for Tethys and Enceladus.

Harris (1961) gave $V_0 = 11.77$ for Enceladus, but Franz and Millis (1973, 1975) find it to be brighter: 11.2 near western elongation and 11.7 near eastern elongation (Table 9.5). These recent results are especially intriguing since they indicate that Enceladus may be 0.5 mag fainter on its leading side than on its trailing side—exactly opposite to the trend for the other inner satellites. Since the observations are limited mostly to elongations, it is impossible to be sure that maximum brightness occurs at western elongation, and not at inferior conjunction. Franz and Millis (1975) believe that there is evidence in their data that Enceladus has a substantial phase coefficient.

Harris' value, $V_0 = 11.27$, for Tethys agrees well with more recent determinations (Table 9.5). With a single exception, available data indicate that the leading side of Tethys is slightly brighter than its trailing side. McCord *et al.* (1971) were the first to detect small brightness variations related to orbital position for this satellite. Subsequently Franz and Millis (1973) announced an amplitude of 0.3 mag in V, and a slightly non-synchronous rotation period (1.8860 days compared with the mean synodic period of 1.8875 days). Considering the number of observations, the uncertainty in each measurement, and the fact that solar phase angle effects were not corrected for in the Lowell data, this conclusion was not compelling and has recently been abandoned (Franz and Millis, 1975). Blair and Owen (1974) found no evidence of orbital brightness variations (to ± 0.2 mag) in their observations using the McDonald area scanner.

The Mauna Kea (MKO) observations of Noland *et al.* (1974b) suggest that the leading side of Tethys is brighter than the trailing side. The formal least-squares fitted lightcurve amplitudes are 0.1 to 0.2 mag (for various colors)—a result which can be reconciled with the other observations referred to above.

The phase coefficients of Tethys are not well determined from the Mauna Kea data and should be considered uncertain. Typical values are 0.02 mag/deg, in agreement with recent data of Franz and Millis (1975).

Dione is the only one of inner satellites for which adequate photometry is available. The agreement among observers as to the value of V_0 is good (Table 9.5), and all observers agree that the leading side is brighter than the trailing side (McCord *et al.*, 1971; Franz and Millis, 1973, 1975; Blair and Owen, 1974; Noland *et al.*, 1974b). There is disagreement, however, as to the amplitude of the lightcurve. The fragmentary data of McCord *et al.* (1971) indicate an amplitude of at least 0.2 mag, while Franz and Millis (1973) suggest 0.4 mag. Blair and Owen (1974) conclude that the lightcurve is not sinusoidal, and that the

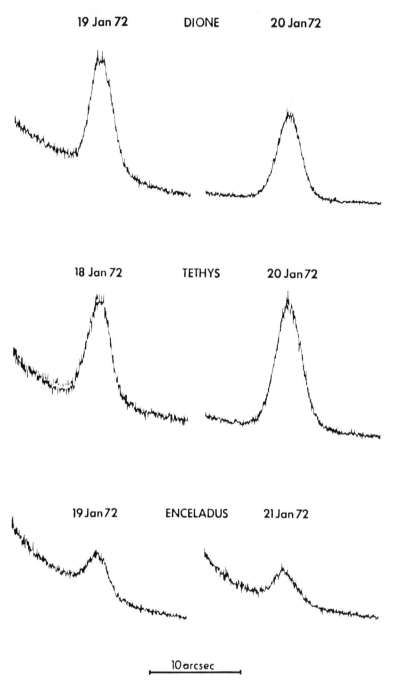

Fig. 9.6. Area scanner profiles across Saturn's satellites obtained by Franz and Millis (1973) in V.

TABLE 9.5
Inner Satellites of Saturn

Satellite	Reference	V_0	$(B-V)$	$(U-B)$
Enceladus	Harris (1961)	11.77*	0.62	
	Franz and Millis (1973; 1975)	11.2 -11.7*		
Tethys	Harris (1961)	10.27*	0.73	0.34
	Franz and Millis (1973)	10.25-10.65*		
	Blair and Owen (1974)	10.28*	0.75	0.32
	Noland *et al*. (1974b)	10.22		
	Franz and Millis (1975)	10.27		
Dione	Harris (1961)	10.44*	0.71	0.30
	Franz and Millis (1973)	10.05-10.40*		
	Blair and Owen (1974)	9.90-10.80*	0.82	0.27
	Noland *et al*. (1974b)	10.38		
	Franz and Millis (1975)	10.46		

* Indicates no correction for solar phase effects

amplitude between $180° \leq \theta \leq 360°$ may be larger than that between $0° \leq \theta \leq 180°$. Between $0 \leq \theta \leq 240°$, their data are well represented by an amplitude of 0.2 to 0.3 mag. However, near $\theta = 270°$ Blair and Owen (1974) find an amplitude close to 0.8 mag, whereas the data of Franz and Millis (1973) show a total amplitude of only 0.4 mag. There are too few Mauna Kea measurements near $\theta = 270°$ to settle this question, but the latest observations by Franz and Millis (1975) favor the smaller amplitude.

The MKO photometry (Noland *et al.*, 1974b) yields phase coefficents for Dione which, although not well determined, are smaller than those for Rhea (see below). Recently Franz and Millis (1975) derived a phase coefficient of 0.033 mag/deg in V—about twice the value quoted by Noland *et al.* (1974b) for this spectral range. This discrepancy remains unresolved. It is possible, however, that the phase coefficients derived by Noland *et al.* for this satellite are systematically low since the analysis assumed sinusoidal brightness variations with orbital phase. As mentioned above, Dione's lightcurve is probably not sinusoidal.

Satellites of Saturn: Rhea

Good photometry of Rhea is now available from several sources. All observers agree that the leading side is bright and the trailing side is somewhat darker; there is only slight disagreement as to the lightcurve amplitude (Table 9.6). The measurements of Harris (1961), McCord *et al.* (1971) and Blair and Owen (1974) are too sparse to define θ_{min} and θ_{max} adequately, but agree with the more extensive observations published by Blanco and Catalano (1971). (N.B., Blanco and Catalano define $\theta = 0°$ to be eastern elongation, contrary to common practice.)

TABLE 9.6
Lightcurve Parameters for Rhea

Reference	Lightcurve Amplitude (Δm)	θ_{max}	θ_{min}	V_0	Comment
Harris (1961)	0.2*	~30	~240	9.76*	
McCord et al. (1971)	~0.3*	90±30	270±30	9.56*	Based on narrow band photometry at 0.56μm
Blanco and Catalano (1971)	0.23*	90°	270°	9.73*	
Blair and Owen (1974)	0.2*	50±50	250±50	9.69*	
Noland et al. (1974b)	0.19±0.02	90±5°	270±5°	9.67	Based on narrow band photometry at 0.55μm

* No correction for solar phase angle effects.

The most extensive photometry of Rhea is that of Noland et al. (1974b) who find a sinusoidal brightness variation of 0.2 mag. At all wavelengths between 0.35μm and 0.75μm, maximum brightness occurs close to $\theta = 90°$ and minimum brightness occurs near $\theta = 270°$. Below 0.5μm, the lightcurve amplitude increases with decreasing wavelength (Table 9.7). In terms of a two-component model of the surface, this result indicates that the contrast between the bright and dark areas increases shortward of 0.5μm.

One of the most significant results of the MKO photometry is a determination of Rhea's phase coefficients. The values range from 0.024 mag/deg in the red to 0.037 mag/deg in the ultraviolet (Table 9.7). Note that these are linear phase coefficients determined near opposition, $\beta(0°–6°)$. Over this range of phase angles non-linear opposition surges are probably important. In fact, Franklin and Cook (1974) do find evidence indicating an "opposition effect" for Rhea at very small phase angles (Figure 9.8).

The large values of the phase coefficients found by Noland et al. are of special interest since Rhea is not a dark object. Morrison and Cruikshank (1974) esti-mate its geometric albedo to be ~ 0.6, which means that the surface texture of Rhea must be very porous to produce such large β's in spite of the importance of multiple scattering within the surface layer (Noland et al., 1974b). High surface porosity is also required to explain the relatively deep negative polarization branch (Bowell and Zellner, 1974) of this high albedo object.

Satellites of Saturn: Titan

A number of important photometric studies of this intriguing satellite (cf. Chapt. 20, Hunten; Chapt. 21, Caldwell; Chapt. 22, Andersson) have been

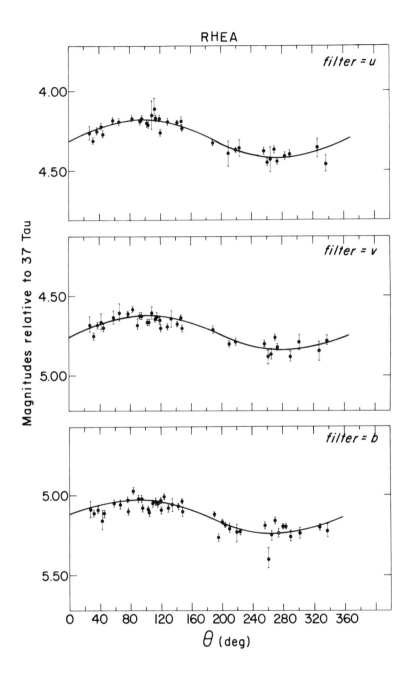

Fig. 9.7. Orbital magnitude variations of Rhea with solar phase angle effects removed. (From Noland *et al.*, 1974b.)

TABLE 9.7

Lightcurve Parameters for Rhea determined by Noland *et al.* (1974b)

λ(μm)	0.35	0.41	0.47	0.55	0.62	0.75
β (mag/deg)	.037±.003	.031±.003	.027±.003	.025±.002	.025±.003	.024±.004
$\triangle$m	0.24±.02	0.20±.02	0.18±.02	0.19±.02	0.19±.02	0.20±.02
θ_{max}	97±5	102±8	90±7	90±5	89±7	93±9
θ_{min}	277±5	282±8	270±7	270±5	269±7	273±9

published (Blanco and Catalano, 1971; Noland *et al.*, 1974b; Franklin and Cook, 1974). A review of much of this material exists already (Veverka, 1974), and only the main points will be touched upon here.

There appear to be no significant brightness or color variations associated with orbital phase, presumably because of Titan's cloudy atmosphere. Phase coefficients obtained by Noland *et al.* (1974b) at wavelengths ranging from 0.35μm to 0.75μm are shown in Figure 9.9. Again, these are linear phase coefficients measured between 0° and 6°, but there is no indication of any non-linear opposition surge close to $\alpha = 0°$. Andersson (Chapt. 22 herein) quotes a phase coefficient of 0.004 mag/deg in V, while the data of Blanco and Catalano (1971) give 0.006—values in agreement with those in Figure 9.9. The wavelength dependence of Titan's phase coefficient is unusual and can be used to discriminate among possible models of Titan's atmosphere (Noland and Veverka, 1974, 1975). Other important information is contained in the satellite's spectral reflectance curve, which is dominated by deep methane absorption bands beyond 0.6μm and by a steep drop in reflectance below this wavelength (Chapt. 11, Johnson and Pilcher).

Evidence has been accumulating that Titan shows long period variations in both brightness and color (Chapt. 22, Andersson; Noland *et al.*, 1974a). Although such variations are not surprising for a cloud-covered satellite, their precise explanation is not yet understood. According to Andersson, Titan has brightened from V = +8.39 to +8.26 between 1956 and 1974, while its (B–V) color appears to have changed from +1.30 to +1.25.

Satellites of Saturn: Iapetus

The photometry of this unusual satellite was reviewed recently by Morrison *et al.* (1975). The bulk of the new data is based on observations by Millis (1973), Noland *et al.* (1974b) and Franklin and Cook (1974) and is compiled in Figures 9.10–9.12 and 11.11.

Morrison *et al.* find that the large orbital brightness variation (Figure 9.10) is explained adequately in terms of a two hemisphere model in which the bright trailing side is about 6 times brighter than the dark leading side. According to

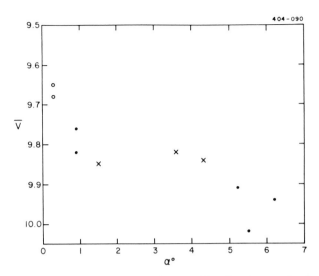

Fig. 9.8. Variation of the V magnitude of Rhea with solar phase angle, for $\theta = 270° \pm 30°$. Dots are from Noland *et al.* (1974b), crosses from Blanco and Catalano (1971), and open circles from Franklin and Cook (1974). (From Franklin and Cook, 1974.)

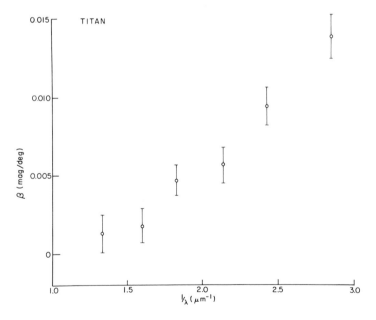

Fig. 9.9. Wavelength dependence of Titan's phase coefficient. (From Noland *et al.*, 1974b.)

their analysis, the linear phase coefficient $\beta(0°-6°)$ is a strong function of longitude and varies from 0.028 mag/deg near $\theta = 90°$ (bright side) to 0.068 mag/deg
near $\theta = 270°$ (dark side). A linear phase law adequately represents most of the
bright side data, although Franklin and Cook (1974) have called attention to a
possible non-linear surge of about 0.1 mag below $\alpha = 2°$. The few dark-side
points suggest a pronounced non-linear surge close to opposition (Figure 9.11).

The linear phase coefficients $\beta(0°-6°)$ published by Noland et al. (1974b) are
hemispherical averages for the dark and bright sides (Figure 9.12); they are
therefore smaller than the extreme values given by Morrison et al. The dark-side
phase coefficients are strongly wavelength dependent; their large values indicate
that the surface texture of the dark side of Iapetus is very porous.

It is interesting that the large orbital brightness variation is associated with
only minor color changes: $\triangle(U-V) \lesssim 0.1$ (Millis, 1973; Noland et al., 1974b;
Morrison et al., 1975), indicating that the shape of the spectral reflectance
curves of the bright-side and dark-side materials may be similar (McCord et al.,
1971; Noland et al., 1974b).

Satellites of Saturn: Hyperion and Phoebe

Harris (1961) gives $V_0 = 14.16$, $B-V_0 = +0.69$ and $U-B = +0.42$ for
Hyperion. Measurements on two nights in B and V by Franklin and Cook (1974)
are consistent with these values. The data are insufficient to look for variations
with either orbital or solar phase angle.

According to Andersson (1974) the magnitude of Hyperion, its colors, and its
phase coefficient are $V = 14.24$, $B-V = 0.78$, $U-B = 0.33$ and $dV/d\alpha = 0.02$
mag/deg, all at $\alpha = 4°$. These colors differ from those of Harris (1961), and
Andersson suggests that this may be because Harris observed mainly the northern hemisphere of the satellite, while the recent photometry has measured mostly
the southern hemisphere. Light variations due to rotation reportedly do not
exceed 0.1 mag (Andersson, 1974).

Observations of Phoebe on four nights in Nov. 1971 by Andersson (1974) are
consistent with a rotational lightcurve of 0.2 mag amplitude and a period of either
11.3 or 21.5 hours. Andersson finds $V = 16.38 \pm 0.05$, $B-V = +0.66 \pm 0.02$,
$U-B = 0.33 \pm 0.03$ and a phase coefficient $dV/d\alpha = 0.10$ mag/deg, all at
$\alpha = 2°$.

Satellites of Uranus

Available observations are few. Harris (1961) measured photoelectric magnitudes for Titania and Oberon, and quoted photographic magnitudes for
Miranda, Ariel, and Umbriel determined by Gehrels (Table 9.8). Andersson
(1974) has obtained new, more extensive photoelectric measurements of Titania
and Oberon (Table 9.8) which agree quite well with those of Harris. Sufficient data to look for solar phase angle effects (opposition effects) are not
available yet.

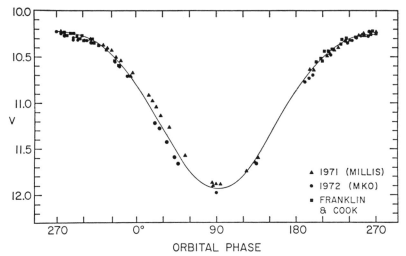

Fig. 9.10 Observed Iapetus lightcurves from 1971–73 compared with the model curve of Morrison *et al.* (1975). Triangles are from Millis (1973), circles from Noland *et al.* (1974b), squares from Franklin and Cook (1974). (From Morrison *et al.*, 1975.)

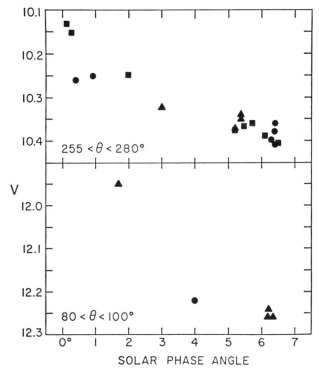

Fig. 9.11. Observed V magnitudes of Iapetus near eastern and western elongation, plotted as a function of solar phase angle. (From Morrison *et al.*, 1975.)

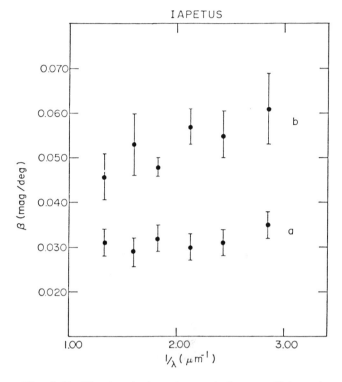

Fig. 9.12. Wavelength dependence of phase coefficients for Iapetus: (a) bright side average; (b) dark side average. (From Noland *et al.*, 1974b.)

Andersson (1974) finds that Titania may possibly show orbital brightness variations, and that Oberon definitely does. For this outer satellite, the trailing side is about 0.1 mag brighter than the leading side.

Satellites of Neptune

The only available measurements are still those of Harris (1961) (Table 9.9). New observations of Triton should be undertaken now that area scanners are available. According to Harris, the leading side of this satellite is 0.2 to 0.3 mag brighter than the trailing side.

SPECIAL TOPICS

Orbital lightcurves: Leading side/trailing side brightness trends

The inner satellites of Jupiter and Saturn tend to have leading sides which are brighter than their trailing sides. The reverse is true for the outermost large

TABLE 9.8
Satellites of Uranus

Satellite	V_0	$(B-V)$	$(U-B)$	Observer	Method*	Reference
V Miranda	16.5	—	—	Gehrels	pg	Harris (1961)
I Ariel	14.4	—	—	Gehrels	pg	Harris (1961)
II Umbriel	15.3	—	—	Gehrels	pg	Harris (1961)
III Titania	14.01	0.62	0.25	Harris	pe	Harris (1961)
	14.01±0.06	0.71±0.11	0.30±0.16	Andersson	pe	Andersson (1974)
IV Oberon	14.20	0.65	0.24	Harris	pe	Harris (1961)
	14.27±0.01	0.68±0.03	0.20±0.06	Andersson	pe	Andersson (1974)

* pg = photographic; pe = photoelectric

TABLE 9.9
Satellites of Neptune

Satellite	V_0	$B-V$	$U-B$	Observer	Method	Reference
I Triton	13.55	0.77	0.40	Harris	pe	Harris (1961)
II Nereid	~18.7*	—	—	Kuiper	pg	Harris (1961)

* Based on a photographic magnitude of 19.5 and an assumed $B-V = 0.8$.

satellites: Jupiter's Callisto and Saturn's Iapetus. In fact, Blanco and Catalano (1974a) have proposed the following general law for the orbital lightcurves of the satellites of Jupiter and Saturn:

$$\theta_{max} = C_1(d)^{-\frac{1}{2}} + C_2$$

where θ_{max} = orbital longitude of maximum brightness, d = distance from planet (in planet radii), and C_1 and C_2 are constants. An analogous relationship was proposed for θ_{min}, the longitude of the lightcurve minimum. The validity of this proposed relation is discussed below; first, we will review the observational facts.

In his dissertation, Johnson (1969) cautioned against speaking loosely of bright sides and dark sides, and showed that the lightcurves of the Galilean

TABLE 9.10
Lightcurve Data for the Galilean Satellites

Reference	θ_{min}	θ_{max}	Remarks
Io			
Harris (1961)	320±30	150±30	NS
Johnson (1969)	320±20	130±20	NS (0.56μm)
Blanco and Catalano (1974a,b)	345*	120*	NS
	315±45	120±45	
Morrison *et al.* (1974)	300±20	100±20	NS
Europa			
Harris (1961)	280±30	100±30	NS
Johnson (1969)	290±20	90±20	NS (0.56μm)
Blanco and Catalano (1974a,b)	280*	45*	NS
	280±20	70±45	
Morrison *et al.* (1974)	280±10	80±10	NS
Ganymede			
Harris (1961)	220±60	60±30	NS
Johnson (1969)	210±30	60±30	NS (0.56μm)
Blanco and Catalano (1974a,b)	240*	60*	NS
	280±20	70±30	
Morrison *et al.* (1974)	270±20	60±10	NS
Callisto			
Harris (1961)	120±40	270±30	NS
Johnson (1969)	170±30	230±30	NS (0.56μm)
Blanco and Catalano (1974a,b)	110*	300*	NS
	90±20	280±20	
Morrison *et al.* (1974)	110±30	270±10	NS

Note: Values of θ_{max} and θ_{min} in V, read from quoted data by present author. NS in the last column indicates that it is clear from the observations that the lightcurve is not sinusoidal. Two sets of values are listed next to Blanco and Catalano. The values followed by the asterisk are those found in Blanco and Catalano (1974a,b). The others were read off by the present author from the graph of Blanco and Catalano. See also Table 16.5 herein.

satellites are definitely non-sinusoidal. He plotted $\theta_0(\downarrow)$, the longitude for which the lightcurve achieves the mean magnitude on its way to minimum brightness, against the satellite's distance from Jupiter. At 0.56μm, he found that $\theta_0(\downarrow)$ increases with distance from Jupiter, from about 60° for Io to about 270° for Callisto. Since the lightcurves are non-sinusoidal, the corresponding longitude $\theta_0(\uparrow)$ does not equal $\theta_0(\downarrow)$ + 180° in general (see also Chapt. 16, Morrison and Morrison).

For simplicity, the following discussion deals only with θ_{max} and θ_{min}, although for a general non-sinusoidal lightcurve all four quantities: θ_{max}, θ_{min}, $\theta_o(\uparrow)$ and $\theta_o(\downarrow)$ are independent. Only observations made in V (or close equivalents) are discussed, and the dependence of θ_{max} and θ_{min} on either wavelength or solar phase angle is not considered. This simplification is excusable as an approximation to reality, but evidence exists that the shapes of some orbital lightcurves change with both λ and α.

The available data for the Galilean satellites are gathered in Table 9.10. To a first approximation, the inner three satellites have brighter leading hemispheres, while Callisto has a brighter trailing hemisphere. Johnson (1969) feels that the data show a progressive shift through 180° in θ_{max}, moving from Io to Callisto (here called the Johnson hypothesis). Morrison and Cruikshank (1974) conclude that the data are equally consistent with $\theta_{max} = 90°$ for Io, Europa and Ganymede, but with $\theta_{max} = 270°$ for Callisto (here called the Morrison-Cruikshank hypothesis). Blanco and Catalano (1974a,b) definitely side with Johnson, but their interpretation of their own data is perplexing to me. In a publication they adopt values of $\theta_{max} = 143°$, $78°$, $50°$ and $280°$ for the Galilean satellites going from Io to Callisto—values which do not agree well with those of Morrison *et al.* (1974). But, the values which I have read from the graphs published by Blanco and Catalano (cf. Table 9.10) do agree.

Since the graphs published by Morrison *et al.* incorporate the earlier data of Johnson and of Blanco and Catalano, the values of θ_{max} and θ_{min} read from them are adopted here, and have been plotted against the satellite's distance from Jupiter in Figure 9.13. It is important to keep in mind that these values may not hold for wavelengths far removed from those included by the V filter.

Figure 9.13 shows that for the Galilean satellites (in V), θ_{min} and θ_{max} do not always occur at 90° or 270°. Thus, the Morrison-Cruikshank hypothesis is not valid in this case. The gradual decrease of θ_{min} with d (Figure 9.13) is consistent with Johnson's view, but the behavior of θ_{max} is apparently complicated. There is no convincing support in Figure 9.13 for the Blanco and Catalano relationship.

Visual observations of Amalthea (J5) by van Biesbroeck (1946) suggest that its leading side may be brighter than the trailing side; this would be consistent with the trend in Figure 9.13.

Values of θ_{max} and θ_{min} (in V) for Saturn's satellites are listed in Table 9.11. Titan shows no regular orbital brightness variations (see for example, Noland *et al.*, 1974b), and there is general agreement that this is due to its atmosphere. All observers also agree that for Iapetus $\theta_{min} = 90°$ and $\theta_{max} = 270°$, to better than 10°.

Reasonable values of θ_{min} and θ_{max} can be derived for Rhea (Table 9.11), but then the situation deteriorates rapidly as the satellites get closer to Saturn and its rings. For Dione the observations are adequate to show that $\theta_{min} \sim 270°$ and $\theta_{max} \sim 90°$; this also seems to be true for Tethys but better observations are required. More observations of Enceladus are also needed to verify that its "trailing" side

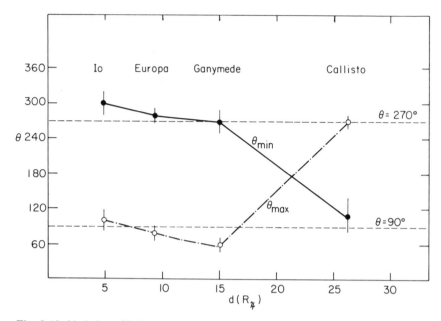

Fig. 9.13. Variation of lightcurve parameters θ_{min} and θ_{max} for the Galilean satellites with distance from Jupiter.

is brighter than its "leading" side (Table 9.11). Brightness differences between the eastern and western ansae of the rings have also been noted (Morrison and Cruikshank, 1974).

The Saturn satellites provide no evidence for the relationship postulated by Blanco and Catalano (1974a). The only values of θ_{min} and θ_{max} that occur seem to be 90° or 270°. Consider Rhea, for example: $\theta_{min} = 270° \pm 5°$ and $\theta_{max} = 90° \pm 5°$, and not 250° and 60° respectively, as required by Blanco and Catalano.

Observations of the satellites of Uranus and Neptune are too scarce to determine θ_{min} and θ_{max}. Harris (1961) suggests that Triton's leading side is brighter than the trailing side by 0.25 mag; Andersson (1974) reports that the trailing side of Oberon may be brighter by 0.1 mag than the leading side.

Orbital Lightcurves: Color Amplitude Trends

For the Galilean satellites the lightcurve amplitudes tend to increase with decreasing wavelength; this color effect is increasingly more pronounced as the satellite's distance from Jupiter decreases. The closer satellites tend to have redder dark sides (Table 9.12; Figure 9.14). For Io, $\Delta(u–y)$ reaches 0.66 mag. If this red color results from Io's interaction with the Jovian magnetosphere, the innermost satellite Amalthea (J5), at only 2.5 R_J, may show even larger color differences—provided that its surface materials are as susceptible to the effects of charged particles as those of Io seem to be (cf. Chapt. 17, Fanale *et al.*).

TABLE 9.11
Lightcurve Data for Saturn's Satellites

Satellite	Reference	θ_{min}	θ_{max}	Remarks
Enceladus	Franz and Millis (1975)	$\sim 90°$ (?)	$\sim 270°$ (?)	More observations needed.
Tethys	McCord et al. (1971)	$\sim 270°$	$\sim 90°$	Not well determined.
	Blair and Owen (1974)	?	?	Not well determined.
	Noland et al. (1974b)	$\sim 270°$	$\sim 90°$	Not well determined.
	Franz and Millis (1975)	$\sim 270°$	$\sim 90°$	Not well determined.
				Conclusion: $\theta_{min} \sim 270°$; $\theta_{max} \sim 90°$ but not well determined.
Dione	McCord et al. (1971)	$\sim 270°$	$\sim 90°$	Not well determined.
	Blair and Owen (1974)	$250° \pm 30°$	$90° \pm 30°$	
	Noland et al. (1974b)	$290° \pm 20°$	$120° \pm 30°$	
	Franz and Millis (1975)	$270° \pm 30°$	$90° \pm 30°$	
				Conclusion: $\theta_{min} \sim 270°$, $\theta_{max} \sim 90°$.
Rhea	Harris (1961)	$\sim 240°$	$\sim 30°$	Not well determined.
	McCord et al. (1971)	$270° \pm 30°$	$\sim 90°$	θ_{max} not well determined.
	Blanco and Catalano (1971)	$270° \pm 10°$	$90° \pm 10°$	Well determined.
	Blair and Owen (1974)	?	?	Not well determined.
	Noland et al. (1974b)	$270° \pm 5°$	$90° \pm 5°$	Well determined.
				Conclusion: $\theta_{min} = 270° \pm 5°$; $\theta_{max} = 90° \pm 5°$
Iapetus	Millis (1973)	$90°$	$270°$	Well established.
	Noland et al. (1974b)	$90°$	$270°$	*Conclusion:* $\theta_{min} = 90° \pm 5°$, $\theta_{max} = 270° \pm 5°$.
	Morrison et al. (1975)	$90°$	$270°$	

By contrast, in the Saturn system the color amplitudes of the lightcurves are small. Millis (1973) finds $\triangle$(U–V) ~ 0.1 for Iapetus, with the dark side redder—in agreement with the data of Noland *et al.* (1974b). For Rhea, $\triangle$(u–y) = 0.05, in the same sense as for Iapetus (Noland *et al.*, 1974b). The lightcurves of Dione and Tethys are sufficiently uncertain to allow only upper limits of $\triangle$(u–y) ≤ 0.1 to be placed on their color amplitudes. It is inappropriate to consider Titan in this context since its surface is probably always obscured by clouds. The conclusion is that in Saturn's satellite system the dark hemispheres of satellites do not appear to get progressively redder as the planet is approached, and the color variations with orbital phase are always small (see also Johnson, 1969; Chapt. 11, Johnson and Pilcher). High quality measurements of Enceladus would be of interest in this context.

In terms of a two-component model of satellite surfaces, one can say that the colors of the dark and bright areas on Saturn's satellites are similar, and that these colors do not change significantly with the satellite's distance from the planet. On the Galilean satellites, however, the dark areas are much redder than the bright areas and this relative redness increases as the satellite's distance from Jupiter decreases (Johnson, 1969). A possibly related fact is that the Galilean satellites have steeper spectral reflectance curves than the satellites of Saturn, excluding Titan (McCord *et al.*, 1971; Johnson and McCord, 1971; Chapt. 11, Johnson and Pilcher). If an explanation is sought in terms of some ubiquitous absorber of ultraviolet and blue light, then this substance is most abundant in the Jovian satellite system—especially close to the planet—but is almost absent in the Saturn system.

Secular Variability of Satellites

As discussed earlier, strong evidence exists that both the color and brightness of Titan change on time scales of years (Chapt. 22, Andersson; Lockwood, 1975; Noland *et al.*, 1974a; Jones and Morrison, 1975). The variations probably are caused by atmospheric fluctuations—changes in the amount of cloud cover, perhaps.

The possibility that satellites with negligible atmospheres may also be variable should be kept in mind. Such variations could be due to changes in the surface albedo caused by the production (or destruction) of surface molecules following perturbations of the satellite's environment. For example, it is possible that the post-eclipse brightening of Io, reported by certain observers (see the review by Morrison and Cruikshank, 1974), is caused by chemical changes in the surface and not by physical changes, such as the sublimation of frost, or variations in surface reflectance with temperature. Radiation belt charged particles could produce substances on the surface during the eclipse which are unstable when exposed to sunlight.

Aside from Titan, Io is the only satellite which is suspected of being variable.

TABLE 9.12
Lightcurve Color Amplitudes for Satellites of Jupiter and Saturn

	Satellites of Jupiter			Satellites of Saturn	
	Distance (R_J)	$\triangle(u-y)$		Distance (R_S)	$\triangle(u-y)$
Io*	5.9	0.66±0.05	Tethys**	4.9	<0.10
Europa*	9.4	0.43±0.04	Dione**	6.3	<0.10
Ganymede*	15.0	0.13±0.04	Rhea**	8.8	0.05±0.04
Callisto*	26.4	0.14±0.05	(Titan)§	20.5	<0.02
			Hyperion†	24.8	?
			Iapetus†*	59.7	~0.1

Note: $\triangle(u-y)$ is the difference in the lightcurve amplitude in u (0.35μm) and y (0.55μm).

 * Morrison *et al.* (1974)   † No data available
 ** Noland *et al.* (1974b)   †* Based on Millis (1973) $\triangle$(U–V)~0.1.
 § Excluded because of atmosphere

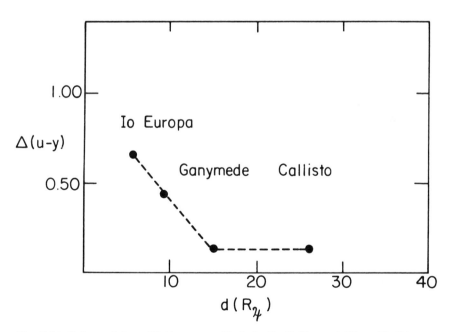

Fig. 9.14. Color variation of lightcurve amplitude for the Galilean satellites with distance from Jupiter.

208   J. VEVERKA

Blanco and Catalano (1974b, 1975) believe that since the observations of Stebbins and Jacobsen (1928) the mean opposition V-magnitude of Io has increased by about +0.2 mag. Whether this apparent change is real, or whether it results from errors in transforming the old observations of Stebbins and Jacobsen to the V system is uncertain (Chapt. 16, Morrison and Morrison).

FUTURE WORK

Satellites of Mars

Spectrophotometry of Deimos, and perhaps even of Phobos, should be possible. This information would be helpful in resolving the question of the surface composition of these satellites (Pollack et al., 1973; Chapt. 14, Pollack).

Satellites of Jupiter

Disk-integrated photometry of the Galilean satellites should be extended by obtaining phase curves at various wavelengths for specific orbital longitudes. Measurements at very small phase angles and at short wavelengths are most important. The results should be combined with the steadily increasing store of photometry of individual surface areas from visual and photographic observations (Dollfus and Murray, 1974; Murray, 1975), spacecraft imaging (Gehrels et al., 1974; Chapt. 18, Gehrels) and mutual events (Murphy and Aksnes, 1973; Vermilion et al., 1974; Greene et al., 1975; Chapt. 3, Aksnes). The aim should be to determine the composition and texture of various regions on the satellites.

Time variations of the Galilean satellites should be looked for carefully, and the controversy concerning the post-eclipse brightening of Io should be resolved by simultaneous observations of the same reappearances using various techniques (Cruikshank and Murphy, 1973; Franz and Millis, 1970, 1974).

Finally, photometry of J5 and J6 is needed.

Satellites of Saturn

The brightness and color of Titan should be monitored regularly using narrow-band filters.

Better and more extensive photometry of Enceladus, Tethys, and even Dione, is needed. An attempt should also be made to measure the brightness of Mimas using area scanners. [Franz (1975) has found with area scanners that Mimas' magnitude is near the new value of Koutchmy and Lamy (1975).]

More observations of Hyperion and Phoebe are needed to establish their rotation periods, rotation lightcurves, and phase coefficients.

Satellites of Uranus and Neptune

Photoelectric photometry of Titania, Oberon, and Triton is feasible. Accurate rotation lightcurves should be obtained, and solar phase effects looked for. In spite of the small excursion in phase angle ($<$ 2 to 3°), the available range covers the region of the opposition effect; such observations would be important in studying the relative surface textures of these satellites.

General Recommendations

In all future work sufficient observations should be obtained to construct phase curves at specific wavelengths, for specific orbital longitudes. This is especially important for satellites with photometrically heterogeneous surfaces. In the future, mean phase coefficients should not be quoted for such satellites.

Two other tasks should receive continuing attention: (1) Spectrophotometry with the highest possible spectral resolution should be extended to the fainter satellites (cf. Chapt. 11, Johnson and Pilcher). (2) Accurate absolute diameters, needed to evaluate geometric albedos and absolute reflectances, should be obtained whenever possible. In this context, the advantages of stellar occultations (O'Leary and van Flandern, 1972; Chapt. 13, O'Leary) and occultations by the Moon (Elliot *et al.*, 1975) should be kept in mind. However, for some satellites, the determination of accurate diameters will have to await direct imaging from spacecraft (Pollack *et al.*, 1973a; Veverka *et al.*, 1974; Chapt. 18, Gehrels).

ACKNOWLEDGMENTS

I am grateful to C. Blanco, S. Catalano, J. Elliot, F. Franklin, J. Goguen, D. M. Hunten, W. Irvine, T. V. Johnson, D. Matson, N. Morrison, R. Murphy, D. Pascu and C. Sagan for helpful discussions. Among numerous colleagues who made available data prior to publication, I am especially indebted to L. Andersson, A. Dollfus and R. Millis. I wish to thank D. Morrison and J. Burns for detailed comments on an earlier version of this manuscript. This work was supported by NASA Grant NGR-33-010-082.

POLARIMETRY OF SATELLITE SURFACES

Joseph Veverka
Cornell University

Available polarization observations of the satellites are reviewed, and the question of what information about satellite surfaces such observations contain is discussed critically.

For satellites with negligible atmospheres, polarization measurements should provide useful information on surface texture and on the opacity of the surface materials. However, they are unlikely to yield specific compositional information.

For Titan, the only satellite known to have an extensive atmosphere, polarimetry indicates the presence of optically thick clouds; it should be possible to deduce the cloud particle characteristics from such data, especially if more extensive observations, covering the full range of available phase angles and wavelengths, are obtained.

Polarization measurements of the brighter satellites have provided useful qualitative information about the nature of satellite surfaces (Dollfus, 1971; Veverka, 1971a, 1973a; Zellner, 1972b, 1973). Unfortunately, at the present time no theoretical understanding exists of the polarization properties of texturally complex surfaces, and thus the full meaning of this information cannot be assimilated. The situation is quite different in the case of Titan where polarimetry indicates the presence of an extensive, and perhaps cloudy, atmosphere which may be optically thick. The latter type of problem has recently become manageable theoretically (cf. Coffeen and Hansen, 1974). As a result, important deductions about Titan's atmosphere, based on polarimetry, are to be expected in the next few years.

For other bright satellites on which polarimetry exists—and which, of course, have at most negligible atmospheres—interpreting the polarization data amounts to understanding the negative branch of the polarization curve (cf. Fig. 10.1b)—a subject on which only empirical and generally semi-quantitative data are available. Nevertheless such measurements provide some information on surface texture, especially when combined with other photometric data. It is unlikely, however, that they contain specific compositional information. At best, one can conclude that the surface material falls into one of several broad categories—such as "bright transparent," "bright opaque," "dark opaque" —which, in themselves, are not directly diagnostic of composition.

To fully appreciate the data reviewed here, it must be realized that polarization measurements of satellites are complicated by a number of factors: (a) Most satellites are faint, and appear close to a bright primary. (b) Phase angle coverage

[210]

TABLE 10.1
Maximum Planetary Phase Angle
Available From Earth

Planet	Maximum Phase Angle
Mars	47°
Jupiter	12°
Saturn	6°
Uranus	3°
Neptune	2°

is limited, especially for the satellites of the outer planets (cf. Table 10.1). (c) Only disk-integrated measurements are possible.

Despite these difficulties, important information about satellite surfaces has been obtained by polarimetry.

POLARIZATION CURVES IN GENERAL

As a result of extensive laboratory work by Lyot (1929), Dollfus (1955, 1961a), and others, a great deal of empirical information about polarization curves exists. These measurements show that polarization curves are sensitive both to the nature of the surface material and to the texture of the surface layer. Experience has shown that information about the texture of a surface layer is much easier to infer from an observed polarization curve than is information about its composition.

Small solar system objects seem to have either loose, particulate surface layers (regoliths), or at least microscopically very intricate surface textures. Accordingly, the polarization properties of such complex surfaces are of primary interest; these are briefly reviewed herein. The notation adopted is similar to that used in the reviews by Dollfus (1961a) and Bowell and Zellner (1974), and is summarized in Figure 10.1. The scattered light is generally partially linearly polarized, with the plane of polarization lying either in the scattering plane (*negative polarization*), or at right angles to it (*positive polarization*).

Realistic theoretical models of polarization from rough, particulate surfaces do not exist. To be useful, model predictions would have to be accurate to ±0.1%; this is a formidable demand in view of the processes involved. Some of these are sketched in Figure 10.1c for an intricate surface layer made up of a variety of particles of different sizes and shapes, most of which are much larger than the wavelength of the incident light.

Surface scattering (1) is a major process (Fig. 10.1c). It depends on the shape, size, and optical properties of the particles. It is affected significantly by shadow-

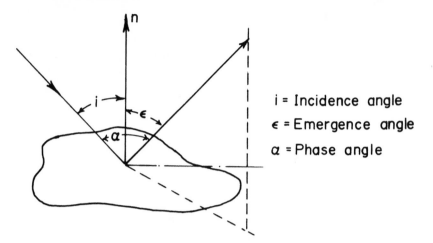

i = Incidence angle

ε = Emergence angle

α = Phase angle

Fig. 10.1a. Scattering geometry and definition of angles.

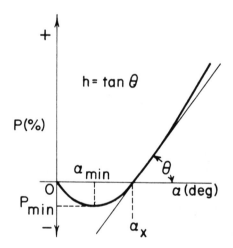

Fig. 10.1b. Schematic diagram of a polarization curve defining the various symbols used in this paper. The phase angle is denoted by α. As α increases beyond α_x, the polarization reaches a maximum value of P_{max} at α_{max} (not shown) and then tends to zero as α approaches 180°.

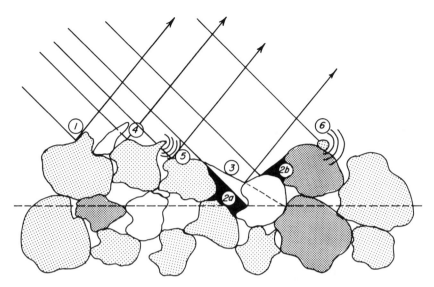

Fig. 10.1c. Idealized representation of some processes involved in the scattering of light from a rough, particulate surface (see text for further explanation).

ing (2) which for a given scattering geometry determines those surface areas that are both illuminated and visible. Multiple scattering of light among particle surfaces (3) is negligible only for very dark surfaces. This is also true for multiple scattering of light which has passed through at least one particle (4). In fact, it is likely that the polarization properties of low opacity surfaces are dominated by processes (3) and (4). Even for very large particles, diffraction "around edges" (5) must be taken into account, and for very small particles such diffraction effects (6) will be dominant.

From this very brief discussion it should be clear why no model of polarization curves exists, and why one must rely heavily on laboratory experience. In summarizing laboratory data it is convenient to separate "opaque" surfaces (such as those the Moon, Phobos, Deimos) from "non-opaque" surfaces (such as those of Rhea, Europa, Io). For a full discussion the reader is referred to a series of papers by Dollfus (1956, 1961a, 1971).

Opaque Materials

Particulate surfaces of opaque materials have a number of important polarimetric properties:

(i) The degree of polarization depends primarily on the phase angle α, and not on i and ϵ separately. Thus, at a given phase angle, the polarization is approximately independent of position on the disk, and the disk-integrated polarization can easily be related to the polarimetric properties of a small surface area.

(ii) A more or less pronounced negative branch is present. Well-developed negative branches ($\gtrsim 0.4\%$) are associated with intricate surface textures (see below).

(iii) Several general correlations between polarization curves and surface reflectance have been made evident for such materials. These are illustrated in Figure 10.2, using the volcanic ash data of Lyot (1929). The inverse correlation between surface reflectance r_n and P_{max} is fairly general and has been exploited by Dollfus and his co-workers in lunar studies (Bowell, 1971; Bowell et al., 1972). A similar inverse correlation between surface reflectance and α_{max}, suggested by the data in Figure 10.2, probably is not always valid since α_{max} appears to be much more sensitive to surface texture than is the surface reflectance.

The "slope-albedo" relationship is very general and well defined (Veverka and Noland, 1973; Bowell and Zellner, 1974). Although highly useful in asteroid studies, it cannot be applied to most satellites since not enough of the polarization curve is observable from Earth to determine the slope h accurately. Phobos and Deimos are the only exceptions, but their surface reflectance can be determined more directly from photometry now that their actual sizes are known from Mariner 9 imagery (Pollack et al., 1973a). Thus, in general, the correlations between surface reflectance and parameters describing the positive branch of polarization curves (P_{max}, α_{max} and even h) are not particularly useful in Earth-based studies of satellites due to the limited range of phase angles available (see Table 10.1).

On the other hand, correlations involving the negative branch parameters (P_{min}, α_{min} and α_x) are of potential importance. Surface reflectance seems to affect both the depth of the negative branch P_{min} and the crossover angle α_x. These trends are discussed more fully later, under *The Negative Branch*.

There appear to be no striking correlations between α_{min} and other surface parameters, except that usually $\alpha_{min} \lesssim \alpha_x/2$.

Non-opaque Materials

Surfaces composed of transparent grains show more complex polarization properties. The polarization usually depends significantly on i and ϵ, separately, and not only on the phase angle α. In such cases, it is difficult to deduce the polarization properties of a surface element from disk-integrated observations. Judging from the available data on such substances (for example, ground glass, snow, water frost, various salts), it seems that negligible negative branches ($\lesssim 0.2\%$; $\alpha_x \lesssim 10°$) are to be expected.

Generally, as the opacity of a transparent powder is increased, the dependence of the polarization on the individual values of i and ϵ decreases, and eventually presumably disappears so that the α dependence alone provides a good charac-

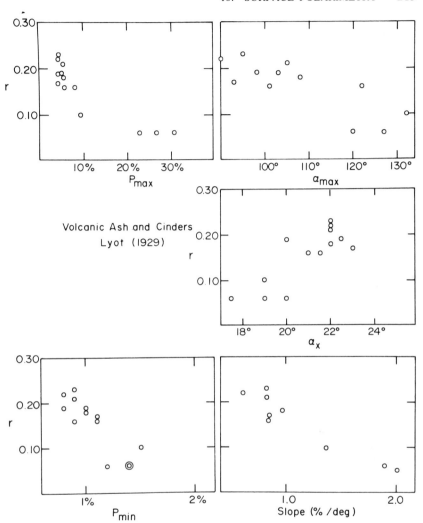

Fig. 10.2. Correlation between the surface reflectance r and various polarization curve parameters (defined in Fig. 10.1b), for unfiltered white light measurements by Lyot (1929) on layers of volcanic ash and cinders. Here, and in Figure 10.4, r represents the bidirectional surface reflectance for i ~ ε ~ 0°.

terization of the polarization curve. At the same time, the prominence of the negative branch is probably enhanced.

Shown in Figure 10.3 are examples of published polarization curves for natural snow (Lyot, 1929) and for laboratory deposits of water frost (Dollfus, 1955). Also of interest in this context are similar measurements by Lyot (1929) on a number of transparent salts including NaCl.

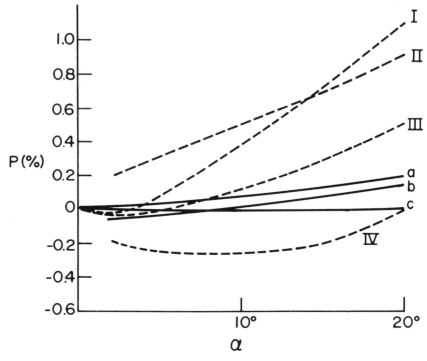

Fig. 10.3. Polarization curves of snow and water frost (Lyot, 1929; Dollfus, 1955). Lyot studied a variety of natural snow surfaces in unfiltered visible light (curves a, b, and c). Dollfus' measurements, also in visible light, refer to laboratory frost layers viewed at $\epsilon = 65°$ (curves I, II, III and IV). The polarization of such layers is strongly dependent not only on α but also on ϵ. Adding absorbers to a frost layer hinders internal multiple scattering and probably enhances both the negative and positive branches of the polarization curve. Polarization curves of frosts are unlikely to be diagnostic of composition since any bright transparent substance will probably give a similar curve.

THE NEGATIVE BRANCH

Various discussions of the negative branch have been given (Lyot, 1929; Dollfus, 1955, 1956, 1961a), but the subject is still poorly understood. It is agreed that texturally complex surfaces are required, but the evidence indicates that mutually free particles are not needed. A large degree of multiple scattering (such as occurs within a surface composed of transparent particles) decreases the importance of the negative branch—other things being equal. However, there are substances such as MgO which are very bright, but which show strong negative polarization at small phase angles (see below).

In this section, the available information on negative branches is reviewed briefly and used to consider the following specific questions: (i) Does a negative branch indicate the presence of a regolith? (ii) Is the depth of the negative branch a good indication of surface albedo? (iii) What can be learned from the wavelength dependence of the negative branch?

The negative branch was discovered by Lyot (1929) who noted that, for opaque powders, surface porosity has an important effect on its depth; as the surface becomes more complex in texture, the negative branch is deepened.

Although single scattering from a flat semi-infinite plane (Fresnel scattering) leads only to positive polarization, there are several ways of producing small amounts of negative polarization with a rough surface:

(a). Multiple scattering from suitably oriented planes can lead to negative polarization (Beckmann, 1968). This sort of mechanism was first suggested by Öhman (1955). However, the special alignments required seem to rule this out as a major mechanism.

(b). Transmission through a particle can give negatively polarized light. The possible importance of this process was suggested by Sagan (private communication, 1974). Although translucent particles probably occur in most surface layers, this is unlikely to be a major mechanism since very dark, opaque powders produce the most pronounced negative branches.

(c). Hopfield (1966) suggested that negative branches result from diffraction around particle edges. Most of this diffracted light is scattered into shadow zones which are invisible at large phase angles. Hopfield qualitatively predicted that the depth of the negative branch should increase with increasing λ/d, where λ is the wavelength of the incident light and d is the mean particle diameter, so that:

$$|P_{min}| \propto \frac{\lambda}{d}. \tag{1}$$

This is, no doubt, the physically most likely explanation of the negative branch. Since the process involves diffraction followed by reflection, the magnitude of the negative branch is expected to be small in all cases.

The most important studies of the negative branch are due to Dollfus (1956; 1961a). In a series of experiments using iron filings (which are opaque at all wavelengths), he showed that surfaces made up of the smallest particles produce the deepest negative branches (in agreement with the earlier conclusions of Lyot). It is likely that these surfaces also have the most intricate surface texture. Inevitably, whenever such a surface was compacted, Dollfus found that the negative branch became shallower. Thus, a complex, porous structure favors the formation of a negative branch. Close proximity among particles is also essential since Dollfus found that no negative branch is observed for a cloud of freely falling particles. This agrees well with Hopfield's concept of the negative branch. Particles must be close enough for the diffracted light from one particle to be scattered back by another, and yet, if the particles get too close together, the structure becomes so closed as to hinder the escape of this scattered component.

Only one scattering of the diffracted light is desirable. If the scattering albedo $\tilde{\omega}_0$ of a typical particle is too high, multiple scattering tends to depolarize the scattered light. This is probably why bright powders tend to have negligible negative branches (Dollfus, 1956, 1961a). There are exceptions, however. Lyot (1929) discovered that MgO, freshly deposited on glass, gives strong negative polarization (-1%) near $\alpha = 1°$. This negative branch is unusually narrow and has not been explained.

The Dollfus measurements on iron filings can be interpreted in another fashion which further supports Hopfield's view of the negative branch: As the particle size decreases, λ/d increases, and the negative polarization increases—as it should according to Hopfield.

In general, the effective $\tilde{\omega}_0$, λ/d, and surface texture will all have an important influence on the negative branch (Pollack and Sagan, 1969). This strong sensitivity of the negative branch to all important surface parameters (particle size, composition, surface texture) makes it an ideal discriminator among various surfaces. It is easy to tell that two surfaces differ in some respect because their negative branches are different. Yet, precisely because of this extreme sensitivity to a variety of parameters, it is almost impossible to interpret such differences uniquely.

The Negative Branch and Surface Texture

What does the presence of a well-developed negative branch tell us about a surface? Dollfus (1956, 1961a) states that negative branches are especially deep for surfaces made up of very fine opaque grains which are allowed to clump into larger aggregates.

Dollfus and his coworkers also found that some lunar rock chips have pronounced negative branches (Geake et al., 1970; Dollfus, 1971; Bowell et al., 1972). Inevitably such rock chips show complex surface grain structure down to a scale comparable to λ when examined with an electron microscope. Thus the presence of a well developed negative branch need not be indicative of a loose regolith (Dollfus, 1971; Veverka, 1973b), but a very intricate surface microstructure is required. A deep negative branch can also be used to rule out a high surface albedo (Fig. 10.4), but a shallow negative branch is by itself ambiguous, since it could be due to "compaction" and not to a high surface albedo.

Variation of P_{min} With Surface Reflectance

Lyot's data (Fig. 10.2) suggest that P_{min} may, in some cases, be related to surface reflectance: the negative branch tends to get deeper for darker samples. However, Egan (1967) has mentioned that this cannot be a rigorous relationship since the negative branch depends not only on surface reflectance but on surface texture. Thus an exact relationship between P_{min} and surface reflectance cannot be expected unless one is dealing with a series of samples having identical

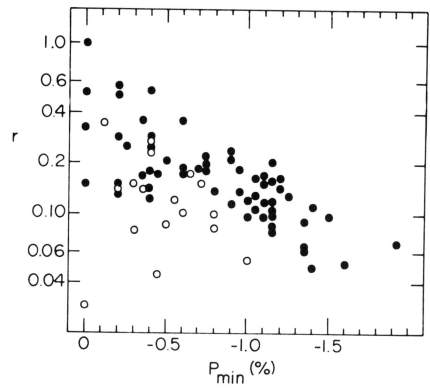

Fig. 10.4. Correlation between the depth of the negative branch (P_{min}) and the normal reflectance of the surface (r) for a variety of samples. From Bowell and Zellner (1974). The solid symbols represent powders or breccias; the open symbols represent solid rock surfaces.

surface texture. Even so, this conjecture that, given a variety of surfaces each made up of similar particle size distributions and having similar textures, there would be a well defined relationship between P_{min} and r_n, has not been tested adequately.

Bowell and Zellner (1974) plotted available laboratory data on a P_{min} versus r_n diagram (Fig. 10.4) and found only a very loose relationship. But these data refer to samples having a wide variety of surface textures. A fair conclusion from Figure 10.4 is that an upper limit on r_n can be estimated from P_{min}. For large values of P_{min} this upper limit can be useful; for example, an object with P_{min} = -1.5% is unlikely to have an albedo greater than 0.2.

One might speculate that many solar system regoliths are sufficiently comparable in texture to allow a good intercomparison of albedos on the basis of measured values of P_{min}. Along these lines, Veverka (1971b) suggested that a single measurement of the depth of a negative branch might be used as a rough discriminator between "bright" and "dark" asteroids. This might prove useful if asteroid regoliths have similar textures.

Zellner (1972b) tried to exploit this relationship in his study of satellites, and at one time proposed that the equation:

$$|P_{min}| \cdot A = 0.16 \qquad (2)$$

adequately represents available asteroid and satellite data. Here A represents the geometric albedo, and $|P_{min}|$ is expressed in percent.

The suggestion that satellite surface textures are sufficiently similar to permit the use of eqn. (2) appears to be only marginally valid. For instance, the geometric albedo of Rhea is certainly $\geqslant 0.5$ and $P_{min} = -0.4\%$ according to Zellner (1972b); yet for the bright side of Iapetus, which most probably has a lower geometric albedo (Morrison et al., 1975; Elliot et al., 1975), $P_{min} = -0.2\%$ (Zellner, 1972b).

At the present time, it seems unsafe to attach much importance to eqn. (2) in deriving satellite albedos. Nevertheless, the depth of the negative branch can be a useful discriminator between "bright" and "dark" surfaces, as demonstrated by Zellner (1972b) in the case of Iapetus.

One final caveat is in order. The lunar highlands and maria have identical negative branches (Lyot, 1929), although they generally differ in reflectance by a factor of two. The wavelength dependence of the negative branch also seems to be the same for the two terrains (Dollfus and Bowell, 1971). Since there is no evidence to suggest that particle size distributions or surface textures are strongly different in the two regions, an over-zealous application of eqn. (2) would lead to the wrong inference that the albedos of the two areas are identical. It is, however, correct to conclude that the albedos are similar, both being sufficiently low for multiple scattering not to be dominant. The reason for the lunar maria and highlands having similar negative branches is not understood at present.

Wavelength Dependence of the Negative Branch

According to Hopfield (1966), the negative branch should deepen as λ increases, since λ/d gets larger. However, for many materials in the visible, as λ increases, so does $\tilde{\omega}_0$. The resulting increase in the importance of multiple scattering tends to counteract the Hopfield effect. Thus the wavelength dependence of the negative branch depends on a balance between these two effects and it is impossible to predict with certainty which of the two will dominate. For lunar regions P_{min} is approximately constant with wavelength, increasing very slightly from -1.0% at $0.325\mu m$ to -1.2% at $1.05\mu m$ (Bowell, 1971; see also Gehrels et al., 1964). A similar trend for lunar-like laboratory samples was reported by KenKnight et al. (1967), but some other measurements (Coffeen,

1965; Egan, 1967) do not show this trend. Dollfus *et al.* (1969) find no significant wavelength dependence of P_{min} for limonite samples between $0.48\mu m$ and $1.05\mu m$, while similar measurements by Egan (1969) show significantly shallower negative branches at $1.0\mu m$ than at $0.5\mu m$. Comparable measurements on other types of materials are lacking, and it is impossible at present to judge whether or not the wavelength dependence of P_{min} on λ is always small, and what—if any—significant information it imparts about the nature of the scattering surface.

For many dark silicate powders, laboratory data suggest that α_x increases with wavelength (KenKnight *et al.*, 1967; Egan, 1967). This is also true for limonite samples measured by Dollfus and Focas (1966) and by Egan (1969). Dollfus and Bowell (1971) found this trend to hold for lunar regions and showed that, on the Moon, the inversion angle increases systematically with wavelength from $20°7$ at $0.325\mu m$ to $26°2$ at $1.05\mu m$—in agreement with older, less extensive measurements by Gehrels *et al.* (1964) and Avramchuck (1964). For all these materials the reflectance increases with increasing wavelength, so that the above trend can also be viewed as a systematic increase of α_x with surface reflectance for surfaces of comparable texture (cf. Lyot's volcanic ash data in Figure 10.2).

Unfortunately too few accurate measurements of the wavelength dependence of α_x exist; in fact, none have been published for transparent materials. It is therefore impossible to assess the generality of the above trend, or to know to what extent it may prove diagnostic of surface properties.

SATELLITE POLARIZATION

Phobos and Deimos

During the 1971 opposition, Zellner (1972a) succeeded in obtaining good polarization measurements of Deimos (Fig. 10.5) and discovered the presence of a deep negative branch ($P_{min} = -1.4\%$). Due to the intense scattered light from Mars, no measurements of Phobos were obtained. However, information about Phobos is available from a few Mariner 9 observations analyzed by Noland *et al.* (1973). Since these measurements were made using a spacecraft television system their absolute accuracy is low in comparison with current Earth-based polarimetry. A single Mariner 9 measurement of Deimos at a phase angle of $74°$ yields $P = 22\pm4\%$, whereas an extrapolation of Zellner's data gives about 16%. This difference very likely represents systematic errors in the Mariner polarimetry which could not be removed in the analysis of Noland *et al.* (1973). The fact that Phobos and Deimos have similar polarizations near $\alpha = 74°$ should hold however, since the systematic errors should be similar in the two cases.

The presence of a deep negative branch in the polarization curve of Deimos indicates a complex surface texture—possibly a regolith (cf. discussion of Chapt. 14, Pollack). A similar conclusion follows for both satellites from the

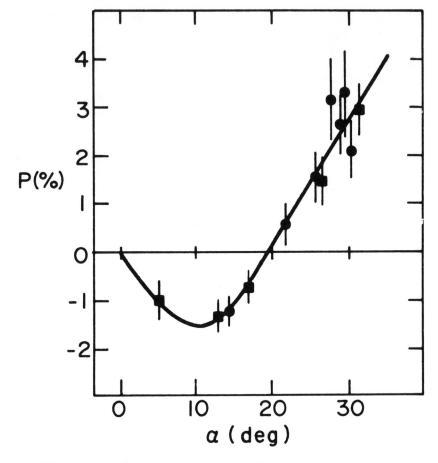

Fig. 10.5. Polarization versus solar phase angle for Deimos in blue light. Circles represent observations at eastern elongation; the squares denote measurements at western
elongation. From Zellner (1972a).

low polarization observed at large phase angles (Noland *et al.*, 1973), since both
objects are very dark ($p_V = 0.05 - 0.07$). It should be recalled that Gatley *et al.*
(1974) have interpreted an eclipse cooling curve of Phobos observed by Mariner
9 in terms of a thin regolith (cf. Chapt. 12 by Morrison). However, none of these
observations can distinguish between a surface layer of loose debris and one of
similar texture in which the grains are effectively welded at their points of
contact. It may be argued on other grounds that the latter alternative is unlikely,
but on the basis of photometry or polarimetry alone, it cannot be rejected.

Io, Europa, Ganymede, and Callisto

The available polarimetric data are gathered in Figure 10.6. Most of the scatter is due to the fact that these satellites have spotted surfaces (Dollfus and Murray, 1974; Morrison *et al.,* 1974; Chapt. 16, Morrison and Morrison; Murray, 1975) so that the observed polarization depends significantly on orbital longitude, and not only on solar phase angle. Large orbital variations in the degree of polarization occur for Callisto and Io, while smaller longitudinal effects have been detected for Ganymede and Europa (see below). In view of this, it is futile to try to define average polarization curve parameters (*e.g.*, α_x, P_{min}) except, perhaps, in the case of Europa. In the future, polarization curves for the Galilean satellites should be constructed for individual orbital longitudes.

For Io, Zellner and Gradie (1975) find orbital variations of 0.4 to 0.5% for $\alpha > 10°$ at $0.52\mu m$. The negative branch is deepest near $\theta = 160°$, and shallowest near $\theta = 300°$. Judging from the large color variation of Io in the ultraviolet (Morrison *et al.,* 1974; Morrison and Morrison, Chapt. 16), even larger variations of polarization with orbital phase might be expected at shorter wavelengths.

Dollfus (1975) studied the orbital variations in Ganymede's polarization at $\alpha = 11°$, and found a polarization maximum of about $+0.2\%$ near $\theta = 0°$ as well as a minimum of 0.0% near $\theta = 180°$.

Callisto's variations in polarization with orbital phase have been studied in greatest detail. On the basis of limited measurements, Veverka (1971a) found some evidence for such variations; recent measurements by Dollfus (1975) and by Zellner and Gradie (1975) are shown in Figure 10.7. The negative branch is deepest near $\theta \sim 90°$, and shallowest near $\theta \sim 270°$; the amplitude is some 0.8 to 0.9% near $0.55\mu m$. Judging from the scatter of points in Figure 10.6, variations of polarization with orbital phase are small for Europa.

Zellner (private communication, 1974) believes the polarization curves of Io, Europa, and Ganymede are consistent with low opacity surface materials in agreement with an earlier suggestion by Veverka (1971a). For Europa and Ganymede this low opacity material is probably water frost (Pilcher *et al.,* 1972; Chapt. 11, Johnson and Pilcher). For Io, it is almost certainly not water frost (Morrison and Cruikshank, 1974). Two candidate materials, sulphur (Wamsteker, 1972, 1974) and an assemblage of evaporite minerals rich in sulphates of magnesium, calcium and sodium, and possibly including NaCl (Fanale *et al.,* 1974a, 1974b; Chapt. 17, herein), are consistent with the polarization data (Zellner and Gradie, 1975). (For a polarization curve of NaCl, see Lyot, 1929).

For Callisto, Zellner (private communication, 1974) feels that a silicate surface similar to that of some "stony" asteroids is indicated, and that the observed orbital variations in polarization are most likely due to differences in surface texture. Dollfus (1975) stresses that the dark (leading) side of Callisto has a negative branch which is similar to that of the Moon, although not quite as deep. The negative branch of the bright (trailing) side, on the other hand, is much

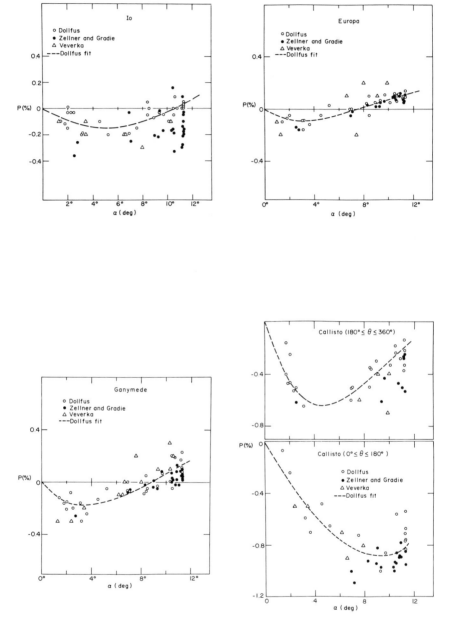

Fig. 10.6. Polarization measurements of Jupiter's Galilean satellites by various observers. The measurements by Zellner and Gradie (1975) were made at 0.55μm, while those of Dollfus (1975) correspond to about 0.5μm. The observations of Veverka (1971a) were made in unfiltered white light.

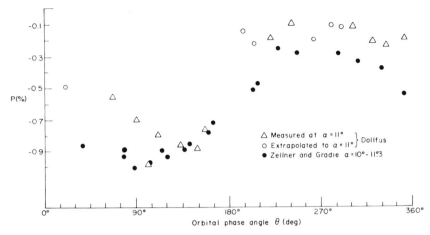

Fig. 10.7. Variations in the polarization of Callisto with orbital phase angle (θ). The solid symbols represent observations at 0.55μm by Zellner and Gradie (1975); the open symbols denote measurements by Dollfus (1975) at about 0.5μm.

shallower, reaching only −0.6% (−1.1% for the Moon), with an inversion angle of only about 13°(23° for the Moon). Dollfus concludes that, like the Moon, the dark side of Callisto is covered with a loose regolith, but that the bright side polarizes light much like some bare, dust-free lunar rock samples. From this fact Dollfus (private communication, 1974) infers that "a chaotic terrain, not regolithized by impacts" pervades the bright side. In view of the sizeable phase coefficient and opposition effect of even the bright side of Callisto (Morrison *et al.*, 1974; Chapt. 9, Veverka), it is difficult to accept this interpretation. There may be other ways of explaining the polarization curve of the bright side: possibly in terms of patches of high opacity "dark-side" material, mixed with patches of some low opacity substance (Veverka, 1971a). The temptation to make statements about the composition of Callisto's outer layer on the basis of polarization data alone is best shunned. It is fair to say that the dark side material has an opacity similar to that of silicate rocks, but actually speaking of "silicate materials" is risky. The polarization data are not that diagnostic.

Finally, the possibility that the polarization of some Galilean satellites may be time-variable should be kept open. Short term brightness changes have been reported for Io, Europa and Ganymede (Morrison and Cruikshank, 1974). Zellner and Gradie (1975) find unexpected scatter in some of their Io polarimetry measurements which could be evidence of variability.

Dione, Rhea, and Iapetus

A series of polarization measurements of these three satellites was made at 0.52μm by Zellner (1972b); the results are sufficient to estimate the depth of the

TABLE 10.2
Polarization Measurements
of Dione, Rhea, and Iapetus

Satellite	$P_{min}(\%)$
Dione	-0.4 ± 0.1
Rhea	-0.4
Iapetus (trailing side)	-0.2
Iapetus (leading side)	-1.3

Note: Measurements were taken at $0.52\mu m$ by Zellner (1972b; Bowell and Zellner, 1974).

negative branch (Table 10.2). However, α_x and h cannot be determined since the maximum range of phase angles is only 6°.

The deep negative branch ($P_{min} \simeq -1.3\%$) found for the dark side of Iapetus is of interest and is consistent with the low albedo and large opposition effect inferred from other sources (Morrison and Cruikshank, 1974; Noland *et al.,* 1974b; Elliot *et al.,* 1975). The small values of P_{min} for Dione, Rhea, and the bright side of Iapetus are consistent with high albedos. See *Discussion* for Zellner's comment and a graph of Iapetus' polarization as a function of orbital phase.

Dione and Rhea have identical values of P_{min} and similar geometric albedos (about 0.6 in the visible, according to Morrison and Cruikshank, 1974). However, the phase coefficients of Rhea are definitely larger than those of Dione (Noland *et al.,* 1974b) suggesting that the texture of Rhea's surface layer is more complex. Why this does not lead to a slightly deeper negative branch for Rhea than for Dione is not clear.

Another puzzle is the low value of $|P_{min}|$ for the bright side of Iapetus: the phase coefficients of the bright side of Iapetus are slightly larger than those of Rhea. (Noland *et al.,* 1974b.) Therefore, one would expect the bright side of Iapetus to have a deeper negative branch than Rhea—unless its albedo is significantly higher. But its value of $|P_{min}|$ is actually smaller than that of Rhea, while its geometric albedo is unlikely to be higher (Morrison *et al.,* 1975; Elliot *et al.,* 1975).

Titan

Disk-integrated polarization measurements of Titan in unfiltered white light were made by Veverka (1973a) during the 1968–69 opposition (Fig. 10.8). The observed polarization was small but always positive. Veverka (1973a) concluded that the available photometric and polarimetric properties of Titan could best be explained in terms of a model in which an opaque cloud deck is overlain by a

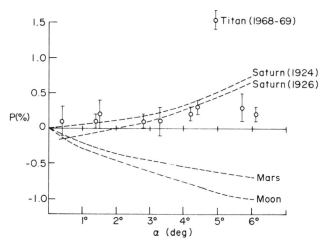

Fig. 10.8. Measurements of Titan in unfiltered white light compared with polarization curves of the Moon, Mars and Saturn. (From Veverka, 1973a.) The Saturn curves are based on measurements by Lyot (1929), made at the center of the planet's disk, and found to be slightly variable from year to year.

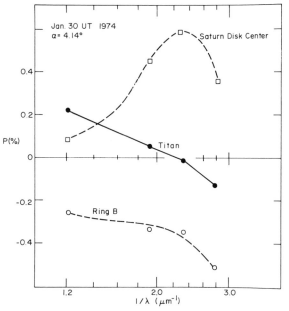

Fig. 10.9. The wavelength dependence of the disk integrated polarization of Titan, compared with similar measurements for the center of Saturn's disk, and for Ring B. Zellner (private communication, 1974).

very thin Rayleigh atmosphere. The optical thickness of this Rayleigh component is now known to be $\leqslant 0.04$ in optical depth at 0.36μm (Caldwell *et al.*, 1973).

New polarization measurements of Titan in three colors (0.36, 0.52 and 0.83μm) were obtained by Zellner (1973) during the 1971-72 opposition. Zellner concluded that his observations "are not consistent with scattering from either an ordinary planetary surface or a pure molecular atmosphere. Apparently an opaque cloud layer with a strong uv-absorbing constituent is needed." No cloud model calculations have yet been carried out to match the polarization measurements.

Veverka (1973a, 1974) noted the similarity between the white light Titan measurements and measurements for the center of Saturn's disk made by Lyot (1929) (Fig. 10.8), and suggested that the atmosphere of Titan near its cloud tops may resemble that of Saturn—a suggestion made previously by McCord *et al.* (1971) on the basis of the similarity in spectral reflectance curves. New polarization measurements by Zellner (Fig. 10.9) prove that this analogy cannot hold in detail. In addition, recent infrared observations (Low and Rieke, 1974b) seem to imply that hydrogen is not a major constituent of Titan's atmosphere as was once believed (Hunten, 1974, and Chapt. 20 herein; Chapt. 21, Caldwell), so that a close analogy between the atmospheres of Titan and Saturn should no longer be expected.

So far, polarization measurements of Titan have produced only qualitative (but very useful) information about Titan's cloud cover. Rigorous analyses of observations at many wavelengths, such as those being obtained by Zellner, should eventually lead to more quantitative data.

FUTURE WORK

Galilean Satellites

Accurate polarization curves at orbital phase intervals of 30° should be obtained for each satellite at several wavelengths. Possible time variations should be looked for, especially in the case of Io. Judging from the steep spectral reflectance curve of this satellite, its polarization is probably significantly wavelength dependent between 0.3 and 0.5μm. Such data might be useful in excluding some candidate surface materials.

Jupiter V

The polarization of Amalthea should be no more difficult to measure than that of Deimos. Such measurements would show the extent of the negative branch and thus indicate whether the surface layer of this rarely studied satellite consists of a low opacity or of a high opacity material.

Titan

Polarization measurements over the available range of phase angles and wavelengths are needed to test competing models of Titan's atmosphere. Available geometric albedos and phase coefficients can be explained through a variety of models (Hunten, 1974, and Chapt. 20 herein; Chapt. 21, Caldwell). Detailed calculations require more complete data than are currently available—especially at short wavelengths ($0.3 - 0.4\mu m$). Zellner's measurements (Fig. 10.9) show that the polarization of Titan is significantly wavelength dependent—a fact that should prove useful in choosing among models.

Since Titan appears to be secularly variable in brightness (Andersson, 1974; Andersson, Chapt. 22; Franklin and Cook, 1974), future polarimetry should be supplemented with photometry so that correlations between brightness changes and possible polarization changes can be studied. It would also be interesting to measure the polarization of Titan in some of the broad methane absorption bands of its spectrum (Coffeen and Hansen, 1974).

Other Satellites of Saturn

Measurements such as those already made by Zellner (1972b) for Rhea and Iapetus should be extended to Tethys and Hyperion. A detailed polarization curve of Iapetus as a function of orbital phase will be helpful in constructing models of this unusual satellite.

Mutual Occultations and Eclipses

Mutual occultations and eclipses of satellites of Jupiter and Saturn are visible from Earth at regular intervals (Brinkman and Millis, 1973; Peters, 1975). Polarization measurements made during mutual occultations would be relatively easy to analyze. For satellites whose surfaces consist of low opacity materials, the degree of polarization near the limbs might be large even at small phase angles (Dollfus, 1961a). Note that, unlike occultation data, measurements obtained during mutual eclipses would not be easy to interpret.

Satellites of Uranus

Measurements of Titania and Oberon should be feasible soon. Even at phase angles of 3.2° the observed polarization can indicate the presence and the depth of a negative branch—data from which some information about surface texture and albedo might be obtained.

Spacecraft Observations

Polarimetry provides an important dimension in studying the surfaces of satellites from spacecraft. The main advantages over Earth-based work are that observations can be made at large phase angles (where the degree of polarization is usually large) and at many points on the disk. A beginning along these lines has already been made for Phobos and Deimos by Mariner 9 (Noland *et al.*, 1973), and for the Galilean satellites by Pioneer 10 (Gehrels *et al.*, 1974; Gehrels, Chapt. 18).

ACKNOWLEDGMENTS

I am indebted to A. Dollfus, B. Zellner and J. Gradie not only for making available some of their recent data before publication but for specific comments. Helpful suggestions were also received from J. Burns, W. Irvine, T. Johnson, D. Matson and D. Pascu. This work was supported by NASA Grant NGR 33-010-082.

DISCUSSION

T. GOLD: The negative branch of the curve is important in allowing one to exclude the possibility of a mixture of light and dark materials in cases where that branch is so prominent that it could only barely be produced by material of the observed (mean) albedo. Any mixture that produces the same mean albedo will produce less of a negative swing.

B. H. ZELLNER: I do not believe that the negative polarization branch is primarily a diffraction effect—at least for darker surfaces. Convincing polari-zation-phase curves complete with deep negative branches have been generated by Milo Wolff (personal communication, 1974) in Monte Carlo computations of multiple Fresnel reflection within a cloud of randomly oriented opaque crystal surfaces. This is strictly a geometrical optics theory ($\lambda \ll d$), and gives P_{min} in excess of 2% for all refractive indices. The essential point seems to be the ability of surface particles to cast microshadows from which only twice-reflected rays can emerge.

The range of P_{min} observed for natural intricate surfaces is presumably due to the polarization-diluting effects of internal transmission and diffraction from structures with $\lambda \approx d$. Thus the observed tendency for albedos to increase with decreasing P_{min} is at least qualitatively explained.

I have comments on two of Saturn's satellites: (1) The wavelength dependence of Titan's linear polarization absolutely rules out Rayleigh scattering and clearly puts constraints on any Mie-scattering models. (2) Figure 10.10 gives the linear polarization of Iapetus in green light as a function of the longitude of the sub-Earth point. The observations were made at the University of Arizona in 1972 and 1973 at solar phase angles between 5.0 and 6.3 degrees; the sub-Earth latitude on Iapetus varied from $-36°$ to $-26°$. The probable error of an observa-tion is on the order of $\pm 0.1\%$.

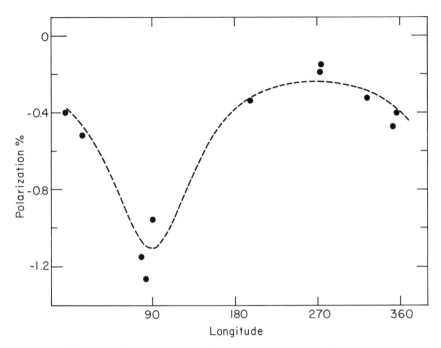

Fig. 10.10. The polarization of Iapetus as a function of orbital phase.

The dashed curve is given by P = 0.38 B⁻¹, where B = 1–0.65 sinθ, representing a sinusoidal light variation of amplitude 1.7 magnitudes. The polarization clearly obeys the Umov law of reciprocity between polarization and albedo, that is, the albedo differences are differences in the unpolarized component only. This result implies that the surface microstructure is similar on the two sides of Iapetus.

SATELLITE SPECTROPHOTOMETRY AND SURFACE COMPOSITIONS

Torrence V. Johnson, Jet Propulsion Laboratory
and
Carl B. Pilcher, University of Hawaii

A review of currently available spectrophotometric data for planetary satellites, excluding Earth's Moon, is presented. These data are interpreted in terms of surface composition with consideration given to the uniqueness of such identifications.

Spectrophotometry, the measurement of the intensity of light as a function of wavelength, is an invaluable tool for investigation of the surfaces and atmospheres of solar system objects. The spectral region which may be investigated from the Earth's surface with optical techniques is usually divided into two regions. In the first, from 0.3 to about 3μ, the spectra of solar system objects are dominated by reflected sunlight. In the second, from 3 to $1,000\mu$, emitted thermal radiation is prevalent. In this chapter we review the existing spectrophotometry of the natural satellites of the solar system (excluding the Moon) in the first region, and discuss the information on satellite surface composition that is implied by these and other data. The properties of thermal radiation from the satellites are discussed in Chapter 12, by D. Morrison.

Two complex topics that have been reviewed elsewhere, the rings of Saturn (Chapt. 19, Cook and Franklin) and the atmosphere of Titan (cf. Chapt. 20, Hunten; Chapt. 21, Caldwell), will be discussed here only briefly. The interested reader is referred to the proceedings of a workshop, edited by Hunten (1974), and an excellent review by Morrison and Cruikshank (1974). The remaining satellites will be discussed in order by their primary.

Many of the data presented here have been measured using UBV or uvby filters; the characteristics of these are shown in Table 11.1. The uvby filters have smaller half-widths than UBV filters, and are used primarily for color determinations. Magnitudes are usually established by converting measurements in the y filter to standard V magnitudes. This can be done to an accuracy of 0.01 mag (1%) by observing several UBV standard stars. Magnitudes determined in this manner frequently will differ from absolute flux measurements obtained with other spectral systems (non-standard filters, interferometric and dispersive spectrometers, etc.) due to differences in standard star selection, spectral resolution, and to unresolved absorptions in the object under study. To facilitate comparison where absolute flux information is not required, spectral data are frequently

[232]

TABLE 11.1
Characteristics of Standard Photometric Systems

Filter	Central Wavelength (Å)	Half-width*(Å)	Ref
U	3600	800	A
B	4200	900	A
V	5400	800	A
u	3500	300	S
v	4110	190	S
b	4670	180	S
y	5470	230	S

* Full-width at half of maximum transmission
A = Allen, 1963, p. 195
S = Strömgren, 1963

scaled to unity at a convenient wavelength; often 0.56μ, near the central wavelength of the V filter, is chosen for photoelectric data.

When geometric albedos (for a definition see Harris, 1961, p. 306; or Chapt. 9, Veverka) are desired, as is often the case in solar system studies, it should be remembered that uncertainties in the values used for the solar flux will produce similar uncertainties in the final albedos. Morrison *et al.* (1974) presented a helpful discussion of their calibration techniques used to derive geometric albedos for the Galilean satellites in the uvby system (see also Chapt. 16, Morrison and Morrison). In general, an uncertainty of at least ±3 to 5% (Arvesen *et al.*, 1969) exists in *all* published geometric albedos due to this uncertainty in the solar flux. Data which are reduced using older solar flux measurements may contain even larger errors.

In discussions of astronomical objects, the question of what constitutes a unique compositional identification often arises. Where the basis for the identification is primarily spectrophotometric data, two relevant criteria may be defined. First, are all spectral features expected to be present (on the basis of the identification) actually observed in the data, considering their spectral resolution and signal-to-noise ratio? Second, if the first criterion is met, can the intensities and shapes of these features be understood in terms of the physical characteristics (e.g., temperature, pressure, proposed surface particle size) of the object under study? If both of these criteria are satisfied, a compositional identification of high reliability can usually be made. In general, the narrower and more numerous the features, the easier it is to unambiguously assign them to a specific chemical species.

In the case of a solid rocky surface, there are two types of absorptions which are of interest: crystal field transitions and charge transfer transitions. Crystal field transitions arise from the 3-d shell electrons of the first transition series of elements, the most important petrologically being Fe^{2+} and Ti^{3+}. When this type of absorption feature is present in a reflection spectrum it can be used to identify the responsible mineral with a high degree of certainty (see R. Burns, 1970; Gaffey, 1974; and Adams, 1974 for discussions of the details of the effects of mineralogy on crystal field transitions). Fortunately, three major rock forming minerals—feldspar, pyroxene and olivine—have such absorption features near $1.0\mu m$. These absorptions are frequently referred to loosely (and incorrectly) as ''iron bands.'' In fact, the reflection spectrum of elemental iron does not exhibit these absorption features. The $1.0\mu m$ features arise from crystal field transitions in 3-d electrons of Fe^{2+} in given crystal structures, and the position and shape of the absorptions are diagnostic of the *mineral structure,* not merely of the presence of iron.

Since crystal field transitions are characteristic of mineral structure, the diagnostic features (i.e., band position and shape) are not altered greatly by changing physical parameters such as particle size, packing, etc. Albedo, *depth* of absorption features and the slope of the reflectance spectrum can be altered by such changes but the center frequency and shape of the absorption remains virtually unchanged (see Hunt and Salisbury, 1970). When such well-defined absorptions exist in solar system reflection spectra, as in the case of the asteroid Vesta, they can be extremely useful in determining surface composition (McCord *et al.,* 1970).

However, absorption features can be suppressed greatly in some cases as a result of mixing with very opaque substances, such as carbonaceous material (see Johnson and Fanale, 1973). Vitrification (which alters the structure of the material) also changes the characteristics of absorption features (cf. Nash and Conel, 1973; Adams and McCord, 1971).

Charge transfer transitions occur between electrons of neighboring ions. The absorptions due to charge transfer are usually located in the blue or ultraviolet portion of the spectrum. The wings of these absorptions, which extend into the visible and near infrared, produce the ''reddish'' slope exhibited by the spectra of most silicates. While these absorptions do depend on the composition and structure of the materials involved, they are not as useful as the crystal field transitions for detailed compositional identifications since the bands are quite broad and much of the ultraviolet portion of these features cannot be observed from Earth.

In summary, when strong crystal field transition features occur in reflection spectra of planetary bodies, they can be reliably interpreted in terms of the crystal structure and mineralogy of the surface materials. In many cases, however, where only charge transfer absorptions are important or absorption features are suppressed by opaque materials (as is the case with carbonaceous compositions), the spectral reflectance must be interpreted in context with other information, such as overall albedo or similarities to meteoritic assemblages. In these cases, while reasonable references concerning probable surface compositions can be drawn (see Johnson and Fanale, 1973, and Johnson *et al.*, 1975), the identification of detailed mineralogy is not as certain as for well-defined absorptions.

Ices and frosts (these terms will be used interchangeably) generally show more numerous and somewhat narrower absorptions than minerals, with most features occurring in the near infrared. Although the intensities of these absorptions are strongly affected by the physical state of the material, the identification of a specific frost as a major component of the surface of an astronomical object can sometimes be made with considerable certainty (see discussions below of the Galilean satellites and Saturn's rings). Finally, spectroscopically active gases usually have a sufficient number of spectral features to make their identification unambiguous if they are present. Exceptions occur when only a few isolated lines are observable or when the signal-to-noise ratio is poor. In these cases some subjective evaluation of the reliability of the identification must be made, usually by considering additional factors such as those based on cosmochemical arguments.

SATELLITES OF MARS

The two satellites of Mars, Phobos and Deimos, are difficult to observe due to their small sizes (and correspondingly faint magnitudes) and their proximity to Mars. Kuiper (1961b) reported early photoelectric photometry from which he found B–V = 0.6 ± 0.1 for both satellites. Zellner and Capen (1974) obtained photoelectric colors for Deimos of B–V = 0.65 ± 0.03 and U–B = 0.18 ± 0.03. The similarity of these colors to those of the Sun (solar values are B–V = 0.64 and U–B = 0.06, H. L. Johnson, 1965) implies that the satellites have flat spectral reflectances. In addition, their geometric albedos are quite low, 0.06–0.07 (Smith, 1970; Zellner, 1972a; Noland *et al.*, 1973; and Zellner and Capen, 1974; cf. Chapt. 9, Veverka), similar to many asteroids (cf. Chapman *et al.*, 1973; Cruikshank and Morrison, 1973; Morrison, 1974c; Zellner *et al.*, 1974; Chapman *et al.*, 1975). Until the satellites' reflectances are better determined, there is no point in making detailed comparisons with the properties of asteroids. However, surface materials with certain types of compositions, such as those of the ordinary chondrites and the surfaces of the Moon and Mars, can probably be ruled out on the basis of the spectral properties cited above. Carbonaceous chondritic material, similar to that suggested for the surfaces of Ceres and Pallas (Johnson and Fanale, 1973; Gaffey, 1974), remains a possibility.

SATELLITES OF JUPITER

We will confine our discussion to the Galilean satellites, since, with the exception of a few photometric measurements of J6 (Andersson and Burkhead,

1970; Andersson, 1974), they are the only Jovian satellites that have been measured photoelectrically.

Visible and Near-Infrared (0.3–1.1μ) Reflectance

Accurate measurements of the satellites' magnitudes as functions of solar and orbital phase angles were first made by Stebbins (1927) and Stebbins and Jacobsen (1928) using a quartz-potassium photoelectric cell. Harris (1961) reduced their data to the UBV system and added some modern observations taken at McDonald Observatory. These combined data established several important facts about the Galilean satellites: (i) They are in synchronous rotation about their primary (i.e., their lightcurves have the same periods as those of their revolutions around Jupiter; cf. Chapt. 6, Peale). (ii) They all show variations in color with orbital phase. (iii) They all have low blue and ultraviolet reflectances, with Io being the extreme example. More recent photometric studies (Blanco and Catalano, 1974b; Owen and Lazor, 1973) have helped to define the satellite's magnitudes and lightcurves with greater precision but have not produced any changes in these basic conclusions (see Chapt. 9, Veverka; Chapt. 16, Morrison and Morrison).

Spectrophotometric studies have revealed much additional information about the Galilean satellites' spectral properties. McNamara (1964) studied the satellites' reflectances in the ultraviolet using twelve narrow-band filters centered from 0.3 to 0.6μ but did not investigate variations with orbital or solar phase angle. More extensive investigations were made by Johnson and co-workers using eighteen to twenty-four filters from 0.3 to 1.1μ (Johnson, 1969; Johnson and McCord, 1970; Johnson, 1971), by Morrison et al. (1974) using the four color (uvby) filter system, and by Wamsteker (1972) using forty-two filters from 0.3 to 1.1μ. The first two of these studies included investigations of the dependence of satellite brightness on orbital and solar phase angle.

The spectral geometric albedos of the leading sides of each of the satellites (data from Johnson, 1971; Johnson and McCord, 1970, 1971; and Morrison et al., 1974) are shown in Figure 11.1 (modified from Morrison and Cruikshank, 1974) and presented numerically in Table 11.2. These data have been scaled from the mean values given by Morrison et al. for the observed geometric albedos (including opposition effects) of the satellites at 0° solar and mean rotational phase angles. The albedos are strongly dependent upon the values adopted for the satellite radii; those used here were taken from Morrison and Cruikshank (1974). It should be noted that to date only the diameters of Io and Ganymede have been determined by the relatively accurate method of stellar occultation (Taylor et al., 1971; Carlson et al., 1973; Chapt. 13, O'Leary). Even using this method, a significant uncertainty remains in the diameter of Ganymede due to the possible presence of an atmosphere on that satellite (Carlson et al., 1973; cf. Chapt. 1, Morrison et al.). Additional uncertainties in

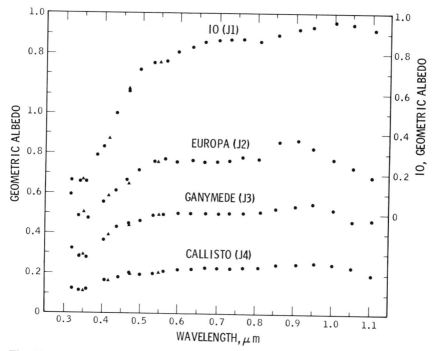

Fig. 11.1. Geometric albedos $(0.3\mu–1.1\mu)$ of the leading sides of the Galilean satellites.

the albedos are introduced by errors in the determination of the satellite's phase functions and, as previously mentioned, the V magnitude of the Sun (see Morrison and Cruikshank, 1974; and Chapt. 9, Veverka, and Table 16.1 for further information). Thus the albedos of Figure 11.1 and Table 11.2 may still undergo considerable revision as these uncertainties are resolved.

The curves of Figure 11.1 are similar, showing decreasing reflectance shortward of 0.55μ and comparatively uniform albedo at longer wavelengths. The 0.319μ turnup in some of the reflectances may be an artifact of the data reduction process (particularly in the solar ultraviolet flux assumed). The data of Wamsteker (1972) are in reasonably good agreement with those shown in Figure 11.1 (see Morrison and Cruikshank, 1974, Fig. 10) despite the fact that Wamsteker's data may refer to different rotational phase angles for different spectral regions. Part, but probably not all, of the differences between the data of Wamsteker and those of the other investigators is due to this averaging process and to Wamsteker's higher spectral resolution. Io's reflectance is most different from those of the other satellites, showing a much steeper decrease in the ultraviolet and an absorption feature, also observed by Wamsteker, centered near 0.55μ. Wamsteker's data suggest the presence of this feature in the reflectances

TABLE 11.2

Geometric Albedos for Leading Sides of the Galilean Satellites

λ(μ)	Io	Europa	Ganymede	Callisto
0.319	0.16	0.59	0.32	0.11
0.338	0.15	0.48	0.28	0.11
0.358	0.14	0.47	0.28	0.12
0.383	(0.29)*	—*	(0.47)*	(0.19)*
0.402	0.33	0.57	0.37	0.15
0.433	0.49	0.62	0.43	0.18
0.467	0.61	0.68	0.45	0.19
0.498	0.72	0.72	0.46	0.20
0.532	0.75	0.77	0.49	0.21
0.564	0.76	0.76	0.49	0.21
0.564	0.76	0.76	0.49	0.21
0.598	0.81	0.76	0.50	0.22
0.633	0.83	0.77	0.50	0.22
0.665	0.86	0.76	0.50	0.22
0.699	0.86	0.76	0.50	0.22
0.730	0.87	0.76	0.50	0.22
0.765	0.87	0.78	0.51	0.22
0.809	0.86	0.77	0.51	0.23
0.855	0.90	0.86	0.53	0.24
0.906	0.92	0.87	0.54	0.25
0.948	0.93	0.83	0.55	0.25
1.002	0.95	0.77	0.52	0.24
1.053	0.95	0.73	0.47	0.23
1.101	0.92	0.69	0.47	0.20
u (0.350)	0.172	0.50	0.30	0.11
v (0.411)	0.382	0.58	0.39	0.16
b (0.467)	0.630	0.64	0.44	0.19
y (0.547)	0.760	0.76	0.49	0.20

Note: Values are from Johnson (1971) and Johnson and McCord (1971), except last four rows which are from the four color filter (uvby) system of Morrison *et al.* (1974).

* Values uncertain due to Balmer transition in standard stars.

of Europa and perhaps that of Ganymede as well. Johnson and McCord (1970) and Johnson (1969, 1971), discussing the 0.55μ feature, concluded that it was not diagnostic of composition. Shallow absorptions in this spectral range are a common characteristic of the spectra of many materials (Hunt and Salisbury, 1970) and can also arise from the combination of surface materials whose spectra exhibit different degrees of short-wavelength absorption (e.g., the two-component model for Io's surface suggested by Wamsteker *et al.*, 1974).

None of the satellite reflectances show any absorption features in the $0.9–1.1\mu$ range that are characteristic of iron-bearing silicates (Johnson and McCord,

1970). In addition, the high albedos of at least the inner three satellites rule out surface materials similar to lunar soils or most meteoritic assemblages. Thus, in the 0.3–1.1μ spectral range there is little diagnostic information concerning the satellites' surface compositions, although some classes of materials (lunar soils, chondritic meteorites, etc.) can be ruled out. There is considerably more compositional information in the near-infrared, as discussed in the following section.

Near-Infrared (1.0–2.5μ) Reflectance

This spectral region, where many volatile materials (e.g., H_2O, CO_2, NH_3, CH_4) exhibit characteristic absorptions, is of major interest in the study of the satellites of the outer planets. The earliest mention in the literature of infrared observations of the Galilean satellites is in an abstract by Kuiper (1957). He states, ''The spectrometer tracings show striking differences between the four Galilean satellites of Jupiter. 1 and 4 roughly resemble the solar and lunar curves among 1 and 2.5μ, but 3 and particularly 2 are markedly different, in the sense that the spectrum beyond 1.5μ is reduced in intensity by a factor of 2–3. This is most readily explained by assuming that 2 and 3 are covered up by H_2O snow. The albedo and color of 2 in visual light are compatible with this hypothesis, while 3, which is darker, may be covered with snow contaminated with silicate dust.'' Unfortunately, these early infrared spectroscopic data were never published. Broadband infrared measurements made by Kuiper in 1956 also showed the low reflectivity of J2 and J3 near 2.0μ. Moroz (1965), using a prism spectrometer, later obtained low resolution ($\Delta\lambda = 1.0$–0.2μ) spectra of all of the Galilean satellites and concluded that the spectra of ''...Callisto (J4) and especially Io (J1) resemble the monochromatic albedo curve of Mars, but spectra of Europa (J2) and Ganymede (J3) resemble spectra of the Martian polar cap and the rings of Saturn. The assumption is thus made that the surfaces of Europa and Ganymede are covered with ice; if not entirely, then at least the most significant fraction.'' [Translation from the Russian by Dale P. Cruikshank.]

These basic features of the satellite infrared spectra were subsequently verified by Gromova *et al.* (1970), Johnson and McCord (1971), and Lee (1972). Observations at even longer wavelengths were made by Gillett *et al.* (1970) and more recently by Hansen (1975). Figure 11.2 shows these data sets adjusted to the new radii and the leading side albedos where possible. Table 11.3 gives the values plotted in Figure 11.2. Lee's Io data show a considerable discrepancy from those of the other investigators that, in light of the existence of a similar discrepancy in spectrophotometric data for Saturn's satellites, may indicate a systematic error in Lee's results. Nonetheless, it is clear from the results of Lee, Gillett *et al.*, and Hansen that Io's albedo remains high out to 5.0μ in contrast to those of the other satellites.

Substantial improvement in the quality of infrared spectral measurements of the satellites was obtained by Pilcher *et al.* (1972) and Fink *et al.* (1973), using

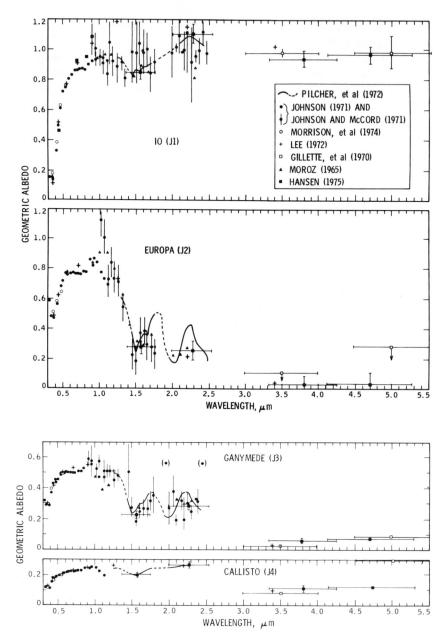

Fig. 11.2. Geometric albedos (0.3μ–3.5μ) of the Galilean satellites. Data are scaled at 0.56μ except for those of Pilcher *et al*. (1972) (scaled near 1.25μ) and Moroz (1965), scaled to match near 1.0μ.

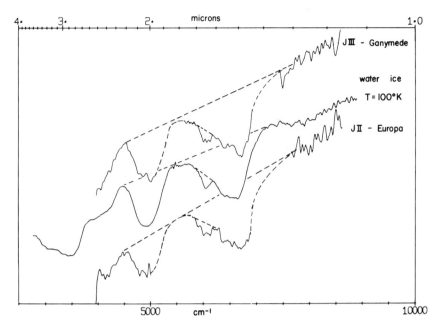

Fig. 11.3. Ratio spectra of Europa and Ganymede with the Moon. A calibration water ice spectrum matching closely the surface temperatures of the satellites is shown for comparison. Data are from Fink and Larson (1975).

Fourier transform spectrometers. These instruments, which permit observations of all spectral elements simultaneously (cf. Vanasse and Sakai, 1967; Bell, 1972), allowed data to be obtained with spectral resolution an order of magnitude better than that previously possible. Ratios of the spectra of the Galilean satellites to that of the Moon are shown in Figures 11.2 and 11.3. The large overall slope in the data of Fink *et al.* is due to the lunar color in this spectral region. Pilcher *et al.* removed this effect from their spectra by multiplying their ratios by the reflectance of the Moon (McCord and Johnson, 1970). These data were scaled to match the broadband data for the infrared geometric albedos of the satellites.

Water frost absorptions in the spectra of Europa and Ganymede are evident in Figures 11.2 and 11.3. Kieffer and Smythe (1974) in a laboratory study set limits of 5 to 28 percent on the abundance of non-water frost materials consistent with these spectra. Both Pilcher *et al.* and Fink *et al.* also indicated the possible presence of weak water frost absorptions in the spectrum of Callisto. From the depths of the frost absorptions Pilcher *et al.* concluded that the following percentages of the satellites' surfaces are frost-covered: Europa, 50 to 100 percent; Ganymede, 25 to 65 percent; Callisto, perhaps 5 to 25 percent. They also

TABLE 11.3
Infrared Albedos of the Galilean Satellites

λ(μ)	Io	Europa	Ganymede	Callisto
	Reference: Johnson and McCord (1971)			
0.906	1.04	0.81	1.60	
0.948	1.01	0.84	0.58	
1.002	0.95	1.12	0.53	
1.053	0.91	1.01	0.57	
1.101	0.84	0.92	0.50	
1.150	1.04	0.84	0.51	
1.199	0.92	0.74	0.51	
1.253	0.88	0.72	0.45	
1.308	0.91	0.55	0.49	
1.349	0.80	—	—	
1.454	0.98	0.23	0.52	
1.504	0.97	0.19	0.27	
1.556	1.05	0.37	0.18	
1.604	0.98	0.38	0.25	
1.658	1.00	0.30	0.27	
1.705	0.85	0.28	0.26	
1.750	0.92	0.24	0.32	
1.792	—	—	0.36	
1.959	—	—	0.57	
1.997	0.97	—	0.27	
2.052	1.02	—	0.38	
2.096	1.10	—	0.19	
2.156	0.99	—	0.33	
2.195	1.12	—	0.21	
2.248	0.91	—	0.30	
2.300	0.96	—	0.25	
2.359	0.99	—	0.33	
2.395	1.12	—	0.31	
2.452	0.97	—	0.58	
	Reference: Moroz (1965)			
0.98	0.94			
1.08	0.96			
1.17	0.96			
1.48	0.82			
1.58	0.89			
1.60	0.99			
1.67	0.84			
2.30	0.81			
2.30	0.88			

λ(μ)	Io	Europa	Ganymede	Callisto
Reference: Moroz (1965)				
0.90		0.85		
1.00		0.90		
1.11		0.90		
1.20		0.80		
1.31		0.62		
1.41		0.42		
1.53		0.31		
1.59		0.29		
1.71		0.37		
2.01		0.24		
2.11		0.22		
2.22		0.27		
Reference: Moroz (1965)				
0.80			0.48	
0.84			0.40	
1.08			0.41	
1.42			0.28	
1.49			0.33	
1.55			0.31	
1.55			0.33	
Reference: Lee (1972)				
0.90	1.04	0.83	0.55	0.24
1.25	1.18	0.73	0.50	0.26
2.20	1.15	0.21	0.31	0.25
3.40	1.00	0.04	0.05	0.09
Reference: Gillett et al. (1970)				
3.50	0.98	0.01	0.02	0.07
5.00	0.98	0.28	0.08	0.21
Reference: Hansen (1975)				
1.57	0.78	0.25	0.21	0.17
2.27	1.15	0.24	0.29	0.27
3.80	0.99	0.02	0.06	0.12
4.71	1.01	0.03	0.07	0.12

concluded that: (1) the albedo of the non-frost material on the surfaces of the outer three satellites increases with increasing wavelength, suggestive of a silicate composition; and (2) the derived percentages of frost surface cover are consistent with the observed visible geometric albedos of the satellites. This last conclusion suggests that the overall differences in albedo and infrared reflectance between these satellites may simply be due to differences in the amount of water frost on their surfaces. Pilcher *et al.* offered support for this conclusion in the form of spectra of the leading and trailing sides of Ganymede, shown here in Figure 11.4. The brighter leading side shows deeper frost absorptions (indicating more exposed surface frost) that are consistent, in this interpretation, with the observed 0.15 mag (15 percent) brightness difference between the two sides. On the other hand, Fink *et al.* concluded that water frost or snow is distributed over the entire surface of Ganymede, based on its low reflectance between 3 and 4 μ where water frost is almost totally absorbing. Images recorded by the Pioneer 10 spacecraft (Chapt. 18, Gehrels) that show greater brightness contrast on Ganymede than on Europa seem to support the conclusions of Pilcher *et al.* regarding the surface distribution of frost (similar results were obtained from groundbased observations of satellite transits by Dollfus and Murray, 1974).

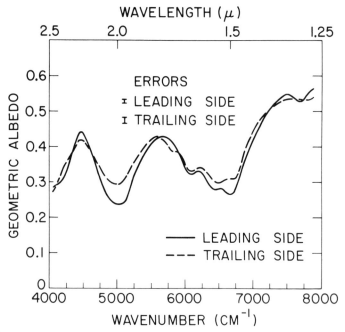

Fig. 11.4. The reflectivities of the leading and trailing sides of J3 scaled to the same value at 5625 cm^{-1}. Both reflectivities are also scaled to the geometric albedos of Johnson and McCord (1971). Each error bar is the average of the standard deviations for all wavelengths (from Pilcher *et al.*, 1972).

Rotational Variation

The correspondence between the periods of the satellites' lightcurves and their orbital periods about Jupiter, and the constancy (since they were first measured in 1927) of the orbital positions of minimum and maximum brightness for each satellite, strongly support the conclusion that the satellites are in synchronous rotation about Jupiter (i.e., the same satellite hemisphere always faces the planet). This is not surprising, since it is expected that tidal braking will lock the satellites into synchronism very early in their lifetimes (cf. Chapt. 6, Peale). Synchronism implies that a satellite's orbital phase angle (measured from superior geocentric conjunction) is always equal to its rotational phase angle. These terms will therefore be used interchangeably. Further, since the inclination of the plane of the satellite's orbit to the ecliptic is small ($\approx 3°$), each satellite presents the same face to the earth at a given orbital phase angle. It is therefore meaningful to compare data taken at a given orbital phase angle even though the observations may have been separated in time by months or years.

The variations in satellite color with orbital phase were studied by Johnson (1969, 1971), using narrow-band filters which spanned the spectral range 0.3–1.1μ, and by Morrison *et al.* (1974), using the uvby system. Figures 9.4 and 16.1–16.4 as well as Figures 11.5 show some of their results. It is apparent that although the largest color variations for all of the satellites occur in the ultraviolet, significant variations occur at other wavelengths for both Io and Europa. To illustrate further the effects of satellite rotation on observed color, Figure 11.6 (modified from Johnson, 1971) shows the ratio of the spectral reflectances of the darker to the lighter hemisphere for each satellite. These ratios can be used with the spectral geometric albedo data of Figure 11.1 to derive the spectral albedo of each hemisphere. The results of this calculation for Io are shown in Figure 11.7, which also shows the values for the two sides from Morrison *et al.* (1974) to be in good agreement with the Johnson data.

The form of most of the variations in color and brightness with orbital phase shows that no Galilean satellite possesses two hemispheres of totally different composition (e.g., a "white" ice hemisphere and a "red" silicate hemisphere). Rather, the differences between the hemispheres of each satellite are qualitatively similar to the differences between mare and upland areas on the Moon (cf. McCord and Johnson, 1970). The observed variations may be understood, for example, in terms of a two-component model in which one component is bright and spectrally neutral, and the other is dark and absorbing in the ultraviolet. Surface mixtures containing different proportions of these two materials can account for the observed color and brightness variations. The non-sinusoidal and varying shapes of the color versus rotation curves (Fig. 11.5) are evidence of a complex distribution of color on the surfaces of the satellites.

The similarity of the ratio curves in Figure 11.6 suggests that the color variations on the different satellites are caused by a similar process or material.

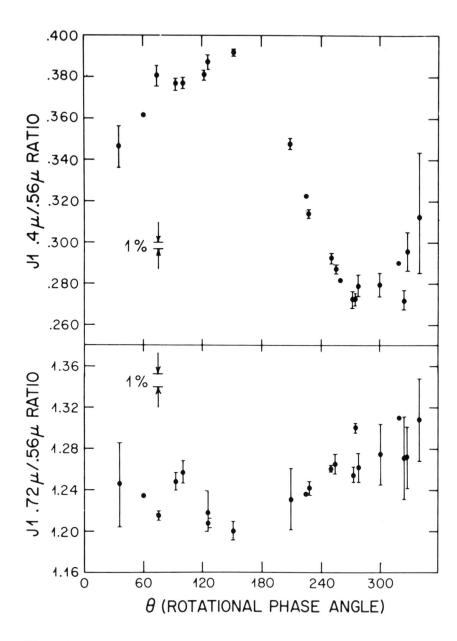

Fig. 11.5 (Above and five pages following). Variation of Galilean satellite colors with rotational phase angle. Data for visual lightcurves and b–y, v–y, and u–y colors from Morrison *et al.* (1974); data for 0.4/0.56μm and 0.72/0.56μm ratios from Johnson (1971).

IO at α = 6°

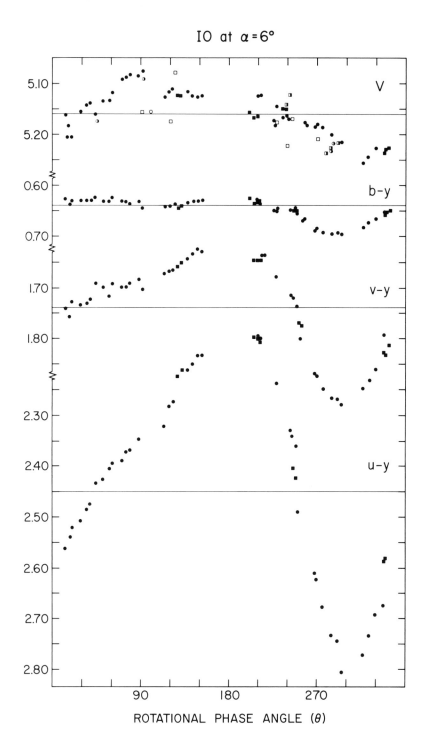

ROTATIONAL PHASE ANGLE (θ)

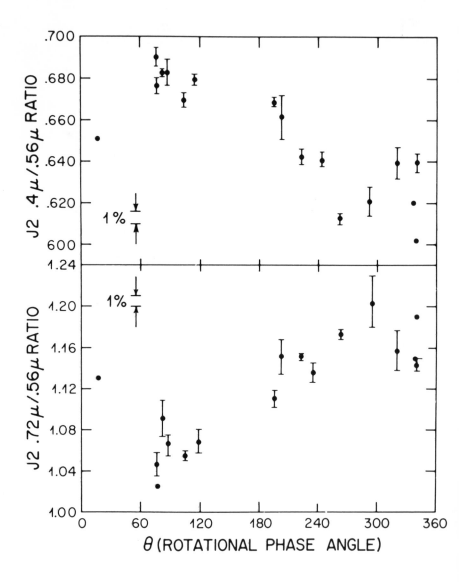

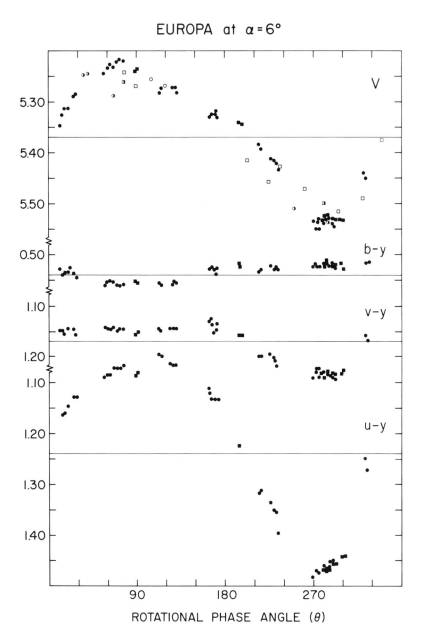

EUROPA at α = 6°

ROTATIONAL PHASE ANGLE (θ)

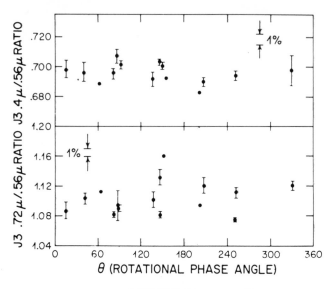

GANYMEDE at α=6°

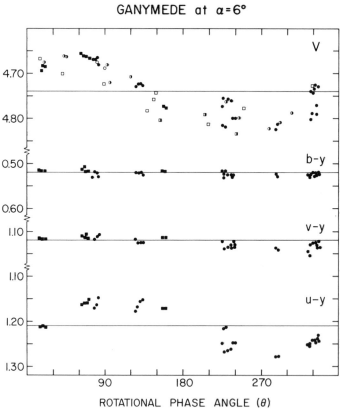

ROTATIONAL PHASE ANGLE (θ)

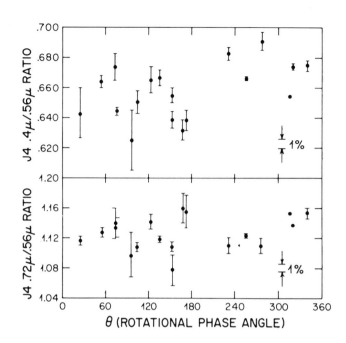

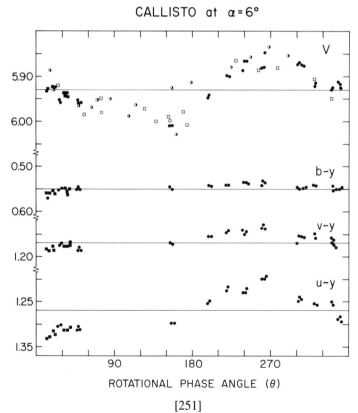

CALLISTO at α = 6°

ROTATIONAL PHASE ANGLE (θ)

[251]

However, any simple leading-side/trailing-side effect is ruled out by the behavior of Callisto, whose leading side is dark and "red" in contrast to those of the other satellites. One suggested coloring process is charged-particle bombardment (Burns, 1968; Axford and Mendis, 1974), producing highly colored free radicals (Binder and Cruikshank, 1964; Veverka, 1971a; Sagan, 1971) and polysulfides (cf. Lebofsky, 1972). However, none of these have been identified in satellite spectra, and many of them may be ruled out on the basis of the existing infrared data. Color center production by radiation damage (Fanale *et al.*, 1974b; Chapt. 17, herein) and sulfur (Wamsteker *et al.*, 1974) have been suggested as well (see the next section). The intensities of these effects could be a function of distance from Jupiter and might thus explain a systematic variation in color with increasing distance from the planet.

Another source of information on the distribution of surface material comes from observations of the satellites in transit across Jupiter. These observations have indicated the presence of dark poles on Io and bright poles on Europa and Ganymede (Lyot, in Dollfus, 1961b; Murray, 1975). Partial confirmation of these results has been provided by direct images (Danielson and Tomasko, 1971) and analyses of mutual event data (Murphy and Aksnes, 1973; Aksnes and Franklin, 1975), and it is expected that images recorded by spacecraft in the vicinity of Jupiter (Chapt. 18, Gehrels) will provide much additional information in the near future. In addition, observations of satellite transits across bands of different colors have suggested that Io's dark poles are redder than the bright equatorial regions (Minton, 1973).

The Surface of Io

The nature of the surface of Io has been particularly puzzling due to the satellite's unusual spectral properties (see Chapt. 17, Fanale *et al.;* Chapt. 16, Morrison and Morrison). Its high visible geometric albedo rules out most silicate surface materials and suggests a frost-covered surface, but its infrared spectrum (Fig. 11.2) shows no frost absorptions. The satellite's ultraviolet albedo, as noted above, is unusually low; the difference between Io's reflectances at 0.35 and 0.60 μ (see Fig. 11.1) is larger than that for virtually any other object in the solar system. Binder and Cruikshank (1964) suggested that this low ultraviolet reflectance might be due to an icy surface containing uv-absorbing free radicals. They conjectured that the free radicals could be formed by the interaction of ammonia and methane in the ice with solar ultraviolet radiation or charged particles from the Jovian magnetosphere. (A similar proposal had been advanced by Rice, 1956, to explain the colors observed on Jupiter.) Variations on this suggestion, including a variety of specific absorbers, have since been put forth by Veverka (1971a; NH_3 and CH_4 impurities in water ice forming species such as NH, CH, CH_3I, $(NH_2NH)_n$, CH_3CS, and polymers of HCN, C_2N_2, and $HCN \cdot NH_3$ under the action of charged particle radiation), Sagan (1971; unspecified free radicals formed by charged particle radiation), and Lebofsky

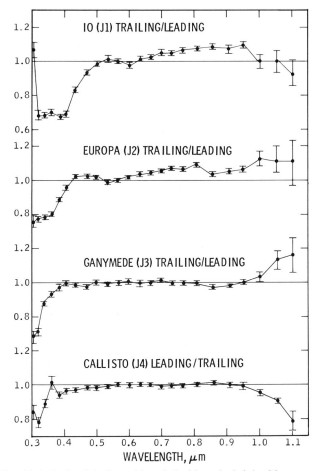

Fig. 11.6. Ratio of the intensities, dark side to the bright side, as a function of wavelength for each of the satellites, modified from Johnson (1971).

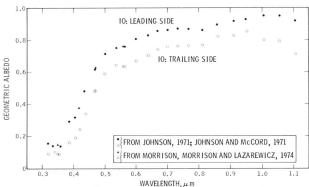

Fig. 11.7. Geometric albedos of the bright (leading) and dark (trailing) hemispheres of Io.

(1972; uv and charged particle irradiated NH_4SH ice). However, as we have noted, Io's spectrum shows no evidence either for surface ice or for any of the suggested absorbers. Kieffer and Smythe (1974) concluded, from comparison of telescopic and laboratory data, that materials other than frosts of H_2O, H_2S, NH_3, NH_4SH, CH_4, and CO_2 must be abundant on Io's surface.

The often suggested icy surface for Io, coupled with the satellite's lack of infrared absorption features, has led to speculation that the normal frost absorptions might be suppressed in Io by the effects of surface microstructure or mixtures with other materials (cf. Morrison and Cruikshank, 1974). A suggestion by McElroy et al. (1974; cf. McElroy and Yung, 1975) that absorptions in an ammonia frost surface might be masked by dissolved sodium is discussed below. The effects of microstructure on frost spectra have been considered both in the laboratory (Kieffer, 1970) and theoretically (Pollack et al., 1973b). In particular, Pollack et al. show that, even for an average particle size of 3 μ, absorptions of up to 20% are still present in the calculated spectrum of H_2O frost. While sub-micron particles due to some peculiarity of Io's environment cannot be ruled out, such a situation seems unlikely in view of the growth characteristics of frost crystals and the fact that Europa and Ganymede do show deep absorptions characteristic of larger particle sizes. Until there is further evidence that some mechanism is acting to mask ice absorptions on Io, it seems reasonable to assume that the satellite does not have a predominantly icy surface (cf. Chapt. 17, Fanale et al.). The lack of absorption near 3μ—a much stronger band than those at 1.4 and 1.8μ—also argues strongly for H_2O free surface materials.

An alternative explanation for the decrease in Io's ultraviolet reflectance, and the smaller decreases in the reflectances of the other satellites, was offered by Johnson (1970; see also Johnson and McCord, 1970). He suggested that the surfaces of all the satellites might consist of low opacity, glassy silicates containing variable amounts of an ultraviolet absorber such as Fe^{3+}. However, this type of surface on Europa and Ganymede is ruled out by their infrared spectra. Johnson discussed the difficulty of finding silicate materials with as low an ultraviolet reflectance as Io, and noted that environmental effects on ice or ferrosilicate surfaces might instead be producing the satellite's spectral characteristics.

The continuing problem of the nature of the surface of Io has led to recent suggestions of some additional surface materials that seem to be in better agreement with the observations. We will first briefly discuss studies of the formation of the Galilean satellites that form a background for these suggestions, and then discuss the newly proposed surface materials, emphasizing the relationships between all of these works.

Lewis (1971a,b; 1974a) developed thermal models of icy satellites from which he concluded that satellites formed at Jupiter's heliocentric distance would be composed of low density ices mixed with higher density rock-like material (sili-

TABLE 11.4
Physical Data

Satellite	Radius(km)	$\rho(\mathrm{g\,cm^{-3}})$
1 Io	1829 ± 10	3.482 ± 0.044
2 Europa	1500 ± 50	$3.4\ \ \pm0.3$
3 Ganymede	$2635^{+\ 15}_{-100}$	$1.93\ ^{+0.3}_{-0.03}$
4 Callisto	2500 ± 150	$1.60\ \pm0.15$
Moon	1738	3.34

From Morrison and Cruikshank (1974),
Anderson *et al.* (1974a,b); cf. Table 1.4

cates, sulfides, hydrous silicates). He proposed that the trend of decreasing satellite density with increasing distance from Jupiter (see Table 11.4) is due to a progression in the relative proportions of the rock-ice mixtures, Io being the most ice-poor satellite and Callisto the most ice-rich. The data given in Table 11.4 (which are considerably more accurate than those available to Lewis in 1971) generally support this conclusion but also suggest that the satellites fall into two classes with regard to bulk composition, Io and Europa being predominately rock-like and Ganymede and Callisto consisting mainly of low-density ices. Pollack and Reynolds (1974) suggested that a large Jovian heat source due to gravitational contraction of the planet during the period of satellite formation could account for this variation in density (see also Graboske *et al.*, 1975; Grossman *et al.*, 1972; Fanale *et al.*, 1974b; Chapt. 17 herein; Chapt. 25, Consolmagno and Lewis; Chapt. 23, Cameron). They assumed, on the basis of Lewis' work, that the non-volatile material in the satellites is rock-like.

On the basis of these cosmochemical arguments and attempts to explain Io's spectral properties Fanale *et al.* (1974b; Chapt. 17 herein), suggested an evaporite-rich surface composition. They argued, following Lewis (1972b), that the original silicate materials which condensed in the vicinity of Jupiter were hydrated silicates similar to carbonaceous chondritic material but that Io retained little or no ice due to its proximity to Jupiter. Fanale *et al.* then reasoned that in an ice-poor Io the concentrations of U, Th, and K are sufficiently high that radioactive heating will raise the internal temperature of the satellite enough to ensure melting of any water ice initially present, and the degassing of most of the water bound in the carbonaceous chondritic material. This will produce a large quantity of liquid water that will percolate toward the satellite's surface, in the process becoming saturated with salts. They suggest that water reaching the surface will be lost to the surrounding space over the age of the solar system, leaving behind a surface composed of a mixture of salts, including components rich in Na^+, Ca^{2+}, Mg^{2+}, and SO_4^{2-}. They showed that such evaporites have reflectances similar to Io's in the infrared but do not show as strong an ultraviolet absorption. However, they demonstrated that the ultraviolet absorption on Io could be due to color centers formed in the salts by charged particle irradiation or caused by production of reduced sulfur from sulfate salts (see below). Interestingly, this "salt-flat" model of Io's surface (formulated before the announcement of sodium emission) provides an abundant source of sodium for the sodium D-line emission that has recently been observed around the satellite (cf. Brown, 1974; Brown and Chaffee, 1974; Trafton *et al.*, 1974; McElroy and Yung, 1975; Bergstralh *et al.*, 1975). Matson *et al.* (1974) proposed that the sodium is sputtered from the surface of Io by proton irradiation and is subsequently injected into a cloud surrounding the satellite where it is observed by resonant scattering of incident sunlight. Besides the evaporite hypothesis of Fanale *et al.*, the only other sodium source proposed is meteoroid infall (Morrison and Cruikshank, 1974).

Wamsteker (1972; see also Wamsteker *et al.*, 1974, and Wamsteker, 1974) has proposed sulfur as a surface constituent of Io. He has shown that sulfur mixed with material having a smaller ultraviolet absorption and no infrared absorptions matches Io's visible and ultraviolet reflectivity well and that the temperature dependence of sulfur's ultraviolet absorption (Sill, 1973) can provide a mechanism for the post-eclipse brightening of Io first observed by Binder and Cruikshank (1964). Fanale *et al.* proposed that one possible source of sulfur is the reduction, by charged particle bombardment, of sulfates present in a salt surface. Comparisons with the spectra of various candidate surface materials and that of Io are presented in Figures 17.4 and 17.5.

McElroy and Yung (1975) inferred a different surface composition for Io from the observations of sodium and hydrogen (Judge and Carlson, 1974) in emission around the satellite. They proposed a surface of ammonia frost containing dissolved sodium. Photolysis of ammonia gas above such a surface would serve as a source of hydrogen, at the same time possibly producing an $NaNH_2$ surface component. They maintained that the dissolved sodium would make the ice electrically conductive, effectively masking its infrared absorptions. It is not clear, however, that this would be an effect of dissolving ionic material in an ice at the temperature of Io. Matson *et al.* (1974) argued against these suggestions, citing the cosmochemical difficulties in retaining the necessary amount of ammonia on Io, the observed high density of the satellite, and the absence of evidence for ammonia frost on any of the other Galilean satellites.

In summary, the surface of Io is composed of some high albedo material— possibly evaporite minerals and/or sulfur—but probably little or no frost of any kind. Europa and Ganymede, on the other hand, have surfaces that are largely covered with water frost. Callisto may have some frost, but another surface component, probably containing silicate and/or carbonaceous chondritic material, must dominate to explain this satellite's lower albedo and lack of absorptions in the 1.0–2.5 μ region. All of the Galilean satellites show a decrease in reflectance in the blue and ultraviolet. The similarity among their colors as well as their color variation with orbital phase suggests that they are all affected by a similar chromophore and/or coloration process.

SATELLITES OF SATURN

Saturn's Rings

The rings of Saturn have been the subject of intensive study for many years. Because of the existence of several excellent reviews (Bobrov, 1970; Cook *et al.*, 1973; Cruikshank, 1974; and Chapt. 19, Cook and Franklin), we will only briefly review the spectrophotometric measurements of the rings.

Franklin and Cook (1965) reported B and V observations of both the A and B rings derived from combined photoelectric and photographic photometry of the entire ring-planet system. Figure 11.8 shows their opposition B-ring magnitudes scaled to the multifilter observations of the isolated B-ring by Lebofsky *et al.* (1970). Table 11.5 (from Lebofsky *et al.*), gives the values of the scaled reflectance. Irvine and Lane (1973) also analyzed photometry in nine filters of the combined planet-ring system to obtain the ring contribution alone. Their data agree with those in Figure 11.8 for wavelengths shorter than 0.7 μ. Discrepancies at longer wavelengths may result from the method used to remove the disk contribution or from the fact that Irvine and Lane measured the entire ring system while Lebofsky *et al.* observed only the ring ansae. Lebofsky *et al.* reported no significant spectral differences between the A and B rings, but their A ring data were not of high quality. Higher resolution spectra of the rings between 0.30 and 0.44μ (Barker and Trafton, 1973a, 1973b) confirmed the low ultraviolet reflectance and showed no evidence for other spectral features.

The curve shown in Figure 11.8 is qualitatively very similar to those of the Galilean satellites but is significantly different from the reflectances of the other satellites of Saturn (see below; Fig. 11.13). As with the Galilean satellites, the high albedo inferred for the ring particles (cf. Morrison and Cruikshank, 1974) suggests frost of some variety, but the visible reflectance is not consistent with pure "white" frosts. Many of the possible chromophores suggested for the Galilean satellites have also been suggested for the rings, including those caused by uv and charged particle irradiation of frosts (cf. Lebofsky *et al.*, 1970).

Again, as in the case of the Galilean satellites, the diagnostic information regarding composition comes from the 1.0–2.5μ spectral region. Kuiper (1952) interpreted his early low-resolution infrared spectra of the rings as indicating the presence of water ice, although no individual absorptions could be identified. Spectra of slightly higher resolution by Moroz (1961) and Shnirev *et al.* (1964) did not result in any alteration of this general conclusion. As in the case of the

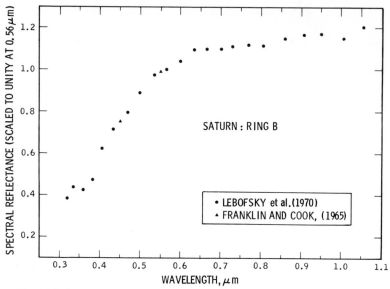

Fig. 11.8. Spectral reflectance of Saturn's Ring B from Lebofsky *et al.* (1970).

TABLE 11.5
Spectral Reflectance of Ring B
(Scaled to unity at 0.56μ)

$\lambda(\mu)$	R_λ
0.319	0.39
0.338	0.43
0.358	0.43
0.383	0.47
0.402	0.62
0.433	0.71
0.467	0.80
0.498	0.89
0.532	0.97
0.564	1.00
0.598	1.04
0.633	1.09
0.665	1.10
0.699	1.10
0.730	1.12
0.765	1.12
0.809	1.11
0.855	1.15
0.906	1.17
0.948	1.18
1.002	1.15
1.053	1.21

From Lebofsky *et al.* (1970)

Galilean satellites, a great improvement in spectral resolution was not obtained until Fourier transform interferometer/spectrometers were applied to the observations. Using such an instrument, Mertz and Coleman (1966) found an absorption feature near 1.66 μ that they attributed to paraformaldahyde. Kuiper *et al.* (1970) obtained higher quality interferometric spectra of the rings which Pilcher *et al.* (1970) showed were due to water frost. Cruikshank and Pilcher (1974) have since confirmed the identification of water frost as the main ring constituent. Pollack *et al.* (1973b) concluded from the depths of the absorption bands in the spectrum of Kuiper *et al.* (1970) that the ice scattering centers have a mean radius lying between 25 and 125 μ.

In summary, the major surface constituent of the ring particles seems to be water frost. However, the nature of the material absorbing in the visible and ultraviolet is unknown. A point worth noting is that the ring spectrum, in the ultraviolet and visible as well as the infrared, is much more similar to the spectra of the Galilean satellites, particularly that of Europa, than it is to the spectra of other satellites of Saturn.

Titan

Titan, the only satellite known to have a substantial atmosphere, has been the subject of intensive study in recent years. Details of current research are found in Morrison and Cruikshank (1974), the proceedings of a workshop on Titan (Hunten, 1974), and the reviews in this book (Chapt. 20, Hunten; Chapt. 21, Caldwell).

Low dispersion spectra taken by Kuiper in 1952 showed the difference in ultraviolet and visible continuum intensities between Titan and other Saturn satellites that is responsible for Titan's frequently noted orange color. Earlier, in 1944, Kuiper had discovered numerous absorption features at wavelengths longer than 0.6μ in Titan's spectrum, two of which he identified as the 6190 Å and 7250Å bands of methane. UBVRI colors published by Harris (1961) showed the steep slope of Titan's reflectance in the blue and the influence of methane absorption on the reflectances in the R and I filters. More recent UBV studies have been reported by Blanco and Catalano (1971). No variation in color with rotation was reported by any of these workers.

Narrowband spectrophotometry $(0.3–1.1\mu)$ by McCord *et al.* (1971) revealed a striking similarity between the scaled spectral reflectance of Titan and that of Saturn's disk (Fig. 11.9). This similarity led McCord *et al.* to suggest that a chromophore which is present in Saturn's atmosphere is also present on Titan, either in the atmosphere or on its surface. They attributed the differences between the ultraviolet reflectances of Titan and Saturn to a greater amount of Rayleigh scattering in Saturn's atmosphere. (Note that the point at 0.3μ which showed a slight upturn for Titan has been deleted since this filter was found to have red leak.) The low ultraviolet albedo of Titan has been confirmed at higher resolution by Barker and Trafton (1973a, 1973b). Caldwell *et al.* (1973)

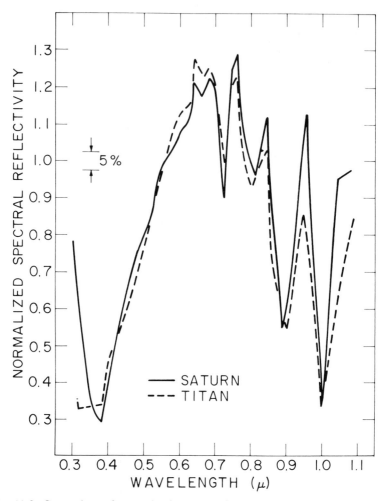

Fig. 11.9. Comparison of spectral reflectances of Saturn (equatorial region) and Titan from McCord *et al.* (1971).

extended the ultraviolet data to $\sim0.2\mu$m and showed that Titan's very low ultraviolet albedo requires an absorber higher in Titan's atmosphere to suppress Rayleigh scattering (see also Danielson *et al.*, 1974 and Chapt. 21, Caldwell).

Recent six-color measurements by Noland *et al.* (1974b) and scanner data with 50 Å resolution by Younkin (1974) are in reasonably good agreement with the earlier work (Fig. 11.10; Table 11.6). Most of the differences between the various data sets of Figure 11.10 are attributable to differences in spectral resolution. It should be noted that the R point from Harris (1961) is high compared to the other measurements. This behavior is repeated in his observations of the

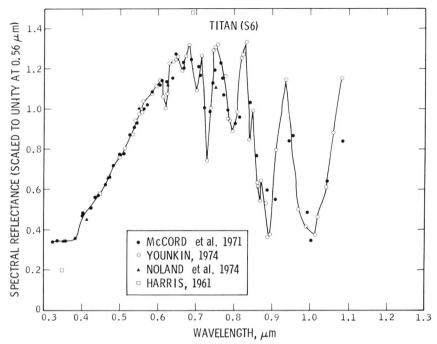

Fig. 11.10. Spectral reflectance of Titan. Lines have been drawn between Younkin's data points. Data scaled to unity at 0.56μ.

other Saturn satellites (Fig. 11.13), suggesting the presence of some systematic error in his R, and possibly in I, measurements.

Research presented at the NASA Titan workshop (Hunten, 1974) has helped us to a better understanding of the satellite's atmosphere. Trafton (1974b; see also Trafton 1972a,b) obtained a refined estimate of the methane abundance and reported a detection of hydrogen and perhaps other atmospheric constituents from high resolution spectral observations. A cloudy atmosphere is suggested both by polarimetric data (Veverka, 1973a, 1974, and Chapt. 10 herein) and atmospheric models (Lewis and Prinn, 1973; Danielson *et al.*, 1974; Caldwell, Chapt. 21). It seems likely that the similarity between the spectral reflectances of Titan and Saturn is due to the presence of similar chromophoric constituents (other than methane) in the clouds on both bodies. However, Titan's much lower geometric albedo suggests that more of the absorber must be present in its atmosphere than in Saturn's.

Iapetus

Iapetus is distinguished among the natural satellites of the solar system by the large amplitude (~2 mag.) of its lightcurve (cf. Chapt. 9, Veverka). A small

TABLE 11.6
Spectral Reflectance of Titan (Scaled to unity at 0.56μ)

McCord et al. (1971)		Younkin (1974)		Noland et al. (1974b)	
$\lambda(\mu)$	R_λ	$\lambda(\mu)$	R_λ	$\lambda(\mu)$	R_λ
0.319	0.343	0.5000	0.752	u(0.350)	0.34
0.338	0.345	0.5124	0.802	v(0.4110)	0.45
0.358	0.347	0.5264	0.870	b(0.4670)	0.65
0.383	0.353	0.5370	0.935	r'(0.6239)	1.12
0.402	0.472	0.5490	0.971	i'(0.7500)	1.10
0.402	0.489	0.5556	1.000		
0.424	0.504	0.5652	1.043		
0.433	0.560	0.5840	1.086		
0.444	0.569	0.5914	1.119		
0.461	0.625	0.6024	1.140		
0.467	0.655	0.6100	1.144		
0.480	0.715	0.6140	1.065		
0.498	0.774	0.6190	1.004		
0.507	0.777	0.6240	1.075		
0.522	0.875	0.6280	1.230		
0.532	0.905	0.6370	1.234		
0.544	0.931	0.6424	1.248		
0.564	1.000	0.6500	1.263		
0.568	1.026	0.6630	1.191		
0.581	1.084	0.6730	1.266		
0.598	1.112	0.6800	1.324		
0.607	1.146	0.7020	1.097		
0.620	1.135	0.7124	1.263		
0.633	1.153	0.7274	0.741		
0.642	1.274	0.7364	1.007		
0.665	1.235	0.7450	1.295		
0.668	1.202	0.7500	1.309		
0.684	1.250	0.7530	1.324		
0.699	1.211	0.7724	1.165		
0.703	1.168	0.7834	0.946		
0.720	1.004	0.7890	0.946		
0.730	0.989	0.7950	0.892		
0.730	0.983	0.8070	0.986		
0.743	1.174	0.8172	1.270		
0.749	1.198	0.8300	1.338		
0.762	1.228	0.8400	0.842		
0.765	1.155	0.8470	0.993		
0.768	1.060	0.8600	0.633		
0.782	0.993	0.8640	0.611		
0.802	0.927	0.8670	0.540		
0.809	0.955	0.8700	0.640		
0.843	1.034	0.8804	0.529		
0.855	0.765	0.8872	0.363		
0.884	0.599	0.8926	0.374		
0.906	0.550	0.9100	0.791		
0.943	0.836	0.9358	1.140		
0.948	0.869	0.9700	0.500		
0.990	0.485	0.9880	0.417		
1.002	0.345	1.0120	0.378		
1.053	0.635	1.0200	0.460		
1.083	0.834	1.0400	0.608		
		1.0600	0.885		
		1.0800	1.151		

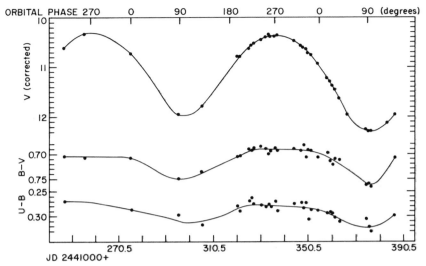

Fig. 11.11. Photometric measurements of the brightness and color (UBV system) of Iapetus in 1971–1972. The formal error in most of the points is ± 0.01 mag. From Millis (1973).

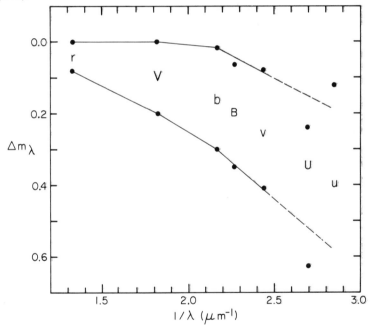

Fig. 11.12. Color of Iapetus relative to the Sun for wavelengths between 0.3 and 0.8μm, normalized to unity at 0.55μm. The UBV data are from Millis (1973); and the uvbyri data are from Noland et al. (1974b). The upper curve is the mean color of the bright material; the lower curve is the mean color of the dark ($p_V = 0.04$) material.

variation in U–B and B–V colors as well was first reported by Harris (1961) and later measured more precisely as a function of orbital phase by Millis (1973). His results, shown in Figure 11.11, showed that the variation might be somewhat larger in magnitude than that reported by Harris and that it occurred in phase with the satellite's lightcurve. Measurements using uvbyri filters, reported by Noland et al. (1974b) and discussed by Morrison et al. (1975) showed similar results.

The spectral reflectances of the bright and dark sides of Iapetus from the measurements of Noland et al., Millis, and McCord et al. (1971) are shown in Figure 11.12. McCord et al. measured the satellite at orbital phase angles near 165° and 340° and thus never obtained a spectrum of the dark side alone (visible only near 90° orbital phase). Their data are therefore shown in Figure 11.12 only for the bright side of the satellite. It can be seen that the dark side is considerably redder than the bright, a result also inferred by McCord et al. from their observations. The spectral reflectance of the bright side of Iapetus is consistent with, but not diagnostic of, some type of surface frost. The spectral reflectance and low albedo (geometric albedo ≈ 0.04; Morrison et al., 1975; Zellner, 1972b) of the dark side are consistent with a surface composed of dark silicates or possibly carbonaceous material. A measurement of the spectral reflectance of the dark side of Iapetus between 0.8 and 1.1 μ, where many iron-bearing silicates show characteristic absorptions, would be extremely useful in determining its surface composition.

Other Satellites of Saturn

Figure 11.13 (from Noland et al., 1974b), shows the normalized spectral reflectances of Titan, Rhea, Dione, Tethys, and Iapetus taken from the narrow-band measurements of McCord et al. (1971), the uvbyri measurements of Noland et al. (1974b), and the UBVRI observations of Harris (1961). As noted earlier in the case of Titan, the R albedos from Harris appear to be systematically higher than the other measurements. As discussed by McCord et al., the reflectances of Rhea, Dione, and Tethys are all consistent with frost surfaces on these satellites, but no positive identification can be made on the basis of existing data. In addition, the albedos of these satellites (cf. Morrison and Cruikshank, 1974; Morrison, 1974a; Chapt. 9, Veverka), while poorly determined compared with those of the Galilean satellites, are sufficiently high for frost to seem a reasonable surface constituent. No significant variations with orbital phase in the color of these satellites has been reported.

Although these satellites are too faint to be measured in the infrared using current interferometric spectral techniques, broadband measurements are possible. Johnson et al. (1975) reported broadband measurements of Rhea at 1.6 and 2.2 μ (H and K filters) that are shown, along with data for Ganymede and two laboratory frosts, in Figure 11.14. Rhea's low reflectance at these wavelengths, similar to that of Ganymede, is indicative of a frost covered surface. Virtually no

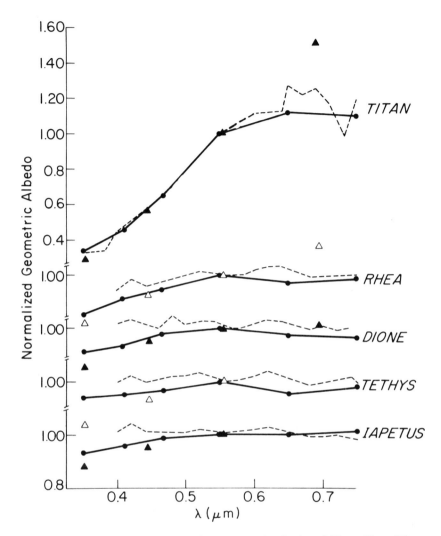

Fig. 11.13. Wavelength dependence of the geometric albedos of Titan, Rhea, Dione, Tethys and Iapetus (at $\theta = 0°$), normalized at 0.55μ (y filter) from Noland *et al.* (1974b). Data of Harris (1961, triangles) and McCord *et al.* (1971, dashed curve), similarly normalized, are also shown.

common rock material (with the exception of highly hydrated minerals) shows such a low infrared reflectance. It is interesting to note that it was exactly this type of evidence that led Kuiper to correctly conclude in the late 1950s that Europa and Ganymede had water frost on their surfaces.

The only other satellite in this system for which photoelectric data are avail-

RHEA (S5)

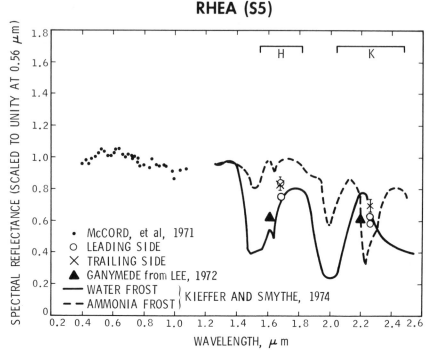

Fig. 11.14. The spectral reflectance of Rhea (scaled to unity at 0.56μm) is shown. Data from 0.4–1.1μm are taken from McCord *et al.* (1971). Data for three observations of Rhea at 1.68 and 2.26μ are plotted, circles for eastern elongation (leading side) and crosses for western elongation (trailing side). Triangles are data at 1.62 and 2.2μ for Ganymede (J3) from Lee (1972). Solid curve is the reflection of water (H_2O) frost and the dashed curve for ammonia (NH_3) frost scaled at 1.25μ to match the trend of the short wavelength data. Both frost curves are taken from Kieffer and Smythe (1974). The half power points of the H and K infrared bandpasses are 1.55–1.81μ and 2.04–2.48μ respectively. Representative error bars (estimated error) are indicated for the data points for the trailing side of Rhea. From Johnson *et al.* (1975).

able is Enceladus. Broadband observations (UBV system) were reported by Harris (1961) and by Franz and Millis (1973; 1975). The B–V color of this satellite is similar to those of the other inner satellites of Saturn.

[*Note added in proof:* Gautier *et al.* (1976), using Fourier interferometer/ spectrometer data, confirm the suggestion of H_2O frost on Rhea's surface and indicate that Iapetus, Dione, and Tethys also have frost covers.]

SATELLITES OF URANUS

The satellite system of Uranus consists of five regular satellites (U5, Miranda; U1, Ariel; U2, Umbriel; U3, Titania; U4, Oberon) about which very little is

TABLE 11.7
Magnitudes and Colors of Uranus' Satellites (from Harris, 1961)

Satellite	V_0	B − V	U − B
Miranda (U5)	16.5	—	—
Ariel (U1)	14.4	—	—
Umbriel (U2)	15.3	—	—
Titania (U3)	14.01	0.62	0.25
	(14.01±0.06)*	(0.71±0.11)*	(0.30±0.16)*
Oberon (U4)	14.20	0.65	0.24
	(14.27±0.01)*	(0.68±0.03)*	(0.20±0.06)*

* Andersson (1974)

TABLE 11.8
**Magnitudes and Colors of
Neptune's Satellites***

Satellite	V_0	B − V	U − B
Triton (N1)	13.55	0.77	0.40
Nereid (N2)†	18.7		

* From Harris (1961)
† Based on a photographic magnitude of 19.5 determined by Kuiper and an estimated B − V of 0.8.

known. Photoelectric data for Titania and Oberon reported by Harris (1961; Table 11.7) show the colors of these satellites to be about the same as those of the inner satellites of Saturn. In the absence of any measurements of the satellites' radii, and hence any knowledge of their albedos, no firm statements can be made regarding their surface compositions except to rule out highly colored materials such as lunar soils. For example, the colors of the satellites are consistent with both high albedo frosts and low albedo carbonaceous material. Although they are difficult to observe, it should be possible to obtain spectrophotometric measurements of these objects in the 0.3–1.1μ region using existing instrumentation. It may also be possible in the near future to obtain broadband near-infrared data similar to those reported for Rhea by Johnson *et al.* (1975).

SATELLITES OF NEPTUNE

As is the case for the Uranian system, we know little concerning the optical properties of Neptune's two irregular satellites, Nereid and Triton. The magnitude of Nereid has only been measured photographically (Harris, 1961). The colors of Triton, measured photoelectrically by Harris (1961; Table 11.8), are similar to those of the inner satellites of Saturn and Uranus, but the absence of any measurements of Triton's albedo introduces the same uncertainties into discussions of its surface composition as were mentioned in the discussion of the satellites of Uranus.

ACKNOWLEDGMENT

We gratefully acknowledge the help given us by all our colleagues in collecting the material for this review; many participants in IAU Colloquium No. 28 kindly provided pre-publication versions of their papers. Special thanks are due to D. Morrison and D. Cruikshank for their cooperation and for their earlier satellite review article which greatly aided preparation for this paper. This work represents one phase of research carried out at the Jet Propulsion Laboratory, California Institute of Technology, under contract NAS 7-100, sponsored by the National Aeronautics and Space Administration.

12

RADIOMETRY OF SATELLITES AND OF THE RINGS OF SATURN

David Morrison
University of Hawaii

Brightness temperatures at wavelengths between 8 μ and 3.7 cm are tabulated for satellites of known size; except for Titan, they are consistent with blackbody emission values. For Titan a more complex thermal radiation spectrum with three separate regimes is identified. Radiometric brightness measurements, when used in conjunction with photometry, are shown to determine satellite sizes and albedos; this technique is calibrated by using satellites and asteroids with radii evaluated from other methods. Eclipse radiometry indicates the thermophysical properties of surface layers; such observations show that the uppermost coverings of Phobos, Callisto, and Ganymede have remarkably low thermal conductivities. Infrared and radar detections of the rings of Saturn are surveyed; the important constraints that they provide on ring particle sizes and bulk composition are outlined.

Most studies of the surfaces and atmospheres of the planets and satellites deal with reflected sunlight (such studies are reviewed in Chapt. 9 by Veverka and Chapt. 11 by Johnson and Pilcher); however, it also is important to study that component of solar energy that is absorbed and reradiated at longer wavelengths. In this review, the term "radiometry" will denote observations of thermal radiation, both microwave and infrared. For the satellites, the wavelength beyond which thermal emission dominates over reflected sunlight varies from about 5 to 10 μ, depending on the temperature and infrared albedo of the object. This distinction between emitted and reflected radiation is not valid in stellar studies, of course, and observations called "radiometry" when made of objects in the solar system would usually be called "photometry" when made of stars.

Ground-based radiometric observations must be made at wavelengths for which the atmosphere of the Earth is at least partially transparent. The transparency of the atmosphere is even more important in radiometry than it is in visible photometry, because the atmosphere serves as a major source of emitted thermal radiation as well as an absorber of radiation from astronomical sources. Most infrared radiometry of satellites and planets has been obtained in the two broad atmospheric windows centered at wavelengths of approximately 10 and 20μ. Short of about 8.0μ, the atmosphere is virtually opaque down almost to 5.0μ due to absorption bands of CO_2 and H_2O. The 10μ window extends from 8 to about 13.5μ, where the wings of the 15μ CO_2 band again produce total opacity. Within the 10μ window the average transmission is nearly 90%, about the same as in the visible, except within a weak absorption band due to O_3 between 9.3 and 10.0μ.

[269]

The 20μ window opens up at 16μ and extends out to approximately 26μ. This window is not nearly as transparent as that at 10μ. Since the main source of opacity is the grouping of H_2O lines distributed throughout the window, there is a substantial advantage to obtaining 20μ observations from dry sites. The average transmission between 17 and 24μ varies, according to the dryness of the atmosphere, between a maximum of about 80% and a lower limit for practical observations of perhaps 40%. There is no sharp cutoff in transparency at the long-wave limit of the 20μ window, and at dry sites significant transmission (up to 25%) exists at windows out to 45μ. This transmission has been exploited by several groups to obtain broadband radiometric measurements with an effective wavelength of about 34μ. Beyond 45μ, the atmosphere is opaque for more than two octaves. The next windows are in the submillimeter, at about 350 and 450μ, and near 1 mm, but these have not been utilized for measurements of satellites. Beyond 1 mm one enters the radio part of the spectrum, where several windows at millimeter wavelengths give almost perfect atmospheric transparency beyond 1 cm. Several radiometric measurements discussed in this chapter have been made at microwave wavelengths of 2.8 to 21 cm.

Planetary satellites were beyond the reach of radiometric techniques until the development in the early 1960s of cooled detectors—bolometers and photo-conductors—of much higher sensitivity than the thermopiles and other devices used during the preceding half century (cf. Pettit, 1961; Sinton, 1961). The first successful radiometry of satellites consisted of observations of the Galilean satellites in the 10μ window by Murray et al. (1964a) at Hale Observatories and by Low (1965) at the University of Arizona. Murray et al. were able to measure the brightness temperatures of Ganymede and Callisto only. Low reported temperatures for all four Galilean satellites and Titan and an upper limit for the temperature of the rings of Saturn. These temperatures were derived from calibrations that were uncertain by as much as 40% in the flux, however; in addition, neither paper gives the radii assumed in the calculation of the brightness temperatures, so that these first infrared measurements are now of historical interest and will not be further discussed here. The first reliable measurements of radiation from satellites in the much lower frequency microwave domain were not made until 1974.

Most radiometric flux measurements of satellites are reported as a brightness temperature, defined as the temperature of an isothermal black disk of the same angular size as the satellite that radiates an equivalent flux density within the bandpass to which the photometer is sensitive. This terminology is unfortunate, for it suggests undue physical significance to the data. Since the satellite is not isothermal and does not radiate as a blackbody, the brightness temperature must not be interpreted as the subsolar surface temperature, the disk-averaged surface temperature, or any other physical temperature. In addition, to convert the measured flux density to a brightness temperature, it is necessary to know the angular size of the object. For the satellites, the sizes are generally not well known, and any error in the size is carried into the brightness temperature. One should realize that, at best, brightness temperature is a derived quantity convenient for specifying the disk-averaged flux density within a spectral passband, provided the assumptions required for its derivation are clearly understood. It is in this spirit that I use the term in this chapter.

In the discussion of infrared radiometry in this review, I will not attempt to place the results obtained by different observers on the same absolute flux scale. Before about 1971, there existed differences of 25% or more among the flux densities adopted for standard calibration sources by various groups of observers; in addition it is frequently unclear from many published reports just what standards were used or what flux scale was adopted. However, during the past four years a reasonable consensus has developed among observers, as indicated by comparisons of the standard sources published by, *e.g.*, Morrison and Simon (1973), Becklin *et al.* (1973), and Low and Rieke (1974a). Most of the observations reviewed here have been published within the past four years, and the flux scales used in the 10μ and 20μ spectral regions agree among observers to within 10% and have absolute errors thought to be less than 15%. For most satellites, these uncertainties are smaller than those associated with the solid angle of the source, so that the dominant source of error in the brightness temperature usually arises from the assumed diameter and/or the random uncertainties in the measurements themselves.

RADIOMETRY OF SATELLITES OF KNOWN SIZE

The diameters of two satellites, Io and Ganymede, have been measured with high precision from photoelectric timings of stellar occultations (Taylor, 1972; O'Leary and van Flandern, 1972; Carlson *et al.*, 1973; Chapt. 13, O'Leary). Two others, Phobos and Deimos, have been measured on high-resolution images obtained by the Mariner 9 television cameras (Pollack *et al.*, 1972, 1973a; Chapt. 15, Duxbury). Ground-based visual estimates of diameter have been made in addition for Europa, Callisto, and Titan (cf. Dollfus, 1970). As discussed by Morrison and Cruikshank (1974) (cf. Chapt. 1, Table 1.4, Morrison *et al.*), these visual sizes for the two Galilean satellites are probably accurate to ± 5%, but the visually measured diameter of Titan might be subject to much larger errors as a result of possible limb darkening. The photometry of a lunar occultation of the latter satellite by Elliot *et al.* (1975) appears to confirm the existence of substantial limb darkening and indicates a radius of 2,900 km, about 15% larger than that obtained by Dollfus (1970). In this chapter, I also adopt a slight revision in the radius of Europa, from 1,550 ± 150 km to 1,500 ± 100 km, based on preliminary reductions of occultations of this satellite by Io (Aksnes and Franklin, 1975; Vermilion *et al.*, 1974). In all, then, seven satellites have reasonably well-determined sizes, and six of them have also been observed radiometrically.

Modern determinations of infrared flux densities from the Galilean satellites in the 10μ and 20μ atmospheric windows have been given by Gillett et al. (1970), Hansen (1972), Morrison et al. (1972), and Morrison and Cruikshank (1973, 1974), and microwave measurements of Callisto and Ganymede have been reported by Berge and Muhleman (1974) and Pauliny-Toth et al. (1974). Infrared observations of Titan have been published by Allen and Murdock (1971), Morrison et al. (1972), Gillett et al. (1973), Low and Rieke (1974b), and Knacke et al. (1975); Briggs (1974a) has published a microwave temperature. Finally, observations of Phobos with the radiometer on Mariner 9 have been made by Gatley et al. (1974). The brightness temperatures derived from these observations and the radii assumed in the calculations are summarized in Table 12.1. In the case of Phobos, which is the only satellite observed over a wide range of phase angles, the temperature is given for zero phase. In most cases, the temperatures are the same as those given by the original observers, but for Titan all of the published temperatures have been modified to correspond to the new diameter obtained by Elliot et al. (1975).

In general, the brightness temperatures given in Table 12.1 are consistent with approximately blackbody emission from surface elements in equilibrium with the insolation for Phobos and the Galilean satellites, while for Titan the apparent emission spectrum is much more complex. We will discuss the radiometry of Titan later (see also Chapt. 20 by Hunten and Chapt. 21 by Caldwell).

Sufficiently accurate broadband radiometry can, in principle, be combined with the known radii and geometric albedos of the Galilean satellites to derive either the photometric phase integral q or the thermal emissivity in the forward direction. To understand this procedure, recall that the phase integral is defined at any wavelength as the ratio of the Bond albedo A_λ to the geometric albedo p_λ. The "bolometric" phase integral q is then defined as the ratio (A/p), where the unsubscripted quantities indicate the average or bolometric values. Given the radius, and the visible and near-infrared photometry, we can calculate p for these satellites, while the radiometric flux density is proportional to $(1-A)$ and to the emissivity. Thus, either q or the emissivity can be found, but not both. Hansen (1972) has utilized this approach to derive values of the phase integral on the assumption that the emissivities of the Galilean satellites must, averaged over all angles, be near unity at 10μ and must show approximately the angular dependence observed for the Moon. Since the thermal conductivities of the satellites are very low (see under *Eclipse Radiometry*), it is possible to ignore radiation from the un-illuminated hemisphere. Hansen finds that the observed brightness temperatures are a few degrees lower than those calculated with a lunar value for the phase integral; therefore, the value of A must be larger than expected, and q must be larger than the lunar value. The method is most useful for objects of high albedo, such as Io and Europa, for which Hansen finds values of q in the range 0.8 to 1.0.

Measured Brightness Temperatures of Satellites

λ[μ]	Δλ[μ]	Phobos (10.9 km)	Io (1820 km)	Europa (1500 km)	Ganymede (2635 km)	Callisto (2500 km)	Titan (2900 km)	
8.4	0.8		149±3	134±3	145±3	160±3	143±5	Gillett et al. (1970)*
8.4	0.8						134±3	Gillett et al. (1973)
8.8	1.0	296±15						Low and Rieke (1974b)
10.2	4.3						122±2	Gatley et al. (1974)
10.3	1.3		137±3	130±3	142±3	152±4		Low and Rieke (1974b)
10.6	5.0						121±2	Hansen (1972)
10.6	5.0							Low and Rieke (1974b)
11.0	5.0							Morrison (1975b)
11.0	2.0		138±4	129±4	145±4	153±5	131±2	Gillett et al. (1970)*
11.0	2.0		139±3	131±3	142±3	157±3	125±2	Gillett et al. (1973)
11.6	0.8						129±1	Low and Rieke (1974b)
12.0	2.0						123±2	Gillett et al. (1973)
12.4	4.0						126±2	Allen and Murdock (1971)
12.6	1.0						98±2	Low and Rieke (1974b)
17.0	2.0						90±2	Low and Rieke (1974b)
17.8	1.0						90±2	Knacke et al. (1975)
18.4	1.0						95±2	Knacke et al. (1975)
19.0	1.0						90±2	Low and Rieke (1974b)
21.0	6.0						91±2	Knacke et al. (1975)
21.0	8.0						91±2	Low and Rieke (1974b)
21.0	8.0		124±4	120±4	132±5	142±6		Hansen (1972)
21.0	10.0		128±5	121±5	138±5	151±7		Morrison et al. (1972)
21.0	10.0		130±3	121±3	143±4	155±5		Morrison (1975b)
21.5	7.3	297±15						Gatley et al. (1974)
22.5	5.0						89±2	Low and Rieke (1974b)
24.5	1.0						84±4	Low and Rieke (1974b)
34.0	12.0						80±4	Low and Rieke (1974b)
2.8 cm					55±14	88±18		Pauliny-Toth et al. (1974)
3.7 cm						101±25		Berge and Muhleman (1974)
3.7 cm							87±26	Briggs (1974a)

* Calibration and assumed radius not given

TABLE 12.1b
Titan Narrowband Measurements
(Gillett *et al.* 1973)

$\lambda(\mu m)$	$T_B(°K)$
8.0	158±4
9.0	130±6
10.0	124±3
11.0	123±3
11.5	128±2
12.0	139±2
12.5	129±2
13.0	128±2

A second way of estimating the phase integral for satellites of high albedo has been suggested by Morrison (1974c). If one assumes that the emissivities in both the 10μ and 20μ bands are equal and near unity, then a radiometric color temperature, which is sensitive to the value of A, can be derived from the ratio of the fluxes at these two wavelengths, independent of the directionality of infrared emission from the satellites and of their angular size. It can be shown that this ratio for most asteroids implies essentially blackbody radiation, with A < 0.2. The same argument applied to the Galilean satellites gives q = 0.6 ± 0.3 for Io, q = 0.9 ± 0.2 for Europa, and q = 1.0 ± 0.3 for Ganymede. Again it should be emphasized that these conclusions are meaningful only if the satellites radiate as black bodies. Chapter 9 by Veverka presents further discussion on p, q and A.

Although the surface materials postulated for the Galilean satellites are expected to have emissivities that are not strong functions of wavelength in the 10μ and 20μ spectral regions, there is some observational evidence of significant spectral structure. From observations made on one night in 1973, Hansen (1974) reported a substantial emission peak near 12μ on Io, Ganymede, Callisto, and the asteroid 1 Ceres. Sinton (1974) observed Io with a somewhat different filter set and found no 12μ features, but he suggested a possible smaller feature near 10.5μ. Hansen (1972) also suggested on the basis of earlier broadband radiometry that there is a deficiency in the 20μ flux from Io, and possibly Europa, relative to that in the 10μ band. Such a deficiency would, of course, result in too high a color temperature and therefore in an anomalously small value for the phase integral if the method described in the previous paragraph were applied, and thus may explain the low value of q obtained for Io by Morrison (1974c). However, in general, I conclude that the quality of the data is such that none of the Galilean satellites has been shown to have an emissivity that is significantly different from unity. [Hansen (1976) has since withdrawn his previous report of a 12μ emission feature on Ceres and the Galilean satellites.]

The observations by Hansen (1972), primarily at 10μ, and by Morrison and Cruikshank (1973) at 20μ, suggested variations in radiometric brightness with orbital phase of up to $\pm$ 0.5 mag for the inner three Galilean satellites. No significant variations are expected for Callisto, due to its low albedo. These variations can also be used to derive mean phase integrals for the satellites. Since the amplitude of variation in the visible is proportional to the variation in p, while the radiometric amplitude is proportional to the variation in $(1-A)$, it is possible, if the radius is known and the assumption is made that q does not vary greatly with wavelength or with position on the satellite, to solve directly for q.

Figure 12.1 shows the 20μ flux curves obtained for Io, Europa, and Ganymede by Morrison (unpublished). It is clear that for orbital phase angles 90° $< \theta <$ 360°, the radiometric curves are shifted in phase by half a cycle with respect to the photometric lightcurves (cf. Figs. 16.1–3, Morrison and Morrison), as would be expected if the variations in both spectral regions were due to variations of albedo with longitude on spherical satellites. These observations can be used to solve for the phase integrals of Io and Europa, based on a model similar to that described for Iapetus by Morrison *et al.* (1975), in which it is assumed that (i) the value of q is the same at all longitudes, and (ii) the albedo distribution can be represented by two uniform hemispheres, one dark and one light. If the diameters are constrained to be within the error bars shown in Table 12.3, the values of the phase integrals (defined without the opposition surge) are q = 0.9 ± 0.4 for Io and q = 1.1 ± 0.3 for Europa. The analysis of the 20μ lightcurve of Iapetus by Morrison *et al.* (1975) discussed in the next section yields for the bright side of this satellite q = 1.0 to 1.5. Most of the uncertainty in the results for all three satellites is due to the uncertainties in the radiometric observations.

The 20μ flux curves illustrated in Figure 12.1 and used to derive values of q for Io and Europa show that both satellites remain near their minimum radiometric brightness for 0° $< \theta <$ 90°. If the radiometric curves were complementary to the photometric lightcurves, the 20μ flux would drop from near its mean value at $\theta = 0°$ to a minimum near $\theta = 90°$. Anomalous thermal behavior is further indicated by the 20μ measurements made in the hour after the end of eclipse (Morrison and Cruikshank, 1973), illustrated in Figure 12.2. There it is clear that the flux recovers to only 70–80% of its pre-eclipse value for Io and to only about 50% for Europa. No satisfactory explanation for this peculiar behavior has been offered. Such an effect could be produced if (i) the satellite develops a substantial heat sink during eclipse, so that its surface does not reach thermal equilibrium with the insolation for many hours after the eclipse, or (ii) q is much smaller on the part of the satellite that faces us during the quarter of an orbit after eclipse than elsewhere. Neither of these possibilities seems very satisfactory.

For satellites with lower albedos, none of the above methods of determining the phase integral and hence the Bond albedo is very sensitive. However, we

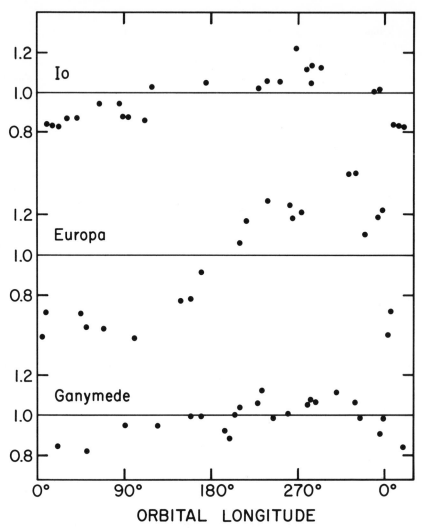

Fig. 12.1. Variation of 20μ brightness of Io, Europa, and Ganymede as a function of orbital longitude. From a preliminary reduction of unpublished observations by D. Morrison.

might expect the phase integrals of darker surfaces such as those of Callisto, Ganymede, and the dark side of Iapetus to be intermediate between those of Io and Europa on the one hand and the Moon and Mercury on the other. Also, Veverka (1970) has suggested upper bounds on q of ~ 0.8 for Ganymede and ~ 0.5 for Callisto from their phase coefficients (cf. Chapt. 9, Veverka). Table 12.2 summarizes all of the determinations of q and A for the Galilean satellites

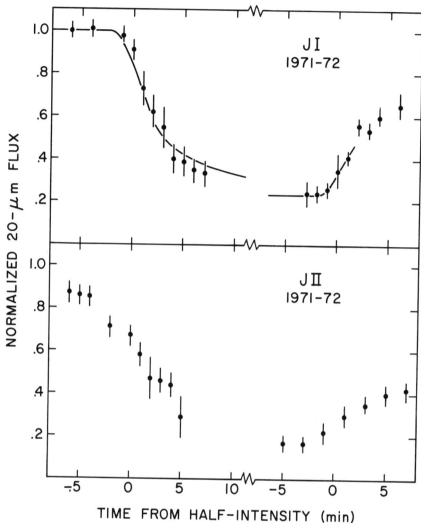

Fig. 12.2. Average 20μ eclipse curves for Io and Europa. The data have been aligned to place the half-intensity of the photometric lightcurves at time t = 0. Note that neither satellite recovers to its pre-eclipse brightness. From Morrison and Cruikshank (1973).

discussed in this paper. It is clear from these results that, while these phase integrals are not determined with high precision, the radiometric studies generally agree on the range of values appropriate to these satellites.

Phobos is so dark ($p_v \simeq 0.06$; cf. Chapt. 9, Veverka) that it can be treated as a black body for purposes of calculating its brightness temperature, as shown by Gatley et al. (1974). In Figure 12.3 the Mariner 9 observations at 10μ are compared with a model in which the illuminated surface is black and no thermal radiation is emitted by the unilluminated surface. This model is consistent with the observations, which are not of course of extremely high precision. Note that Phobos is the only satellite for which a radiometric phase curve has been obtained.

The first radio detection of a satellite was reported by Gorgolewski (1970), who obtained a brightness temperature for Callisto at 3.5 mm of 255 ± 80 °K. This value is substantially higher than the anticipated average subsurface temperature of about 100 °K, but the uncertainty is too great to warrant much interpretation. Much more reliable microwave measurements have recently been made of this satellite by Berge and Muhleman (1974), who obtained $T_B = 101 ± 25$ °K at a wavelength of 3.75 cm, and by Pauliny-Toth $et\ al.$ (1974), who obtained $T_B = 88 ± 18$ °K at 2.82 cm. These values agree with each other and with simple theory. Pauliny-Toth $et\ al.$ also measured Ganymede at 2.82 cm, obtaining $T_B = 55 ± 14$ °K. This temperature is significantly below the expected value of ~ 90 °K, but no explanation for such a low value is apparent, and judgment should probably be deferred until the measurement has been repeated.

All of the satellite radiometry discussed so far is consistent with approximately blackbody or greybody thermal emission mechanisms. However, the wavelength dependence of the brightness temperature of Titan is in striking contrast to this behavior, and its discovery has contributed greatly to the recent interest in this satellite (cf. Hunten, 1974, and Chapt. 20 herein; Chapt. 21, Caldwell). The gross features of the spectrum—a high brightness temperature near 10μ but low near 20μ—were initially interpreted to indicate a large greenhouse effect, based primarily on pressure-induced opacity in molecular hydrogen at wavelengths longer than about 15μ (cf. Morrison $et\ al.$ 1972; Pollack, 1973; Sagan, 1973). However, an alternative model in which the excess 10μ flux originates in a temperature inversion high in the atmosphere has been proposed by Danielson $et\ al.$ (1973) (cf. Chapt. 21, Caldwell) and supported with additional observations by Gillett $et\ al.$ (1973). Since the details of these opposing theories are discussed in Chapter 20 by Hunten and in Chapter 21 by Caldwell, only a few brief comments on the observations are appropriate here.

Table 12.1a lists all of the broadband and intermediate band radiometric measurements of Titan. There exist in addition eight flux measurements in the 8μ to 13μ band obtained by Gillett $et\ al.$ (1973) with a spectrophotometer with a resolution of almost 0.15μ; these data are listed in Table 12.1b. All of the observed infrared flux densities of Titan are plotted in Figure 12.4 (cf. Figs. 20.3, 21.1; Table 20.1).

Figure 12.4 suggests at least three separate regimes for the thermal radiation from Titan. The first, and that which is responsible for most of the radiated energy, extends from the start of the 20μ atmospheric window (about 16μ) to longer wavelengths. Most of the observed brightness temperatures in the region

TABLE 12.2
Phase Integrals and Bond Albedos for the Galilean Satellites

Parameter	Io	Europa	Ganymede	Callisto	Reference
p_V	0.63±0.02	0.68±0.09	0.43±0.02	0.17±0.02	Morrison and Morrison (Chapt. 16)
q	<0.6	<1.2	<0.8	<0.5	Veverka (1970)
q	0.9±0.2	0.9±0.2	1.2±0.4	—	Hansen (1972)
q	0.6±0.3	1.0±0.2	1.0±0.3	—	Adopted, Morrison (1974c)
q	0.9±0.4	1.1±0.3	—	—	This chapter
q	0.9±0.2	1.0±0.2	1.0±0.3	0.8±0.4	Adopted (this chapter)
p	0.62±0.02	0.58±0.08	0.38±0.02	0.16±0.02	Adopted (this chapter)
A	0.56±0.12	0.58±0.14	0.38±0.11	0.13±0.06	Adopted (this chapter)
$1-A$	0.44±0.12	0.42±0.14	0.62±0.11	0.87±0.06	Adopted (this chapter)
T_{max} (°K)	141±11	139±12	154±6	167±3	Adopted (this chapter)

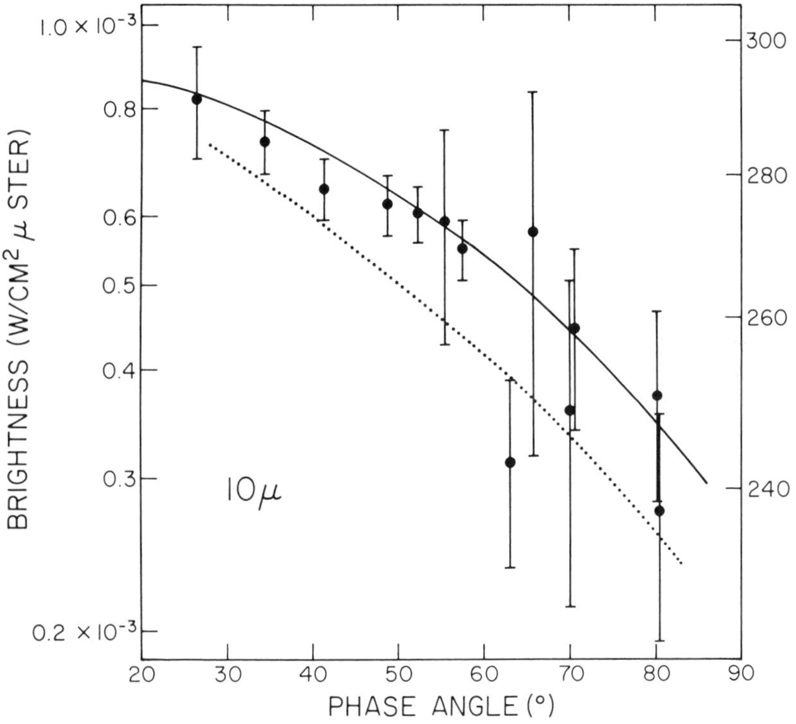

Fig. 12.3. Brightness temperature at 10μ of Phobos as a function of phase angle from Mariner 9 observations. The solid curve corresponds to the predicted brightness of a totally absorbing satellite with zero thermal conductivity. The dotted curve represents an average of the predictions for a totally absorbing surface with a thermal inertia of 2.5×10^{-3} cal cm^{-2} sec$^{-1/2}$ (°K)$^{-1}$. From Gatley *et al.* (1974).

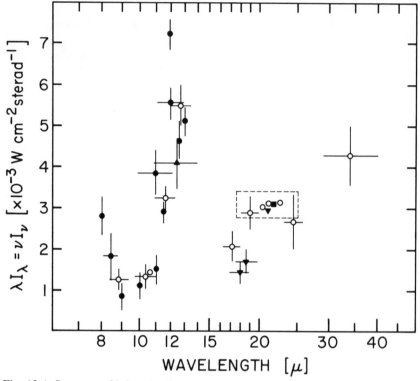

Fig. 12.4. Summary of infrared radiometry of Titan. All specific intensities are adjusted to correspond to a diameter of 5800 km. Approximate bandwidths and error bars are illustrated for most points. In the 20μ window, the dashed box delineates the region within which several broadband results fall. The scales of the figure are such that area under the curves is proportional to emitted energy.

are in the vicinity of 90 °K, and the radiation can therefore be characterized as approximately Planckian. The second region lies near the long-wave end of the 10μ window (*i.e.*, 11μ to 13μ) and may extend out to near 16μ. There is a very strong and narrow peak in the radiated flux near 12μ, and the highest resolution data suggest that several emission lines may be present. Since the 12μ peak is too narrow to be due to Planckian radiation, there seems to be little doubt that molecular emission from a hot atmosphere is responsible for much of this radiation. The third region extends from 9μ shortward and is observed up to the atmospheric cutoff just short of 8μ. The narrowband observations clearly suggest molecular radiation here also, and the obvious candidate is the 7.7μ band of CH_4, which is known to be present in the atmosphere of Titan (Danielson *et al.*, 1973). The 8μ brightness temperature of Titan (~155 °K) is the highest measured at any wavelength, but as is clear from Figure 12.4, radiation in this band plays a fairly minor role in the radiation balance of the satellite. This temperature does set a lower limit, however, on the physical temperature in the atmosphere where the band originates. As one proceeds toward still shorter wavelengths, the next spectral window is at 5μ, where the apparent brightness temperature of about 160 °K (Low and Rieke, 1974b; Knacke et al., 1975) is presumably due primarily to reflected solar radiation and not to thermal radiation from the satellite.

All of the observations illustrated in Figure 12.4 have been made at wavelengths where the atmosphere of Titan, if sufficiently massive, might well be expected to have substantial opacity. Thus, unlike the case of the other satellites, there is no necessity to associate any of these measured brightness temperatures with the surface temperature. At centimeter wavelengths, however, the atmosphere should be transparent, so that the detection of Titan at 3.7 cm by Briggs (1974a) assumes a particular importance. His observation, adjusted to the new occultation diameter of 5800 km (Elliot et al., 1975), yields a brightness temperature of 87 ± 26 °K, and for plausible emissivities the corresponding mean surface temperature is 102 ± 34 °K. This value is in entirely satisfactory agreement with the expected equilibrium temperature of Titan, computed on the assumption that the atmosphere is sufficiently massive to result in a nearly isothermal surface (cf. Leovy and Pollack, 1973) of 85 ± 3 °K. Although the measurement is insufficiently precise to rule out a substantial greenhouse effect, resulting in a surface temperature approaching 140 °K, it appears to exclude some of the models discussed by Sagan (1973) and Pollack (1973), in which surface temperatures in excess of 150 °K are postulated. In the dispute between greenhouse models and temperature inversion models for the atmosphere of Titan, the weight of the radiometric evidence, while not yet conclusive, now tends to favor the latter. The greatest potential for further radiometry of this satellite now appears to be in continued spectrophotometry in the 10μ band with the highest possible spectral resolution and in efforts to improve upon the precision of the measured microwave brightness temperature.

RADIOMETRY OF SATELLITES OF UNKNOWN SIZE

If the diameter of a satellite is not known, the brightness temperature cannot be derived from a radiometric measurement. However, as was originally demonstrated by Allen (1970) and Matson (1971), it is possible to use a measurement of the radiometric brightness together with simultaneous visible photometry to derive the size of such an object. This photometric/radiometric technique was first applied to satellites by Murphy et al. (1972), who obtained radii for Rhea and Iapetus from 20μ radiometry. It has since been described in detail and applied to several satellites by the author and his colleagues (Morrison, 1973a, 1974a; Jones and Morrison, 1974; Morrison et al., 1975).

The principle of the photometric/radiometric technique is simple. The fraction A of the total sunlight striking the body is reflected while the fraction (1–A) is absorbed. If the illuminated surface is everywhere in equilibrium with the insolation, all of the absorbed energy will be reradiated in the thermal infrared, while the reflected component can be measured by conventional photometric techniques. A simultaneous measurement of both the reflected and the emitted radiation thus yields the Bond albedo and hence the diameter. However, in practice it is possible to measure only that part of the reflected and thermal radiation which is directed toward the observer, that is, the visible and infrared brightness at approximately zero phase. In order to solve for the diameter, it is necessary to model the relationship between the geometric and Bond albedos (given by the phase integral q) and between the infrared emissivity at zero phase and that averaged over all angles. Since systematic errors can be introduced by modelling these effects and by any inconsistencies between the photometric and radiometric flux scales, calibration of the technique against objects of known diameter is required.

The Galilean satellites have provided the best calibration for determination of radiometric diameters (cf. Morrison, 1973a; Jones and Morrison, 1974). Unfortunately, even these satellites are insufficiently well understood to serve as really satisfactory calibrators. Only two, Io and Ganymede, have well determined sizes, and both of these objects may have enough atmosphere (Chapt. 13, O'Leary) to modify their photometric and radiometric properties relative to those of airless satellites and asteroids (cf. Cruikshank, 1974). Io, in addition, appears to have a unique surface composition and may be substantially influenced by its interaction with the Jovian magnetosphere (cf. Morrison and Burns, 1976; and Chapt. 11, Johnson and Pilcher; Chapt. 17, Fanale et al.). Finally, the albedos of these satellites are fairly high, whereas the radiometric technique has the least ambiguities when applied to dark objects. The advantage of low albedo arises from the fact, pointed out by Morrison (1973a), that the technique yields the *geometric* albedo directly for dark objects and is insensitive to the value of the phase integral. This situation obtains because, as A approaches zero, the thermal flux becomes insensitive to changes in A, while the photometric brightness remains proportional to the geometric albedo. In contrast, for objects of high albedo the application of this technique requires a knowledge of q, and in fact it is just this sensitivity to q that was used by Hansen (1972) to estimate the phase integrals of Io and Europa. Callisto has low albedo, large distance from Jupiter, and high infrared brightness; if its diameter were better known, it would probably provide the best calibration.

Very little is known about the radiometric phase curves of satellites, and there is no familiar parameter analogous to q to express the relationship between emissivity normal to the surface and averaged over all angles. Aumann and Kieffer (1973), however, do define such an "anisotropy index," which they call

Q, for infrared emission. Initial applications of the photometric/radiometric technique simply assumed that the effective emissivity was unity at small phase angles. However Jones and Morrison (1974) noted that infrared studies of the Moon clearly show departures from this assumption, as indicated for instance by the fact that the 10μ brightness temperature at the center of the full Moon is about 12 °K higher than that expected for a black body. They therefore introduced in their thermal models a free parameter, T_0, defined as the subsolar *brightness* temperature at 1 AU from the Sun, and they selected the value of T_0 that gave the most accurate diameters for the Galilean satellites. Adjusting this parameter allows, in effect, for variations in the total emissivity and in the thermal limb darkening as well as in the ratio between normal and average emissivity. They concluded that $T_0 \cong 400$ °K gave the most accurate results, although using data available for the Galilean satellites they could not exclude values as large as $T_0 = 408$ °K. Fig. 12.5 illustrates the dependence of the derived diameter on both T_0 and q for a light object (Io) and a dark object (1 Ceres). An independent check of these models is provided by the diameter of Iapetus measured from the photometry of the 30 March 1974 lunar occultation (Elliot *et al.*, 1975). As discussed by Morrison *et al.* (1975), the occultation diameter is significantly smaller than the radiometric value obtained with $T_0 = 400$ °K, suggesting that a more accurate solution for this satellite will be obtained with $T_0 \geq 408$ °K. In addition, a comparison of diameters obtained for asteroids by the polarimetric method and the radiometric method (Chapman *et al.*, 1975; eqn. 10.2 gives the first method) shows that for these objects (which have lower albedos than most satellites) the agreement between the two methods is better with $T_0 = 408$ °K. Thus, there may be more peaking of the emissivity near zero phase angles for dark objects than for light ones, just as dark objects tend to have larger photometric phase coefficients. Indeed, it should probably not surprise us if the attempt by Jones and Morrison (1974) to find a single radiometric parameter that would apply to all airless objects proves to be an oversimplification, just as it would be a mistake to assume that all of these objects have the same photometric phase integral.

In spite of the uncertainties in calibration discussed above, the photometric/ radiometric method is still a very useful way to determine the diameters and albedos of satellites and asteroids. If we use the formulation by Jones and Morrison (1974) and adopt values of T_0 anywhere in the range 400 to 410 °K, the diameters obtained should be correct to ± 10%. This conclusion applies, of course, only to objects with surfaces essentially in equilibrium with the insolation; the presence of substantial atmospheres, rapid rotation, or nonsolar sources of energy will render it invalid. These restrictions do not limit the application of the technique to most satellites, however, and only for Titan and perhaps Amalthea (J5) is it clearly inappropriate. [Rieke (1975a) has derived a color temperature for J5 of 155 ± 15 °K from six infrared brightness measurements.]

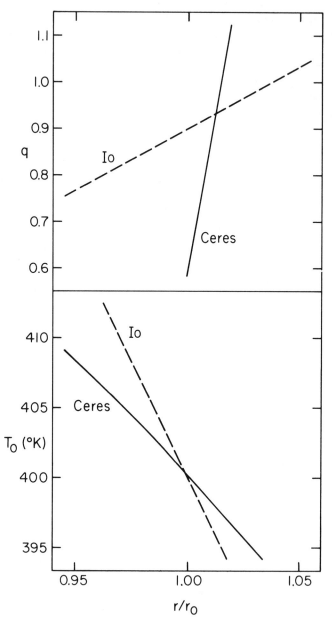

Fig. 12.5. Dependence of the radius of an object, as calculated by the radiometric/photometric technique, on the phase integral q and the degree of peaking of the thermal radiation, given by the temperature coefficient T_0. Io, a high-albedo object, and Ceres, a low-albedo object, are considered. From Jones and Morrison (1974).

TABLE 12.3
Calibration and Results of the Photometric/Radiometric Technique

Parameter	Io	Europa	Ganymede	Callisto	1 Ceres	4 Vesta	Iapetus (dark side)	Rhea	Dione
D (km)	3640±20	3000±200	5720±50	5000±300	931±50[a]	490±25[a]	1400±250[b]	1500±200[b]	950±250[b]
V(1,0)	−1.69[c]	−1.37[c]	−2.09[c]	−0.96[c]	3.70[d]	3.55[d]	2.37[e]	0.11[e]	0.82[e]
p/p_V	0.98	0.86	0.89	0.97	0.9	1.1	1.0	1.0	1.0
q	0.9	1.0	1.0	0.8	0.6	0.6	0.6	1.0	1.0
$V-m_{10}$	4.70[f]	3.91[f]	5.69[f]	7.26[f]	9.95[g]	8.38[g]	—	—	—
D_{10} (km)	3763	3020	5239	4416	959	498	—	—	—
$V-m_{20}$	9.29[f]	8.81[f]	10.26[f]	11.72[f]	12.76[g]	11.24[g]	12.4[h]	8.9[i]	9.2[i]
D_{20} (km)	3588	2850	5275	4969	1034	515	1600	1700	1275
$\bar{D}$ (km)	3676	2935	5257	4692	996	506	1600	1700	1275
ΔD (%)	+1	−2	−9	−7	+7	+3	—	—	—

a. Polarimetric diameter (Zellner et al., 1974)
b. Occultation diameter (Elliot et al., 1975)
c. Morrison and Morrison (Chapt. 16)
d. Zellner et al. (1974); Chapman et al. (1975)
e. Noland et al. (1974b)
f. Morrison (1975b)
g. Morrison (1974c)
h. Morrison et al. (1975)
i. Morrison (1974a)

Table 12.3 shows the diameters of the Galilean satellites, the best observed asteroids (1 Ceres and 4 Vesta), and three satellites of Saturn (Iapetus, Rhea, and Dione), obtained by the radiometric method with $T_0 = 408$ °K, and compares these results with diameters derived by other means. In all cases, photometric V magnitudes extrapolated linearly to zero phase have been used, so that the opposition surge is not taken into account. The values of the phase integral and the geometric albedo are therefore not strictly correct but correspond to the unprimed quantities discussed for the Galilean satellites by Morrison and Morrison (Chapt. 16, herein). The sources of the data are given in the footnotes to the table. This table summarizes all of the published radiometry of satellites of unknown diameter (unknown at the time the observations were made, in any case), and gives the data on which a calibration of the photometric/radiometric technique can be based.

Because of the great difference in albedo between its leading and trailing faces, radiometric studies of Iapetus yield more than the size of this satellite. Murphy et al. (1972) first measured the brightness of Iapetus at 20μ at points near both eastern and western elongation and showed that the darker side emitted more thermal radiation than the brighter, consistent with the photometric variations being due to albedo differences between the two faces. More extensive observations, sufficient to yield a rough radiometric flux curve, have been reported by Morrison et al. (1975) and are illustrated in Figure 12.6. Once the radius is established from the 20μ flux at eastern elongation, the amplitude of the flux curve can be used to derive the phase integral q as discussed for the Galilean satellites in the previous section. Morrison et al. (1975) derive (for the model with $T_0 = 408$ °K) a value of q = 1.3 corresponding to the solid curve in Figure 12.6 and q = 1.1 for the dashed curve. Both values refer to the bright side only, since the amplitude is insensitive to the value of the phase integral of the darker face. If the radius is 800 km and q $\simeq$ 1.2, the Bond albedo of the bright side of this satellite is about 0.5. The Bond albedo of the dark side cannot be measured, but if the phase integral is similar to that of the Moon, then A $\simeq$ 0.04. (The photometric lightcurve of Iapetus is shown in Fig. 9.10.)

ECLIPSE RADIOMETRY

Infrared radiometric observations of the thermal response of the surface to changing insolation can be used to study the thermophysical properties of the uppermost surface layers of the satellites. Earth-based observations cannot explore the diurnal temperature variations (since the maximum phase angle, even for the Galilean satellites, is not even 12°; cf. Table 10.1) but they can be used to investigate the temperature changes resulting from eclipses. The first measurements of the thermal emission of a satellite during eclipse were made in 1963 by Murray et al. (1964b), who observed the initial cooling and final heating phases

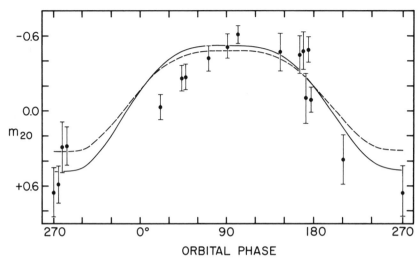

Fig. 12.6. Brightness variations of Iapetus at 20μ as a function of orbital (rotational) phase, as observed from Mauna Kea in 1971–73. The two theoretical curves are calculated from a model for the albedo distribution that fits the visual lightcurve. The solid curve, which provides the best fit, corresponds to a phase integral for the bright material of $q = 1.3$; the dashed curve is for the same albedo distribution but $q = 1.1$. From Morrison *et al.* (1975).

of an eclipse of Ganymede in the 10μ band. Although they could not detect the satellite during mid-eclipse, they were able to conclude from the rapid temperature changes they observed that Ganymede must have a thermal conductivity at least as low as that of the Moon. The first radiometry throughout an eclipse was obtained in the 20μ band by Morrison et al. (1971), who showed that the conductivity of Ganymede was substantially lower than that of the Moon. Subsequently, Hansen (1973), Morrison and Cruikshank (1973), and Morrison and Hansen (1975) made extensive ground-based eclipse observations of all four Galilean satellites, while Gatley et al. (1974) observed an eclipse of Phobos with the Mariner 9 radiometer.

The easiest eclipses to observe and to interpret are those of the outer two Galilean satellites, Ganymede and Callisto. Both are very bright at 10μ and 20μ, and the eclipses take place sufficiently far from the planet that scattered thermal radiation does not contaminate the observations, which can be continued throughout the eclipse. Furthermore, the interpretation does not require taking into account nonsolar sources of energy, whether infrared radiation from Jupiter or impacts with energetic charged particles in the Jovian magnetosphere. Figure 12.7, taken from Morrison and Cruikshank, shows observations at 10μ and 20μ of the eclipse of Callisto on 11 August 1972. Since the satellite passed near the edge of the planet's shadow, the rise and fall in insolation took place relatively

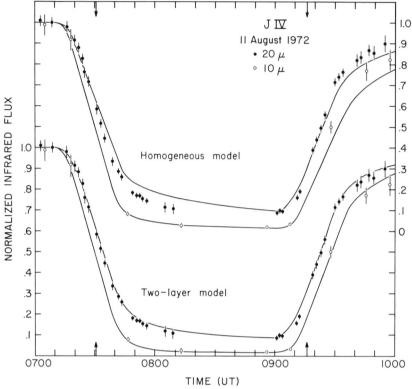

Fig. 12.7. Radiometry of an eclipse of Callisto at 10μ and 20μ plotted with the predictions of the best fitting homogeneous and two-layer surface models. From Morrison and Cruikshank (1973).

slowly, each requiring about 25 minutes, as compared to 4 minutes for a typical eclipse of Io. In the upper part of the figure are illustrated eclipse curves in the two passbands, generated by a homogenous thermal model and constrained to fit the observed minimum residual brightness at 20μ, which is 0.09 ± 0.01 of the pre-eclipse value. Such a model is clearly inadequate. In the lower part of the figure, results are shown for a two-layer model, in which the thermal inertia [defined as $(K\rho c)^{1/2}$, where K is thermal conductivity and ρc is heat capacity per unit volume] of the upper layer is much less than that of the lower one. This two-parameter model (thermal inertia and thickness of the upper layer) fits all of the observations satisfactorily. Since the cooling of a typical surface element during eclipse has five times more effect on the radiated flux at 10μ than at 20μ, the agreement with the model at both wavelengths precludes the exposure on the surface of any significant amount of material of high thermal conductivity and consequent smaller temperature drop during eclipse. From their observations of several eclipses of Ganymede, Hansen (1973) and Morrison and Cruikshank (1973) have derived essentially identical values of the thermal parameters for this satellite, while Morrison and Hansen (1975) confirm the solutions obtained earlier for Callisto.

For the inner two Galilean satellites, eclipse radiometry is more difficult to obtain, in part because the cooling and heating curves must be observed for different eclipses and then combined with proper normalization, and the observations and interpretations are less satisfying. Figure 12.8 illustrates a composite 10μ eclipse reappearance curve of Io obtained at the Hale 5m telescope by Hansen (1973). While the same sort of two-layer model fits the data, the residual flux level just before third contact is nearly an order of magnitude higher than was the case for Callisto and Ganymede. In consequence, Hansen finds that the thermal inertia of the uppermost layer of Io is much larger than that of the outer two satellites. On the other hand, Morrison and Cruikshank, observing only at 20μ, do not find the pre-third-contact brightness of this satellite to be anomalous, and they derive a thermal inertia for its surface similar to that for Callisto and Ganymede (see Figure 12.2). No thermal model that involves only heat conduction in a solid surface is compatible with both sets of observations, and the discrepancy remains unresolved.

Leaving aside the anomalous behavior of Io, the eclipse radiometry provides some useful clues to the nature of the surfaces, at least for Callisto and Ganymede. Table 12.4 summarizes the results of fitting models to the observations. Even though the thermal models used are crude, they show that the uppermost few millimeters of the surface must consist of a material of remarkably low thermal conductivity, nearly an order of magnitude lower than that on the Moon, and that below this layer there must be a fairly rapid transition to a material of much higher conductivity, and, probably, higher density. Further, there can be no significant exposure of this underlying material; Morrison and Cruikshank (1973) set the upper limit at 1% of the surface for Callisto and 5% for Ganymede. It is difficult to understand how such a structure could be maintained in an environment of repeated meteoroidal bombardment, unless the surface material consolidates and re-fuses below the surface following impact. Morrison and Cruikshank suggested a model in which volatiles play an important role: the low conductivity layer is a low density frost, and the high conductivity layer is ice or an ice-dust-rock matrix. Such a model is consistent with the spectroscopic detection of water ice on Ganymede and Europa (Pilcher *et al.*, 1972; Fink *et al.*, 1973). It is not apparent, however, why the thermophysical properties of Callisto and Ganymede should be so similar when their albedos and infrared spectra are so different; in addition there are other reasons to be skeptical about the suggestion of large amounts of frost on Callisto, particularly in view of the absence of spectroscopic features due to this material (cf. Johnson and Pilcher, Chapt. 11).

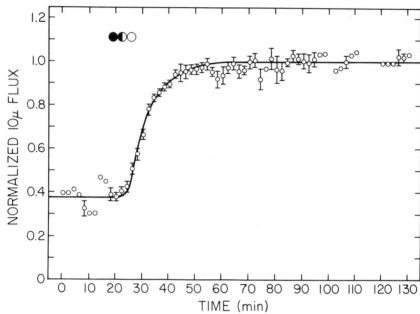

Fig. 12.8. Radiometry of the heating phase of an eclipse of Io at 10μ, plotted with the prediction of a best fitting two-layer surface model. From Hansen (1973).

TABLE 12.4

Thermal Models of the Galilean Satellites from Eclipse Radiometry

Parameter	Io	Europa	Ganymede	Callisto	Reference
Thermal inertia = $(K\rho c)^{1/2}$	1.3 ± 0.4	<4	1.4 ± 0.2	1.0 ± 0.1	1
$(10^4$ erg cm^{-2} s$^{-1/2}$ °K$^{-1})$	3.8 ± 0.3	1.4 ± 0.4	1.2 ± 0.3	—	2
Thermal parameter = $(K\rho c)^{-1/2}$	3200 ± 1000	>1000	3000 ± 400	4200 ± 400	1
(cal^{-1} cm^2 s$^{1/2}$ °K)	1100 ± 100	3000 ± 1000	3400 ± 700	—	2
Mass of upper layer	0.10 ± 0.04	—	0.15 ± 0.03	0.11 ± 0.02	1
(g cm^{-2})	0.11 ± 0.03	0.10 ± 0.05	0.19 ± 0.03	—	2

1: Morrison and Cruikshank (1973)
2: Hansen (1973)

On 12 December 1972 an eclipse of Phobos was successfully observed at 10μ and 20μ from Mariner 9 (Gatley et al., 1974). The 10μ observations are illustrated in Figure 12.9. Like the Galilean satellites, Phobos exhibited very rapid heating and cooling and an extremely low flux level throughout most of the eclipse. The best fitting value of the thermal inertia is $(K\rho c)^{\frac{1}{2}} \leq 2 \times 10^4$ erg $cm^{-2} s^{-\frac{1}{2}} {}^\circ K^{-1}$, a value nearly as low as that of the satellites of Jupiter. Gatley et al. note that this thermal inertia requires a dust layer more porous than that on the Moon, a result that is perhaps not unexpected in view of the much smaller acceleration of gravity on Phobos. They set a lower limit of a few millimeters on the thickness of this insulating layer and also an upper limit of 10% on the fraction of the surface that is dust-free. The results of Gatley et al. are also exhibited in Figure 14.9.

No eclipses have been observed for the satellites of Saturn; such events take place only when the Sun is near the orbital plane of the satellite. In addition the observations would be impossible to obtain with present ground-based equipment, as a result of the closeness of the satellites to the planet at the time of eclipse and due to the extremely low flux levels involved. However, the thermal emission from the rings has been measured just past eclipse; these and other radiometric observations of the rings are discussed next.

RADIOMETRY OF THE RINGS OF SATURN

Infrared radiometry of the rings was first attempted in the mid-1960s, when the Earth was near the ring plane and the rings presented only a small solid angle. Low (1965) set an upper limit to the 20μ brightness temperature of the combined A and B rings of 80 °K in about 1964, when the tilt angle $|B|$ was perhaps $10°-15°$ (no actual dates of observation were given). A subsequent publication (Aumann et al., 1969) attributes to Low an upper limit of 60 °K at the same wavelength, probably obtained in 1965–66, when $|B| < 8°$. The first infrared detection of the rings was made at a wavelength of 12μ in 1968 and 1969 by Allen and Murdock (1971), when the tilt angle was $|B| \simeq 17°$. For the combined A and B rings they obtained a brightness temperature of 83 ± 3 °K. At about the same time Low obtained a nearly identical temperature at 20μ, which was never published (cf. Armstrong et al., 1972).

The next measurements of the infrared brightness temperatures of the rings were not made until 1972, when the rings were nearly fully open with respect to Earth ($|B| > 24°$). In this work, which was done with the Mauna Kea 2.2m telescope and a photometer aperture of 5 arcsec diameter, the temperatures were obtained for the ansae of the A and B rings separately and so cannot be compared directly with the earlier observations. Murphy (1973), observing in 1972 in a broad spectral band from 17μ to 26μ, obtained $T_B = 94 \pm 2$ °K for ring B and $T_B = 89 \pm 3$ °K for the fainter A ring. Morrison (1974b), observing with the

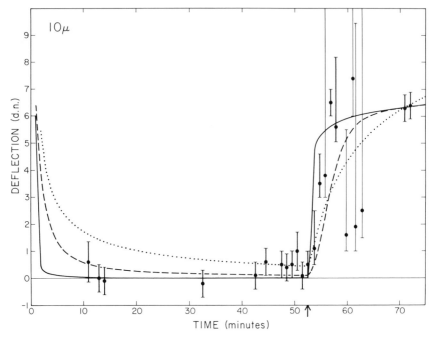

Fig. 12.9. Observations of the eclipse of Phobos on 12 December 1972 obtained at 10μ from Mariner 9. The curves represent models in which the surface has a thermal inertia of 2.5×10^{-4} (solid curve), 7.7×10^{-4} (dashed curve), and 2.5×10^{-3} (dotted curve) cal cm^{-2} $s^{-\frac{1}{2}}$ $°K^{-1}$. The data obtained at about 56 min (4 min. after re-emergence of the satellite from the shadow) by themselves demonstrate the low thermal conductivity of the surface material. From Gatley *et al.* (1974).

same photometer, found temperatures for ring B of 96 ± 2 °K in the same 20μ band and 92 ± 2 °K in the 8μ to 14μ band, and confirmed that the A ring had a lower surface brightness at 20μ. Murphy also reported an anomalously high brightness temperature for the inner (C) ring of 89 ± 4 °K, but Morrison failed to confirm this result and argued that the signal which seemed to be coming from the C ring is probably due to response of the photometer to radiation from the disk and ring B at the edges of the 5 arcsec field. Finally, a measurement in 1973 at a wavelength near 35μ by Nolt *et al.* (1974) yields a temperature for the combined A and B rings of 90–95 °K.

The infrared observations now extend over nearly the full range of tilt angles and clearly show a large variation in the surface brightness of the rings at 20μ and probably also at 10μ. The upper limit of 60 °K obtained in 1966 represents a decrease in 20μ surface brightness of a factor of about 100 over that observed in 1972 and 1973. In a review of these observations, Murphy (1974) has reduced all of the brightness temperatures to equivalent values for the B ring alone; these

TABLE 12.5

Adjusted B-Ring Brightness Temperatures, T_B. Adapted from Murphy (1974).

Observation Year	Observer	$\lambda(\mu m)$	$T_B(°K)$	B'
1964	Low (1965)	10	<85	~9°
1965	Low (unpublished)	20	<64	~5°
1969	Allen and Murdock (1971)	12	86±3	17°
1972	Murphy (1973)	20	94±2	26°
1973	Morrison (1974b)	11	92±3	26°
1973	Morrison (1974b)	20	96±3	26°
1973	Nolt et al. (1974)	35	92–97	26°

results are given in Table 12.5. Unfortunately, the data are not well distributed with respect to inclination to the orbit plane (the angle B') and include a variety of observing techniques, spectral bands, and modes of calibration. It is therefore not possible to deduce the functional form of the dependence of temperature on B', although the observations are certainly consistent with a simple picture in which $T_B^4 \propto \sin B'$. Several models for the variation of brightness with tilt are discussed in the review paper on the rings by Pollack (1975).

At the distance of Saturn from the Sun, the equilibrium temperature of a surface normal to the insolation is given approximately by $127 [(1-A)/E]^{1/4} °K$, and the 20μ brightness temperature of a nonrotating sphere is approximately $118 [(1-A)/E]^{1/4} °K$. For a collection of spheres that is not optically thick, of course, the apparent brightness temperature would be even lower. It thus appears that, if the ring particles have an infrared emissivity of near unity, the observed brightness temperature of 95 °K requires that A < 0.5. If there is significant shadowing or if the particles radiate a substantial part of the absorbed energy from their un-illuminated hemispheres because of rapid rotation, no value of A greater than zero is consistent with the assumption of unit emissivity. However, these conclusions are obviously at odds with the fact that the rings are known to have high geometric albedos (cf. Bobrov, 1970; Cook et al., 1973; Chapt. 19, Cook and Franklin). It follows that the emissivities must be peaked strongly toward small phase angles and/or there must be a source of heat in addition to the insolation. These possibilities have been discussed by Aumann and Kieffer (1973), who conclude that significant additional heating (tens of percent of the absorbed insolation) is provided by infrared radiation from Saturn. Their models suggest that for a large, but probably reasonable, peaking of the emissivity, Bond albedos as great as about 0.5 to 0.6 could be consistent with the observed brightness temperatures, a result also supported by Morrison (1974b). Even in extreme cases, however, it appears that the Bond albedo is smaller than the geometric albedo, indicating a value of q for the particles of less than unity. A more detailed treatment of this problem by Pollack (1975) reaches similar conclusions.

The difference between the brightness temperatures of the B ring measured at 10μ and 20μ by Morrison (1974b) has not been explained. It would be useful to make additional measurements at both wavelengths, and with higher spectral resolution as well, as the rings close during the next few years. To my knowledge, there exist no emission spectra of the rings in either the 10μ or the 20μ atmospheric windows.

The ring particles pass through the shadow of Saturn once each orbit, and measurements of their change in temperature at the time of eclipse can provide information on the thermal properties of the particles. Because the maximum phase angle of Saturn is 6°, it is not possible to measure the temperature of the particles *during* eclipse, and the only available data are measurements of the brightness temperature of the rings a few arcsec beyond emergence from eclipse. Morrison (1974b) found that the 20μ temperature 5 arcsec past the shadow, averaged over a 5 arcsec beam, was 2.0 ± 0.5 °K lower than at the equivalent position before eclipse. Aumann and Kieffer (1973) have modelled the eclipse cooling and reheating of the ring particles and shown that the temperature at this point is sensitive to the size of the particles, which must according to these observations be either less than 500μ or more than 2 cm in radius. Since the smaller range of sizes would suggest lifetimes that are short in comparison to the age of the solar system (cf. Chapt. 19, Cook and Franklin), Morrison (1974b) interprets his observations to favor the size greater than 2 cm. Westphal (private communication, 1974) has made similar observations (but with higher spatial resolution) at 10μ with the Hale 5m telescope, and a preliminary reduction of his data confirms this magnitude for the post-eclipse temperature drop.

Morrison (1974b) also reported that, at the time of his observations, the east ansa was brighter than the west ansa by about 10% at 20μ. This difference should not be a result of the eclipse cooling, and no explanations that require that the ring particles be in synchronous rotation seem very plausible either (cf. Fig. 6.1, Peale). Asymmetries in the ansae have also been noted in reflected light, in the visual region (cf. Bobrov, 1970) and in the near infrared (Cruikshank and Pilcher, 1974). None of these phenomena has been explained satisfactorily, and if these observations are all correct, they represent one of the more peculiar properties of the rings awaiting theoretical analysis.

The rings of Saturn, unlike the other satellites discussed in this review, have been the object of numerous radiometric studies at microwave frequencies as well as in the infrared. Recent interferometric observations, in particular, clearly separate the radiation of the rings from that of the disk at a variety of wavelengths. Even from single-dish observations, however, it is possible to set some useful limits on the microwave brightness temperature of the rings. Two techniques are available for this purpose. The first is to observe with the same equipment over a period of several years, during which the apparent solid angle of the rings varies as the tilt angle B' changes; the varying component of the flux can then be assigned to the rings. The second technique is to look for any excess in integrated radiation over that expected from the disk alone, based on either a model atmosphere for Saturn or on the assumption that the ratio of the flux from Saturn to that from Jupiter is independent of wavelength.

Observations of Saturn from 1965 through 1969 ($0 < |B'| < 17°$) at a wavelength of 3.3 mm by Epstein *et al.* (1970) show no systematic variation with B', suggesting an upper limit to the brightness temperature of the rings in 1969 of about 20 °K at this wavelength. Epstein and his collaborators have continued these observations, but no post-1969 data have been published. From an analysis of the millimeter-wave spectrum of Saturn, Wrixon and Welch (1970) concluded that none of the observations published in the 1960s indicated a detection of the rings at these wavelengths. More recently, Janssen (1974) has critically reviewed the observations at wavelengths between 1 mm and 2 cm, normalizing the Saturnian temperatures to values obtained with the same equipment for Jupiter, to search for evidence of radiation from the rings. With the one exception discussed below, he finds that these microwave data are all consistent with the hypothesis that the rings are non-emitting. At 1.4 mm, however, Rather *et al.* (1974) find an apparent brightness temperature for Saturn that is 45 ± 15 °K larger than the value they find for Jupiter. Since both the model atmospheres and the observations at longer millimeter wavelengths suggest that Saturn should have a temperature ~ 0.95 that of Jupiter, they attribute the excess to the rings; correcting for solid angle factors at the time of the observation (in 1973), they obtain a 1.4 mm brightness temperature for the combined A and B rings of 35 ± 15 °K. Thus, the non-interferometric observations suggest that only at the shortest millimeter wavelengths can the rings be detected, and even here the brightness temperature is much lower than that observed in the infrared.

Interferometric observations have recently set much more stringent upper limits on the brightness of the rings at wavelengths longer than 3 cm. In the first such study published, Berge and Read (1968), observing when the Earth was near the ring plane, set an upper limit to the brightness temperature of ~ 10 °K at a wavelength of 10 cm. From observations made in 1970 and 1971 at 21 cm, Briggs (1973) and Berge and Muhleman (1973) set upper limits of 6 and 10 °K, respectively. Most recently, Briggs (1974b) has established upper limits of about 15 °K at both 4 and 11 cm from observations obtained near the maximum value of B', at the same time that temperatures greater than 90 °K were being measured in the infrared, while Cuzzi and Dent (1975) obtained the first positive detection of the rings at a wavelength of 4 cm, finding a brightness temperature of 15 ± 3 °K. Note, however, that this radiation is not necessarily due to thermal emission from the rings, as will be discussed below.

At sufficiently high spatial resolution, the presence of the rings can be sought not only from their thermal emission but also from their effect on the radiation from Saturn that must pass through them in order to reach the Earth. Observations of both effects will yield the transparency as well as the brightness temperature of the rings. That the rings are not perfectly transparent was first shown by Briggs (1974b), who found that interferometrically derived equatorial diameters for the microwave-emitting disk of Saturn are 2–3% greater than that of the visible disk. This apparent increase near the equator is presumably due to a low brightness near the poles, which in turn is the result of obscuration by an optically thick, cold ring. Briggs discussed observations made with the NRAO three-element interferometer at frequencies (wavelengths) of 1420 MHz (21 cm), 2695 MHz (11 cm), and 8085 MHz (3.8 cm). He modelled the expected range of brightness distribution from both the disk and the rings in order to derive two basic parameters of the rings: g, the microwave opacity relative to the visible opacity, and T_R, a measure of the equivalent radio brightness of the particles if they are opaque. Basically, the degree of obscuration of the disk by the B ring determines g, while the brightness temperature of the whole ring system is sensitive to both parameters. At each frequency, there are ranges of both g and T_R that provide acceptable fits to the data, as illustrated in Figure 12.10. Also shown are lines of constant brightness temperature. Both the 3.8 cm and the 11 cm observations indicate that the radio opacity must be nearly as great as the visible opacity and that T_R is very low. At 21 cm, the data suggest that the opacity decreases, although this result, unlike that at shorter wavelengths, is somewhat sensitive to the choice of limb-darkening parameters for Saturn.

Another recent observation is important to the interpretation of these results. In January 1973, Goldstein and Morris (1973) succeeded in measuring the reflectivity of the rings at a wavelength of 12.6 cm with the JPL Goldstone 64m radar telescope. They obtained a surprisingly strong signal, corresponding to a backscattering cross section of approximately 60% of the projected geometrical cross section of the combined A and B rings, or nearly a factor of 10 greater reflectivity than is usually found in radar studies of planetary surfaces. The combination of large radar cross section and low brightness temperature at the same microwave frequencies is unusual, because experience with other bodies in the solar system as well as the usual Fresnel laws governing the interaction of radiation with a dielectric surface predict that the emissivity should be high and the reflectivity low—exactly opposite to the situation for the rings.

The properties of the rings at wavelengths near 12 cm can be summarized as follows: low brightness temperature ($T_B \leqslant 15\,°K$), high opacity to radiation from the disk ($g \cong 1$), and very high backscattering efficiency. Several authors (e.g., Goldstein and Morris, 1973; Pollack et al., 1973b; Pettengill and Hagfors, 1974) have proposed models for the ring particles that might be consistent with these properties, and a comprehensive treatment of the implications of these observations has been given by Pollack (1975) and an alternate view in Chapter

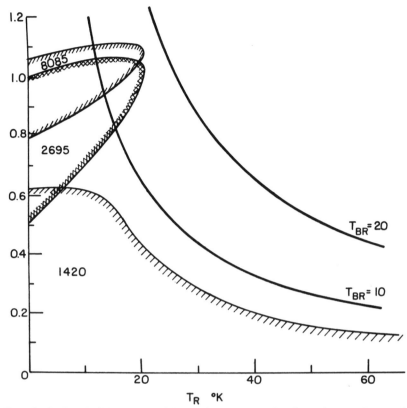

Fig. 12.10. Permissible ranges of T_R and g (see text for discussion) at frequencies (wavelengths) of 8085 MHz (3.8 cm), 2695 MHz (11 cm), and 1420 MHz (21 cm). T_{BR} is the effective ring brightness temperature. From Briggs (1974b).

19 by Cook and Franklin. In this chapter I will only summarize some of the theoretical implications, primarily following Pollack's (1975) analysis.

Three models seem to be capable of explaining the high radar backscattering efficiency of the rings. The first, suggested by Goldstein and Morris (1973), is that of large (>1 m), rough spheres. If they are composed of silicates or ice, it is marginally possible, by optimizing the gain factor in the model, to reproduce the observed reflectivity. On the other hand, if the particles are metallic (and thus have very high dielectric constant), it is easy to generate the required backscattering efficiency. The second model, that of smooth, transparent spheres, was proposed by Pettengill and Hagfors (1974). In this approach, the high reflectivity is produced by internal reflection and refraction in ice spheres large with respect to the wavelength; the mechanism is very efficient, so that only a relatively small geometric cross section of such smooth spheres is required. A third model was proposed by Pollack *et al.* (1973b), in which the role of multiple scattering is stressed. Their "bright-cloud" model generates the required radar reflectivity by scattering among many small particles, each of a few centimeters diameter and each with a high single-scattering albedo. It is analogous at microwave frequencies to the multiple-scattering clouds in the Earth's atmosphere, which are strongly backscattering in the visible. Each of these models makes different predictions as to the microwave transparency and emissivity of the rings, and as shown by Pollack (1975), the radiometric observations are now sufficient to distinguish among them.

We have seen that the thermal emission from the rings at microwave frequencies is extremely low, whereas the physical temperature is known from infrared radiometry to be $\sim 100\,°K$. The effective emissivity of the rings must therefore be very low, due either to a small optical depth or to low emissivity of the individual ring particles. In the model of large rough particles, such low emissivity requires metallic composition; large rough silicate or ice particles are definitely excluded by the low microwave brightness temperatures. The low temperatures are consistent with both other models, however. In the model of Pettengill and Hagfors, both the optical depth and the particle emissivities are low and brightness temperatures of $\sim 1\,°K$ are predicted. In the Pollack *et al.* model, the single-scattering albedos are high and hence the particle emissivities are low, yielding predicted brightness temperatures of a few degrees. In both of these latter models, the microwave properties of the particles depend on their size relative to the wavelength, so both can, with proper choice of particle size, be made consistent with the brightness temperature at 1.4 mm of $35 \pm 15\,°K$. This wavelength-dependence of emissivity does not apply to the metallic particle version of the Goldstein and Morris model, however, and the detectability of the rings at millimeter wavelengths appears to exclude this possibility as well.

At 3.8 cm and 11 cm, Briggs' observations (1974b) indicate optical depths for the rings near unity, as discussed above and illustrated in Figure 12.10. However, the large ice-sphere model of Pettengill and Hagfors predicts very small optical depths throughout the microwave spectrum. The only model that is consistent with both large opacity and low emissivity is that of Pollack *et al.*, in which the individual particles in the bright cloud are highly reflective and low in emissivity, while the total optical depth is substantial.

The radiometric and radar results considered together not only exclude the models proposed by Goldstein and Morris (1973) and by Pettengill and Hagfors (1974), they also provide constraints on the optical properties, and hence on the size and composition, of the particles postulated in the bright-cloud model of Pollack *et al.* (1973b). Compare with the discussion of Cook and Franklin in Chapter 19. We have already noted that, in any model, the particles must be at least a few centimeters in diameter in order to reflect 12 cm radar efficiently. Silicate particles even 2 or 3 cm in diameter, however, will radiate at greater than the observed levels for wavelengths of both 4 cm and 3 mm. Water ice, in contrast, has optical properties such that particles of the required size do not begin to have appreciable emissivity until wavelengths near 1 mm are reached. That the *surface* of the ring particles is primarily ice was already known from infrared reflection spectroscopy (Pilcher *et al.*, 1970), but the evidence discussed here is the first to show the *bulk* composition of the particles.

The optical properties of water ice at temperatures near 100 °K are fairly well known, so that it is possible to determine what ranges of particle sizes are consistent with the observed microwave opacities and brightness temperatures. A lower limit of a few centimeters is set by the radar reflectivity and by the substantial optical depths measured at wavelengths of 3.8 and 11 cm. If, as suggested in Figure 12.10, the optical depth is substantially less at 21 cm than at shorter wavelengths, an upper limit of about 6 cm is set on the diameters. Less restrictive, and more certain, upper limits are implied by the upper limit to the brightness obtained at 3 mm, which requires a mean diameter less than 30 cm. The bright-cloud model with particles composed of water ice also provides an explanation for the ring temperature of 15 ± 3 °K found at 3.8 cm by Cuzzi and Dent (1975), as originally pointed out by Cuzzi and van Blerkom (1974). If the single-scattering albedo is greater than 0.95, a brightness this great is expected due to scattered thermal radiation from Saturn. In summary, Pollack's analysis leads to a model in which the rings are composed of a cloud of particles, composed primarily of water ice and with characteristic diameters of a few centimeters, of sufficient thickness to have total optical depth near unity in the B ring at centimeter wavelengths. The majority of the particles are apparently something between snowballs and large hailstones, although of course these observations do not exclude the possibility of occasional, much larger objects that might represent a substantial fraction of the total mass of the rings.

It is clear from the above discussion that even these initial microwave radiometric (and radar) studies of the rings are crucial in determining the size and composition of the ring particles, even though much theoretical analysis remains to be done. In future studies, the infrared radiometry can also be expected to aid in the determination of the photometric properties of the particles.

PROSPECTS FOR FUTURE WORK

In this chapter I have reviewed radiometric observations of nine planetary satellites and the rings of Saturn. All of these have been measured at 20μ, and most are also accessible at a variety of other infrared wavelengths extending from 8μ to 34μ. In addition, radiometric studies at microwave frequencies have been carried out for Callisto, Ganymede, Titan, and most particularly the rings of Saturn. All of these results have been obtained since 1964, and most have been published since 1971. Radiometry of satellites is a young field, and there are prospects for much important work at both infrared and radio wavelengths in the near future.

In the infrared, improvements can be expected in spatial resolution, spectral resolution, and detection of faint signals. For the Galilean satellites, there are two high priority problems. The first is to search for departures from grey-body emission in the 10μ and 20μ regions, such as those suggested by Hansen (1974). Filter-wheel spectrometers, such as those now in use or under construction at several observatories, are entirely adequate to this task. The second is to re-examine the anomalous behavior of Io and Europa during and after eclipses as reported by Hansen (1973), Morrison and Cruikshank (1973), and in this chapter. In order to improve on the existing observations and to assure that the data are not being contaminated by scattered radiation from Jupiter, these observations should be made with large-aperture instruments, such as the Hale 5m or Mayall 4m telescopes, or with the NASA 3m National Infrared Telescope now under construction. Elsewhere among the Jovian satellites, it may be possible with present facilities and techniques to measure the 10 and 20μ radiation from J6 and J7 (cf. Morrison and Cruikshank, 1974), and thus to determine the sizes and albedos of these two satellites. [A successful detection has been accomplished by Cruikshank (1976b).]

A variety of important infrared observations can be expected among the satellites of Saturn, although for the most part these require large-aperture telescopes both to reach low flux levels and to eliminate contamination by radiation from the planet and its rings. Spectrophotometry of Titan has already proved of crucial importance for the study of the atmosphere of this satellite, and observations of higher spectral resolution are needed, especially within the 10μ band, where there appears to be a great deal of spectral structure. Parenthetically, similar infrared spectrophotometry of all the Jovian planets would be very interesting. The 20μ radiometry of Iapetus, Rhea, and Dione obtained by the Hawaii observers should be verified and extended by others. It would be interesting to obtain improved radiometric diameters for the inner satellites to be compared with those derived from lunar occultation photometry. Finally, the rings should provide a rich harvest of radiometric results during the next few years. Spectrophotometry at 10μ and 20μ is obviously needed, particularly in view of the discrepancy between broadband 10μ and 20μ temperatures reported by Morrison (1974b). High spatial resolution, such as is obtainable only with the largest telescopes, is needed in order to measure the separate brightness temperatures of rings A and B and to search for radial and azimuthal structure in the rings. High-resolution radiometry immediately after eclipse, such as that being carried out by Westphal with the Hale telescope, will contribute to an understanding of the thermophysical properties of the ring particles. As the tilt of the rings changes, it will be useful to monitor the changing brightness temperature of the rings at a variety of wavelengths. A determination of the dependence of temperature upon illumination angle will provide important input data to any comprehensive model of the photometric properties of the rings.

Microwave studies of the satellites have only begun, and with the completion of the upgrading of the NAIC Arecibo 300m telescope a whole range of additional programs will become possible. Briggs (1974c) has discussed the capabilities of this instrument, which can be used as either a single-element telescope or in combination with an auxiliary 30m antenna to form a two-element interferometer, for observations of faint sources in the solar system. Because of the necessity of separating the satellite signals from those of the planets, most observations will probably be carried out in the interferometer mode. The Galilean satellites will be observable at wavelengths from 4 cm to nearly 20 cm, while Titan and the largest asteroids can be studied from 4 cm to about 21 cm. Also J6 and J7 may be detectable, depending on their size, at 4 or 5 cm. In addition, of course, the Arecibo facility will permit extensive radar studies of the Galilean satellites and of many asteroids.

In view of the great contribution now being made to our understanding of the rings of Saturn by microwave interferometric radiometry and radar, it seems certain that these techniques will continue to be applied. The application of interferometric techniques at millimeter wavelengths, where the rings are expected to have non-zero brightness temperatures, would be extremely interesting. Even without interferometry, it may be possible to establish the brightness temperature of the rings at 3 mm, and such an observation would provide a useful discriminant among models, particularly in defining the size distribution of the ring particles.

ACKNOWLEDGMENTS

I am grateful to many colleagues for useful discussions and for making results available in advance of publication, particularly F. H. Briggs, D. P. Cruikshank, O. L. Hansen, D. L. Matson, R. E. Murphy, J. B. Pollack, and J. A. Westphal. I thank N. D. Morrison for a critical reading of an early version of this chapter and T. J. Jones for assisting with some of the calculations. This research was supported in part by NASA grant NGL 12-001-057.

STELLAR OCCULTATIONS BY PLANETARY SATELLITES

13

Brian O'Leary
Princeton University

Calculations have shown that several occultations of stars by the large satellites of the outer planets could be observed each decade with existing equipment at terrestrial telescopes (O'Leary, 1972). It is possible, in principle, to detect thin atmospheres and to determine an occulting satellite's radius and shape to great precision. Partially successful results already have been obtained for two Galilean satellites. The May 1971 occultation of Beta Scorpii C by Io determined an upper limit of 10^{-4} mb for a surface pressure on that satellite (Taylor et al., 1971). Assuming a hydrostatic fluid triaxial figure distorted from a sphere by rotation and tides raised by Jupiter, a mean radius of 1818 ± 5 km for Io was derived from timings of six occultation events (O'Leary and van Flandern, 1972). In June 1972 Ganymede occulted the eighth magnitude star SAO 186800. Observations of four occultation events showed non-abrupt light intensity changes, indicating the presence of an atmosphere whose surface pressure is greater than 10^{-3} mb (Carlson et al., 1973). The radius derived from fitting the occultation durations as chords to a model disk is 2635 ($+15$, -100) km. A close pass of Callisto to the fifth magnitude star 21 Capricorni occurred in 1973; had this resulted in an occultation, about twenty observatories in western North America would have obtained data. G. E. Taylor of the Royal Greenwich Observatory in England provides predictions for occultations involving the Galilean satellites, Titan and many asteroids. A systematic program of occultation predictions and observations is urged in order to improve our knowledge about the gross physical properties of planetary satellites prior to future spacecraft encounters.

Occasionally a solar system body occults a star of comparable brightness. Though these events are extraordinarily rare for the brighter planets, stellar occultations by the fainter planets, satellites, and asteroids are relatively frequent (O'Leary, 1972). Prediction accuracies have improved, and recent success with two of Jupiter's Galilean satellites shows that it is possible to determine atmospheric and geometric properties of solar system bodies with great precision by using simple photoelectric equipment at small and moderate aperture telescopes. This chapter reviews recent work in the field, observational techniques, and future opportunities.

THE 1971 IO-BETA SCORPII C OCCULTATION

Photoelectric observations of the occultation of the fifth magnitude star Beta Scorpii C by the satellite Io (J1) on May 14, 1971, were made in Florida, Jamaica, and the Virgin Islands (Taylor *et al.*, 1971). Some visual observations

[302]

also were made. This was the first occasion on which accurate observations of such a rare event have been obtained, although several visual observations of the occultation of a star by Ganymede (J3) were made in 1911. The predictions for Beta Scorpii C were issued only a few weeks before the event; the predicted track of the occultation was uncertain, owing primarily to the uncertainty in the declination of Jupiter. It has been estimated that, on the average, Io will occult a star as bright as β Scorpii C only once per millenium.

Photoelectric lightcurves and accurate timings of four disappearances and two reappearances were obtained at four sites: Gainesville, Florida; Tampa, Florida; Kingston, Jamaica; and St. Thomas, Virgin Islands. None of the lightcurves showed any signs of an atmosphere on Io: in all cases the curves were flat just before and after occultation with abrupt changes in intensity at immersion and emersion. The presence of even a tenuous atmosphere would be indicated by a gradual change in intensity during a period of several seconds before and after occultation, caused by the defocusing of differentially refracted starlight through the gaseous medium. The observations showed that the light intensity changed by no more than a few percent during these phases, leading to an upper limit of $\sim 10^{-4}$ mb surface pressure for N_2 and CH_4 and $\sim 10^{-3}$ mbar for H_2. Because molecular hydrogen would most likely have escaped from a primordial atmosphere, 10^{-4} mb is the probable upper limit for any atmosphere on Io, according to these observations.

The track of Io's shadow over the Earth is shown in Figure 13.1. The analysis of the timing data for four disappearances and two reappearances indicates that the apparent path of the star with respect to Io, as seen from the observing sites, was as shown in Figure 13.2. The six observations can then be fitted to a model disk by a least squares technique, weighting each observation according to its estimated error (Taylor *et al.*, 1971; Taylor, 1972; O'Leary and van Flandern, 1972). If Io is assumed to be a perfect sphere, the mean radius is 1829.5 ± 2.0 km. Unfortunately the precision in timings after separation of the chords is not sufficient to empirically determine Io's apparent ellipticity, the error being larger than the derived value.

On the other hand, rotation and tides raised by Jupiter both act to distort Io from a perfect sphere (O'Leary and van Flandern, 1972). Given its internal density distribution, a precise figure can be derived if we assume that Io acts as a fluid in hydrostatic equilibrium. The assumption that Io acts as a "relaxed" fluid is almost certainly a reasonable one (Johnson and McGetchin, 1973); its orbit about Jupiter has negligible eccentricity and very low inclination, indicating very stable rotational and tidal configurations over long periods of time. It comes as no surprise that photometric data show a synchronous rotation period (Chapt. 6, Peale). It is, therefore, highly probable that Io has taken on a permanent triaxial figure in which the A-axis of moment of inertia lies along a line to Jupiter, the B-axis is tangent to Io's orbit, the C-axis is orthogonal to A and B and coincident

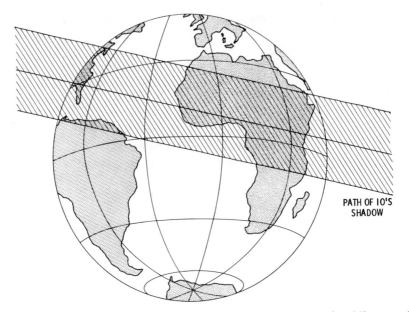

Fig. 13.1. The May 14, 1971, occultation of Beta Scorpii C by Io. (After O'Leary and van Flandern, 1972)

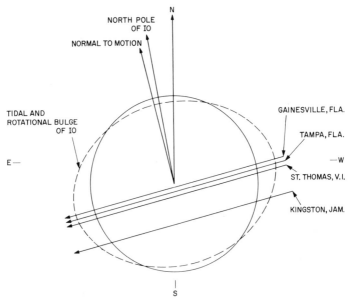

Fig. 13.2. The apparent path of the star with respect to Io as seen from each of the four observing sites at which photoelectric results were obtained. (After O'Leary and van Flandern, 1972)

with its axis of rotation, and C > B > A (Kaula, 1968; Chapt. 6, Peale). Deviations from hydrostatic equilibrium are expected to be very small compared to the magnitudes of rotational and tidal bulges.

O'Leary and van Flandern (1972) calculated the rotational and tidal figures of Io assuming that the satellite is a homogeneous fluid in hydrostatic equilibrium with a rotational axis orthogonal to its orbit. Were there an appreciable density gradient with depth (such as is the case with the outer planets) Io's triaxial ellipticity would be some tens of percent less than the homogeneous case. On the other hand, by analogy with the Moon, whose mass and mean density are close to those of Io, one might expect a nearly homogeneous mass distribution within.

The results of our calculations indicated that the radii along the three principal axes of Io are a = 1829.2 km, b = 1814.8 km, and c = 1809.9 km. The mean radius for Io is then $\overline{R}$ = (abc)$^{1/3}$ = 1817.9 ± 2.5 km. This can be contrasted with R = 1829.5 ± 2.0 km assuming a spherical shape, which is actually closer to a measure of the equatorial radius at the time of observation. The mean density thus derived is 3.54 ± 0.06 g/cm^3, the largest uncertainty being in the mass of Io (see Table 1.4 for Pioneer value).

Any strong concentration of mass toward the center of Io could increase the value of $\overline{R}$ by a few kilometers, but in no case would it exceed the value derived for a sphere; such a central concentration is unlikely on cosmogonic grounds (cf. Chapt. 25, Consolmagno and Lewis) and thus it seems improbable that $\overline{R}$ exceeds 1823 km.

Five of the six residuals of the observation exceed their mean errors based on the uncertainty of each occultation timing. The Tampa observation is 2.6 times its mean error, which, according to the observer, was conservatively estimated. Surface irregularities might be a possible explanation for this. Indeed, if the analogy with our own Moon is applicable here also, surface irregularities of a few kilometers are to be expected although stress relief would rapidly smooth a surface of ice (Johnson and McGetchin, 1973). The sign of the residual is in the sense that the immersion occurred over a lowland about 3 km below Io's mean limb profile. For our Moon, the chances of encountering a feature 3 km higher or lower than its mean limb profile in six random samplings of its limb is about 30%. From this, and the excess of the residuals over their mean errors, O'Leary and van Flandern (1972) suggest that the occultation timings indeed detected surface irregularities on Io. As in the case of the Moon, it is possible that the mean surface radius of Io is 2 to 3 km less than the mean limb radius derived above because intrinsically low elevations are rarely seen in projection on the limb.

The combined uncertainties in the degree of central concentration (which would increase $\overline{R}$ by a few kilometers, at most, from the homogeneous case considered herein) and the presence of surface irregularities (which would decrease $\overline{R}$ by a few kilometers, at most) suggest $\overline{R}$ = 1818 ± 5 km as the most probable value and error for the mean radius of Io.

THE 1972 GANYMEDE-SAO 186800 OCCULTATION

On June 7, 1972, Ganymede occulted the eighth-magnitude star SAO 186800 (Fig. 13.3). Successful photoelectric observations obtained at Lembang, Java (Indonesia), and Kavalur, India, show non-abrupt immersions and emersions (Figs. 13.3 and 13.4). By contrast, the occultation of Beta Scorpii C by Io was observed to be instantaneous to within the time resolution of the instruments. In the absence of an atmosphere on Ganymede, the intensity change should have been more rapid than the ~ 0.05 second integration time of both observations. The presence of an atmosphere would produce a more gradual lightcurve through refraction.

Reduced Lembang data are available from several minutes before the onset of the event until several minutes after its termination. The tracing of the data in Figure 13.3 shows clearly that the occultation was in fact observed. The time of mid-occultation was approximately one minute from that predicted, well within the accuracy of the prediction. Moreover, photographs taken from Lembang confirm that the events occurred at the times indicated by the photoelectric records. Immersion appears quite gradual, lasting perhaps several seconds. Emersion, while less clear, is also non-abrupt; this interpretation is made a little more difficult by the presence of a noise spike near emersion (Fig. 13.4).

The Kavalur immersion curve (Fig. 13.4) provides additional evidence that the falloff is not instantaneous and suggests a falloff time of approximately 0.5 second, although it is consistent also with considerably longer falloff times. The emersion curve is suggestive of an intensity rise taking several seconds, but this may be due in part to fluctuations of Jovian scattered light (Jupiter's limb was only 20 arcsec away) resulting from telescope guiding oscillations which occurred with time scales of the order of several seconds.

The data of Figures 13.3 and 13.4 are complicated by scattered light fluctuations, occasional noise spikes, and some data dropouts. It must be remembered that the intensity drop was only of the order of 5 percent, that the star was itself only eighth magnitude, and that there was some scattered light from Jupiter. Nevertheless, the data are of sufficient quality to determine the occultation radius and to support the inference that the intensity changes are nonabrupt. Thus it appears that Ganymede does possess at least a modest atmosphere.

Carlson *et al.* (1973) fit the two sets of observations as chords to a model disk (Fig. 13.5) and found a discrepancy of about 5 seconds between the absolute times for the two observatories. This discrepancy is too large to be explained entirely by the gradual nature of the events, yet too small to erode our confidence that the occultation was in fact observed at both locations. There is the possibility of an error in the setting of the clock at one observatory or the other. Also difficult to interpret completely is the suggestion that the immersion and emersion recorded from Lembang were more gradual than those from Kavalur. It is

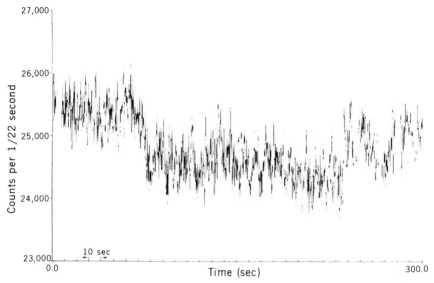

Fig. 13.3. Photoelectric lightcurve of the occultation of SAO 186800 by Ganymede from Lembang, Java. Note that the time scale is compressed relative to the Kavalur data (see Fig. 13.4) and that each 10-second interval contains 220 data points. (After Carlson *et al.*, 1973)

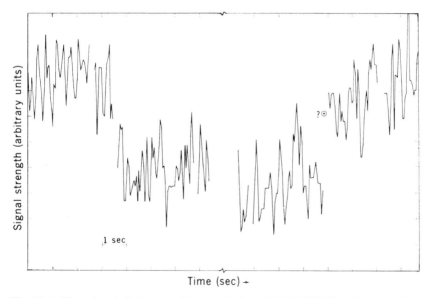

Fig. 13.4. Photoelectric lightcurve of the occultation of SAO 186800 by Ganymede from Kavalur, India. (After Carlson *et al.*, 1973)

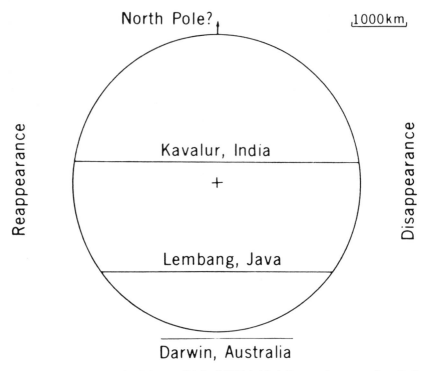

Fig. 13.5. Apparent path of the star SAO 186800 behind Ganymede as seen from India and Java on 7 June 1972. (After Carlson *et al.*, 1973)

clear that deriving a meaningful scale height and composition of Ganymede's atmosphere from the shape of these lightcurves is impossible.

However, it is possible to set a lower limit to Ganymede's surface pressure and to determine the radius at that level. The data suggest a surface pressure greater than $\sim 10^{-3}$ mb (attributable to the lack of any noticeable abrupt event in the photoelectric record). Infrared observations suggest an upper limit of less than ~ 1 mb (Hansen, 1972). Further analysis of the data may narrow these limits. Spacecraft radio occultations may shed additional light in the near future on the nature of Ganymede's atmosphere.

An analysis was made by using the four observed times (which were all given equal weight) in conjunction with ephemerides of Jupiter and Ganymede. The equations of condition contained three unknowns: corrections to the right ascension, declination, and adopted semi-diameter of Ganymede. Assuming a mean molecular weight of 28 (molecular nitrogen) and a temperature of 100°K, the scale height would be about 20 km and the solid surface would lie about 140 km below the occultation layer. The figure would be even greater for constituents of lower molecular weight or for a warmer atmosphere. Thus the diameter of Ganymede is 5270 (+ 30, ~ − 200) km, and the mean density 2.03 (−0.03, ~ + 0.2) g cm⁻³.

Because of its slow (synchronous) rotation rate (the period of rotation is 7.155 days), the equatorial flattening of Ganymede should be very small. If we make the reasonable assumption that Ganymede is a homogeneous fluid body in hydrostatic equilibrium, the equatorial radius would exceed the polar radius by less than 1 km and the permanent tidal bulge directed toward Jupiter would be only 2 km greater than in an orthogonal direction (O'Leary and van Flandern, 1972). Therefore, the assumption that Ganymede is spherical is well within the accuracy of these observations.

FREQUENCIES OF STELLAR OCCULTATIONS BY PLANETARY SATELLITES

It is clear that stellar occultations by satellites are a powerful tool for detecting the presence or absence of thin atmospheres, and in determining accurate radii and thereby mean densities. In principle, it is also possible to determine shapes and topography if there are a sufficient number of observers located at disparate sites. How often may we expect such events to happen? Occultation frequency calculations have been carried out for those fourteen satellites of outer planets whose mean angular diameters exceed 0.1 arcsec (O'Leary, 1972).

These calculations are subject to a number of assumptions, and the final answers are not rigorously accurate. Nevertheless, they indicate the objects that are most susceptible to occultations and, therefore, the direction that prediction and experimentation ought to take. The frequencies are computed first by determining the area of the sky swept out by each object each year. Two general cases are considered: (1) occultations observable in the night sky above a given site on the Earth and (2) occultations observable from somewhere on the Earth. As seen from a given site, the area swept out is the product of the occulting object's mean angular motion and its mean angular diameter; as seen from some region on the Earth—which is just as likely to be New Guinea as Mt. Palomar—the area is the product of the object's mean angular motion and the sum of its angular diameter and twice its geocentric parallax. The mean motion and mean parallax for the outer planetary systems are listed in Table 13.1. Once the area has been determined, one simply multiplies this by the mean number of eligible stars per unit area that are brighter than the appropriate magnitude (Allen, 1963), to obtain the frequency of occultations.

The calculations are subject to the following assumptions:

1. The near ultraviolet (about 3600 Å) is the best region for observing occulta-

TABLE 13.1
Preliminary Data for Calculation of Stellar Occultation
Frequencies for Satellites of the Outer Planets
(After O'Leary, 1972)

Satellite	Mean motion (deg/yr)	Mean geocentric parallax (arcsec)*	Galactic latitude (deg)†
Jupiter	50	3.5	12
Saturn	28	2.0	29
Uranus	12.5	1.1	55
Neptune	8.2	0.6	25

* Mean geocentric parallax is expressed in Earth diameters.
† Galactic latitude is for 1 January 1972.

tions. This is because most bodies in the solar system resemble late-type stars in having very red color indices. Thus, in most cases, observing in the ultraviolet minimizes the brightness of the planet with respect to that of the occulted star.

2. Counts of stars in a given magnitude range per square degree of sky are taken for photographic (blue) magnitude. Transforming these counts to the ultraviolet requires an adjustment in the ultraviolet − blue (U − B) color index which is negligible for most stars, that is, those bluer than spectral type KO and redder than AO. This transformation has been neglected.

3. Star counts correspond to those at intermediate galactic latitudes, that is, the "mean" values compiled by Allen (1963). Table 13.1 indicates a list of the galactic latitudes of the outer planets on 1 January 1972 (the other planets move fast enough to yield a "mean" result during the course of a year or two). With the exception of Jupiter and Pluto, the outer planets are situated at intermediate galactic latitudes. Averaged over the next several decades, these star counts should be well within a factor of 2 of the mean values.

4. The mean angular motion of a planet in the sky is obtained from its movement in right ascension and declination, given in the *American Ephemeris and Nautical Almanac,* over a representative time span of 1 or 2 years (Table 13.1). Each satellite has an added motion that is sometimes an appreciable fraction of that of its parent planet—in the case of Io, almost 50 percent. On the other hand, the satellite moves fastest while it is in conjunction with the planet, and these are the times when it is poorly visible or not visible at all. We assume that these effects roughly cancel out and that each satellite's motion in the sky during periods of observation is the same as that of its planet.

5. Mean apparent diameters and mean geocentric parallaxes are assumed: time-variant deviations are very small for satellites of the outer planets.

6. It is assumed that one out of five occultations in line with a given site will be observable in the night sky over that site. This takes into consideration a factor of 2 because only one hemisphere can be seen, another factor of 2 because observations can only be made at night, and another 20 percent for poor observing conditions, such as twilight and low altitude.

The results are shown in Table 13.2. In comparing observations from a given site on the Earth with observations from somewhere on the Earth, we see that occultations can be seen much more frequently when one is willing to mount an expedition to perform the observations. This gain is important for satellites, which are smaller than the Earth; then the baseline provided by geocentric parallax "buys" the observer more space to move into the object's cylindrical shadow from the star.

The case is considered in which the intensity drop is 10 percent or greater (i.e., the star is, at most, 2.5 magnitudes fainter than the satellite in U; the appropriate stellar magnitudes are listed in Table 13.2, Column 7). Any drop of less than 10 percent in intensity is assumed to be a marginal observation because of signal-to-noise problems in the photometry; it is especially difficult to interpret in the case of a satellite with an atmosphere, for which the intensity changes gradually during immersion and emersion. (Note that the Ganymede-SAO 186800 occultation involved a 5 percent intensity drop but that was probably still sufficient to deduce nonabrupt events, i.e., an appreciable atmosphere.) There are possible ways around the signal-to-noise problem for brighter satellites seen through telescopes with large apertures. In observing the occultation of Sigma Arietis by Jupiter, Baum and Code (1953) used a spectral band-pass of 10 Å centered in the K line of ionized calcium. Because the star totally lacked this absorption, which cuts out most of the reflected sunlight from Jupiter, it was possible to gain a factor of about 4 in the intensity ratio of star to planet.

Column 8 of Table 13.2 lists the telescope apertures necessary to produce 100 counts per second for observations made through a wide-band U filter with a photomultiplier with high quantum efficiency. This arrangement corresponds to 10 percent r.m.s. fluctuations in the photon statistics for the satellite for 1-second integration times, a time resolution that normally translates into spatial resolutions on the satellite of about 10 km (or one scale height). These figures are conservatively estimated from theoretical considerations and from the direct experience of observing Pluto at the Mt. Palomar 200-inch telescope. In effect, these apertures are considered the minimum values necessary for getting meaningful results from occultations; they can serve as a rough guide in deciding whether a given occultation should be observed with a given apparatus, at home or in the field. A practical lower limit of 3 inches in aperture was set because at smaller apertures there are serious problems with scintillation noise. We see that small-aperture telescopes (about 20 inches) are adequate for observations of all objects except the satellites of Uranus and Neptune.

The occultation frequencies in columns 4 and 6 of Table 13.2 are given per

TABLE 13.2
Frequency of Occultations of Stars by Satellites (After O'Leary, 1972)

	1	2	3	4	5	6	7	8
			Observable from a given site		Observable from somewhere on earth		Faintest photographic magnitude of occulted star	Minimum telescope aperture (inches)
Satellite	U magnitude (mean)	Mean diameter (arcsec)	Area swept in sky (deg²/year)	Occultations $N/5$	Area swept in sky (deg²/year)	Occultations N		
Ganymede	+ 6.3	1.4	0.019	0.04	0.065	0.7	9	3
Europa	+ 7.0	0.8	0.011	0.05	0.057	1.4	10	3
Callisto	+ 7.3	1.2	0.016	0.10	0.063	1.9	10	3
Io	+ 7.7	0.9	0.012	0.13	0.059	3.0	10	3
Titan	+10.8	0.75	0.0058	1.2	0.021	22	13	7
Rhea	+11.2	0.20	0.0015	0.4	0.017	24	14	9
Tethys	+11.7	0.15	0.0012	0.4	0.017	29	14	11
Dione	+11.8	0.13	0.001	0.4	0.017	31	14	11
Iapetus	+12.4	~0.1	~0.0008	~0.6	~0.016	~ 50	15	15
Enceladus	+13	~0.1	~0.0008	~0.9	~0.016	~ 90	15	20
Mimas	+13	~0.1	~0.0008	~0.9	~0.016	~ 90	15	20
Titania	+15	~0.1	~0.0003	~2	~0.004	~130	17	50
Oberon	+15	~0.1	~0.0003	~2	~0.004	~130	17	50
Triton	+15	~0.15	~0.0003	~3	~0.0025	~ 80	17	50

Note: In columns 4 and 6, N is the number of stars occulted in 10 years. The factor of 5 takes into account that one of every four occultations will be visible in the night sky above a given site and that an additional 20 percent of occultations will be invisible because of poor conditions, that is, low altitude and twilight. In column 5, the larger area swept out is due to the geocentric parallax of the object (see Table 13.1). The minimum telescope aperture is the approximate aperture, based on the mean U magnitude of the planet, that will produce 100 photon counts per second for observations through a wide-band U filter with a phototube of high quantum efficiency. The aperture is given in inches.

decade, which is a convenient time scale for a systematic research program. With the possible exception of the faintest satellites, these figures should be correct to within a factor of 2 when integrated over several decades. In order to minimize confusion in the interpretation of Table 13.2, the freedom of pursuing a given occultation at a suitable field station (column 6) is allowed.

The results for the large satellites of the outer planets are most interesting. Although the occultation of Beta Scorpii C by Io was a fortuitous one, observable encounters with fainter stars are relatively frequent. It is possible that by the end of the decade, occultations involving all four of Jupiter's Galilean satellites will be observed. With proper planning, we should have answers to questions about thin atmospheres on the Galilean satellites as well as detailed information about their diameters, shapes, and ephemerides.

The situation with Saturn's satellites is even more promising because their fainter magnitude allows more candidate-stars. Two or more events occur each year for each of the seven largest satellites, but a moderate-aperture telescope is required to observe most of these events, which involve thirteenth to fifteenth magnitude stars. From a given observatory, on the average, one event could be observed each decade for each satellite (Table 13.2, column 4). Since we know that Titan has an atmosphere, it is obviously of interest to probe it by occultation techniques, and the statistics predict frequent enough events for this satellite.

The two largest satellites of Uranus and Neptune's satellite Triton are even better candidates; their predicted occultation frequencies are similar to those of Pluto. (Unfortunately, Triton is difficult to observe because its separation from Neptune never exceeds 20 arcsec.) Thus, we see that the large satellites of the outer planets are prime candidates for observations of stellar occultations. In principle, a systematic attack on the problem should yield information about the physical properties and positions of these poorly understood objects prior to, and concurrent with, spacecraft missions to the outer solar system.

The problem remains of the mechanism and accuracy of prediction. Occultations of stars brighter than ninth magnitude by the seven brightest planets, the major Jovian and Saturnian satellites, and 34 of the largest asteroids are predicted by G. E. Taylor (private communication, 1974). These predictions are based on a combination of planetary ephemerides with a magnetic-tape version of the *Smithsonian Astrophysical Observatory Star Catalogue*. On the other hand, column 6 of Table 13.2 indicates that the most frequent events occur with far fainter objects, Pluto and the satellites of Saturn, Uranus, and Neptune. Predictions for these must be performed by photographing the sky ahead of the planet in question and measuring the plates immediately; this method has already been applied to Saturn's rings and Titan with stars down to twelfth magnitude. The need for continuous monitoring of the sky ahead of Saturn, Uranus, Neptune, and Pluto, in relation to stars brighter than the photographic magnitudes listed in column 7 of Table 13.2, is apparent.

A major source of inaccuracy in the predictions are errors in star positions and in the ephemerides of the planets and satellites. These errors are typically a few tenths of an arcsec in the direction perpendicular to motion, and in certain instances, as in the 1968 occultation by Neptune, they can exceed 1 arcsec. Satellites whose angular diameters are of the order of 0.1 arcsec would seem hardly worth the mounting of an expedition. However, there are ways to refine the predictions which will not be discussed here. The Galilean satellites and Titan, with angular diameters of the order of 1 arcsec, are relatively immune from the problem of inaccurate ephemerides.

However, there have been missed opportunities because of uncertainties in ephemerides. The best example involved Callisto and the fifth magnitude star 21 Capricorni on May 26, 1973. The path went north of that predicted, owing to errors in Jupiter's ephemeris and lack of time to refine the predictions; the advance notice in this case was less than one week. Twenty to thirty observatories were prepared to make photoelectric measurements following a crash effort to inform observers and a successful occultation would have surely resulted in obtaining several chords across the disk.

In conclusion, I urge that a systematic and expanded system of occultation predictions be initiated to include stars as faint as those listed in column 7 of Table 13.3 for each of the satellites considered. In view of the short lead times (of the order of a few weeks) for the fainter and more frequent events, there should be a system of cooperation among observatories, whose staffs should be alert for predictions in IAU circulars and equipped to observe each event. The more observations there are of an event, the better is the probability of success. Moreover, an object's diameter can be obtained only if there exist two or more chords across the disk, as observed from different sites on Earth in timing the event. The full potential of this undertaking can be realized only through cooperation and not through competition.

DISCUSSION

GORDON E. TAYLOR: The present situation is that we use the SAO star catalog for most predictions though we do photograph the sky ahead of Saturn, for likely occultations by Titan and Iapetus of stars to 11 mag, and ahead of Pluto for stars to 17 mag. Ideally we would like a star catalog for a band about 3° wide around the ecliptic, including stars to 15 mag, that is, a catalog of one-half million stars.

BRIAN O'LEARY: As an immediate practical solution to the prediction problem, it would seem fruitful to extend the sky photography ahead of Saturn for satellite occultations of stars to 15 mag and to examine the sky ahead of Uranus and Neptune for satellite occultations of stars to 17 mag. An ad hoc working group on extending predictions claims it is possible for Royal Greenwich Observatory to extend their Saturn prediction capability to 15 mag and for the University of Texas to use existing Palomar plates of the sky ahead of Uranus and Neptune to search for eligible stars to 17 mag. Refinements of predictions of possible occultations might also require taking new plates from some southern site. Once these predictions are refined, we will find from Table 13.2 above that there should be several observable occultations per year as seen from somewhere on Earth and one or more observable events from a particular site. It therefore would seem to be very worthwhile to initiate the extended prediction program to encompass Triton and the satellites of Saturn and Uranus.

CLARK CHAPMAN: On a related topic, I would like to mention the problem of measuring asteroid diameters by these techniques. As pointed out, one difficulty with the stellar occultation technique for measuring diameters is that the event is observable on Earth only from within the cylindrical "shadow" of the object, which is of course, nearly the same size as the object itself. For small objects correct predictions of the position of the narrow path of the cylinder across the Earth is difficult and the probability of a successful observation is low. As demonstrated by Elliot *et al.* (1975) the lunar occultation technique (in which the darkened limb of the Moon covers the object under study) is probably superior, in a practical sense, for small nearby objects such as asteroids. David W. Dunham, Donald R. Davis, and I have issued predictions during the past year for lunar occultations of asteroids, which occur quite frequently and are observable from wide areas on the Earth. To date I am not aware of successful photoelectric observations of these events. [O'Leary (cf. O'Leary *et al.*, 1976) has been partially successful in organizing an effort to see Eros occult a star.] I should point out that the prediction effort is *not* being pursued regularly any longer. But if there is interest in observing such events, they are not hard to predict. These observations have the potential of accurately calibrating the indirect techniques of radiometry and polarimetry currently used to "measure" the diameters of these small bodies.

PART IV

OBJECTS

PHOBOS AND DEIMOS

James B. Pollack
NASA-Ames Research Center

Ground-based and spacecraft observations of Phobos and Deimos are reviewed and the satellites' origin is discussed. Both bodies are heavily cratered, with crater densities close to the saturation level. The largest impact events may have knocked chips off these bodies and caused extensive fracturing of their surfaces. The surfaces are at least 1.5 billion years old and may date back to the early history of the solar system. The Martian satellites display large deviations from sphericity. As a result of tidal processes, they are in synchronous rotation. Several independent lines of evidence show that they have regoliths. Despite some provocative arguments, their internal strengths and the nature of their interior are poorly known at present. Photometric measurements suggest that they are made of either carbonaceous chondritic material or a basalt. Sinclair (1972a), Born and Duxbury (1975) and Shor (1975) apparently have successfully determined Phobos' secular acceleration. Their value of approximately 1×10^{-3} degrees/(year)^2 implies that the interior of Mars has a low Q (~100), which in turn may indicate that a portion of the Martian interior is experiencing partial melting. The low inclination of the satellites' orbits indicates that they are not captured bodies, but were formed as part of the same process that resulted in Mars.

In 1969, the same year that man first walked on Earth's Moon, space-age observations of the next-nearest satellites were initiated. The tiny Martian satellite, Phobos, was found projected against the disk of Mars on one of the Mariner 7 photographs. Analysis of this photograph by Smith (1970) yielded the first direct determination of that body's dimensions and albedo. High resolution images of both Phobos and Deimos were obtained by the TV experiment aboard the Mariner 9 spacecraft, which went into orbit about Mars on November 14, 1971, and remained operational for about a year. Additional useful observations of Phobos were made with the Mariner 9 infrared radiometer and ultraviolet spectrometer. Studies of these Mariner data have yielded information about the Martian satellites' cratering history, surface microstructure, surface composition, shape, rotation rate, orbits, and internal structure. More details on the shape and rotation of Phobos and Deimos are given in Chapter 15 by Duxbury.

Ground-based observations also have provided important information about Phobos and Deimos. Measurements of their positions over the past century have led to accurate determinations of their orbits, including estimates of Phobos' secular acceleration. Finally, additional insight into the composition and microstructure of their surfaces has been gained from recent polarimetric and photometric studies.

[319]

In this review, we assess the current state of our knowledge about the physical properties of Phobos and Deimos. The sections of this chapter are organized according to physical characteristics. Each section contains a discussion and analysis of all the observational data that bear upon a particular subject. The chapter is concluded with a discussion of several models of the origin of these bodies.

CRATERING HISTORY

In this section, we make use of the Mariner 9 TV images to discuss the influence of meteoroid bombardment on the satellites' surface morphology and to obtain estimates of their ages (Pollack *et al.*, 1972; Pollack *et al.*, 1973a; Masursky *et al.*, 1972). All Mariner 9 satellite images are published in Veverka *et al.* (1974); many are presented in this chapter and in Chapter 15 by Duxbury.

As illustrated in Figures 14.1 and 14.2 for Phobos and Figures 14.3 and 14.4 for Deimos, both bodies are heavily cratered. We will assume throughout this chapter that the craters are the result of meteoroid impact; it is difficult to see how vulcanism could occur on such small bodies. The craters differ greatly in their state of preservation, as illustrated by two craters seen near the right-hand, central portion of Figure 14.4. The smaller crater appears much more well-defined than the larger one. This diversity in crater morphology probably is the result of erosion due to saturation bombardment, a point discussed in more detail below.

At least some of the craters have well-defined rims (Masursky *et al.*, 1972). The rims may have been formed either as the result of the uplifting of bedrock or as the result of the shock lithification of a powdered material. In the latter case, some contribution may be expected from fragmental ejecta despite the low surface gravity, as discussed below in the section dealing with the satellites' regoliths.

The well-defined large crater shown in the bottom right-hand part of Figure 14.1 and the top of Figure 14.2 measures about 5 km across, while the somewhat more subdued crater appearing at the top left-hand side of Figure 14.1, and the bottom left-hand part of Figure 14.2 has a diameter of about 8 km. Since Phobos' diameter is only about 20 km, we might suspect that the largest impact events have caused extensive damage and that the satellite barely escaped from being totally destroyed in a hostile meteoroid environment.

Figure 14.5 provides some further information about the possible effects of meteoroid bombardment. The long, linear feature in shadow represents the boundary region in which the altitude of the surface undergoes a marked change. This linear feature ends near the rim of the 8 km crater, which appears in the bottom, left-hand side of the figure. This relationship is further illustrated in Figure 14.6 (cf. Duxbury, Fig. 15.8), which shows a map of the major features of Phobos found from the collection of Mariner 9 photographs. In the coordinate

Fig. 14.1. Mariner 9 photograph of Phobos obtained on orbit number 80 at a phase angle of 83°. Sub-spacecraft point is 65° S, 356° W. The illuminated area is approximately 23 km high and 13 km wide. North is at the top. (IPL roll 937, 103305)

Fig. 14.2. Mariner 9 photograph of Phobos obtained in orbit number 131 at a phase angle of 45°. Sub-spacecraft point is 36° S, 66° W. The illuminated area is approximately 19 km high and 21 km wide. North is at the top. (IPL roll 1114, 175715)

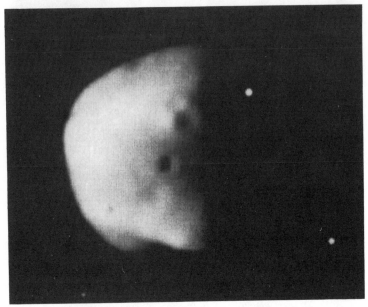

Fig. 14.3. Mariner 9 photograph of Deimos obtained on orbit 25 at a phase angle of 68°. On this version south is at the top. The predicted sub-spacecraft point is 14° S, 20° W. (IPL roll 752, 144007)

system used for this map, the pole is taken as a perpendicular to the orbital plane and 0° longitude passes through the sub-Mars point. The 8 km crater appears near 60° W longitude at 0° latitude, while the linear feature displays an arcuate geometry on the map, beginning near the lip of the 8 km crater and following a great circle around half the satellite. Pollack *et al.* (1972) pointed out that laboratory cratering studies carried out by Gault and Wedekind (1969) could provide an explanation for the linear feature as well as the irregular edges shown, for example, at the bottom of Figure 14.3. When the energy of an impacting object approaches but lies somewhat below the energy needed to totally disintegrate its target, the impact not only results in the formation of a crater but also fractures the surface and breaks off parts of the target.

In Figure 14.7 is illustrated the number density of craters on Phobos and Deimos (in logarithmic increments) as a function of crater diameter. Shown for comparison are the crater counts for the lunar uplands and several parts of Mars, which range from among the most heavily to the least heavily cratered areas on that planet. The number density of craters on the satellites is comparable to that found for the lunar uplands. Because the crater density of kilometer-sized craters on the lunar uplands is close to the saturation limit, we conclude that a similar situation holds for Phobos and Deimos. At the saturation limit, a maximum equilibrium value of crater density is reached at which the number of new craters of a given size that are created in a certain time interval balances the number of

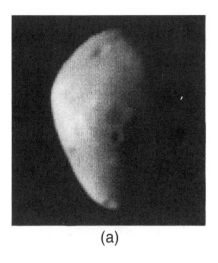

(a)

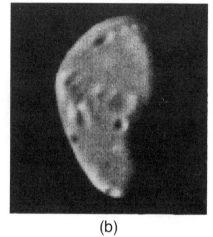

(b)

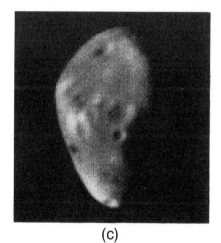

(c)

Fig. 14.4. Three versions of a Mariner 9
photograph of Deimos obtained on revolution 149 at a phase angle of 65°. The predicted sub-spacecraft point is 28° N, 355°
W. The illuminated area is approximately
12 km high and 7 km wide. North is at
the top. (Stanford AIL product 020501,
STN 0156)

324   J. B. POLLACK

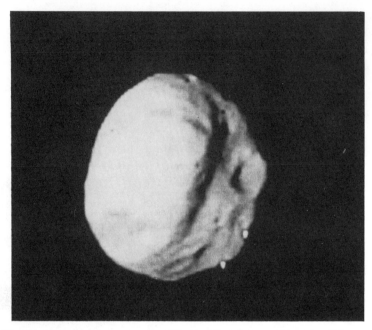

Fig. 14.5. Mariner 9 photograph of Phobos obtained on orbit number 129 at a phase angle of 19°. The predicted sub-spacecraft point is 28° S, 19° W. The illuminated area is approximately 22 km high and 20 km wide. North is at the top. (IPL roll 1114, 172829)

older craters of that size, which are destroyed as a result of the new impacts. This conclusion is also supported by the morphological evidence discussed above.

Before we can further compare the crater densities on Phobos and Deimos with those on the Moon and Mars, we must estimate the variation in the size of a crater produced on the four bodies by a meteoroid of the same size and impact velocity. Prior comparisons have implicitly assumed that there would be no variation in crater dimensions. Gault and Moore (1965) have defined two useful limits for the scaling relationships for impact craters. For sufficiently small craters the diameter D_c of the crater will be determined by the conversion of the kinetic energy of impact into the energy needed to fracture the target material. In this limit D_c scales as the diameter of the impacting object D_i and is independent of gravity g. At the other extreme the size of the crater is determined by the work required to lift material out of the crater. For this case $D_c \propto D_i^{3/4} g^{-1/4}$. Suppose first that kilometer-sized craters on all four bodies of interest lie within the gravity scaling limit. In that event we find that a meteoroid of a fixed size impacting at a fixed velocity produces a crater that is about 3½ times larger on the Martian satellites than on the Moon. The corresponding scaling factor is 4½

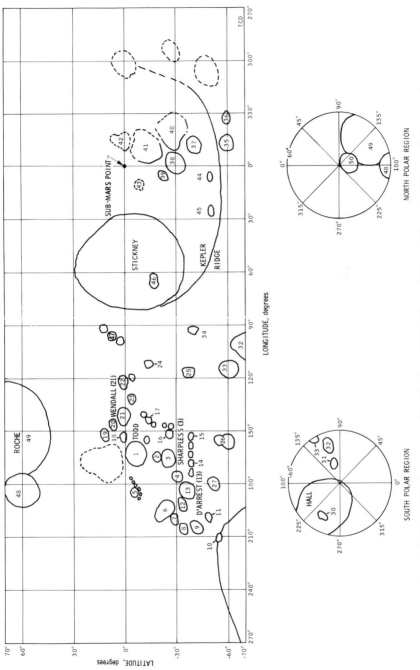

Fig. 14.6. Map of Phobos, prepared by Duxbury (1974; Chapt. 15 herein) from the Mariner 9 photographs.

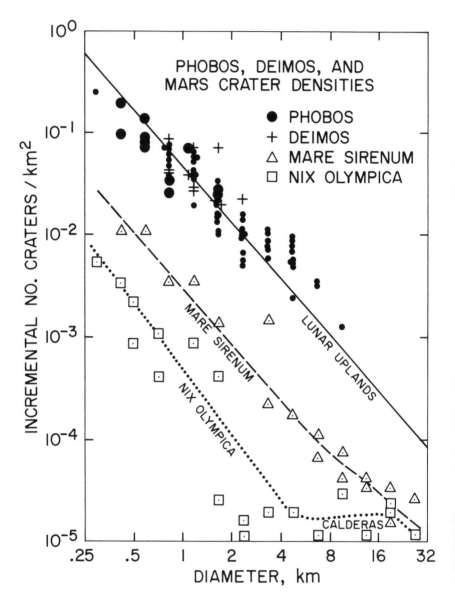

Fig. 14.7. Crater densities on Phobos and Deimos compared with those in the lunar highlands and in two regions on Mars. From Pollack *et al.* (1972).

for a comparison between the Martian satellites and Mars. In the section dealing with the internal structure of the satellites, we find a modest amount of evidence that the binding energy of the satellites is determined principally by their internal strength and not by their self-gravity. If this is the case, a scaling factor closer to unity should apply in the comparison of craters on the Martian satellites with those on the Moon and Mars.

We now proceed with a comparison of the crater density on Phobos and Deimos with that on Mars. In the gravity scaling limit we should compare the incremental number of craters of a given size on the satellites with the number of Martian craters of a size that is a factor of 4½ times smaller. For this case we find from Figure 14.7 that the number of impacting events recorded on Phobos and Deimos is comparable to that recorded on one of the most heavily cratered areas on Mars. If we allow for the fact that the crater density on the satellites is close to the saturation limit, while that found on Mare Sirenum is not, we conclude that even on the most heavily cratered areas on Mars, some kilometer-sized craters have been obliterated by a process other than cratering. This conclusion is strengthened by the discussion above which indicates that a scaling factor closer to one is perhaps more appropriate, as can be seen from Figure 14.7.

A lower limit to the age of the satellites can be obtained by comparing their crater density with that of selected areas on the Moon. Such an estimate has been made by Pollack *et al.* (1972), where the known age and crater density of the Apollo 12 site on the Moon was used as a standard of comparison. Based on the best available numbers at that time, the present frequency of meteoroid bombardment at Mars was assumed to be twenty times that at the Moon. Also the crater diameter scaling factor was set equal to one. With the above input data, a lower bound of 3.4 aeons was derived for the age of the satellites. We note that because the crater density on the satellites shows saturation effects only a lower bound on the age can be found from the observed crater density.

We now update this age calculation to allow for both a revision in the scaling factor and in the relative frequency of bombardment. Wetherill (1974) has presented arguments indicating that the present relative frequency of bombardment at Mars as compared with the Moon is much closer to unity than a factor of 20. If the crater size scaling factor is set equal to its upper bound of 3½, we obtain a minimum lower bound of about 1.5 aeons for the age of the satellites. With the use of a somewhat more likely value for the scaling factor, we find that the satellites date back to the beginning of the solar system.

In summary, the surfaces of Phobos and Deimos are heavily cratered, with a number density of craters close to the saturation value. The larger cratering events may have knocked chips off these bodies and created extensive fracture zones. The age of the surfaces, as implied by the crater density, is at least 1.5 billion years, and they may date back to the origin of the solar system.

SHAPE, SIZE, AND SPIN

We now will discuss estimates that have been made of the dimensions and rotation rate of the Martian satellites (see also Chapt. 15, Duxbury). Toward the end of this section we will indicate the influence of tides raised by Mars on the evolution of their spin vectors.

As mentioned earlier, Smith (1970) obtained the first direct determination of Phobos' size from an image of Phobos that was detected on one of the Mariner 7 photographs of Mars. From a careful analysis of this image, which was only several resolution elements across, Smith estimated that the projected area was 18 by 22 kilometers across.

Accurate determinations of the three-dimensional figure of Phobos and Deimos have been obtained from the Mariner 9 photography (Pollack *et al.*, 1973a; see also Chapt. 15, Duxbury). At closest approach, Mariner 9 came within 5,000 km of the satellites and obtained pictures of them with a resolution element size of several hundred meters. The surfaces were modeled as triaxial ellipsoids. Furthermore, the bodies were assumed to be in synchronous rotation with one axis of the ellipsoid in the direction of the Mars-satellite line, a second axis also in the orbital plane, and a third one perpendicular to the orbital plane (cf. Chapt. 6, Peale). Computer-generated overlays of various trial ellipsoids were made for each high resolution photograph. The ellipsoid axes were adjusted to obtain the best fit to the outline of the satellites shown in the various photographs. About 30 pictures of Phobos and 10 of Deimos were used in this manner.

Table 14.1 shows the values for the principal axes so found (cf. Table 15.2, Duxbury). Note that the axes are radii and not diameters. The range in viewing geometry among the photographs permitted an accurate determination of all three axes for Phobos but good determinations only for the two smallest axes of Deimos. These results indicate that both bodies show large departures from sphericity, a conclusion which is consistent with their small sizes and extensive cratering history. As we will see in the section dealing with their internal structures, their gravitational fields may be small compared with their internal strengths.

We have mentioned above that, in performing the shape determination, the satellites were assumed to be in synchronous rotation. This assumption was confirmed by checking that certain surface features, such as the centers of craters, always appeared at about the same predicted latitude-longitude location. Furthermore, the largest axis was found to be in the Mars-satellite direction, while the smallest one was determined to be in the direction perpendicular to the orbit plane. Such a configuration is expected for synchronous rotation (see Chapt. 6, Peale).

Burns (1972) and Pollack *et al.* (1973a) have carried out calculations to show that it was quite reasonable for both satellites to be in synchronous rotation.

TABLE 14.1
Principal Axes of Phobos and Deimos*
(Adapted from Pollack *et al.*, 1973a)

Satellite	Largest Axis (km)	Intermediate Axis (km)	Smallest Axis (km)	Volume (km³)	Mass (10¹⁸ g)†
Phobos	13.5±1	10.7±1	9.6±1	5810	17.4
Deimos	7.5$^{+3}_{-1}$	6.0±1	5.5±1	1040	3.1

* The axes are radii, *not* diameters. On the average the largest axis points toward Mars, the smallest axis lies normal to the orbit plane, and the intermediate axis is perpendicular to the other two (Chapt. 6, Peale).
† A density of 3 g/cm³ has been assumed.

Evolution to this state is caused by the frictional dissipation of solid body tides raised on the satellites by Mars. In general, the final spin state need not be the synchronous value, but can be some simple fraction of that value. However, when the fractional difference between the moments of inertia is larger than a factor that depends on the orbital eccentricity, the synchronous value will always be realized as the end state (presented by Peale in Chapt. 6). The small values of the orbital eccentricities and the large differences in the moments of inertia, as implied by the values for the principal axes, assure that this condition is met by several orders of magnitude for both satellites.

In order for the Martian satellites to achieve synchronous rotation, they must be tidally despun in a time less than the age of the solar system. This time scale depends upon geometrical factors, the mass of Mars, a dissipation factor for the satellites and the rigidity of their interior (Chapt. 6, Peale). With a range of plausible values for the dissipation factors and a rigidity typical of solid rock, it has been shown (Pollack *et al.*, 1973a) that the despin time for Deimos lies between about 10⁶ and 10⁸ years. If the interior of Deimos has less coherence than that of solid rock, even smaller time scales result. The corresponding time scale for Phobos is about a factor of 100 smaller, chiefly because it is closer to Mars. Thus, in a time much less than the age of the solar system, tidal forces raised by Mars cause an evolution of both satellites to a synchronous rotation state.

In summary, estimates of the principal axes of Phobos and Deimos are given in Table 14.1. Both satellites show large departures from sphericity and have synchronous rotation periods. This latter result is consistent with the end state expected from tidal evolution and the time scale required for them to be despun is much less than the age of the solar system.

REGOLITH

We will summarize several pieces of evidence which indicate that the Martian satellites have regoliths, then describe the spatial extent of the regoliths, and finally, discuss how they may have been produced. The term regolith refers to a top layer of loose, individual, small particles such as the layer of fine soil that is present on our Moon, where it was generated by repeated meteoroid impacts.

The presence of a regolith on both satellites is suggested by photometric, polarimetric, and thermal measurements. We first consider the photometric results. The brightness of the limbs of the satellites have been studied (Pollack *et al.*, 1973a) as a function of the phase angle. For dark surfaces, such as those of the Martian satellites, the limb brightness depends almost entirely on the phase angle and should have little dependence on the angles of incidence and reflection. The limb brightness of both bodies showed a very similar behavior, with the brightness decreasing rapidly with increasing phase angle. Comparison of the observed photometric function with those for a number of laboratory samples led to the conclusion that the surface layers were texturally complex (Chapt. 9, Veverka). The photometric behavior was consistent with the presence of a regolith, but did not uniquely require a regolith.

Polarimetric measurements of the satellites have been obtained at large phase angles from the Mariner 9 photography and at smaller phase angles from ground-based observations (cf. Chapt. 10, Veverka). Using Mariner 9 photographs of Phobos and Deimos taken with a triplet of polarization filters, Noland *et al.* (1973) found that both bodies had a large positive polarization (20–25%) at phase angles ranging from 74° to 81°. Positive polarization refers to the situation where the largest E vector lies perpendicular to the Sun-planet-Earth plane. Again, this result is consistent with the presence of a regolith, but may not uniquely require a regolith (see Chapt. 10, Veverka).

Zellner and Capen (1974) measured the polarization of Deimos from the ground at phase angles between 0 and 45°. Figure 14.8 summarizes their results as well as the Mariner 9 results, which were discussed above. The filled circles are the polarization values for Deimos; the open circles are the data for Phobos. The presence of a deep negative branch in the polarization curve for Deimos provides additional evidence in favor of a regolith. By analogy, since Phobos is very similar to Deimos in its photometric and polarimetric behavior, it, too, probably has a regolith. The ground-based measurements of Deimos' polarization are shown in Figure 10.5.

Perhaps the strongest piece of evidence showing the presence of a regolith is supplied by the thermal infrared observations of Phobos obtained by the Mariner 9 infrared radiometer experiment (cf. Fig. 12.9, Morrison). Gatley *et al.* (1974) observed Phobos at wavelengths of 10 microns and 20 microns as the satellite passed through the shadow cast by Mars. Their results at a wavelength of 10

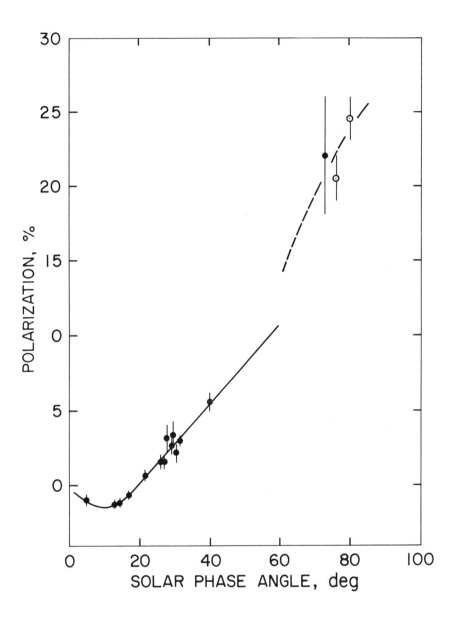

Fig. 14.8. Polarization of Phobos and Deimos as a function of phase angle. The solid points refer to observations of Deimos, while open circles represent corresponding measurements of Phobos. From Zellner and Capen (1974) and Noland *et al.* (1973).

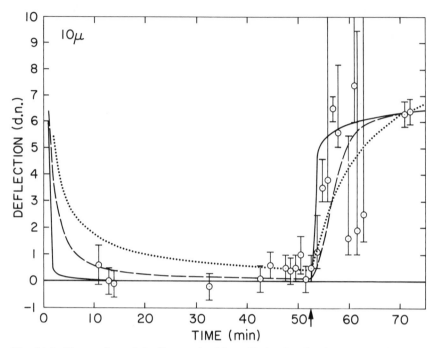

Fig. 14.9. Observations of the flux received from Phobos in a bandpass centered near 10 microns. The measurements were obtained during and after Phobos passed through the shadow of Mars. The solid dots with their accompanying error bars represent individual observations. The solid, dashed and dotted curves represent calculated fluxes from surfaces characterized by thermal inertias of 2.5×10^{-4}, 7.7×10^{-4} and 2.5×10^{-3} cal cm^{-2} sec$^{-1/2}$ °K^{-1}. From Gatley *et al.* (1974).

microns are shown in Figure 14.9. Very similar data were obtained at 20 microns. The dots with their associated error bars show the observed flux values, while the lines represent the predictions of theoretical cooling models that differ in the value of the surface's thermal inertia. The solid, dashed, and dotted curves are characterized by thermal inertias of 2.5×10^{-4}, 7.7×10^{-4}, and 2.5×10^{-3} cal cm^{-2} sec$^{-1/2}$ °K^{-1}, respectively. Phobos was within the umbra of Mars' shadow from the start of the observations until a time of slightly greater than 50 minutes, which is marked by an arrow along the time axis.

The data at both wavelengths require the model with the lowest thermal inertia. Models having even lower values for the thermal inertia lead to acceptable predictions. If nominal values of density and specific heat are assumed, a thermal conductivity of about 10^{-6} cal sec^{-1} cm^{-1} °K^{-1} or smaller is implied by these results. Such low thermal conductivities are realized only for powdered surfaces. By way of contrast, a texturally complex, porous, rocky surface, which might be compatible with the photometric observations, would have a thermal conductivity much higher than the upper bound deduced above. The thermal conductivity of porous materials is determined by the fractional area of thermal contact between its solid elements. In the case of porous rocks, the fractional area of thermal contact is never very small and therefore neither is its thermal conductivity. Only in the case of powders can this fractional area be extremely small, leading to very low values of the thermal conductivity. We conclude that both satellites possess regoliths.

The regolith must extend to at least the depth penetrated by the diurnal thermal wave. Using the above upper limit to the thermal conductivity, Gatley *et al.* (1974) found a value of 1 mm for this minimum depth. Thus, we have only a very weak constraint on the depth of the regolith.

Except in areas of topographic variation, such as the inside of craters, Phobos appears photometrically quite homogeneous on the Mariner 9 photographs. This result, perhaps, implies that its regolith covers all parts of its surface. However, as shown in Figure 14.4, Deimos has some photometric irregularities, as illustrated by the bright streaks near the left-center of the picture (see also Chapt. 15, Fig. 15.4, Duxbury). The bright streaks may have been produced by a recent meteoroid impact that either exposed an underlying rock surface or excavated deeper, bright, powdered material, in much the same way that bright rays are generated on our Moon.

Because Phobos and Deimos are so small, the presence of a regolith is perhaps surprising. One might have expected that the powdered rock produced by meteoroid impacts would easily escape from these bodies since the escape velocity for these satellites is only about 10 meters/second. However, it has been pointed out by Pollack *et al.* (1973a) and by Soter (1971) that though the ejecta from a meteoroid impact will escape from the satellite, almost all of the ejected material will have too low a velocity to escape from Mars' gravitational potential. They will go into orbit about Mars and in a reasonably small period of time (100 to 10,000 years) be recaptured by their parent bodies. Subsequent impacts may cause some material to repeat this cycle. Material close to the surface may have been recycled many times and thus the surface layers may consist of a very fine powder.

Recent laboratory studies of hypervelocity impacts into unconsolidated targets lead to a significant revision of the above model of the maintenance of the satellites' regoliths. Stöffler *et al.* (1975) found that ejecta from impact craters formed in quartz sand travelled approximately three to four orders of magnitude less far from the crater than material derived from impacts into solid basalt. The prior analysis discussed above made use of the results for solid targets.

The new data imply that about 99.9% of the ejecta resulting from impacts into the satellites' regoliths will not escape from the satellites (S. Soter, private communication, 1974). Hence the above recycling of material from the satellites into orbit about Mars and back again is applicable only if impacts reach well-consolidated material beneath the regolith, that is, perhaps only to the larger impact events. One might even argue that the satellites formed initially with a regolith, derived from the last stages of satellite formation, and that subsequent hypervelocity impacts have led to little recycling (J. Burns, private communication, 1974). Such a picture of the non-evolution of the satellites' regoliths is applicable only if the initial regolith is deeper than the average depth of subsequent meteoroid excavation. Pollack *et al.* (1973a) estimated from the excavated volume of the observed crater population that the average depth of meteoroid excavation was several hundred meters.

Thus, if the satellites are composed entirely of a collection of small particles, recycling has played only a minor role in the regolith's subsequent history. However, if initially the satellites formed with a regolith less than several hundred meters deep and with a coherent, rock-like interior, meteoroid impacts significantly deepened the regolith to an average depth of several hundred meters and recycling permitted the accumulation of most of the mass ejected by large craters. Finally, we note that the effect of impact events into a regolith is to generate both larger particles through shock lithification and fusion and at the same time smaller particles by fracture (Stöffler *et al.*, 1975).

We conclude that both Phobos and Deimos possess regoliths. The regoliths may have been generated partially from the powdered ejecta produced in meteoroid impacts, which escaped from the surfaces of these bodies, went into orbit about Mars, and were subsequently recaptured. However, the importance of this process depends critically on the depth of the satellites' soil layers at the time of their formation. If initially the soil layers were deep, recycling was unimportant.

INTERNAL STRUCTURE

The degree of consolidation of the satellites' interiors may provide clues as to their evolutionary histories. We can consider two extreme possibilities. The interiors may be made up of a loose collection of small particles, as are the tops of their surfaces. In this case, gravitational forces are responsible for holding the bodies together. Alternatively, their interiors may be made up of well-consolidated rocks, in which case the structural strength of the rocks may play a more important role than the gravitational forces. Below we discuss arguments advanced earlier (Pollack *et al.*, 1973a) to distinguish between these two possibilities, consider a new argument based on their shape, and discuss the cosmogonic implications of these alternative, structural models.

The linear feature, shown in Figures 14.5 and 14.6 and which was discussed in the section dealing with cratering, offers strong evidence in favor of the well-consolidated model. In that section we interpreted the linear feature as being the result of a fracturing of the surface caused by the impact event that created the 8 km crater. While it is easy to see how such a localized dissipation of part of the impact energy can occur in crystalline material, it is difficult to imagine how such a feature could be produced in a loose aggregate of soil. Put another way, if the structural strength of the material is more important than gravity, an initial weakness, which is a zone of a large gradient in the structural integrity, can be propagated over a considerable distance, and lead to a near-discontinuity in the amount of mass spalled off at the position of the linear feature. If gravity is more important, a more likely scenario would be a fairly smooth variation of the amount of mass ejected with position on the surface. In analyzing the likely failure modes of a weakly-coherent material, a clear distinction should be made between the effects of tensile forces, which are important for the situation of interest, and those due to compressive forces. For example it is quite possible to create grooves in a sand-like material by exerting a linear compressive force, since the compressive strength of this material is much greater than its tensile strength.

A distinction between the two structural models was also attempted by Pollack *et al.* (1973a) on the basis of past cratering history of the satellites. The binding energy for the two models was compared with the impact energy of the largest meteoroid that was likely to strike each satellite over the age of the solar system. This comparison indicated that the largest meteoroid would have very likely totally disintegrated both Phobos and Deimos if they were made of unconsolidated material. However, equating the kinetic energy of impact to that of the fragments resulting from a total disintegration of a satellite, we find that the mean velocity of the fragments would be quite small (~ 10 m/sec). Therefore the fragments would travel in very similar orbits about Mars and might re-form the satellite in a fairly short time (S. Soter, private communication, 1974). We conclude that attempts to infer the satellites' structural strength from their cratering history may be less useful than originally thought.

The shape of the satellites offers another means of assessing their internal strength. Soter and Harris (1976) have proposed that the triaxial figure of Phobos may be due to the combined effect of Phobos' rotation, its gravitational field, and the tidal component of Mars' gravitational field. By matching the observed axes given in Table 14.1 with those of equipotential surfaces, they were able to obtain estimates of the mean density of Phobos. They found a value of 3.8 ± 0.5 g-cm^{-3}. Furthermore, because Phobos' distance from Mars is changing rapidly (see below), Phobos had to adjust to its present configuration within about the last ten million years in order to have a figure that lies within the error bars of Table 14.1.

In order for Phobos to assume the shape of its gravitational equipotential surface, its interior tensile strength must be small compared with its gravitational field. Using formulae (Pollack *et al.*, 1973a) for the binding energy due to gravity and structural strength, we find that the above requirement implies that Phobos' tensile strength is less than about 10^6 dynes/cm^2. Soter and Harris interpret their result as implying that Phobos may be composed entirely of a collection of small particles.

Above we have seen that there exists apparently contradictory evidence concerning Phobos' internal strength. On the one hand, the existence of the long linear feature implies that Phobos' internal strength is more important than its gravitational field, while Soter and Harris' analysis of its figure leads to the opposite conclusion. We now examine these arguments in a more critical fashion to see if they can be reconciled. Consider first the argument based on the linear feature. We might try to dismiss this argument. It could be claimed that while there is experimental evidence that such a feature can be produced in a crystalline material, no experiment has been performed to show that it cannot be produced in a ball composed of sand grains. Alternatively, one could point out that the strength needed to generate this feature, whose height is about a kilometer, need not exceed the upper bound to the strength obtained from the shape argument. In particular, tensile strengths between about 10^5 and 10^6 dynes/cm² might satisfy both results.

Let us next consider in further detail Soter and Harris' arguments based on Phobos' shape. We could argue that by pure coincidence they were able to match the observations with an equipotential surface. After all, they have two adjustable parameters—Phobos' mean density and its mean radius. This argument is not very persuasive because there are three observational quantities that are being matched and because the mean density they derive is not unreasonable. However, it is worth noting that Deimos' shape should be very close to that of a sphere according to their model and therefore the differences shown in Table 14.1 must be dismissed as arising from observational error, a not altogether impossible suggestion. Also, the mean density they derive, even allowing for its error bars, appears to be larger than what one might expect for a body that is pure regolith (Keihm and Langseth, 1975). This latter point, if accepted, does not negate Soter and Harris' explanation of Phobos' shape, but merely their inference about its internal structure. Finally, the presence of many well-defined craters on Phobos may be difficult to understand within the context of Soter and Harris' model. Almost all of these craters formed at a time when Phobos was much further from Mars than it is today. During these earlier epochs it should have had a much more spherical shape than it currently does. Even allowing for the effects of shock lithification in a thin zone near the surface of the craters, it is hard to see how the subsequent readjustments of the shape of Phobos by several kilometers would not have severely distorted and perhaps destroyed most of its craters.

The above discussion indicates that no definitive statement can be made as yet concerning the internal strength of the satellites, although we have a slight preference for models of intermediate or large strength. However, it is of interest to consider what strengths we might expect from a variety of materials. A collection of particles in a vacuum exhibits a cohesion derived from weak van der Waals electrostatic forces. Measurements of the mechanical properties of the lunar regolith indicate that its cohesion is about 5×10^3 dynes/cm² (Scott and Robertson, 1968). Such a material is certainly consistent with the upper bound on the tensile strength found from the shape of Phobos but may be too small to account for the linear feature. Little quantitative information is available about the tensile strengths of carbonaceous chondrites, which are welded together by a combination of compressive forces, low temperature welding, and variable amounts of moderate higher temperature-induced welding. Type I carbonaceous chondrites are very friable, while later types are stronger. Finally, solid igneous or metamorphic rocks typically have tensile strengths of about 10^8 to 10^9 dynes/cm^2. However, the development of cracks due to a history of extensive meteoroid bombardment will greatly lower these values. Thus, conceivably an interior composed of heavily fractured rock could be weak enough to conform to the equipotential surface. In view of the density derived from the shape argument, such a model may be the preferred one.

We conclude this section by considering the relationship between the internal structure of the satellites and their origin. First, suppose they are composed of a collection of small particles. In that case we could argue that they formed with their current dimensions and that their gravitational fields were too weak to consolidate them.

If their interiors contain a substantial amount of more well-consolidated, rock-like material, one could argue that they were originally part of a much larger object, whose self-gravity was large enough to form well-consolidated rocks. Subsequently, this parent body was fragmented, perhaps by meteoroid impact, leading to the formation of the present satellites. Alternatively, we might imagine that the satellites formed with their present dimensions and that some source of energy, other than gravity, caused them to melt, transforming the primordial, unconsolidated material into one having the strength of solid rock. Among the possible energy sources that might be important in the early history of the solar system are short-lived radioactive substances (Fish *et al.*, 1960) and the early solar wind (Sonett *et al.*, 1968). Similar hypotheses have been discussed to explain the rock-like strength of many meteorites.

In summary, the internal strength and degree of consolidation of the satellites' interior are not well known at present, although there is some indication that much of their interior contains highly fractured, rock-like material. In view of this uncertainty it is premature to draw any strong inferences concerning their origin.

COMPOSITION

Photometric observations have provided clues as to the composition of the surfaces of the satellites. However, at present no definitive identification has been made. Below, we discuss determinations of their albedo at visual wavelengths, observations of the shape of their spectral reflectivity curves, and the implications of these results for composition.

Estimates of the geometric albedo of the Martian satellites have been obtained by combining spacecraft measurements of their geometrical dimensions with ground-based observations of their integrated brightness. Using a Mariner 7 photograph, Smith (1970) estimated that Phobos' geometric albedo was about 0.065. Values of 0.05 ($^{+.03}_{-.01}$) and 0.06 ($^{+.02}_{-.01}$) were found by Pollack *et al.* (1972) for the geometric albedos of Phobos and Deimos, respectively (see the summary given in Chapt. 9 by Veverka). These results were derived by combining the Mariner 9 size information with Kuiper's (1961b) ground-based photoelectric observations. Finally, Zellner and Capen (1974) derived geometric albedos of 0.062 ± 0.012 and 0.070 ± 0.031 for Phobos and Deimos, respectively. This last set of values is probably the most accurate since Zellner and Capen made use of several magnitude determinations, including their own, and corrected each of those observations to 0° phase angle, making use of their observed value for Deimos' phase coefficient. Zellner and Capen also obtained a value of 0.065 for Deimos' geometric albedo from their determination of the slope of the positive branch of the satellite's polarization curve (cf. Chapt. 10, Veverka). Thus both satellites have a very low albedo, a result that supplies some compositional information as discussed below.

The above results indicate Phobos and Deimos have a similar visual albedo. The most accurate determination of the ratio of the two satellites' reflectivity stems from photometric analysis of the Mariner 9 photographs, which was discussed in the section dealing with the satellites' regoliths. The ratio of the reflectance of Deimos to that of Phobos is 1.05 ± 0.05 (Pollack *et al.,* 1973a). Thus the two Martian satellites have almost identically the same reflectivity at visual wavelengths.

We next discuss the variation of the Martian satellites' reflectivity with wavelength (cf. Chapt. 11, Johnson and Pilcher). According to Kuiper's (1961b) photoelectric observations, the two satellites have a (B–V) value of 0.6 ± 0.1 magnitudes. To within his observational uncertainty, Phobos and Deimos show the same flat reflectivity curve between the blue and yellow region of the spectrum. Zellner and Capen (1974) have obtained the best current estimates of the UBV magnitudes of Deimos. They find that Deimos has (B–V) and (U–B) values of 0.65 ± 0.03 and 0.18 ± 0.03 magnitudes. The corresponding magnitude differences for the Sun are 0.63 and 0.14. These results imply that Deimos' reflectivity declines only very slightly (~5%), if at all, from 5500 Å to 3500 Å.

The Mariner 9 ultraviolet spectrometer has obtained measurements of Phobos' reflectivity between a wavelength of 2100 Å and 3500 Å. (Lane, private communication, 1974; Pollack *et al.,* 1976b). In marked contrast to its longer wavelength behavior, the reflectivity curve has a noticeable red-ward slope, decreasing by almost a factor of two between the long-wavelength and short-wavelength boundaries.

We now turn to the compositional implications of the above results (cf. Chapt.

12, Johnson and Pilcher). That Phobos and Deimos have almost identical geometric albedos in the visible and very similar (B–V) colors strongly suggests that their surfaces are made of the same material. Carbonaceous chondritic material and basalt are the only two substances that are abundant in the solar system and that show albedo and color characteristics similar to those found for Phobos and Deimos (Pollack *et al.*, 1972). In the former case absorption is chiefly the result of the carbon present in the material, while in the latter case iron oxide is the source of the darkening. Distinction between these two possibilities is most important since with the first possibility the satellites would be composed of undifferentiated material, while in the second case the opposite would be true.

The similarity in the surface composition of Phobos and Deimos is not the result of their environment, but reflects their intrinsic chemical similarity. For example, studies of the Moon indicate that meteoroidal material constitutes only a percent or so of the lunar surface material. Also the amount of ejecta from impact events on Deimos that ends up being captured by Phobos will be far less than the quantity of ejecta from Phobos that is recaptured by that body.

In summary, both satellites are extremely dark at visual wavelengths and have a fairly flat reflectivity curve throughout most of the visible portion of the spectrum. However, in the near ultraviolet, Phobos' reflectivity declines markedly towards shorter wavelengths. Both satellites are very likely made of the same material. Carbonaceous chondritic material and basalt are the best compositional candidates.

ORBITS

We will discuss our current knowledge of the satellites' orbits, point out their significance for inferring certain properties of Mars, and finally consider the very important problem of the secular acceleration of Phobos' orbit. Ground-based observations of the Martian satellites' orbital position have been made since their discovery in 1877 by Asaph Hall (1878). An increasingly sophisticated theory has been developed by such people as Sharpless (1945), Wilkins (1966), Sinclair (1972a) and Shor (1975) who make use of these observations to determine a number of parameters, including the orbital elements. Sinclair's and Shor's theories include an allowance for the secular and periodic variations in the orbital elements due to the combined effect of the oblateness of Mars and the attraction of the Sun.

At the time the Mariner 9 spacecraft went into orbit about Mars and began its observations of Phobos and Deimos, Wilkins' theory provided the best predictions of the satellites' positions. While the ephemerides based on this theory were quite accurate, they alone were not accurate enough to permit obtaining close-up photographs of the satellites with the Mariner 9 TV camera. Hence satellite photographs were first obtained at relatively large distances from the satellites and used to refine the satellites' ephemerides. Using the revised ephemeris, the satellite group of the Mariner 9 television team obtained photographs of Phobos and Deimos from distances as small as 5000 km (Pollack *et al.*, 1972, 1973a). The most important corrections to Wilkins' theory for Phobos were an advance of 2.8° (about 600 km) in its mean longitude and a 0.3° decrease in the inclination of its orbital plane relative to the Earth's equatorial plane. The most important correction for Deimos was a decrease of 0.3° (about 150 km) in its mean longitude.

Analysis of the satellites' observed positions have not only been useful for evaluating their orbital elements, but they have permitted a determination of a number of properties of Mars. These properties include Mars' mass, the J_2 moment of its mass distribution, its oblateness, and the location of its axis of rotation (cf. Burns, 1972). Recently, studies of the radar tracking of the Mariner 4, 6, 7, and 9 spacecraft have led to a considerable refinement in most of these values (cf. Born, 1974).

From his studies of the motions of the satellites of Mars, Sharpless (1945) obtained a positive value for the secular acceleration of Phobos, which implied that the satellite would crash into Mars in only another several times 10^7 years. This result led to a plethora of explanations for such a large secular acceleration, including the suggestion by Shklovskii that the Martian satellites were artificial bodies, created by a technologically advanced civilization (Shklovskii and Sagan, 1966). However, the Mariner 9 photographs of Phobos and Deimos appear to rule out this imaginative hypothesis. Sharpless' ephemeris is in clear conflict with the most recent observations of Phobos' position; it incorrectly predicts its current longitude by almost 10° (Burns, 1972). Wilkins (1967, 1968) and Sinclair (1972a) have redetermined the value of Phobos' secular acceleration, using more refined models and more extensive observational data than Sharpless used. They obtained several, quite disparate values, depending on the set of data employed, and they concluded that the observations were not sufficiently accurate to permit a determination of Phobos' secular acceleration.

However, more recently both Shor (1975) and Born and Duxbury (1975) have obtained positive determinations of Phobos' secular acceleration. Shor made use of an even larger collection of ground-based measurements than did Sinclair, including some early Russian data, which were unavailable to Sinclair. Shor found a value of $(1.43 \pm 0.15) \times 10^{-3}$ degrees/(year)^2 for the secular acceleration. (Note that because of the manner in which Shor and Sinclair define "secular acceleration," the first derivative of the mean motion, which is used in the tidal equations below, equals *twice* the "secular acceleration.") In contrast to Sinclair, Shor determined that this value was relatively insensitive to the omission of some of the data.

Born and Duxbury (1975) compared the position of Phobos found on the Mariner 9 photographs with the predictions of Sinclair's and Shor's theories. In processing the Mariner 9 TV data, they made use of an ephemeris model that included the solar perturbations as well as those due to the zonal and tesseral harmonics of Mars' gravity field. They found that a much better fit to the observed positions of Phobos was given by the theories that included a secular acceleration term than ones that did not. For example, at the beginning of the Mariner 9 observations, the observed longitude of Phobos differed by 2° from that predicted by Sinclair's theory without a secular acceleration term. By contrast the longitudes predicted by Sinclair's theory with a secular acceleration of 0.96×10^{-3} degrees/(year)^2 and Shor's theory with a secular acceleration of 1.43×10^{-3} degrees/(year)^2 differed by only 0.3° and 0.4°, respectively, from the observed value. The observed longitude was known to about $\pm 0.05°$ and the predicted longitudes are accurate to several tenths of a degree. Thus, the theories with a secular acceleration term agree with the Mariner 9 observations to within one sigma of their formal uncertainties.

In conclusion, the excellent analyses of Shor (1975) and Born and Duxbury (1975) have provided positive estimates of Phobos' secular acceleration. However, in view of the difficulty of determining this parameter, it would perhaps be wise to await further studies before we consider that Phobos' secular acceleration has definitely been detected. Below we assume that these determinations are correct and explore their exciting implications. In the calculations below, we make use of both the secular acceleration values quoted above.

We first consider the implications of Phobos' secular acceleration for the state of the interior of Mars. According to Burns (1972), by far the most likely cause of Phobos' secular acceleration is the tidal torque exerted by Mars, which in turn arises from the tide raised by Phobos on Mars (cf. Chapt. 7, Burns). We will assume that this is the correct explanation. Aside from known geometrical factors, the secular acceleration is proportional to $(mk_2)/Q$, where m is the mass of Phobos and k_2 and Q are the Love number and the specific dissipation factor of Mars, respectively; high Q means low dissipation (Burns, 1972; Chapt. 7 herein). Assuming that the mean density of Phobos is 3 g-cm^{-3} and making use of dimensions of the satellite obtained from the Mariner 9 photographs, we obtain the estimate of Phobos' mass shown in Table 14.1. The value chosen for the mean density is consistent with that of the best compositional candidates of the previous section. The uncertainty in this number is probably no more than $\pm 20\%$. With this value for Phobos' mass, we find from equations given in Chapter 7 (Burns) that the ratio Q/k_2 for Mars equals approximately 780 and 1,150 for Shor's and Sinclair's accelerations, respectively.

We next estimate a value for k_2 by using an appropriate terrestrial analog. Aside from well-determined quantities, the Love number k_2 depends upon the value of the rigidity (Chapt. 6, Peale; Chapt. 7, Burns). The rigidity varies somewhat with composition and undergoes modest increases with increasing pressure and decreasing temperature. Here we use the terrestrial rigidity at a depth having the same pressure as that at a mean depth beneath the Martian surface. We choose the mean depth for Mars as one-quarter its radius—850 km—which was derived by finding the volumetrically averaged radial distance from the center of Mars. The corresponding "equivalent depth" for the Earth is 320 km. According to Gray (1972) the rigidity of the Earth equals 8×10^{11} dynes/cm^2 at this location. Using this value we find that k_2 equals 0.093 for Mars and therefore Q for Mars equals about 75 and 110 for Shor's and Sinclair's accelerations, respectively. The uncertainties in the mass of Phobos and the rigidity of Mars lead to uncertainties of about $\pm 35\%$ in the above values for Q. Smith and Born (1975) have independently arrived at estimates of Q ranging from 50 to 150.

These values of Q are comparable to the smallest value ($\sim$80) found at any depth beneath the surface of the Earth (Kaula, 1968; Chapt. 7, Burns). The region of low Q for the Earth coincides with a zone of partial melting. At greater depths in the Earth, where the local temperature lies increasingly below the melting point temperature, Q steadily increases and reaches values of about 2,000 in the lower mantle (Kaula, 1968). Therefore, we suggest that the comparatively low value for the Q of Mars may be due to the presence of an extensive zone of partial melting within the interior of Mars. This result is consistent with estimates of the ages of the youngest volcanic units on the Martian surface, which apparently have been formed very recently (Carr, 1974). With a more detailed analysis than given here, the secular acceleration of Phobos may provide an important constraint on models of the Martian interior.

We next turn to a consideration of Phobos' past and future orbital evolution, applying equations given by Burns (1972; cf. Burns, Chapt. 7 herein). For the purpose of the present set of calculations, we assume that Q has been constant with time. Making use of Phobos' present secular acceleration and mean motion, we extrapolate its orbital decay into the future and find that it will crash into Mars in only another 3.4×10^7 and 5.1×10^7 years for Shor's and Sinclair's accelerations, respectively. The present secular acceleration also implies that Phobos was much further from Mars in the past. We find that 4.6 billion years ago, at the time of its birth, Phobos had a semimajor axis that was between 2.00 (Sinclair) and 2.13 (Shor) times its present value a_p. These past values may be compared to a value of 2.18 a_p and 2.50 a_p for the stationary orbit position and the present semimajor axis of Deimos, respectively. The stationary orbit position refers to the distance at which the orbital period of a satellite equals the rotation period of Mars. Satellites located closer to Mars than this distance spiral into Mars as a result of tidal evolution and those at larger distance spiral outward (Chapt. 7, Fig. 7.1, Burns). Because of its smaller mass and greater distance from Mars, Deimos has undergone only a negligible change in its orbital distance. The above calculation indicates that in the early epochs of the solar system the orbits of Phobos and Deimos were closer together and that both were close to the stationary orbital position.

In summary, the orbits of the two Martian satellites have been accurately defined by an extensive set of ground-based measurements spanning almost a century. Analysis of the Mariner 9 photographs and several studies of the ground-based data have led to an improvement in the value of several of the orbital elements. Up until recently, this positional information also yielded the most accurate determination of several properties of Mars.

Shor's (1975) and Born and Duxbury's (1975) studies of Phobos' motion have led to a positive determination of Phobos' secular acceleration, with this parameter equal to about 1×10^{-3} degrees/(year)^2. Making use of this value, we find that for tidal forces to be responsible for the secular acceleration the ratio of Q/k_2 for Mars must be about 1,000, which in turn implies that a portion of Mars' interior may be a zone of partial melting. We also find that Phobos will crash into Mars in about 4×10^7 years and that 4.6 billion years ago Phobos was much closer to both the stationary orbital position and to Deimos.

ORIGIN

The terrestrial planets have far fewer natural satellites than the outer Jovian-type planets. Furthermore, it has been argued that our own Moon may have been tidally captured by the Earth (cf. Singer, 1968; cf. Chapt. 27, Wood). Thus, Phobos and Deimos may be the only presently existing primordial satellites of the inner solar system (cf. Chapt. 23, Cameron). Two types of hypotheses (Hartmann et al., 1975) have been advanced concerning the origin of Phobos and Deimos. One model draws attention to the nearness of Mars to the asteroid belt and suggests that like perhaps our Moon, the two satellites are captured bodies. If this is true, then there may presently be no satellites of the inner solar system which formed about their parent planets (cf. Burns, 1973; Ward and Reid, 1973). The second model postulates that the Martian satellites did, in fact, form in the same vicinity as Mars and that they have always been satellites of that planet.

Singer (1971) has studied the orbital evolution of bodies captured by Mars. He finds that it is possible for such bodies to evolve to the present positions of the Martian satellites and to have their eccentricities decreased to very small values, which are comparable to those of the Martian satellites at the present time. However, he points out that tidal capture of objects of the small mass of Phobos and Deimos is extremely improbable and that time scales of almost an order of magnitude larger than the age of the solar system are required for the tidal evolution to take place (Chapt. 7, Burns). Instead he advocates that Mars captured a much larger object, which through tidal evolution crashed into Mars. Shortly before this occurred, he speculates Phobos and Deimos were split off.

Perhaps the strongest objection to the capture theory is the difficulty of evolving the captured bodies into orbits having very low inclinations (Burns, 1972). The orbital planes of Phobos and Deimos are inclined by only about 1° and 2° to Mars' proper plane (Sinclair, 1972a). According to Burns (1972; Chapt. 7

herein) tidal forces would cause only a negligible change in the inclination of a captured body's orbit. Unless some way is found to overcome this objection, we must conclude that Phobos and Deimos have always been satellites of Mars. This raises an interesting speculation. Could the Earth have originally had satellites of the size of Phobos and Deimos, which were subsequently destroyed by collision with the Moon, after that body was captured and tidally evolved through the location of these bodies?

Burns (1972) has pointed out that Deimos lies close to the stationary orbit location (the distance at which the orbital period equals Mars' rotational period). Also, as discussed in the section on orbits, Phobos too may have been initially close to this position and evolved to its current location through tidal processes. Then the question is raised as to whether the Martian satellites formed preferentially near the stationary orbital position. Gold (1972, 1975b) has suggested a mechanism that would favor the formation of satellites near this position. He believes that dust particles in orbit about a magnetized planet will accumulate near the stationary position because particles located elsewhere are acted on by Lorentz forces that drive them to that position.

In summary, there are two competing theories of the origin of the Martian satellites. One hypothesis postulates that they are captured bodies; the other that they formed as part of the same process that formed Mars. The low inclinations of Phobos and Deimos argue in favor of the latter model.

ACKNOWLEDGMENTS

I am very grateful to Steven Soter, Joseph Burns, Torrence Johnson, Verne Oberbeck, and Richard Young for their helpful comments on the manuscript and to Alan Harris, George Born, Donald Gault, and Theodore Bunch for several very useful discussions. This work was supported by the Office of Planetology Programs, National Aeronautics and Space Administration.

DISCUSSION

THOMAS GOLD: The mechanical strength of an object cannot be judged from the shapes into which it breaks. Weaker impacts on weaker materials can always generate the same shapes again from solid rock down to a weakly cohesive powder. An absolute calibration can perhaps be obtained from the appearance of some circular craters that imply that material was excavated by hot gas generated at the impact, and removed at speeds appropriate to hot gas. This requires certain minimum normal pressures in the crater at the time, and perhaps this consideration can be used to define the strength of the rock.

CLARK CHAPMAN: I am not persuaded that the presence of a large crater and associated fracture on Phobos proves that it has "rock-like strength." Certainly there must be *some* slight degree of internal strength, but the important question is whether Phobos must be a fragment of a much larger object or alternatively whether it might have accreted at approximately its present size, cemented only slightly by intergrain forces (cf. Hartmann *et al.*, 1975). A very weakly cohesive body will require, of course, a smaller impact (*i.e.*, a lower velocity or a smaller impacting particle) to produce the same effect as on a stronger body. But even a fragile sandcastle can be fractured.

FRASER FANALE: To explain the coherence apparently required for the linear feature, you invoke derivation from a larger object or early intense heating. I wonder whether simple, low temperature intergrain cementation brought about perhaps by leaching at ice-silicate boundaries and salt deposition later between grains might give the required coherence. Carbonaceous chondrites, which make plausible candidate materials for Martian moons (in view of their low albedos) may be sufficiently coherent. In any event, this mechanism would not require intense heating.

PHOBOS AND DEIMOS: GEODESY

15

Thomas C. Duxbury
Jet Propulsion Laboratory

Results of geodesy studies of Phobos and Deimos based on Mariner 9 imaging data are presented. This analysis includes a review of the surface coverage and high resolution pictures obtained and the determined sizes, shapes, topographies, and librations of the two Martian satellites. Also, exploration of Phobos and Deimos by missions such as Viking is discussed.

The Mariner 9 mission to Mars opened a new frontier in the exploration of natural satellites. The Mariner 9 spacecraft was successful in gathering and transmitting to Earth scientific data on Mars and its satellites for nearly a year while in orbit about Mars. Close satellite encounters allowed the viewing of the satellites of another planet for the first time at sub-kilometer resolution. Well over one-half of Phobos was viewed while only about one-half of Deimos was seen. Results of processing these satellite data for determining satellite size, shape, topography and libration are presented (see also Chapt. 14, Pollack, for interpretation of this data). Significant work still remains to be done with the Mariner 9 data, particularly in the area of mapping the surfaces of the satellites.

Follow-on missions orbiting Mars such as Viking should allow the completion of the surface coverage of Phobos while extending but not completing the surface coverage of Deimos at high resolution. Furthermore these missions should give valuable data on the questioned secular acceleration in the longitude of Phobos and give a measure on the masses of both satellites.

SATELLITE SURFACE COVERAGE

Over 50 high resolution (narrow angle) TV pictures of Phobos and Deimos were obtained during the circumplanetary orbital phase of Mariner 9. An atlas of these pictures is given by Veverka *et al.* (1974). These pictures were taken to maximize the surface coverage within viewing constraints imposed by the relative spacecraft and satellite orbits and by spacecraft pointing limits of the TV cameras. To discuss the satellite surface coverage, it is useful to define a body-fixed coordinate system and mean surface for the satellites. Since the satellites

[346]

are in synchronous rotation (cf. Chapt. 6, Peale), an **x** axis from which longitudes will be referenced was chosen to be along the satellite centered-Mars position vector. The satellite north pole (**z**) was chosen to be parallel to the satellite orbit angular momentum vector making the satellite equatorial plane in the satellite orbit plane. With both Phobos and Deimos having prograde orbits whose inclinations are within a few degrees of Mars' equator, the north poles of Phobos and Deimos are within a few degrees of the north pole of Mars. A mean satellite surface in the form of an ellipsoid was assumed in defining longitudes (position west) and latitudes of surface points.

Mariner 9 sub-spacecraft points relative to body-fixed coordinates at the times of the high resolution satellite pictures are given in Figure 15.1. Table 15.1 lists these sub-spacecraft points by the revolution number of Mariner 9 about Mars. It can be seen that good southern hemisphere coverage of Phobos was obtained. Limited northern hemisphere coverage at significantly poorer resolution was obtained only near the end of the mission. Poor coverage of the sub-Mars point and no coverage of the eastern equatorial region of Phobos was obtained.

Deimos surface coverage was severely restricted since the orbit of Mariner 9 was inside the orbit of Deimos with a relative orbit inclination of over 60°. Close encounters (less than 10,000 km) generally occurred as Mariner 9 passed through the orbit plane of Deimos. This is reflected in Figure 15.1, showing the sub-Mariner 9 points on Deimos to be only in the region near the sub-Mars point.

Figures 15.2, 15.3, and 15.4 are representative of the high resolution coverage of Phobos and Deimos. The longitude-latitude grids were drawn using the previously defined coordinate system and reference surface, assuming synchronous rotations and ellipsoidal shapes. Grid lines are drawn at 45° longitude intervals and at ±40° and ±80° latitudes. North poles are indicated by "N," south poles by "S," sub-Mars points by "V," anti-Mars points by "A," sub-spacecraft points by " + ," sub-solar points by the large "*," and terminators by the series of small "*." Marked differences in Phobos surface resolution between the southern hemisphere pictures taken within 7500 km and the northern hemisphere pictures taken outside of 11,000 km are quite evident. Three large distinct craters of 8 km, 7 km and 5 km are seen on Phobos near the sub-Mars point, the north pole and the south pole, respectively. Other large craters of similar size are indicated but are more subdued, possibly being obliterated by smaller and more recent impacts. The rougher appearance of the northern hemisphere relative to the southern hemisphere is possibly explained by the larger magnification of the lower resolution northern hemisphere Phobos pictures. Further Mariner 9 images of the Martian satellites are Figures 14.1 to 14.5.

Only the single northern hemisphere Deimos picture shows significant surface detail. A major surface feature is visible at the south pole with six large craters on the front side. It is noted that the shape of the smaller Deimos is less accurately represented by an ellipsoid than is the shape of the larger Phobos.

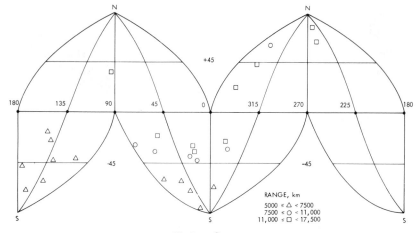

a. Phobos Coverage

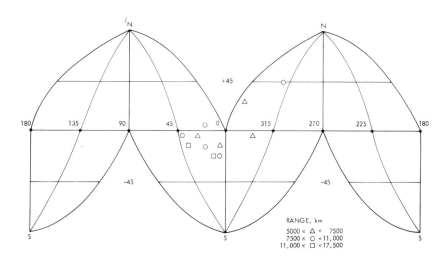

b. Deimos Coverage

Fig. 15.1. Mariner 9 sub-spacecraft points when high resolution pictures were obtained. *Above:* Phobos coverage; *Below:* Deimos coverage.

TABLE 15.1
Sub-spacecraft Points for Narrow Angle Satellite Pictures Listed by the Revolution (Rev) Number of Mariner 9 About Mars.
See Veverka *et al.* (1974) for the photos.

Rev	Satellite	Longitude, deg	Latitude, deg	Range, km
25	Deimos	20	-14	8800
27	Phobos	142	-25	7200
31	Phobos	341	-26	14500
34	Phobos	160	-70	5700
35	Deimos	35	-13	12300
41	Phobos	109	-41	7400
43	Phobos	152	-18	7400
48	Phobos	53	-63	7200
53	Phobos	24	-33	11900
57	Phobos	131	-61	6000
63	Deimos	7	-12	7800
73	Deimos	41	-4	10100
73	Phobos	134	-41	6500
77	Phobos	344	-30	8100
80	Phobos	356	-67	7000
87	Phobos	79	-57	7200
89	Phobos	172	-47	5800
111	Deimos	27	-1	7200
117	Phobos	86	-28	10200
121	Deimos	13	-21	15300
129	Phobos	17	-31	12500
131	Phobos	66	-35	10400
133	Phobos	83	-82	6100
145	Phobos	52	-21	15300
149	Deimos	356	28	5500
150	Phobos	344	23	14500
159	Deimos	3	-19	10100
161	Phobos	21	-40	10000
171	Phobos	333	43	13600
197	Deimos	335	-1	7400
207	Phobos	47	-67	7000
221	Phobos	22	-41	9800
430	Phobos	341	61	10700
437	Deimos	323	45	8600
444	Phobos	251	79	13100
675	Phobos	241	68	14000
676	Phobos*	94	33	17300

*Note: This picture was the last picture returned to Earth by Mariner 9.

[349]

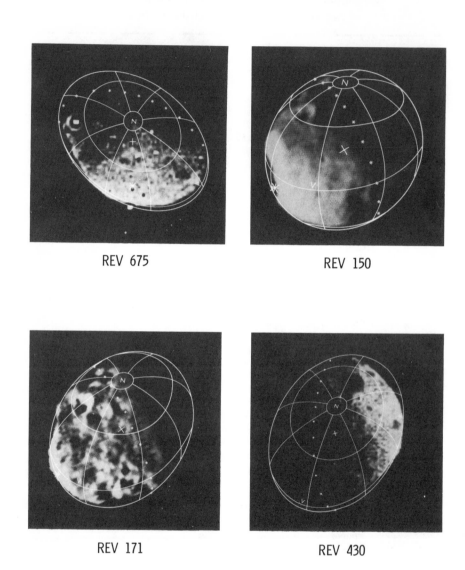

REV 675

REV 150

REV 171

REV 430

Fig. 15.2. Northern hemisphere viewing of Phobos.

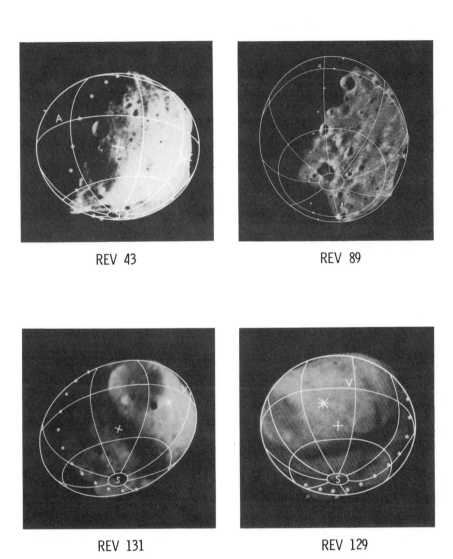

REV 43   REV 89

REV 131   REV 129

Fig. 15.3. Southern hemisphere viewing of Phobos.

[351]

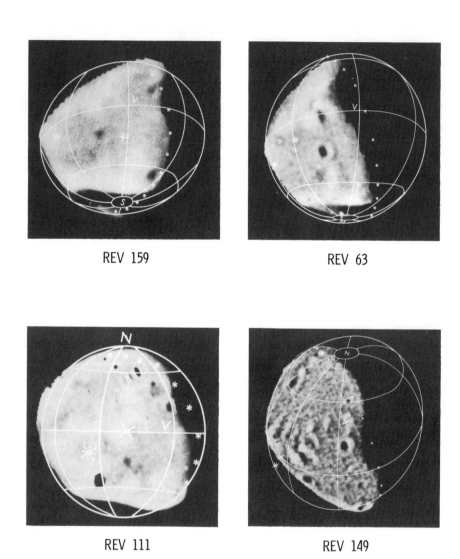

REV 159

REV 63

REV 111

REV 149

Fig. 15.4. Viewing of Deimos.

SATELLITE SIZE AND SHAPE

The three radii of the ellipsoids (Fig. 15.5) which define the mean surfaces of Phobos and Deimos are listed in Table 15.2 and their relative sizes are shown in Figure 15.5. The first entry in Table 15.2 comes from ground-based photometric data (cf. Chapt. 9, Veverka; Table 14.1, Pollack) and assumes a geometric albedo of 0.15 for the satellites, similar to that of Mars. The Mariner 7 entry was determined from a picture of Phobos in transit across Mars (Smith, 1970). Phobos then produced an image area of tens of picture elements (pixels) compared to the Mariner 9 satellite image areas of thousands of pixels. Figure 15.6 shows a magnification of the Mariner 7 Phobos image together with a limb contour giving the size of Phobos. Since Phobos was in transit, only the two smaller radii were observable with the largest radius nearly along the line of sight to Phobos. Mariner 9 entries in Table 15.2 were initially determined by matching computer-drawn overlays to the lit limbs of the satellites as shown in Figures 15.2, 15.3, and 15.4. The most accurate determination of radii was obtained for Phobos by using cartographic techniques to process 38 landmarks in nine pictures for determining the mean surface (Duxbury, 1974). Mariner 9 data yielded satellite geometric albedos of ~ 0.05 bringing the Earth-based determined radii in line with the Mariner data (Pollack *et al.*, 1972). Satellite masses of 1.74 $\times$ 10^{16} kg for Phobos and 3.11 $\times$ 10^{15} kg for Deimos are obtained using the volumes defined by the mean surfaces and assuming a mean density of 3.0 g-cm^{-3}.

SATELLITE TOPOGRAPHY

The determination of surface topography not only requires complete surface coverage but also overlapping coverage to yield stereo viewing. Only about 70% of Phobos and 50% of Deimos had coverage, a portion of which was at poor resolution. Very poor stereo coverage of Deimos was obtained; however, stereo coverage of the Phobos southern hemisphere was obtained. The cartographic processing of this coverage found surface height variations of as much as 20% of the local mean radius. The long linear feature (Kepler Dorsum) had a height of over 1.5 km relative to its base. Both satellites appear to have a saturated crater density (Pollack *et al.*, 1973a). Also, TV (Pollack *et al.*, 1973a), infrared (Gatley *et al.*, 1974; cf. Chapt. 12, Morrison), and polarimetric (Zellner, 1972a; cf. Chapt. 10, Veverka) observations support the existence of a regolith on both satellite surfaces (cf. Chapt. 14, Pollack).

SATELLITE LIBRATION

The amplitude, period and phase of libration can yield information on the recent impact history and internal structure of a satellite (see Chapt. 6, Peale).

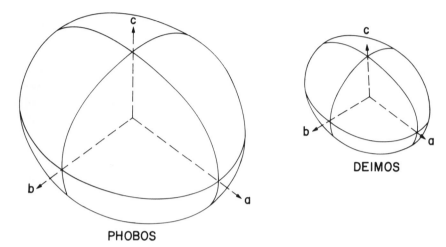

Fig. 15.5. Relative sizes of the mean surfaces of Phobos and Deimos defined by an ellipsoid of radii a, toward Mars; b, in the orbit plane; and c, normal to the orbit plane.

TABLE 15.2
Radii, in Kilometers, of Phobos and Deimos

Source	Phobos			Deimos		
	a	b	c	a	b	c
Earth-based*	7.0	7.0	7.0	4.0	4.0	4.0
Mariner 7	—	11.0	9.0	—	—	—
Mariner 9						
Overlays	13.5	11.5	9.5	7.5^{+3}_{-1}	6.1 ± 1.0	5.5 ± 1.0
Cartography	13.5 ± 0.5	10.8 ± 0.7	9.4 ± 0.7	—	—	—

*Pre-Mariner value which assumed geometric albedos of 0.15. The actual albedos are about 40% of this (Chapt. 9, Veverka).

For a satellite in synchronous rotation, the sum of the in-plane optical and forced libration angles relative to true anomaly and measured positive about the satellite north pole (Eckhardt, 1967) is

$$\theta = -2e\sin M + \frac{2e\sin M}{1 - 1/3Y} \quad \text{for } Y < 1 , \tag{1}$$

where the first term is optical libration, the second term is forced libration, e is the satellite orbital eccentricity, M is the satellite mean anomaly and

$$Y = (B - A)/C , \tag{2}$$

where A, B and C are satellite principal moments of inertia. The libration is zero at the satellite periapsis and apoapsis while being maximum at $M = 90°$ and $270°$. For a spherical, homogeneous satellite ($Y = 0$), eqn. (1) reduces to the standard optical libration caused by a satellite's rotation rate being constant throughout its orbit while the rate of change in true anomaly varies throughout an eccentric orbit. However, for an ellipsoidal satellite, restoring torques from the planet cause the forced libration in addition to the optical libration. Assuming a uniform density, ellipsoidal satellite with radii a, b and c, Y can be expressed as a function of these radii as

$$Y = (a^2 - b^2)/(a^2 + b^2) \tag{3}$$

which is only dependent on the two radii in the satellite orbit plane (Fig. 15.5). For Phobos (Table 15.2 and e = 0.015)

$$\theta = -5.1 \sin M \quad \text{(deg)} \tag{4}$$

with about 2° from optical libration and 3° from forced libration, and for Deimos (Table 15.2 and e = 0.001)

$$\theta = -0.2 \sin M \quad \text{(deg)}. \tag{5}$$

The orbital eccentricities used were obtained from processing the Mariner imaging data (Born and Duxbury, 1975; cf. Table 1.4). Small differences in the observed amplitude of in-plane libration would probably be caused by errors in the assumed moments of inertia. However, large differences in amplitude, period or phase would indicate a free oscillation, caused by a major impact in the last thousands or millions of years which has not been dissipated by planetary tidal restoring torques (see Chapt. 6, Peale; Burns, 1972).

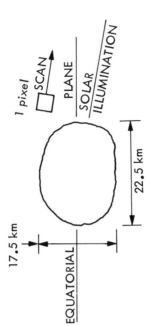

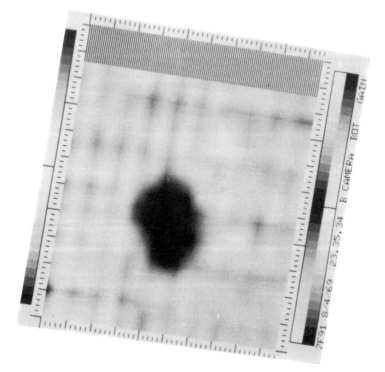

Fig. 15.6. Mariner 7 picture of Phobos while in transit across Mars. From Smith (1970)

A limited cartographic effort which processed 38 landmarks in 9 Phobos pictures established the existence of an in-plane libration for Phobos. A maximum libration angle of about 5° was observed (Fig. 15.7) which is in good agreement with the predictions. This libration can also be inferred from Figures 15.2 and 15.3 by observing the positions of a few surface features relative to the latitude-longitude grids which were produced assuming no libration. Care must be taken, however, since errors in the assumed size and shape of the mean surface used for the overlays and the fact that the surface features are not on the mean ellipsoidal surface will give the appearance of a libration. The 9 pictures processed were taken during a 3-month period and were within ±30° of either Phobos periapsis or apoapsis. Therefore, a period and phase for libration were not determined. Also the amplitude of libration could be larger than the observed 5°, since very limited orbital coverage was obtained.

Out-of-plane librations of less than a few degrees were indicated by the cartographic effort with a measurement error of about ±1°. Since no out-of-plane forced libration should be expected, this observed difference could possibly reflect the lack of good stereo coverage and represent errors in the determined locations and radii of the 38 control points. Improved knowledge of the librations could be obtained from existing Mariner 9 satellite pictures not already processed and by future exploration. A map of surface features on Phobos produced by processing the 38 control points is shown in Figure 15.8 (also Fig. 14.6).

FUTURE EXPLORATION OF PHOBOS AND DEIMOS

Future spacecraft exploration of Phobos and Deimos should attempt to complete the high resolution surface coverage of both satellites. Phobos surface priorities should be the sub-Mars and north polar regions with both of these areas being potential landing sites for follow-on missions. Overlapping coverage to yield stereo viewing should be included. Also, this coverage should be obtained throughout the satellite orbits and with sufficient satellite true anomaly sampling to observe the amplitudes, periods and phases of libration. These imaging data could be taken using color and polarizing filters to simultaneously yield photometry data. Complete satellite orbital coverage with accurately known TV pointing data is also important for satellite ephemeris improvements (cf. Chapt. 3, Aksnes; Chapt. 5, Pascu). Many hundreds of satellite pictures would be required to satisfy these objectives.

Another opportunity for spacecraft exploration of the Martian satellites is the Viking 1976–77 mission. The Viking TV cameras are ideally suited for satellite photography. Rather than having a wide and a narrow angle camera, the Viking orbiters each have two narrow angle cameras with 1.7° × 1.5° fields of view and about six arcsec pixel angular resolutions. These cameras are not boresighted, making possible a properly exposed satellite picture with one camera and a star background picture with the other camera to yield precision TV pointing data.

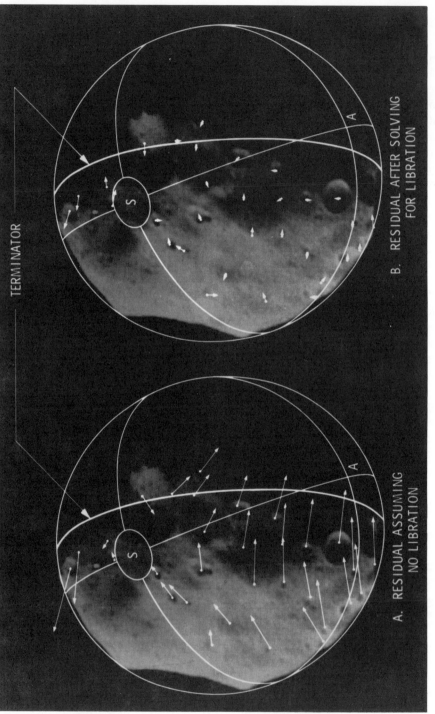

Fig. 15.7. Magnified landmark position residuals when (a) assuming no libration and (b) including a 5° in-plane libration.

The Viking spacecraft allow tens of satellite encounters within ranges of a few thousand km, whereas the closest ranges on Mariner 9 were about 5000 km. Completing the high resolution surface coverage of Phobos should be possible, while viewing the backside of Deimos at high surface resolution will probably still not be possible. However, some northerly and southerly surface coverage as well as improved front side coverage of Deimos should be permitted. The major concern relating to Viking imaging of the satellites is that the satellites are not of prime importance to Viking planners. Viking's prime objectives were to place landers on the surface of Mars. Up to the time of landing, the orbiter's prime function has been in landing site surveillance and certification. After landing, the orbiters are still to be primarily focusing on Mars. It is anticipated that only about 50 satellite TV pictures will be obtained during the orbiter phase with the remaining thousands of pictures supporting the primary objectives of landing and Mars coverage.

This limited number of pictures should be sufficient to complete the Phobos surface coverage and obtain the maximum Deimos coverage possible within the orbiter trajectory and TV pointing constraints. However, stereo coverage will be severely restricted as well as coverage of the satellites throughout their orbits. This will seriously degrade any libration analysis and cartography efforts. Also, the two-camera approach with stars in one camera and a satellite in the other camera will have to be curtailed to maximize satellite images in the allocated picture budget.

Satellite pictures are especially obtained during the approach to Mars when the large distances make the satellite images appear as point sources. While not being valuable for surface analysis, these images together with star images in the same picture make these pictures extremely valuable for satellite ephemeris improvement (Born and Duxbury, 1975). The star images should enable satellite position fixes to be made to a 5 km accuracy level throughout the satellite orbits. Accurate satellite longitude fixes are a key to answering the questioned secular acceleration in the longitude of Phobos (cf. Chapt. 14, Pollack). These same approach pictures can also be used to search for new satellites of Mars (Chapt. 5, Pascu).

Finally, the close satellite encounters offer a unique opportunity to determine the masses of the satellites. Close flybys of a few thousand km would accelerate the orbiters which in turn would be detected in the Earth-based doppler tracking of the orbiters. The accuracy of determining the gravitational constant of a small satellite from doppler data is approximated by (Anderson, 1971)

$$\sigma_{GM}^2 = \frac{16BV^3}{\pi} h\sigma_z^2 , \qquad (6)$$

where σ_{GM} is the one-sigma uncertainty of the satellite gravitational constant in units of km³/sec², B is the closest approach distance, V is the flyby velocity, and

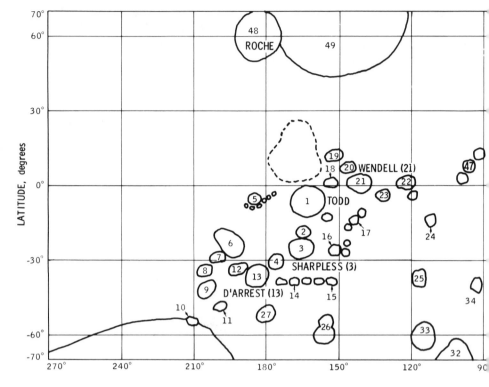

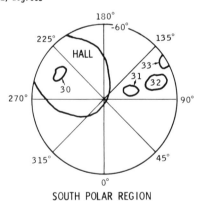

SOUTH POLAR REGION

[360]

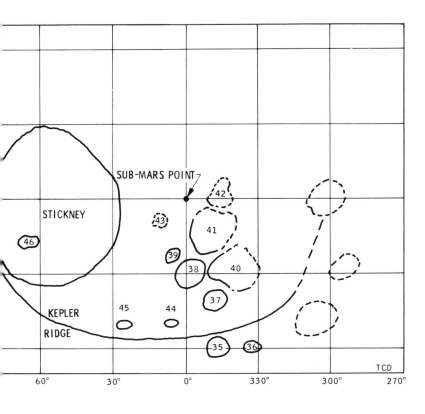

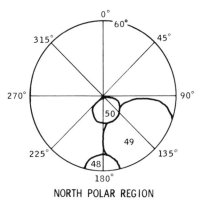

NORTH POLAR REGION

Fig. 15.8. Map of the observed surface features of Phobos. The dashed lines represent features which may exist but were not clearly identified.

σ is the range rate error for doppler data sampled at a time interval h. For a relative flyby velocity of 2 km/sec at 1000 km and a σ_z of 1 mm/sec for a 60 sec sample interval.

$$\sigma_{GM} = 0.00078 \text{ km}^3/\text{sec}^2 , \qquad (7)$$

which is about 70% for Phobos and 400% for Deimos for one flyby. Multiple encounters and long period effects will aid in the mass determination. Therefore, satellite mean density calculations will probably be limited by errors in the satellite volumes because of the lack of sufficient stereo coverage. A knowledge of mean densities will give important information on internal composition (Chapt. 14, Pollack).

SUMMARY

Valuable satellite geodesy information has been deduced from the Mariner 9 imaging data of Phobos and Deimos. Some of the Mariner 9 data still remains untouched as far as satellite geodesy is concerned. The Viking mission offers an opportunity to complete the Phobos coverage while increasing the Deimos coverage. Potentially limited satellite pictures from Viking, however, probably will severely restrict libration and cartographic efforts. Mass determinations of Phobos and Deimos by Viking appear to be feasible.

ACKNOWLEDGMENTS

The author wishes to thank Mike Wolf and the JPL Image Processing Laboratory for the picture processing support. This paper presents one phase of research carried out at the Jet Propulsion Laboratory, California Institute of Technology, Pasadena, California, under NASA Contract No. NAS7-100.

PHOTOMETRY OF THE
GALILEAN SATELLITES

David Morrison
and
Nancy D. Morrison
University of Hawaii

The Galilean satellites have been studied photometrically by Stebbins (1927), Stebbins and Jacobsen (1928), Harris (1961), Johnson (1971), Blanco and Catalano (1974b), and Morrison et al. (1974). From these observations, we derive the dependence of the magnitudes and colors on both solar and orbital phase angles. The rotational variations are best defined by the most recent observations; as a consequence of the complex albedo distributions on the satellites, the lightcurves cannot be represented by simple analytic expressions. The recent data also establish the mean magnitudes on the V photometric system to within ± 0.02 mag and permit us to derive accurate geometric albedos. The dependence of magnitude on solar phase angle, and particularly the behavior at very small phase angles, is best shown by the extensive observations from 1926 and 1927. We use the newly determined rotational curves to reduce these data and express the phase effect in terms of a linear curve for α>6° and a quadratic curve for α<6°. The phase dependence of these satellites is compared with that observed for other small bodies in the solar system; particularly interesting are the anomalously large phase coefficient and opposition effect of Io.

Although many physical and dynamical investigations of the four Galilean satellites of Jupiter have been carried out in the past few decades, only in the 1970s has effort been devoted to refining the pioneering broadband photoelectric studies carried out 50 years ago by Stebbins (1927) and Stebbins and Jacobsen (1928). Harris (1961) transformed these early observations to the UBV system and added some new data of his own, and his paper has long been the standard reference for the photometry of these satellites. In this book Veverka (Chapt. 9) summarizes the basic ideas of photometry and reviews the results for all satellites. Johnson (1971) obtained extensive observations in 1969 with a 24-color spectrophotometer, but his emphasis was on deriving spectral information rather than on refining the broadband photometry. Most recently, however, Blanco and Catalano (1974b) have presented a new set of UBV data obtained in 1971, and Morrison *et al.* (1974) obtained photometry in 1973 on the *uvby* system with an average precision of ± 0.01 mag. Some of the conclusions reached by these authors have been conflicting, but the last two papers approach a consensus concerning the photometric properties of the satellites. This chapter will summarize the results of Morrison *et al.* (1974) and use them together with all previously published photometry to derive optimum values for the magnitudes, colors, and albedos and to find the dependence of these parameters on solar and

[363]

orbital phase angles. [Extensive new photometry has been published (Millis and Thompson, 1975) and a general review on the Jovian satellites written (Morrison and Burns, 1976).]

As noted by all photometric observers of the Galilean satellites, there is no straightforward way to separate the effects of solar and orbital phase angles on the magnitudes and colors. It is necessary to proceed iteratively, first assuming a nominal phase dependence, then obtaining a first-order rotation curve from which to derive a more refined phase law, etc. (cf. Chapt. 9, Veverka). In the following discussions, we present what is in our judgment the most self-consistent interpretation of all of the photometric data, without describing the many trials used in reaching our final conclusions.

PHOTOMETRIC SYSTEMS, MEAN MAGNITUDES, AND ALBEDOS

Each of the major photometric studies of the Galilean satellites has used a different photometric system, and in order to compare the data we must transform them to a common system. In conformity with most astronomical photometry, we use the V magnitude of the UBV system. Harris (1961) observed on this system, but, since he never published his individual observations, we cannot make much use of his data in this analysis. Blanco and Catalano (1974b) also observed on the UBV system. Morrison et al. (1974) observed on the uvby system, but since they transformed their y magnitudes to V magnitudes, their results should be directly comparable to those of Blanco and Catalano. We rely primarily on these two recent groups of observations, which agree to within their observational accuracy of ± 0.01 mag, to define the mean magnitudes of the Galilean satellites.

The other two data sets are not as easily transformed to V magnitudes. Johnson (1969, 1971) published individual observations for only the one of his 24 narrowband filters that is centered at 0.56 μ, near the central wavelength of the V band. He did not transform these observations to the UBV system but published the ratio of the brightness of the Galilean satellites to that of the G5 star o Vir, for which he derived a magnitude at 0.56 μ of 3.99 from observations of the ratio of the brightness of this star to that of α Leo. Johnson assumed that the magnitudes of the satellites derived in this way were close to true V magnitudes. We observed o Vir on four nights, however, and we obtain V = 4.15 ± 0.02 and use this value to transform Johnson's data. Johnson (private communication, 1974) informs us that he believes most of the 0.16 mag difference between his magnitudes at 0.56 μ and true V magnitudes is a result of his failure to transform the narrowband magnitude difference between o Vir and α Leo into a difference in V. If we assume that the magnitudes of the satellites transform to V exactly as does that of o Vir—that is, if we neglect the color dependence of the transformation—we derive magnitudes from Johnson's data in satisfactory agreement with those obtained by Blanco and Catalano and by Morrison *et al.*

In contrast, the early observations by Stebbins (1927) and Stebbins and Jacobsen (1928) were made in an extremely broad band centered near 0.45 μ. Harris (1961) transformed these magnitudes to the UBV system, but we have not attempted to use this transformation directly. Instead, we have simply found that for Europa, Ganymede, and Callisto, all of which have similar mean colors and small rotational amplitudes in $b-y$ and B–V, we can match recent V magnitudes closely if we subtract 0.74 from the magnitudes as originally published. For Io the situation is complicated by its red color and strong color variation, and there appears to be no satisfactory way to transform these observations to V magnitudes. The mean magnitudes (with rotational effects removed), however, agree with the recent data if we subtract 0.91 from them.

We derive two values for the absolute V magnitude for each satellite reduced to R = $\triangle$ = 1 AU, according to the two conventions in common use for this quantity. In this discussion we define V(1,0) to be the V magnitude at $\alpha = 0°$ *as obtained by linear extrapolation from magnitudes observed at 6° < α < 12°.* This is the definition used in studies of asteroids (Gehrels, 1970). We define V′(1,0) *as the V magnitude as would actually be observed at $\alpha = 0°$,* that is, the magnitude that includes the opposition effect (Chapt. 9, Veverka). This definition has been used in most previous studies of the satellites. V(1,0) is well determined observationally for many objects, and values obtained by different observers can be compared. However, V′(1,0) is more physically meaningful, and it is required for a determination of the true geometric albedo (cf. Harris, 1961). In order to derive the albedos, we adopt a magnitude for the Sun of $V_\odot = -26.77$ (cf. Gehrels *et al.*, 1964) and use the radii as summarized by Morrison and Cruikshank (1974), similar to those in Table 1.4.

The colors of the satellites are best determined by multifilter spectrophotometry, as reviewed by Johnson and Pilcher in Chapter 11, herein. However, multicolor photometry of high precision can usefully supplement the spectrophotometry. Harris (1961), Owen and Lazor (1973), and Blanco and Catalano (1974b) have all published rotation curves in the U–B and B–V colors of the UBV system. However, the most accurate color rotation curves appear to be those of Morrison *et al.* (1974), and we will limit our discussion of the mean colors to these data.

Table 16.1 lists the mean V magnitudes and *uvby* colors of the satellites and the corresponding albedos for $\alpha = 6°$ and for $\alpha = 0°$. For the albedos we use the symbol p, which denotes geometric albedo, although we remind the reader that only the value of p derived from V′(1,0) is a true geometric albedo. The uncertainties given for the albedos include the stated uncertainties in the radii but not the uncertainties in the magnitude of the Sun.

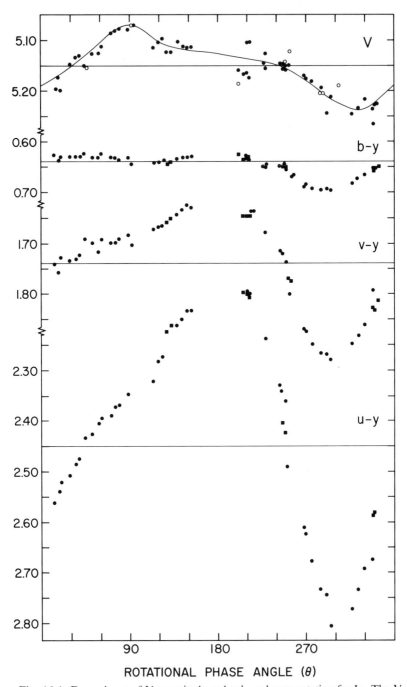

Fig. 16.1. Dependence of V magnitude and *uvby* color on rotation for Io. The V magnitudes have been corrected for dependence on phase as indicated in Table 16.2. Filled circles are from Morrison *et al.* (1974) and open circles are from Blanco and Catalano (1974b), all for $\alpha > 3°$. The color indices, which have been normalized to be zero for Vega, are from Morrison *et al.* (1974); circles are for $\alpha > 6°$, squares for $\alpha < 6°$.

EUROPA at α = 6°

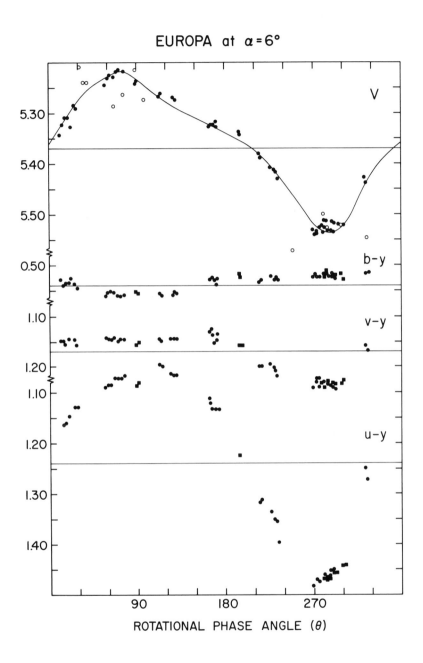

Fig. 16.2. Same as Figure 16.1, for Europa.

ROTATIONAL VARIATIONS

The variations of magnitude with rotation are best determined from observations made over a limited range in solar phase angle. Such data were obtained by Morrison *et al.* (1974) for Io and Europa on five nearly consecutive nights in June 1973. For Ganymede and Callisto, with their longer periods of rotation, it is necessary to synthesize the lightcurves from data obtained over a wider range of phase angle; therefore, errors in the phase dependence assumed for these two satellites show up as increased scatter in the rotation curves. We conclude that for greatest accuracy, rotation curves for the inner two satellites should be constructed only from the recent observations by Morrison *et al.* (1974) and by Blanco and Catalano (1974b). We restrict ourselves to observations obtained at $\alpha > 3°$, where uncertainties in the size of the opposition effect are unimportant. In contrast, the best results for the outer satellites will be obtained from a combination of the observations by these authors with those by Stebbins (1927) and Stebbins and Jacobsen (1928), again limited to $\alpha > 3°$. Johnson's (1971) observations are omitted because of their apparently greater scatter. For the rotation curves in the *uvby* colors, of course, we are restricted to the observations by Morrison *et al.* (1974).

The rotational variations of the Galilean satellites in V, $b-y$, $v-y$, and $u-y$ are illustrated in Figures 16.1 through 16.4. All of the plots are to the same scale to facilitate comparison. We have corrected all of the points plotted in Figures 16.1 through 16.4 to $\alpha = 6°$, using the phase curve parameters given in Table 16.2. In general, the observational errors are comparable in size to the symbols for the colors and perhaps twice as large for the V magnitudes. Evidently, the curves are complex, and the color curves do not agree in the details of shape with the lightcurves. In Table 16.3 we list the magnitudes at intervals of 10° in longitude that define the nominal rotation curves. Further discussion of rotational variations follows Figures 9.4 and 11.5.

VARIATIONS WITH SOLAR PHASE

The phase dependence of the magnitudes of objects without atmospheres has two parts. At phase angles greater than about 6°, the magnitude changes linearly with phase; this rate of change we will call the *phase coefficient*. At smaller phase angles, however, there is typically an "opposition effect," such that the brightness increases much more rapidly very near opposition. We will call the difference between the magnitude at $\alpha = 0°$ that is actually observed and that obtained by extrapolation of the linear phase law the *opposition surge*. These terms are discussed in more detail in Chapter 9 by Veverka.

We model the opposition effect by fitting a parabola through the points at $\alpha < 6°$, constrained so that the curve and its first derivative are continuous with

TABLE 16.1
Mean Magnitudes and Albedos of the Galilean Satellites

	Io	Europa	Ganymede	Callisto
$V(1,6°)$	-1.55 ± 0.02	-1.33 ± 0.02	-1.96 ± 0.02	-0.77 ± 0.02
$V(1,0)$	-1.68 ± 0.03	-1.37 ± 0.03	-2.08 ± 0.03	-0.95 ± 0.03
$V'(1,0)$	-1.85 ± 0.03	-1.46 ± 0.03	-2.15 ± 0.03	-1.20 ± 0.03 (L)
				-1.08 ± 0.05 (T)
$b-y$ (6°)	0.64 ± 0.01	0.54 ± 0.01	0.52 ± 0.01	0.55 ± 0.01
$v-y$ (6°)	1.91 ± 0.01	1.34 ± 0.01	1.29 ± 0.01	1.34 ± 0.01
$u-y$ (6°)	3.87 ± 0.03	2.66 ± 0.02	2.63 ± 0.02	2.69 ± 0.02
Radius (km)	1820 ± 10	1550 ± 150	2635 ± 25	2500 ± 150
$p_V(6°)$	0.56 ± 0.01	0.62 ± 0.12	0.39 ± 0.01	0.14 ± 0.02
$p_V(0°)$	0.63 ± 0.02	0.64 ± 0.12	0.43 ± 0.02	0.17 ± 0.02
$p_{V'}(0°)$	0.72 ± 0.02	0.69 ± 0.12	0.46 ± 0.02	0.22 ± 0.02 (L)
				0.20 ± 0.02 (T)
$p_b(6°)$	0.45 ± 0.01	0.55 ± 0.10	0.34 ± 0.01	0.12 ± 0.02
$p_v(6°)$	0.25 ± 0.01	0.47 ± 0.09	0.30 ± 0.01	0.11 ± 0.01
$p_u(6°)$	0.10 ± 0.01	0.35 ± 0.07	0.22 ± 0.01	0.08 ± 0.01

Note: For the Sun, we have taken $V_\odot = -26.77$, $b = -26.36$, $v = -25.74$, and $u = -24.74$ (cf. Morrison *et al.*, 1974). The radii are from Morrison and Cruikshank (1974). See also Table 1.4 herein. (L) and (T) refer to leading and trailing sides.

the straight line at $\alpha = 6°$. Thus three independent parameters describe the complete curve: V at $\alpha = 6°$, $dV/d\alpha$ for $\alpha > 6°$, and V at $\alpha = 0°$. This is the number of parameters used in the more traditional parabolic fit of a single curve over the range $0° < \alpha < 12°$ (cf. Stebbins, 1927; Harris, 1961), but we find that the parameters we have chosen represent the data better and facilitate comparison with other objects.

In order to derive the phase curves for these satellites, we must properly allow for the rotational variations, which exceed the phase variations for all but Callisto. In most cases, we used the mean V-magnitude rotation curves indicated by the solid lines in Figures 16.1 through 16.4 to make this correction. However, the observations of Io by Stebbins and by Stebbins and Jacobsen require special treatment. The rotational amplitude obtained by these authors is greater than that shown in Figure 16.1 because of the shorter effective wavelength of their photometry, but it matches closely the amplitude observed in the b filter. We therefore used the b lightcurve to correct these observations. In principle, the other satellites should be treated similarly, but in practice the amplitude in $b-y$ is small enough for all but Io that no correction to the V-magnitude curves is required.

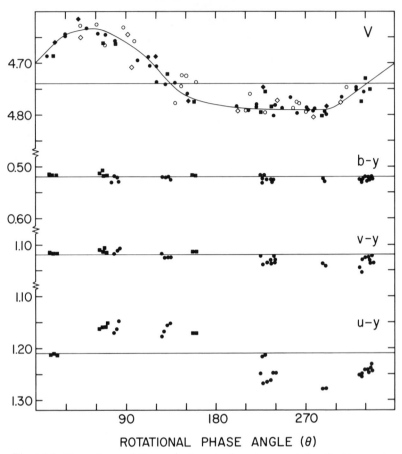

Fig. 16.3. Dependence of V magnitude and *uvby* color on rotation for Ganymede. The V magnitudes have been corrected for phase as indicated in Table 16.2. Circles are from Stebbins (1927) or Stebbins and Jacobsen (1928); diamonds from Blanco and Catalano (1974b); and squares from Morrison *et al.* (1974), all for $\alpha > 3°$. Only one point per night is plotted. Open symbols have larger internal errors. Color indices are as described for Figure 16.1.

Figure 16.5 illustrates the phase curves, corrected for rotation as described above. Note that only the magnitudes from the 1920s have been adjusted in zero point to match the other curves; the observations by Morrison *et al.*, Blanco and Catalano, and Johnson (when reduced with V = 4.15 for o Vir) all agree naturally with each other. Examined separately, the magnitudes from each source except Johnson appear to have a mean scatter less than ± 0.02 mag. Reference is made to the discussion following Figure 9.5.

For each of the three inner satellites, a single phase relationship represents all

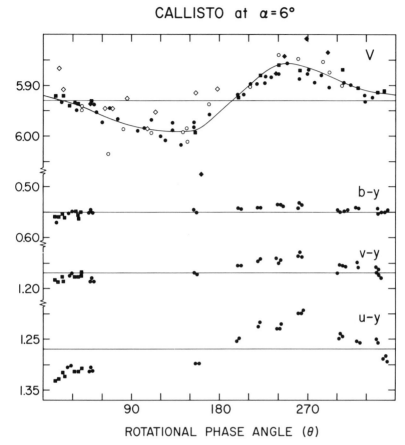

Fig. 16.4. Same as Figure 16.3, for Callisto.

of the observations. However, Stebbins and Jacobsen (1928) found that the leading and trailing sides of Callisto appear to show different opposition surges. We have therefore plotted the two sides of this satellite separately in Figure 16.5. The magnitudes from 1927, taken at $180° < \theta < 360°$ for $\alpha < 3°$, clearly fall below the curve defined by the other data. Unfortunately, since no subsequent observations include this part of the orbit near opposition, this effect is unverified. In view of the high accuracy of all of the Stebbins and Jacobsen photometry, however, there is no reason to doubt this interesting result; further, polarimetry by Veverka (1971a), Gradie and Zellner (1973), and Dollfus (1975) also show surprising differences between the two sides of this satellite (cf. Veverka, Chapt. 10 herein).

According to our best judgment, the solid curves in Figure 16.5 represent the

TABLE 16.2
Dependence of Magnitude on Solar Phase Angle*

	Notes	Io	Europa	Ganymede	Callisto (L)	Callisto (T)
$V\,(\alpha = 0°)$		4.85 ± 0.03	5.24 ± 0.03	4.55 ± 0.03	5.50 ± 0.03	5.62 ± 0.05
$V\,(\alpha = 6°)$		5.15 ± 0.02	5.37 ± 0.02	4.74 ± 0.02	5.93 ± 0.02	5.93 ± 0.02
$\dfrac{dV}{d\alpha}\,(\alpha > 6°)$	a	0.021 ± 0.003	0.004 ± 0.002	0.019 ± 0.002	—	0.028 ± 0.004
$\dfrac{dB}{d\alpha}\,(\alpha > 6°)$	b	0.024 ± 0.004	0.009 ± 0.003	0.018 ± 0.002	0.032 ± 0.003	0.027 ± 0.005
$\dfrac{dm}{d\alpha}\,(\alpha > 6°)$	c	0.022 ± 0.003	0.006 ± 0.003	0.018 ± 0.002	0.030 ± 0.003	0.030 ± 0.003
$A\,(\alpha < 6°)$	d	0.0780	0.0373	0.0420	0.1133	0.0733
$B\,(\alpha < 6°)$	d	-0.0047	-0.0026	-0.0020	-0.0069	-0.0036
ΔV	e	0.17 ± 0.03	0.09 ± 0.03	0.07 ± 0.03	0.25 ± 0.04	0.13 ± 0.05

a: Based on V magnitudes from Blanco and Catalano (1974b) and Morrison et al. (1974).

b: Based on broadband magnitudes from Stebbins (1927) and Stebbins and Jacobsen (1928).

c: Based on a combination of all data; leading (L) and trailing (T) sides of Callisto not solved for separately.

d: Where $V(\alpha) = V(0) + A\alpha + B\alpha^2$.

e: $\Delta V \equiv V(6) - V(0) - 6\,dm/d\alpha$ (i.e., the opposition surge).

*Compare these results to those of Table 9.3.

TABLE 16.3

Nominal V-Magnitude Rotation Curves

$\theta(°)$	Io	Europa	Ganymede	Callisto
0	0.040	−0.010	−0.020	0.000
10	0.030	−0.040	−0.040	0.004
20	0.016	−0.070	−0.060	0.010
30	0.002	−0.086	−0.076	0.014
40	−0.012	−0.116	−0.080	0.020
50	−0.030	−0.130	−0.086	0.030
60	−0.044	−0.140	−0.084	0.040
70	−0.062	−0.150	−0.080	0.048
80	−0.076	−0.150	−0.070	0.054
90	−0.080	−0.136	−0.060	0.060
100	−0.074	−0.120	−0.046	0.064
110	−0.060	−0.106	−0.030	0.066
120	−0.056	−0.092	−0.012	0.068
130	−0.048	−0.080	0.010	0.070
140	−0.034	−0.070	0.030	0.070
150	−0.030	−0.060	0.042	0.068
160	−0.028	−0.050	0.050	0.066
170	−0.026	−0.040	0.056	0.050
180	−0.022	−0.030	0.060	0.030
190	−0.020	−0.020	0.064	0.014
200	−0.016	−0.010	0.066	−0.002
210	−0.012	0.004	0.068	−0.020
220	−0.010	0.020	0.070	−0.038
230	−0.006	0.044	0.070	−0.054
240	−0.002	0.070	0.070	−0.062
250	0.008	0.090	0.070	−0.066
260	0.018	0.120	0.070	−0.066
270	0.030	0.150	0.070	−0.060
280	0.044	0.162	0.070	−0.054
290	0.060	0.164	0.070	−0.048
300	0.070	0.150	0.064	−0.040
310	0.080	0.130	0.050	−0.030
320	0.086	0.090	0.038	−0.020
330	0.084	0.050	0.020	−0.010
340	0.074	0.024	0.010	−0.006
350	0.056	0.006	−0.004	−0.002

optimum three-parameter fits to the data. To determine the phase coefficients, we have obtained least squares fits to various combinations of the data in the ranges $5° < \alpha < 12°$ and $6° < \alpha < 12°$. We have treated the observations from the 1920s, which have an effective wavelength near 0.45μ, separately from the more recent observations (all near 0.56μ) in order to determine the wavelength dependence of the phase coefficient. Each of these data sets, however, yields similar phase coefficients, so that it seems justified to plot them together in

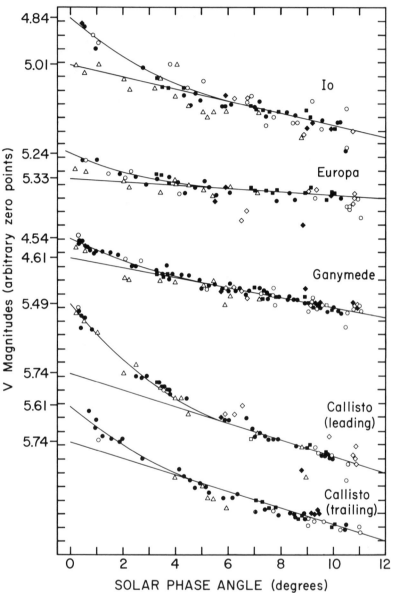

Fig. 16.5. Dependence of mean V magnitudes on solar phase angle. The points are nightly averages and have been corrected for rotational variations as indicated in Table 16.3. Circles are from Stebbins (1927) and Stebbins and Jacobsen (1928); triangles from Johnson (1969, 1971), diamonds from Blanco and Catalano (1974b), and squares from Morrison *et al.* (1974). Open symbols have larger internal errors. The coefficients that define the mean phase-dependence curves are given in Table 16.2.

Figure 16.5 and to quote a single phase coefficient for each satellite based on all of the observations analyzed together. The parabolic part of the curves ($\alpha < 6°$) was adjusted by eye to give the best fit to the opposition surge. However, it is clear that other possibilities, characterized for instance by slightly different slopes in the linear part ($\alpha < 6°$) but passing through the same points near the extreme values of α, are also allowed by the data. The parameters of these curves are given in Table 16.2 together with errors derived from the least squares fits (for the phase coefficient) or from eye estimates (for the opposition surge). Also given are values of the phase coefficients as determined separately for the data at 0.46μ and 0.56μ.

INTERPRETATION AND SUMMARY

The literature on photometry of the Galilean satellites contains a number of determinations of the absolute visual magnitudes and the geometric albedos (cf. Chapt. 9, Veverka). In Table 16.4 we compare our values for V(1,0) and V′(1,0) with those already published. In cases where the authors fitted only a parabola in the phase angle to the observations, we obtained V(1,0) by computing the magnitude at $\alpha = 6°$ from their V′(1,0) and their quadratic coefficients A and B, and then extrapolating back to $\alpha = 0°$ using the linear phase coefficients derived in this paper. Thus, in effect, we are comparing the magnitudes obtained by the various investigators at $\alpha = 6°$.

The agreement among the various observers is as good as can be expected, given the variety of photometric systems used. As noted above, Johnson's untransformed magnitudes are not readily compared with the other values. Also, Harris' transformation for the unfiltered observations from the 1920s may introduce some errors into the magnitudes attributed to Stebbins and Jacobsen in the table. We suspect that such transformation difficulties probably account for the apparent decrease with time in the brightness of Io, although of course we cannot exclude a real change in the satellite. The most recent values agree well, however. In addition, we have shown that the individual observational points are themselves in excellent agreement, and a part of those differences in the opposition magnitudes that do remain in Table 16.4 are attributable to differences in analysis rather than in the data. There is no evidence for variations in the solar constant or for other long-term changes in the brightness of the satellites, with the possible exception of Io noted above. [Possible secular variations have been suggested by Lockwood (1975b).]

The phase coefficients of the Galilean satellites can be compared with those measured for other airless bodies in the solar system (cf. Veverka, Chapt. 9, herein). The coefficients of Io and Ganymede ($dV/d\alpha \simeq 0.020$) are similar to that of the average asteroid (0.023), according to the compilation by Gehrels (1970). The coefficient of Callisto (0.030) is similar to that of the Moon (0.027); the only objects listed by Gehrels with significantly larger phase coefficients are

TABLE 16.4

Comparison of V Magnitudes With Other Observations

	Io		Europa		Ganymede		Callisto‡	
	V(1,0)	V'(1,0)	V(1,0)	V'(1,0)	V(1,0)	V'(1,0)	V(1,0)	V'(1,0)
Stebbins and Jacobsen* (1928)	−1.88	−1.99	−1.43	−1.53	−2.10	−2.16	−1.00	−1.23
Harris (1961)	−1.79	−1.90	−1.43	−1.53	−2.10	−2.16	−1.01	−1.20
Johnson (1971)†	−1.77	−1.82	−1.50	−1.57	−2.23	−2.26	−1.07	−1.29
Blanco and Catalano (1974b)	−1.71	−1.86	−1.38	−1.47	−2.09	−2.12	−1.01	−1.24
This Paper	−1.68	−1.85	−1.37	−1.46	−2.08	−2.15	−0.95	−1.20

* As transformed to the UBV system by Harris (1961)
† Untransformed
‡ Leading side

2 Pallas (0.036), Mercury (0.038), and 1 Ceres (0.050). The phase coefficient of Europa (0.006) is the smallest reliably measured for any body without an atmosphere, but it agrees with the phase coefficient calculated by Veverka (1973c) for a smooth, snow-covered planet. In general, these comparisons are consistent with the expectation that darker objects, where multiple scattering in the surface material is less important, will have larger phase coefficients (Gehrels, 1970; Veverka, 1971c); however, it appears that the Galilean satellites tend to have larger coefficients, relative to their geometric albedos, than do the other measured satellites and asteroids (cf. Veverka, Chapt. 9 herein). The very large value for Io is particularly anomalous. This satellite, in spite of its high geometric albedo (virtually identical to that of Europa in the visible), has a phase coefficient of the sort usually associated with dark, rocky materials. In these photometric properties, as in so many other respects (cf. Chapt. 17, Fanale *et al.*), Io appears to be unique among the satellites.

The sizes of the opposition surges of these satellites are not very well correlated with either albedo or phase coefficient. The value for the leading side of Callisto (0.25 mag) is among the largest observed, but that of the trailing side of this satellite appears to be about 0.1 mag smaller. Both Europa and Ganymede, in spite of their otherwise very different photometric properties, have opposition surges of about 0.1 mag. Io is again the most peculiar object, with its opposition surge of about 0.2 mag.

The rotational light and color curves illustrated in Figures 16.1 through 16.4 are complex, and it is not our intention to attempt to interpret them in terms of the surface distribution of albedo features. However, a few patterns deserve brief comment. In all cases the satellites appear to brighten more rapidly than they darken as they rotate, and in most cases they also become blue more rapidly than they redden. However, the dramatic variation of Io in the ultraviolet exhibits a slow rise and very rapid decline. Clearly, the lightcurves in different wavelengths have different shapes as well as amplitudes. We therefore conclude that, although in general regions that are dark in the visual are also red in color, there is no one-to-one correspondence between visual albedo and color.

The most striking individual albedo feature indicated in the rotation curves is centered at longitude 300° on Io. This feature is exceedingly red. From the narrowness of the dip in the lightcurves, it is apparent that the feature is substantially smaller than the hemisphere in which it lies. Therefore, when we look at Io at $\theta = 300°$ we must be seeing other, more neutral surface as well as the red feature, and the true u–y color index of the spot must be at least one magnitude. It is tempting to speculate that this dark red spot on the trailing hemisphere consists of the same material as the dark red polar caps which were discovered on Io when seen at $\theta = 180°$ (Minton, 1973; Murray, 1975).

The rotation curves illustrate a monotonic decrease in the longitudes (θ) of the maximum and minimum V magnitudes and of the crossover points where the curves pass through the mean values, as noted previously by Johnson (1971) and

TABLE 16.5
Longitudes (in degrees) of V_{max}, V_{min}, and Crossover V_0
for the Galilean Satellites

	Io	Europa	Ganymede	Callisto
θ_{max}	90 ± 10	75 ± 10	55 ± 10	255 ± 15
$\theta_0(\downarrow)$	240 ± 15	205 ± 5	125 ± 5	0 ± 20
θ_{min}	315 ± 15	285 ± 10	250 ± 40	140 ± 20
$\theta_0(\uparrow)$	30 ± 10	355 ± 10	345 ± 10	200 ± 5

Blanco and Catalano (1974b). A similar effect is seen in Saturn's system. The effect is most clearly shown in the longitudes of the crossover. In Table 16.5 we tabulate these longitudes for the V magnitudes (cf. the discussion by Veverka, Table 9.10 herein); values for the colors have been given by Morrison et al. (Table VII, 1974). The most obvious manifestation of this trend is, of course, the reversal of Callisto's lightcurve relative to those of the inner three satellites.

In this chapter, we have summarized the observational results of the past 50 years from broadband and intermediate-band photometry of the Galilean satellites, with emphasis on the V magnitudes. The data are mutually consistent and suggest no long term changes in the photometric properties of these satellites. We note that, in spite of several recent sets of high quality observations, the long series of observations obtained in 1926 by Stebbins and in 1927 by Stebbins and Jacobsen, with a 12-inch telescope and what is by our standards exceedingly primitive photometric equipment, remains the best source for determining the phase curves of all of the satellites. It is beyond the scope of this paper to undertake the interpretation of these photometric observations, but we urge others to consider the boundary conditions on the surface composition and microstructure of the Galilean satellites imposed by this substantial body of photometric data.

ACKNOWLEDGMENTS

We thank T. V. Johnson, C. B. Pilcher, and J. Veverka for useful discussions. This research was supported in part by NASA grant NGL 12-001-057.

IO'S SURFACE AND THE HISTORIES OF THE GALILEAN SATELLITES

17

Fraser P. Fanale, Torrence V. Johnson,
and Dennis. L. Matson
Jet Propulsion Laboratory

We consider the problem of the chemical evolution of Io's surface in the context of the general problem of the histories of the Galilean satellites. Any satisfactory hypothesis explaining Io's unusual optical (and other) properties must also satisfy certain fundamental cosmochemical constraints, including those imposed by density differences among the Galilean satellites and implied conditions in the circum-Jovian cloud during accretion. Resulting constraints on the bulk compositions of the satellites imply differences in internal thermal history among them; in turn, these imply differences in degassing history, hence surface chemical evolution. Surface evolution is also affected by the interaction of each satellite with its space environment: specifically, we discuss changes in surface composition and optical properties brought about by exospheric escape (e.g., dehydration) and interaction with the intense Jovian magnetospheric proton flux (e.g., sputtering and F-center development). In the context of these constraints, surface compositional models must explain the optical properties of the satellites. These include their albedos, spectral signatures—particularly the absence of ice bands in Io's infrared spectrum—and polarization. Also included is the puzzling observation that Io has dark polar regions. Finally, it is necessary to explain the intensity, orbital phase dependence and spatial distribution of the recently discovered Na D-line emission which surrounds Io.

After considering current data and various compositional hypotheses, we conclude that Io's properties can best be explained if it is postulated that the surface of Io is largely covered by "evaporite" salts produced by defluidization of Io's interior, migration of salt-saturated solutions to Io's surface and subsequent H_2O loss to space. Laboratory reflectance studies show that evaporites constitute a good match to Io's spectrum in the infrared, in contrast to ices or frosts, the presence of which does not seem likely in view of the absence of near infrared ice bands in Io's surface spectrum. Likely coloring agents in the blue include elemental sulfur, which may be produced from sulfates by proton irradiation or other processes, and F-centers produced by irradiation with magnetospheric protons. Preferential irradiation of material in the polar regions may account for Io's peculiar dark polar caps. Within the limits of theoretical and observational constraints, other materials, including montmorillonite-like phyllosilicates may be present on Io's surface, but these materials do not exhibit fully satisfactory optical properties, nor do they offer special advantages in explaining Io's Na D-line emission or dark poles. Other materials, such as frosts or high-temperature silicates, seem unlikely to be present in large quantities. Io's surface seems to represent the end result of a surface dehydration process. On Europa's surface, this dehydration is not complete, and it appears that "clean" H_2O ice has been added to it lately at a faster rate than the rate of loss. Ganymede and Callisto (especially Callisto) appear to have very thick ($\geqslant 100$ km) ice crusts overlying huge ($\geqslant 600$ km) liquid H_2O mantles. These crusts may be contaminated with initial primordial or meteoritic silicate. They appear to be thick enough that subduction and recrystallization, resulting in purification of the ice, did not occur globally. Implications of these models for the history of conditions in the near-surface environment of the Galilean satellites are discussed and future astronomical, laboratory and spacecraft tests are suggested.

[379]

Our study models the histories of the Galilean satellites and, in particular, that of Io. The following constitute *direct observational constraints:* (1) the densities of Io and the other Galilean satellites (see Table 1.4), (2) the optical properties of these satellites (see Chapts. 9 and 10 by Veverka, 11 by Johnson and Pilcher, and 12 by Morrison) and (3) the observed Na D-line emission from and around Io. Models for the histories of the Galilean satellites and the evolution of their surfaces not only must explain or be consistent with these direct observations but must also satisfy important *cosmochemical precepts.* Specifically, we will examine the compatibility of models for the evolution of Io's surface with estimates of the temperature and pressure in the vicinity of the incipient Jovian system implied by astrophysical models of the solar nebula, and the local perturbation induced in these parameters by the newly formed primary—Jupiter. The preceding considerations imply certain constraints on the bulk composition of Io and the other satellites. In turn, these limits on bulk composition imply internal thermal histories, differentiation histories, and degassing histories that are different for each of the Galilean satellites. We will examine hypotheses concerning the composition and evolution of the satellites and their surfaces in the context of the observational evidence and cosmochemical precepts that we have listed above.

PRE-ACCRETION HISTORY

The temperature and pressure in the pre-planetary nebula at Jupiter's distance from the Sun just prior to planetary accretion may be estimated from the work of Cameron (1963) and Lewis (1972a). Cameron estimated temperatures and pressures in the nebula as a function of heliocentric distance and time. Cameron (1973) and Cameron and Pine (1973) have subsequently put forth new astrophysical models of the solar nebula. Lewis (1972a; 1974b) found that the uncompressed densities (hence mean atomic weights) of the inner planets can be explained in terms of condensation from the nebula, assuming an adiabat which may be understood as approximately representing conditions at an imaginary moment when gas-dust separation occurred. If, in fact, this adiabat is correct, it would have predictable consequences for the bulk chemical and (initial) mineralogical composition of objects in the asteroid belt and beyond. One of these consequences, pointed out by Lewis (1972a), is that carbonaceous chondrite-like material would have condensed in the asteroid belt. This prediction was subsequently confirmed by evidence that many main belt asteroids may consist of such material (Johnson and Fanale, 1973; Gaffey, 1974). In addition, Lewis predicts that objects condensing at much greater than asteroidal distances from the Sun will consist of carbonaceous chondritic material plus water ice. At still greater heliocentric distances, objects would consist of carbonaceous chondritic material, plus water ice, plus $NH_3 \cdot H_2O$, etc. (Lewis, 1972b; 1974a).

Based on Lewis' best adiabat, the last assemblage would correspond to solid objects accreting at Jupiter's heliocentric distance. This model certainly does not explain all the observations, including apparent differences in composition between the Earth and the Moon. Even more puzzling is the fact that Vesta, a member of the main asteroid belt in a regular orbit, has an apparently achondritic (refractory) surface composition, as implied by the observations of McCord *et al.* (1970). Such exceptions notwithstanding, the list of features of the solar system that are explained by Lewis' (1972a) model, including the fact that the Earth's uncompressed density is actually higher than that of Venus, is impressive—especially considering the simplicity of his model.

However, it now appears that temperatures (and possibly pressures) near the forming Jupiter may have been substantially raised by the proximity of that body. This was first suggested by Kuiper (1952) to explain the observation that the densities of the satellites of Jupiter appeared to increase with increasing proximity to the primary (cf. Cameron, Chapt. 23 herein). Recently an astrophysical model for the early history of Jupiter, considered as a low mass star, has been developed by Graboske *et al.* (1975). Their model describes the Jovian surface luminosity as a function of time. From this, and from the Galilean satellite densities, Pollack and Reynolds (1974) estimated the temperature and pressure at which the satellites formed and calculated the time interval between Jupiter's formation and satellite formation. One conclusion they drew was that none of the Galilean satellites—even the outermost, Callisto (J4)—would have incorporated any ice but H_2O ice. Thus one might expect Jupiter's satellites to have accreted from carbonaceous chondritic material and ice, with a greater proportion of ice at greater distances from Jupiter. It is possible that the material which formed Io might have condensed at temperatures sufficiently high that Io formed from anhydrous silicates. However, Ganymede (at $\sim 15\ R_J$) would have had to condense at a temperature no higher than about 160°K (Pollack and Reynolds, 1974) to have incorporated the large amounts of ice which account for its present density of ~ 1.9 (Carlson *et al.*, 1973; Anderson *et al.*, 1974a, 1974b; Table 1.4 herein).

Figure 17.1 shows the temperature and pressure in the circum-Jovian cloud as a function of time and distance from Jupiter. We assume that the radiant energy from the early luminous Jupiter decreased as the square of the distance from its surface and that the blackbody equilibrium temperature throughout the circum-Jovian cloud was proportional to the one-fourth power of the radiation received. We neglect such effects as nebula opacity, convection, albedo differences, etc. Then we would expect that, in the vicinity of Io, at about 5.9 Jovian radii, material accreted at temperatures less than 300°K. Therefore, based on meteorite thermometry (e.g., see Du Fresne and Anders, 1962) and the condensation sequence of Lewis (1972a), the assumption that Io accreted largely from at least partly hydrated silicates appears sound. On the other hand, based on the high

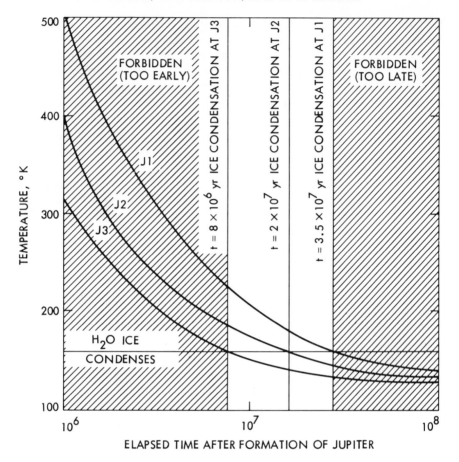

Fig. 17.1. Schematic representation of possible surface temperatures on J1, J2, and J3 as functions of time. We assume (1) the time-luminosity history of Jupiter's surface is that of Pollack and Reynolds (1974); (2) identical and constant albedos for all the objects; (3) a completely transparent gas cloud; and (4) the absence of atmospheric greenhouse effects.

densities of Io (3.5 g-cm^{-3}) and Europa ($\sim$ 3.1 g-cm^{-3}), it appears that no large amount of H$_2$O ice was incorporated into Io. In the next section we will show that the maximum thickness of the possible H$_2$O mantle on Europa is only 75 km.

The preceding analysis has been rather simplistic. For one thing, it assumes that the thermal history of pre-satellite Jovicentric dust was determined only by radiation from a point source. Note, however, that if one scaled the dimensions of the Jovian system to those of the solar system, all of Jupiter's large inner satellites would lie well inside the orbit of Mercury. Early in the history of the Jovian system and prior to either satellite accretion or separation of gas and dust, Jupiter constituted an extended source, rather than a point source, of radiation.

Thus, the existence of a large proportion of ice in Ganymede does not absolutely preclude exceedingly high temperatures in the region where Io formed. Moreover, we have not considered the finite opacity of the gas in the cloud. Possibly this factor caused the gradient of temperature with heliocentric distance to be sharper than we have assumed. Finally, we have not considered the possible effects of vertical thermal gradients in the circum-Jovian cloud analogous to those considered by Cameron (1973) for the solar system. Despite all these uncertainties, the balance of evidence suggests that Io incorporated substantial amounts of low temperature hydrated silicates but little or no ice and that differences in evolutionary history between the satellites are primarily due to differences in the bulk initial ratio of H_2O ice to silicates as suggested by Lewis (1971a). In any event, we feel safe in concluding that Io did not accrete from a reservoir of high temperature anhydrous silicates as did Earth's Moon. We will now examine the effect of these differences in initial composition upon the subsequent internal thermal and differentiation histories of the Galilean satellites with an eye to explaining the unique properties of Io's surface.

SATELLITE INTERNAL DIFFERENTIATION HISTORY

What effect would the pre-accretion history of the Galilean satellites, particularly the fact that they initially incorporated vastly different proportions of ice, have had on their subsequent internal differentiation history? We believe the effect would have been profound. Lewis (1971a; cf. Chapt. 25 by Consolmagno and Lewis) pointed out that the thermal gradient $\partial T/\partial z$ in satellites with "solar" ice-to-silicate ratios in thermal steady state may be represented by the relationship:

$$-K\frac{\partial T}{\partial z} = \frac{S\rho r}{3} \; , \qquad (1)$$

where K = thermal conductivity (erg cm^{-1} sec^{-1} °K^{-1}), S = heat production rate per gram (erg gm^{-1} sec^{-1}) and ρ = density (g-cm^{-3}). For a present-day satellite with Ganymede's radius and with solar proportions of silicates and ice, Lewis uses these average values: K = 2.2×10^5 erg cm^{-1} sec^{-1} °K^{-1}, S = 1.7×10^{-8} erg/g-sec, and ρ = 1.9 g-cm^{-3}. But Lewis (1971a) noted that this treatment would probably not be applicable to Io. For Ganymede, eqn. (1) yields an average gradient of $-\partial T/\partial z \sim 1.7 \times 10^{-14}$ r (Lewis, 1971a) or about 1.25 °K/km. To perform the calculation for an ice-free Io, we use S = 6.8×10^{-8} erg/g-sec (the present rate of heat generation in chondrites) and ρ = 3.5 g-cm^{-3}. It is hard to estimate the conductivity that should be used for Io. The major mineral in Type 1 carbonaceous chondrites appears to have a crystal structure very much like that of montmorillonite (Bass, 1971; Fanale and Cannon, 1974).

However, the Type 2 and 3 carbonaceous chondrites (Io has a density similar to Type 3 carbonaceous chondrites) have matrices that consist largely of fine grained olivine. Therefore we will use the conductivity of olivine as an approximation. The conductivity of dunite at $0°$ C is 4.8×10^5 erg cm^{-1} sec^{-1} °K^{-1} and at $200°$C it is 3.2×10^5 erg cm^{-1} sec^{-1} °K^{-1} (see Clark, 1966). The conductivity of pure olivine at $500°$C is 4.0×10^5 erg cm^{-1} sec^{-1} °K^{-1} and at $1000°$K it is 3.2×10^5 erg cm^{-1} sec^{-1} °K^{-1} (see Toksöz et al., 1972). Hence considering other uncertainties in the calculation, we assume a temperature-independent conductivity of 4×10^5 erg cm^{-1} sec^{-1} °K^{-1} or 1×10^{-2} cal cm^{-1} sec^{-1} °K^{-1}. If the rocky material is largely serpentine, as suggested by Lewis (1974a), K might be as low as 2.4×10^5 erg cm^{-1} sec^{-1} °K^{-1} (Clark, 1966), but this would not greatly alter our results. Thus the relationship for Io becomes $-\partial T/\partial z = 2.0 \times 10^{-13}$ r. Since r = 1830 km for Io, the average gradient should be $-\partial T/\partial z = 3.7$ °K/km. Our point is that this gradient is similar to that expected for the outermost portion of a hypothetical initially cold and chondritic Moon as calculated by MacDonald (1959). In such an object, melting of mafic rocks and minerals would be expected to occur at a depth of ~ 400 km, and hence, although extensive normal igneous extrusion might not occur, extensive degassing of water would be expected throughout such a body.

We have developed models for the thermal history of the Galilean satellites using a programmed solution to the inhomogeneous equation of heat conduction in spherical coordinates with time-dependent heat sources. The program (Conel, 1974) is based upon a solution for a radially symmetric distribution of heat sources given by Lowan (1933). The boundary condition at the outer surface is the so-called linearized radiation boundary condition or "Fourier's Problem of the third kind." We assume that, at the outer boundary, the body radiates to a medium at a constant temperature of $130°$K. In these models we employ the best currently available values for the radii and masses of these objects. The masses are based upon the Pioneer 10 results of Anderson et al. (1974a, 1974b) and the radii are essentially those compiled by Morrison and Cruikshank (1974) (see Table 1.4). As a further input to these models, we have calculated the ice-silicate ratios for J1, J2, J3, and J4 based upon the assumption that they presently consist not of a mixture of ices and low density silicates, as might be expected on the basis of the initial condensation sequence, but rather of ice and high density silicates ($\rho \sim 3.5$ g-cm^{-3}). This assumption will be validated by the results of our models which indicate that ice-silicate separation occurred throughout most of the Galilean satellites early in their history, and that the subsequent evolutionary path of the resulting silicate cores would include autometamorphism at temperatures greater than those which would be necessary (Mason, 1973) to produce Type 3 carbonaceous material ($\rho \sim 3.5$ g-cm^{-3}) from initially Type 1 carbonaceous material ($\rho \sim 2.3$ g-cm^{-3}). Moreover, we will show that such intense metamorphism is predicted for all reasonable sets of assumed initial conditions;

TABLE 17.1

Selected Model Parameters for the Galilean Satellites

Parameters	J1	J2	J3	J4
r: radius (km)	1.83×10^3	1.55×10^3	2.64×10^3	2.5×10^3
m: mass (g)	8.80×10^{25}	4.81×10^{25}	1.47×10^{26}	1.05×10^{26}
ρ: density (g/cm^{-3})	3.5	3.1	1.9	1.6
X_m: mass fraction of silicate (g·g^{-1})	1.00	0.95	0.67	0.53
X_v: volume fraction of silicate	1.00	0.87	0.37	0.25
Z_I: predicted maximum thickness of H_2O mantle (km)	0.0	7.5×10^1	7.4×10^2	9.3×10^2
Z_R: predicted minimum radius of silicate core if completely differentiated (km)	1.83×10^3	1.43×10^3	1.89×10^3	1.57×10^3
C: specific heat (cal gm^{-1} °K^{-1})	0.20	0.25	0.46	0.57
D: thermal diffusivity $K\rho^{-1}c^{-1}$ (cm^{2}sec^{-1})	1.44×10^{-2}	1.29×10^{-2}	1.14×10^{-2}	1.08×10^{-2}
U_0: average uranium concentration (g·g^{-1})	1.10×10^{-8}	1.05×10^{-8}	7.36×10^{-9}	5.84×10^{-9}
β: potassium/uranium ratio	1.00×10^5	1.00×10^5	1.00×10^5	1.00×10^5

Note: Masses are derived from the work of Anderson *et al.* (1974 a,b). Radii are essentially those given by Morrison and Cruikshank (1974), or in Table 1.4. Mass fractions, volume fractions, specific heats, uranium contents, thermal diffusivities, maximum thickness of H_2O mantles, and minimum radii of silicate cores are predicted on the basis of X_m and X_v which are, respectively, the mass fraction and volume fraction of silicate in the objects. The latter two parameters are estimated on the basis of m, r and the assumption that each object consists of two end members: H_2O and silicate. It is assumed that the silicate portion of each satellite has a density at present which is close to 3.5g-cm^{-3}, regardless of its original density. This assumption is defended in the text.

that is, our reasoning is not circular. Table 17.1 gives the input parameters for our thermal history models plus some other useful parameters.

Using the reasoning just stated, we have calculated volume (X_v) and mass (X_m) fractions of silicate for all the satellites. Then, based on X_m, we have calculated the specific heat (C), thermal conductivity (K), thermal diffusivity (D) and average uranium content for each of the assumed ''homogeneous'' objects. We feel safe in neglecting radiative conduction because, for the olivine lattice, it now appears to be less important than lattice conductivity at least to 1700°K

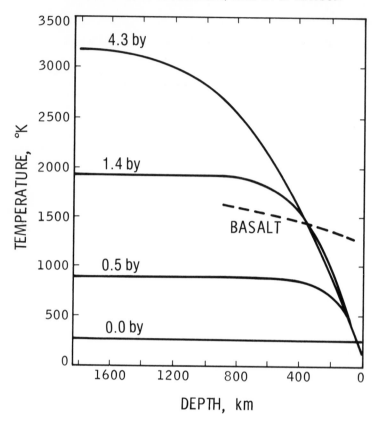

Fig. 17.2. Internal thermal profiles for Io at 0.5, 1.4 and 4.3 billion years after its formation. These computer plots are generated using heat transfer parameters and assumptions discussed in the text. Also plotted as a dotted line is the basalt solidus for Io.

(Shankland, 1970). The K/U ratio β is assumed to be chondritic, and a surface temperature T_s of 130°K is used (see Morrison and Cruikshank, 1974). Finally, although this is not needed for the simple (homogeneous) thermal models we present here, we have calculated the present thickness of the H_2O mantles resulting from the differentiation suggested by our models (see below) according to the following relationship:

$$\frac{R_1}{R_2} = \left[\frac{\rho_T - \rho_{H_2O}}{\rho_S - \rho_{H_2O}} \right]^{1/3}, \tag{2}$$

where R_1 = the radius of the silicate core, R_2 = radius of the entire object, ρ_S = density of the silicate core, ρ_{H_2O} = density of the H_2O mantle and ρ_T = mean density of the whole object.

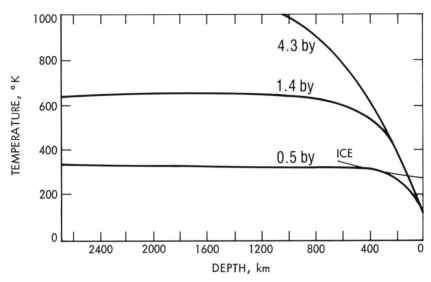

Fig. 17.3. Internal thermal profiles for Ganymede at 0.5, 1.4 and 4.3 billion years after its formation. the curves for 1.4 and 4.3 billion years may be regarded as fictional in that extensive ice melting would cause the thermal history to follow a different track than that which we project. The ice solidus is shown as a solid line.

Our resulting thermal models for J1 and J3 are given in Figures 17.2 and 17.3, respectively. In Figure 17.2, note that, for Io, the basalt solidus is crossed at a depth of ~ 400 km. Also, a temperature of 700°C is reached at a depth of only ~ 200 km. Extensive degassing of chemically bound water and its subsequent migration to the surface would be expected to result from such a thermal profile. Generation of surface igneous extrusions seems somewhat less likely but certainly cannot be ruled out. The presence of molten silicate in Io's interior is definitely predicted. Note, in Figure 17.2, that the temperature at a depth of 200 km was as high or higher 3 billion years ago than it is now. This suggests that most of the release of near-surface chemically bound water may have occurred relatively early in Io's history, and allowed substantial time for loss of the resulting surface ice, as will be discussed later. However, it will also be shown that the amount of H_2O loss is unlikely to have been great enough to have significantly altered the bulk density of Io.

The results for Ganymede (Fig. 17.3) must be regarded as partly fictional, in that differentiation of H_2O and silicate obviously would have occurred throughout much of Ganymede during the first third of its history and would have affected profoundly its subsequent evolution. According to Table 17.1, the silicate core resulting from that differentiation could have a radius of up to 1.9 × 10³ km—or somewhat less because the outermost portion of Ganymede is likely not to have melted and differentiated all the way to the very surface (see Fig.

17.3). The initial ice-to-silicate ratio we calculated for Ganymede may be somewhat too low, because we have assumed a present density of 1.0 for the non-silicate portion. Consolmagno and Lewis, in Chapter 25 herein, have presented detailed models of the initial distribution of ice phases in icy satellites. Their work suggests that the initial average density of the non-silicate portion of a 2,500 km radius icy satellite might be as high as 1.4 g-cm^{-3}. However, our models suggest that melting probably occurred early in Ganymede's history. Based on data compiled by Clark (1966) on the variation of the density of water and ice with temperature and pressure, and on our thermal model, it appears that the current effective mean density of the non-silicate portion of Ganymede is close to 1.15 g-cm^{-3}. This may imply a significant amount of planetary expansion.

Europa, like Io, would achieve very high interior temperatures and any ice initially present would be melted. The maximum thickness of any resulting ice mantle for Europa can be calculated by making the assumption that the silicate portion of Europa has as high a density as does Io. Again, this assumption can be defended since: (1) based on Io's thermal history and the similarity of the total amount of silicate between J1 and J2, the thermal history of Europa (though slightly less severe than Io's owing to the slightly smaller radius of Europa's silicate core) would produce temperatures greater than 600°C throughout most of Europa's interior; (2) the thermal boundary between Type 2 and Type 3 carbonaceous chondrites lies at only $\sim$ 600° to 700°C (Mason, 1973); and (3) the density of the Type 3 carbonaceous meteorites is $\sim$ 3.5 g-cm^{-3}. Moreover, our reasoning is not circular; if the density of the silicate in Europa is less than 3.5 g-cm^{-3}, then the quantity of ice must be lower, and interior temperatures (and resulting silicate densities) even higher than we have assumed.

On the basis of bulk satellite density, we expect a rather thin ($\leqslant$ 75 km) H$_2$O outer shell on Europa and a very thick ($\sim$ 740 km) H$_2$O outer shell on Ganymede. How thick would their solid ice crusts and liquid H$_2$O mantles be? If our thermal model for Ganymede as depicted in Figure 17.3 is correct, it seems possible that the ice crust on Ganymede might be over 100 km thick. From Table 17.1, the total H$_2$O layer on Europa (water plus ice) can be only $\sim$ 75 km thick, vs. 740 km for Ganymede. We have not constructed a detailed thermal model for Europa. However, we may consider the two dimensional problem of transfer, through a "thin" layer of ice, of a heat flow equal to that from Europa's silicate core. In this case

$$t = (T_m - T_S) \, K_i/Q \, , \tag{3}$$

where t = ice thickness, T_m = ice melting point, T_S = surface temperature, K_i = thermal conductivity of ice, and Q = surface heat flow (originating in the silicate core). Thus, we calculate that Europa should have an ice crust with a maximum thickness of 40 km. Obviously, if Europa has a total outer H_2O shell much thinner than this, it would not be underlain by a zone (mantle) of liquid H_2O. Similarly, we would expect that Callisto would have an even thicker ice crust than Ganymede, because the silicate core mass of Callisto (hence the rate of heat generation) is much less than that of Ganymede whereas their surface areas (through which the heat must flow) are roughly the same.

The observations are as follows: Europa has a reflectance spectrum similar to that of "clean" H_2O ice while Ganymede has a lower albedo, and Callisto has a much lower albedo still than Ganymede (see Tables 1.4 and 16.1; Chapt. 11, Johnson and Pilcher). This implies successively increasing amounts of non-icy material in the ice crusts of J2, J3 and J4, respectively (Pilcher *et al.*, 1972). Thus, we are tempted to speculate that there is a possible relationship between the thickness of the ice crusts (hence their stability in the face of meteorite bombardment, convective overturn or internal degassing) and the ability of each satellite to "purify" their surfaces of original non-icy material.

We conclude that Io experienced a thermal history not unlike that which would be expected for an initially cold and chondritic Moon. This resulted in extensive degassing of chemically bound water in Io and, conceivably, even the generation of surface igneous magmas: Europa probably experienced an internal thermal history that was only slightly less intense and is presently covered with a small ($\leqslant$ 75 km) H_2O layer. Ganymede and Callisto may have very thick ($\geqslant$ 100 km) ice crusts overlying huge ($\geqslant$ 700 km) liquid H_2O mantles. Their ice crusts may never have been subducted, melted and "purified" of original silicates. But in both Ganymede and Callisto, most of the silicate has aggregated together in a silicate core within which temperatures developed that produced core densities similar to those of Type 3 carbonaceous chondrites and to that of Io.

CHEMICAL EVOLUTION OF IO'S SURFACE

Transport of Solutions to the Surface

What would be the consequences of the internal differentiation histories just described for satellite surfaces? Hot water liberated from silicates and percolating through the outer layers of Io would quickly become saturated with soluble salts. The fine grain size of meteoritic materials in general and carbonaceous chondrites in particular (most of the matrix grains in the latter are phyllosilicate grains < 0.1 μm in diameter), together with our laboratory results, suggests that saturation would occur almost immediately if the silicate-to-water ratio were sufficiently high.

What would be the composition of the soluble salts carried in the upward migrating aqueous solutions on Io? The composition of salts in terrestrial sea water (Table 17.2, column 2) provides some intuitively attractive but potentially misleading evidence, since the salts in sea water have experienced a rather

TABLE 17.2
Partial Chemical Composition (expressed as weight percent) of Dissolved Salts in Sea Water, River Water, and Two "Evaporites" Produced by Leaching Basalt and the Orgueil Meteorite

1	2	3	4	5	6
	Sea Water (Sverdrup *et al.*, 1942)	River Water Less Cyclic Salts (Sverdrup *et al.*, 1942)	Basalt Leach Evaporite (this work)	Orgueil Leach Evaporite (this work)	Enrichment (depletion) Factor (Orgueil Evaporite ÷ Total Orgueil)
Na	30.6	2.6	24	11	18
Ca	1.2	20.7	8.4	6.8	7.8
Mn	—	—	0.05	0.7	4.1
K	1.1	2.0	4.3	<2	<4
Ni	—	—	—	2.9	2.9
Mg	3.7	3.0	3.6	25	2.6
Al	—	<0.1	0.03	0.25	(0.42)
Si	—	5.8	0.70	3.7	(0.35)
Fe	—	<0.1	<0.003	0.08	(0.004)
$SO_4^=$	7.7	11.4	16	34	—
Cl^-	55	0	14	—	—

Note: Cations are listed in order of decreasing degree of enrichment in the Orgueil evaporite relative to the total composition of the meteorite (which is approximately similar to average solid solar system material).

different history from what we would expect on Io. First, the ions in sea water are present in relative concentrations (κ) determined not only by their relative present and past supply rates to the ocean, but also their relative rate of removal from the ocean. That is: $\kappa = \tau \cdot T_{1/2}/0.693$ where τ = input flux, $T_{1/2}$ = half life in oceans, and $T_{1/2}/0.693$ = mean residence time.

Thus the ocean is a temporary reservoir in transient equilibrium with both sources and sinks, and the mean residence times of ions in the ocean are <<4.5 billion years. We should also consider the total composition of soluble salts that would occur in a closed system evaporite deposit. Another possibly important consideration is that the composition of the source material leached in Io to provide the evaporite deposits is likely to be very different from the composition of the rocks that have been eroded on Earth to provide terrestrial evaporites and sea water salts. For example, terrestrial salts would not be as rich in S as those produced by leaching meteoritic material owing to the concentration of S in Earth's interior.

Table 17.2, column 3, gives the composition of river water, less cyclic or windblown oceanic salt. This represents a somewhat closer approximation to the total evaporite composition. The difference between columns 2 and 3 is profound, and results partly from removal of Si, Ca, and Al and Fe from river water (e.g., in deltas) before supplying the ocean, coupled with their rapid removal from the ocean by clays and carbonates. Chemically analogous reservoirs containing hydroxides or carbonates may be hidden below the surface on Io.

Further difficulties in estimating the possible composition of Io's surface are introduced by the fact that evaporite deposits are highly stratified by nature, and contain a highly variable mineral composition. For example, most evaporite deposits on Earth contain mainly calcium carbonate, calcium sulfate and (of course) sodium chloride. But when 98% of the water has evaporated, other salts begin to crystallize (e.g., see Mason, 1958). Continued isothermal evaporation at 25°C leads to the crystallization of bloedite (Mg $\cdot$ Na_2 $(SO_4)_2$ $\cdot$ XH_2O) and eventually epsomite (Mg SO_4 $\cdot$ XH_2O). From then on, the sequence of crystallization depends upon whether the previously separated salts can react with the solution. These considerations play an important part in interpretation of the composition of the salt found in meteorites and in the estimation of Io's surface composition, as will be shown. Needless to say, considerable gardening and mixing of evaporite material on Io's surface may have taken place as the result of meteorite impact.

Let us now examine some evidence we have obtained in the laboratory suggesting the composition of a "total" evaporite that might be obtained by the leaching of igneous rocks or meteoritic material by hot water in a closed system followed by evaporation or sublimation of the aqueous salt solution. Column 4 shows the composition of an "evaporite" we obtained by leaching a finely ground sample of fresh basalt in sub-boiling water for one hour. Differences in composition between the basalt leach evaporite and river or sea water salts are major and may be attributed to many causes, including the fact that our laboratory leaching procedure was extremely mild and barely "scratched the surface." That is, more extensive leaching might produce salt with cations in relative abundances somewhat closer to those in the rock. Also, a "light" leach might have incorporated an undue contribution from microcrystals of groundwater-deposited evaporite minerals.

Column 5 gives results of analysis of a leach evaporite from the Orgueil meteorite. This was obtained by leaching a very small ground sample (70 mg) for 1 hr. in sub-boiling water. The yield of evaporite resulting from this mild procedure was 3.4 mg or 5% by weight of the original Orgueil sample. The results are subject to some interpretative problems. In particular, the composition may be strongly affected by redissolving the natural evaporite salt that occurs in veins and pores of Orgueil. Similar salts occur in other carbonaceous chondrites. Table 17.3 shows that epsomite, bloedite and gypsum ($CaSO_4$ $\cdot$ $2H_2O$) have been identified in carbonaceous chondrites by Du Fresne and Anders (1962). These workers attributed the presence of salts to an aqueous solution and deposition in the meteorites' parent body or bodies, probably in a "temporary atmosphere" protected by a permafrost layer. Reviewing the history of studies of the Orgueil

TABLE 17.3
Partial List of Salts Identified in Meteorites (Carbonaceous Chondrites)

Salt	Composition	Meteorite	Class	Amount (wt%)
Epsomite	$MgSO_4 \cdot 7H_2O$	Orgueil	I CC	17
		Tonk	I CC	21
Gypsum	$CaSO_4 \cdot 2H_2O$	Mighei	II CC	One Crystal
Bloedite	$MgSO_4 \cdot Na_2SO_4 \cdot 4H_2O$	Ivuna	I CC	"One Tiny Particle"

Note: See Du Fresne and Anders (1962).

meteorite, Nagy (1966) noted that investigators prior to the turn of the century had also reported the presence of soluble salts of ammonia, potassium, magnesium, and sodium with sulfates and chlorides as anions. The presence of ammonium salts was not subsequently confirmed by other investigators, however. Nagy concluded that "the water soluble material in Orgueil appears to be magnesium sulfate only." But material balance considerations alone would suggest that salts extracted by leaching a carbonaceous chondrite should comprise a more chemically complex assemblage. Leaching of Orgueil, which consists mainly of 0.1 μ grains of phyllosilicates exhibiting close to solar relative abundances of nonvolatile and semi-volatile elements, would hardly be expected to result in an evaporite deposit consisting of pure epsomite (see below). In considering Table 17.3, recall from above that it would be expected that slowly evaporating solutions would deposit primarily NaCl, but that the final stages of their evaporation would produce a series of strata rich in epsomite and bloedite. We suggest that the few small samples of carbonaceous chondrites that are available came from different layers in a chemically stratified interstitial evaporite deposit (or deposits) in the pores in the outer zones of the parent body (or bodies). We also believe that the domain size of these "strata" was much greater than the domain size sampled by any individual meteorite.

The last column in Table 17.2 shows the degree of enrichment of each element in the Orgueil evaporite, that is, the ratio of the abundance of each cation in the Orgueil evaporite to its abundance in Orgueil itself. Note that not only is Na the most enriched element in the Orgueil evaporite, but it is also enriched almost to the maximum possible extent. That is, the Na enrichment factor is 18, whereas (by definition) the maximum enrichment factor permitted by the ratio of "evaporite" to total meteorite mass is 20. This is of importance in understanding the origin and properties of Io's surrounding Na emission, as will be shown later.

This model of evaporite formation allows us to explain why different carbonaceous chondrites contain different salt crystals (Table 17.3) and also allows us to reconcile the complex assemblage of salts indicated by the chemical composition of our Orgueil leach evaporite (Column 5, Table 17.2) with the isolated mineralogical identification of epsomite in Orgueil. Also, we may conclude that any "cap" or surface deposit formed by leaching of average solid solar system material is likely to be very rich in both Na and SO_4. We will now consider whether such a cap or crust might have formed on Io.

Development of Io's Present Surface

It might be argued that a salt deposit would not be formed by supplying salt-saturated aqueous solutions to the surface of Io because the sublimation and planetary escape rate of H_2O would be too low at Io's surface temperature. From the data in Figure 17.3 we can estimate the maximum sublimation rate for H_2O based on the work of Watson et al. (1963). These are the rates at which ice could sublime if the process were utterly irreversible. These "maximum" or irreversible sublimation rates are so high that even at the present surface temperature on Io of $\sim 130°K$ (Morrison and Cruikshank, 1974; Morrison, Chapt. 12 herein) as much as 2 km of H_2O ice would sublime in geological time. This sublimation rate certainly would be sufficient for the development of a salt cap—especially since much higher surface temperatures are estimated to have prevailed in Io's past (Fig. 17.1), and since the sublimation rate is sharply dependent on the temperature (Watson et al., 1963). It should be noted that this approach only applies to the case where the sublimation rate per se is the rate-limiting factor.

However, the inferred Pioneer 10 upper limit on the atmospheric pressure has been given as $\lesssim 10^{-8}$ bars (Kliore et al., 1974; 1975). Since Io's surface temperature is about $130°K$ and since the equilibrium vapor pressure of ice at that temperature is 10^{-13} bars, it is quite possible that Io's atmosphere is saturated with water at present and that the rate-limiting step that determines the H_2O loss rate on Io is the atmospheric loss rate not the sublimation rate. Later we will detail arguments suggesting that the present day upper limit on the mean residence time against escape of atmospheric species may be as low as 1 to 2 years or less. Thus, the minimum loss rate is the total number density in the atmospheric column corresponding to the equilibrium vapor pressure divided by the atmospheric mean residence time. For a surface water pressure of $\sim 10^{-13}$ bars, the number density at the surface would be about 7×10^8 H_2O molecules cm^{-3}. If the atmospheric temperature were similar to the surface temperature, the scale height would be ~ 26 km, and there would be a total of 2×10^{13} cm^{-2} H_2O molecules in the atmospheric column. If this were the only loss mechanism, then the present minimum net loss rate can be estimated to be $\sim 6 \times 10^{-10}$ g cm^{-2} yr^{-1} or only 3 cm of ice in 4.5 AE. This rate would hardly seem sufficient to permit development of a salt cap, even though water at $0°C$ may contain as much as 15% or more dissolved sulfate salt. On the other hand, this estimate of the loss rate is probably several orders of magnitude too low for two reasons. First, simple momentum-transfer probably does not remove species nearly as fast from Io as magnetic field sweeping of primary and secondary ions. Second, as in the "irreversible" case, the absolute rate of water destruction in, and loss from,

Io's atmosphere would increase enormously with increasing surface temperature (hence basal H_2O pressure). This would be crucial earlier in Io's history when surface temperatures, especially on Io, were higher (Fig. 17.1).

The water loss mechanisms we suggest for Io cannot apply with equal effect to Europa: our calculations for Europa show that, integrated over the satellite's history, the amount of H_2O that could have been removed from Europa by the sublimation-sweeping process is $\leq 10^{-2}$ of that which could have been removed from Io. Furthermore, any magnetic field sweeping removal (we have only taken simple momentum transfer sweeping into account) would be less effective on Europa than on Io.

In summary, it is difficult to estimate the amount of H_2O that may have been lost from Io to space during geological time. The thickness of ice lost may have been as low as a few centimeters or as high as a few kilometers depending on the effective mechanisms of exospheric loss and the total atmospheric pressure during Io's history. Almost all mechanisms of loss operate less effectively on Europa. This conclusion (together with the likelihood that Europa probably had a higher initial H_2O content than Io) may account for the data of Pilcher et al. (1972) which shows that "clean" ice covers much of the surface of Europa, but not that of Io (cf. Chapt. 11, Johnson and Pilcher).

Optical Measurements and Models for Io's Surface Composition

We will now consider the relationship between the optical properties of the surface materials on J1, J2, J3, and J4 and their probable surface composition. In the preceding sections, we have traced the pre-accretion history of the material that formed Io, and presented a model for the internal differentiation history of Io. We have shown that enrichment of the surface material in soluble salts, deposited by upward migrating aqueous solutions, is one likely consequence of Io's probable history. Elsewhere (Fanale et al., 1974a,b) we have suggested that Io's optical properties could also be explained if its surface were largely covered with salts. Here, we will review the optical evidence and its bearing not only on the evaporite hypothesis, but also on other hypotheses of surface composition that are, at least to some extent, also compatible with the probable cosmochemical history of Io.

The optical properties of Io have long been difficult to explain—particularly Io's high visual albedo and very low blue and ultraviolet reflectance (Harris, 1961; Morrison and Cruikshank, 1974; cf. Fig. 11.2, Johnson and Pilcher, herein). Figure 17.4 (from Fanale et al., 1974b) shows Io's spherical reflectance as the heavy line. This was derived from the geometric albedo using a phase integral q of 0.7 (Morrison and Cruikshank, 1974; Chapt. 12, herein, Morrison). Io's high albedo, its polarimetric properties (Veverka, 1971a) and the high derived value for the phase integral (Morrison and Cruikshank, 1974) are all satisfied if Io's surface is covered by low opacity, multiply scattering material.

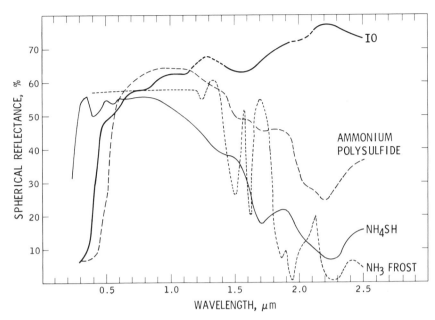

Fig. 17.4. The spectral spherical albedo of Io is shown as the heavy line. Data is taken from Johnson and McCord (1971) in the visible and Pilcher *et al.* (1972) in the infrared. Also shown are spectra of ammonium polysulfide and NH$_4$SH (Sill, 1973) as well as that of ammonia frost (Pilcher *et al.*, 1970). The visible portion of the last is extrapolated. Comparison of the telescopic and laboratory data is not exact. When the laboratory data were taken with an integrating sphere, they are plotted as percent reflectance relative to MgO. Data of Sill (1973) were taken at 45° incident and normal reflectance angles relative to LiF. These data are arbitrarily reduced to spherical reflectance by using a phase integral of 0.7 (which is similar to Io's phase integral).

Also, the polarization-phase curve for Io as measured by Dollfus (1961a), Veverka (1971a), and Zellner (to be published) shows a negative branch of depth 0.2% at a phase angle near 6° with the inversion angle believed to fall between 10° and 15°. This suggests rather translucent surface materials on Io (cf. Veverka, Chapt. 10 herein). Now let us compare Io's spectrum with those of serious candidate materials that have been suggested. The spectra of some of these materials are also shown in Figure 17.4. A common suggestion is that Io's surface albedo is high because Io is largely covered by some variety of ice or frost. This is certainly true for Europa, which has about the same albedo as Io (Table 1.4) and exhibits strong H$_2$O water frost bands in the near-infrared. But Io's near-infrared spectrum shows no evidence of such bands (Pilcher *et al.*, 1972; Fink *et al.*, 1973; Johnson and Pilcher, Chapt. 11 herein). In addition, Io's surface is colored, not white, suggesting that even if ice or frost were present on Io's surface, its visible spectrum would have to be accounted for by some additional, albeit minor, component (cf. Morrison and Cruikshank, 1974; Johnson and Pilcher, Chapt. 11 herein). Finally, Europa exhibits bright polar caps (Kuiper, 1973; Murphy and Aksnes, 1973; Vermilion *et al.*, 1974 [see, however, Aksnes and Franklin, 1975b]) as would be expected for a satellite with ices covering a large portion of its surface whereas Io has dark, "reddish" caps (Dollfus, 1971; Minton, 1973; Murray, 1975). However, a frost surface for Io cannot be entirely ruled out since Io's environment has not been completely matched in laboratory experiments. The possibility of grain size differences or impurities affecting a frost spectrum are discussed by Johnson and Pilcher in Chapter 11 and by Kieffer and Smythe (1974). McElroy and Yung (1975) have suggested that an electrically conducting ammonia ice might not exhibit deep absorptions, but there are no theoretical or experimental treatments of these types of materials (cf. Chapt. 11, Johnson and Pilcher). We feel that the case for abundant frost or ice on Europa's surface is very strong and that for abundant frost or ice on Io appears very weak. Alternatively, some investigators have suggested other materials, such as NH_4SH and sulfur, possibly derived from H_2S (Kuiper, 1973), ammonia hydrosulfides (Lebofsky, 1972; 1973b) and low opacity silicates or glasses derived from them (Johnson, 1969). Figure 17.4 shows that the ammonia compounds have near-infrared features that do not appear in Io's spectrum (Sill, 1973). Silicates such as alkali feldspar seem unlikely in view of thermal considerations discussed in the previous sections and the petrological unlikelihood of a monomineralic surface consisting of a single high temperature silicate mineral.

Among the more acceptable possibilities, Figure 17.5 shows that sulfur does match the visible spectrum of Io and its high infrared albedo quite well and could be derived from H_2S. However, there are other possible origins for reduced sulfur entirely compatible with the evaporite hypothesis as will be shown. The role of sulfur as a spectrally active component on Io's surface is treated in more detail by Wamsteker *et al.* (1974) and Wamsteker (1974). They suggest that the spectrum of Io may be reproduced by a 60/40 mixture of two spectra: one the spectrum of sulfur and the other the spectrum of some unspecified material which shows a wavelength dependence of its reflectivity similar to the non-H_2O component of the spectrum of the rings of Saturn.

Figure 17.5 also shows the spectrum of natural halite (NaCl). Note that, unlike the frosts, natural halite provides a reasonably good match to Io's near-infrared spectrum. In addition, Zellner (personal communication, 1974) points out that the polarimetric properties of Io are, in fact, identical with those of table salt, as measured by Lyot (1929). The halite curve in Figure 17.5, which shows a fairly uniform albedo from 0.3 to 2.5 μ, provides a poorer match to Io in the visible and ultraviolet. As indicated earlier, we do not expect theoretically that salt on Io's surface would consist entirely, or even dominantly, of NaCl. It is well known that small amounts of impurities produce strong and varied colora-

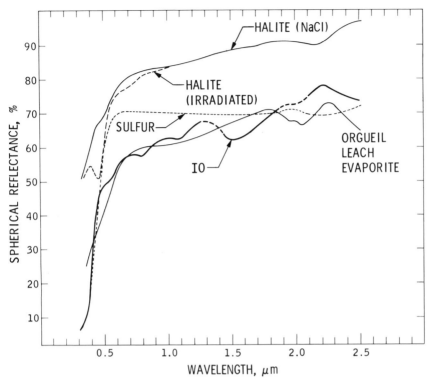

Fig. 17.5. Io's spectral reflectance is plotted with a heavy line, as in Figure 17.4. Also shown is the spectra of selensulfur (Sill, 1973), natural halite, the halite sample after irradiation by protons (Nash *et al.*, 1975) and the Orgueil leach evaporite. Comparison of laboratory and telescopic spectra is accomplished using the same procedure as for Figure 17.4.

tion in some terrestrial halite deposits. Moreover, the high energy ($\geqslant$ 1 MeV) particle flux on Io can easily penetrate the most dense atmosphere that Io could conceivably possess ($\sim 10^{-8}$ bars). [The Pioneer papers in Gehrels (1976) discuss the magnetospheric environment.] On hitting the surface materials, the high energy protons could produce coloration by two mechanisms: color center production and chemical alteration. Salts are especially susceptible to color center production under irradiation (cf. Przibran, 1956). Most previous suggestions of irradiation coloring have focused on production of free radicals, polymers or organic compounds (Binder and Cruikshank, 1964; Sagan, 1971) but many of these possibilities can be eliminated because of the absence of characteristic bands in Io's near infrared spectrum. Color center production would not necessarily suffer from this problem. For example, we irradiated the natural halite sample (Fig. 17.5) with a total dose of $\sim 2 \times 10^{17}$ protons cm^{-2}, having an energy from 2.0 to 7.0 KeV. Immediately after irradiation, the sample exhibited the reflectance shown as a dotted curve in Figure 17.5. The deep adsorption centered near 0.48 μ is due to F-centers in the NaCl lattice. The band depth decreased with time as a result of room temperature annealing. Our results suggested that a greater dose would have produced a deeper band. In any event, irradiation of the halite did not lessen the match to Io's near-infrared spectrum, and it improved the match in the blue.

The reflectance curve for the Orgueil leach evaporite is also shown in Figure 17.5 and provides a better match to Io's curve. Based on chemical considerations discussed earlier, we would expect that any evaporite produced by leaching of meteoritic material would, like the Orgueil evaporite, or the basalt evaporite, be rich in sulfates. Moreover, the ability of prolonged proton bombardment to chemically reduce compounds is well known and would be expected to result in production of small amounts of reduced sulfur in a sulfate-rich assemblage. Thus, while the match of the Orgueil leach evaporite curve to Io's curve in the visible is not perfect, it might be improved by prolonged proton bombardment. For instance, the evaporite material has the optical properties suggested by Wamsteker (1974) for the second (unidentified) component in his model. The match in the near-infrared is already within experimental error of the telescopic data (Pilcher, personal communication, 1974). If this reasoning is correct, then it also provides an explanation for the otherwise perplexing observation that Io has dark, "reddish" polar regions (Dollfus, 1961a; Minton, 1973; Murray, 1975). These areas may receive higher doses of trapped radiation and are also colder than the rest of the satellite, thus may have enhancement and retention of coloration (Fanale et al., 1974b).

One possible component of the surface, the possible presence of which has not yet been suggested, is a talc-like or montmorillonite-like phyllosilicate (sheet-silicate). Cosmochemically, the presence of such material on Io's surface is entirely possible, since the last solid materials expected to form in equilibrium with the nebula prior to the condensation of H_2O ice are tremolite and serpentine or talc (Lewis, 1972a; Lewis, 1974b). Moreover, the matrix material in carbonaceous chondrites (thought to resemble the initial non-icy condensate in the outer solar system) is apparently extremely similar to montmorillonite in its structure (Fanale and Cannon, 1974). The albedos and infrared spectra of talc, montmorillonite and serpentine (Hunt and Salisbury, 1970; Fanale, unpublished data) are compatible with those of Io. There are several serious problems with the phyllosilicate hypothesis, however:

1. The spectral match of these mineral phases to the spectrum of Io in the visible is poor. Also, we have not been able, in preliminary attempts, to improve this match by proton irradiation, as was possible for halite. While a "stain" of sulfur on the phyllosilicate would improve the match in the visible considerably, it would necessitate deriving the phyllosilicate from one source and the sulfur from another, whereas in the evaporite model, the sulfur is originally present as an essential constituent of the major surface component.

2. The only primordial phyllosilicate actually known to occur in the solar system (that in carbonaceous chondrites) occurs in intimate mixture with the black opaque components, carbon and organic compounds, which are expected on a theoretical basis to respectively condense before and after the phyllosilicate during nebula cooling. Obviously there are problems in explaining the high albedo of Io on this basis.

3. The reflectances of all these minerals normally show absorptions due to water at 1.4 and 1.9 μ and, more importantly, deep hydroxyl bands between 2.0 and 2.5 μ (Hunt and Salisbury, 1970). Io's surface spectrum exhibits no such bands.

4. Unlike the evaporite hypothesis, the phyllosilicate hypothesis offers no special advantages in explaining either the source of Na in the cloud surrounding Io (see next section) or Io's dark poles.

Thus, we favor the evaporite hypothesis over the phyllosilicate hypothesis for Io's surface composition. However, we feel that there is greater likelihood that phyllosilicate may be present as a component of Io's surface than other materials which have been suggested.

In summary, a comparison of visible and near-infrared spectra of possible candidate surface materials for Io with spectra obtained for Io's surface suggests that the peculiar optical properties of Io's surface and the differences between the spectra of Io and Europa can best be explained by hypothesizing an evaporite surface on Io, rather than one covered by ices, igneous rocks or other previously suggested materials.

INTERACTION BETWEEN IO AND ITS ENVIRONMENT

A number of poorly understood processes are currently operating upon Io, and over geologic time they may have significantly altered both the optical properties and the chemical composition of its surface. These processes currently produce an assemblage of peculiar manifestations which are observable at optical and radio wavelengths. What implications does our model hold for properties such as the observed sodium D-line emission?

Sodium D-Lines

During the 1972 opposition of Jupiter, Brown (1974) discovered that sodium D-line emission was emanating from Io. Almost immediately he found that the intensity of the line was time variable (Brown, 1974; Brown and Chaffee, 1974). Shortly after the announcement of the original discovery, Trafton et al. (1974) and later Sinton (personal communication, 1973) independently discovered that the emission was not confined to Io but also came from a cloud as large as 20 arcsec in diameter. During the 1974 opposition, the effect was extensively observed and studied (Macy and Trafton, 1975; Bergstralh et al., 1975; see also the extensive discussion of papers in Gehrels, 1976).

Inquiry is directed toward the discovery of the source of sodium and the mechanisms by which it is excited. The half-life of sodium atoms for photoionization is about 10^6 seconds at Jupiter's distance from the Sun. Thus, the sodium cloud about Io must be replenished with fresh material from some source. The only viable source of sodium appears to be the surface of Io itself (McElroy *et al.*, 1974; Matson *et al.*, 1974). The concept of a sodium-rich surface is certainly in accord with our model for Io's surface (see Table 17.2) which, it should be noted, was formulated prior to the D-line observations. Surface material is probably removed by proton or ion sputtering as has been suggested by Matson *et al.* (1974). Sputtering is thought to occur because Io orbits deep within Jupiter's magnetosphere and is subject to an intense flux of protons. (The properties of this interaction are detailed in Nash *et al.*, 1975.) In the 0.4 to 1.0 MeV range alone, the flux observed by instruments aboard Pioneer 10 amounts to 10^7 protons cm^{-2} sec^{-1} (Trainor *et al.*, 1974). At lower energies, the flux is thought to be even higher. Fluxes as high as 10^9 protons cm^{-2} sec^{-1} have been suggested by Carlson (1975) as being necessary in order to explain the Pioneer 10 Lyman α observations in terms of a charge exchange model for the destruction of hydrogen.

A large number of heavy ions are also expected to impact Io's surface. These ions are created both in the sputtering process and by ionization of any atmospheric gases. The Jovian magnetosphere rotates with a period of about 10 hours. Since Io's orbital period about Jupiter is only 1.77 days the relative velocity of the magnetic field with respect to Io is approximately 56 km-sec^{-1}. Any ion near Io which is in this moving magnetic field will be accelerated either into Io or away from it.

Electric fields are also expected to be present since plasma sheaths may occur at the interface between Io and the Jovian magnetosphere. In some of the plasma sheath models, potentials as high as 600 KeV are developed (see Gurnett, 1972; Hubbard *et al.*, 1974). Both the magnitude and the sign of these potentials depend critically on longitude and latitude at Io's surface, Io's position in the magnetosphere, and the electrical properties of Io's surface and ionosphere. Typical characteristic heights of the sheaths vary from 0 to 100 km. Any ion produced within or entering the sheath region will be accelerated and will leave the sheath either by striking the surface or by traveling outward into the magnetosphere. However, many of the outward-bound ions which enter the magnetosphere are returned again to Io because the gyroradius of even a 600 KeV sodium ion (ejected at the equator perpendicular to a field of 0.035 gauss) is approximately 150 km. Heavy ions are 10^3 times more efficient at sputtering than protons but presently we have no reliable way of estimating their flux on Io's surface. It is possible that ion bombardment could be responsible for most of the sputtering.

Once sodium has been sputtered, its history depends upon the density of Io's atmosphere. Unfortunately, the Pioneer 10 radio occultation cannot yet provide a neutral atmospheric density which is not model dependent (cf. Kliore *et al.*, 1974; 1975). If the surface pressure is less than approximately 10^{-11} bars, the sputtered material will travel into the cloud on a ballistic trajectory. If the pressure is 10^{-10} bars or greater, the material will be thermalized in the atmosphere and then ejected chiefly by elastic scattering due to energetic protons together with their secondary collision products.

An atmosphere on Io with a basal pressure of 10^{-8} bars will have roughly 10^{18} molecules cm^{-2} in the total atmospheric column. Assuming a molecular cross section of 10^{-15} cm^2, it would appear that each proton of MeV energy could eject up to several hundred molecules from Io's atmosphere. For a proton flux of 10^8 cm^{-2}-sec (the flux on Io if one includes the protons of energy as low as 0.5 MeV) the resulting atmospheric mean residence time would be $\sim 10^7$ seconds or less than one year. Scattering by protons of ≤ 0.5 MeV energy may also be important in further lowering the mean residence time. Analogous calculations have been performed for Earth's Moon (Herring and Licht, 1959). On Earth's Moon, however, $\mathbf{v} \times \mathbf{B}$ sweeping has been found to be a more important atmospheric removal mechanism than elastic collision (Hodges *et al.*, 1974). This may also be the dominant mechanism for atmospheric escape from the Galilean satellites since the velocities of the magnetic field relative to the satellites are high (Carlson, 1975).

The sodium in the atmosphere and cloud about Io may be excited by several mechanisms. Resonant scattering of sunlight is the most important (Trafton *et al.*, 1974; Matson *et al.*, 1974; Bergstralh *et al.*, 1975). If Io has a surface pressure of 10^{-10} bars or greater, significant excitation by atmospheric processes could occur. Several models involving nitrogen have been suggested and studied by McElroy *et al.* (1974) and McElroy and Yung (1975). Finally, as noted by Nash *et al.* (1975), excitation of sodium atoms also occurs in the process of sputtering itself and in this case, D-lines are emitted within a few millimeters of the surface. The amount of the contribution cannot yet be calculated because of the uncertainties in the proton and ion fluxes. Telescopic spectra of Io have been cited as evidence in favor of an atmospheric or surface contribution. Parkinson (1975) and McElroy and Yung (1975) argue that the D_2 and D_1 lines occur in a ratio which cannot be explained entirely by resonant scattering of sunlight. On the other hand, data presented by Bergstralh *et al.* (1975) showing a strong correlation between orbit phase and emission implies that any constant atmospheric or surface contribution must be small.

Finally, the characteristics of the Na cloud permit a preliminary attempt at an important, if somewhat qualitative, steady state calculation to be made. It is known that emission comes from a cloud extending to ~ 10 Io radii (1.8×10^4 km). It is also likely that a large proportion of the Na atoms are traveling outward with a velocity comparable to the escape velocity (Matson *et al.*, 1974; Parkinson, 1975). In one extreme case, that of an atmosphere where outward radial streaming occurs with uninhibited ballistic trajectories and without a substantial fraction of the atoms returning to Io, it would be necessary to replenish the column—which consists of $\sim 10^{11}$ Na atoms cm^{-2} (Trafton et $al.$, 1974)— approximately every $1.8 \times 10^4/3 = 6 \times 10^3$ sec. Recall from above that the upper limit on the mean residence time implied by the photoionization rate is only $\sim 1 \times 10^6$ seconds and it could be as short as 1×10^5 seconds (Carlson, 1975). Thus, depending on whether escape or photoionization dominates in removing atoms, the supply rate should be sufficient to supply between $1 \times 10^{11}/(6 \times 10^3)$ and $1 \times 10^{11}/(1 \times 10^5)$ or between 2×10^7 and 1×10^6 Na atoms cm^{-2} sec^{-1}. The necessity for a Na source of about this magnitude has also been suggested by Macy and Trafton (1975). Based on laboratory sputtering data, Matson et $al.$ (1974) estimated that a surface with $\sim 10\%$ Na could yield between 1×10^6 and 1×10^7 Na cm^{-2} sec^{-1} if bombarded with protons of flux and energy comparable to that expected near Io's surface. We consider this to be in good agreement with the steady state balance despite all the uncertainties in the calculation. While materials with, say, a factor of 10 lower Na concentration cannot yet be discarded on this basis alone, it is clear that models involving materials with only trace concentrations of Na would encounter serious difficulties in supplying the necessary flux. More significant statements concerning surface composition may be possible based on future observations of other lines as will be discussed in the next section.

Hydrogen Torus

In addition to the sodium cloud, a hydrogen torus was observed by the uv photometer on Pioneer 10 (Judge and Carlson, 1974; Carlson and Judge, 1974), extending some 120° around Io's orbit. The existence of this torus and the absence of similar tori around Europa or Ganymede raise the question of the source of hydrogen on Io. McElroy et $al.$ (1974) suggested that photodissociation of subliming NH_3 could supply the necessary hydrogen, but the rate of supply requires large amounts of ammonia frost on Io's surface. In view of some of the arguments advanced against the presence of so much ammonia frost (Matson et $al.$, 1974), other sources need to be investigated. Some possibilities include the degassing of H_2 and/or NH_3 from an interior heated to very high temperatures by radionuclide decay (see Fig. 17.2), the higher proton flux at Io and the effects of surface composition on hydrogen retention and escape. For instance, ice-covered surfaces (such as those of Europa and Ganymede) may be relatively more efficient chemical sinks for incoming protons and degassed hydrogen than the salt regolith we propose for Io. The latter might become saturated with protons in < 1 my; in which case a hydrogen outflux equal to the proton influx might be sufficient to sustain the torus. McDonough (1975) has suggested that a co-rotating plasma may be both source and sink for the hydrogen. However, he leaves open the possibility that a hydrogen source on Io may be required.

FUTURE WORK ON IO

It is obvious that the next important clues to the source of the sodium about Io will come from observations of its temporal and spatial extent. The new vidicon and other two-dimensional photometry systems can be combined with very high dispersion (e.g., 8 mm/Å) astronomical spectrographs, permitting pictures of Io in each of the D-lines to be obtained. Coupled with spectrographic observations to obtain line profiles and possibly radial velocity distributions much diagnostic data should be obtained and current theories will undoubtedly be subjected to severe testing. Observations of eclipses of Io at many wavelengths, especially for the sodium lines, could be important in determining the relative contribution of any non-resonant scattering processes to the sodium emission.

There are other specific studies which can be done which will immediately advance our understanding of Io. The sputtering process is not very selective and other elements from the surface of Io are probably also present in the cloud about Io. For those elements with resonant lines, it is possible to predict their intensities relative to sodium for different assumed surface compositions. Table 17.4 shows anticipated results for various possible surface compositions. In this calculation it is assumed that sodium is excited by near continuum levels of solar energy while the line in question is excited only by the residual intensity in the core of the appropriate Fraunhoffer line. The calculation is thus a conservative estimate of contrast with the undisturbed solar continuum compared with the observed maximum sodium emission. The actual surface brightness which could be observed from one of these lines depends, of course, on the solar energy available at the frequency of the resonant line. Thus, many of the lines in the ultraviolet will be difficult to observe, even from above the atmosphere, due to the low value of the solar ultraviolet continuum. In a detailed cloud model other factors, such as the relative lifetimes of the different atomic species against ionization, would have to be included for an accurate prediction of line strength; the values given here are only estimates to illustrate the uses of this technique and to suggest further lines of inquiry.

Upon inspection of the predictions in Table 17.4, it is clear that serious searches should be carried out for the resonant lines of Ca, Mg, Si and K. Of particular interest is the Mg line at 2852Å which should be bright enough to observe from Earth satellites or high altitude balloons. Simultaneous ground-based observation of the Na D-lines would permit the determination of the ratio of Mg to Na. This is an important ratio for the determination of surface composition, as it varies widely for the various candidate surface materials. (It will also be important to determine whether or not Mg is significantly fractionated relative to sodium during supply of both elements to the cloud about Io or during loss from the cloud.)

The two very different electron density profiles (nightside and dayside) observed by Pioneer 10 and the role of the atmosphere as a temporary reservoir for

TABLE 17.4
Predicted Line Contrast With Respect to Undisturbed Solar Continuum
(Normalized to Na = 1.00) for Various Materials

Element	Wavelength	Atomic Abundance (and Line Contrast*) for Chondrites		Atomic Abundance (and Line Contrast*) for Basalt		Atomic Abundance (and Line Contrast*) for Orgueil Leach Evaporite	
Na	5896Å	1.00	(1.00)	1.00	(1.00)	1.00	(1.00)
Mg	2852Å	26	(13.)	2.5	(1.3)	2.3	(1.2)
Ca	4227Å	1.5	(0.1)	2.3	(0.2)	0.34	(0.03)
Si	2516Å	31	(0.3)	10	(0.1)	0.26	(0.00)
K	7665Å	0.2	(0.1)	0.3	(0.2)	<0.1	(<0.05)

*Estimates based on atomic abundances and the oscillator strengths as discussed by Greenstein and Arpigny (1962). See text.

escaping material make additional radio occultations high priority objectives. It is becoming clear that many occultations are needed and that this will probably be a task best suited for a Jupiter orbiter mission. Thus, a small, detachable Io-orbiting transmitter may be a justifiable experiment.

The small innermost satellite Amalthea (J5) is an attractive target for a spacecraft imaging experiment. Amalthea is subjected to even more intense proton fluxes than Io and atomic line emissions caused directly by sputtering could be strong (cf. Nash *et al.*, 1975).

Many more laboratory experiments should be done with the proton and ion irradiation of materials of geologic interest. The irradiation effects on the spectral reflectivities of salts and especially sulfates would have direct application to Io, including its dark polar caps. Simultaneously, reduction of sulfate to elemental sulfur by any proton flux could be investigated using electron spectroscopy and other surface analysis techniques. Solid state luminescent phenomena could be important. Will it be possible to construct a theory to compete with the frost theory for Io's eclipse brightening but based only upon solid state effects? For example, one could argue that during the eclipse when the dark polar surfaces are cold, many of the irradiation-produced defects do not anneal and remain "frozen" in the crystal structure until they are exposed to solar uv radiation, whereupon annealing of the defects may stimulate the crystals to luminesce. These questions can be pursued in the laboratory by studying irradiation effects as a function of temperature. The sputtering of ices with protons and heavy ions needs to be investigated. The effects caused by the proton fluxes impinging upon the (icy) surfaces of the other satellites—Europa, Ganymede and Callisto—should also be studied.

[*Note added in proof:* Potassium has also been observed in emission around Io (Trafton, 1975c, and Münch, personal communication, 1976). The presence of ionized sulfur, S^+, has been reported by Kupo *et al.* (1976) on the basis of faint spectral lines present on plates they obtained. Thus, H, Na, and K are known to be present in the cloud about Io, and sulfur is very likely to be there too.

The basal pressure of Io's atmosphere, derived from electron densities determined by radio occultation, may be $\lesssim 3 - 5 \times 10^{-11}$ bars. This value is two to three orders of magnitude less than was given by Kliore *et al.* (1974, 1975). This new pressure estimate comes from a reinterpretation of the Pioneer 10 radio occultation data in light of current knowledge of the low energy particle fluxes in the Jovian magnetosphere (Johnson *et al.*, 1976).

The presence of sodium, potassium, sulfur and several other elements in abundance on Io's surface is predicted by the hypothesis that the satellite's surface consists of evaporite mineral deposits. Thus the new discoveries of K and S^+ emissions are compatible with our model. Of perhaps greater ultimate interest is the failure thus far to detect Ca, Mg, Al, Si, or any of their ions. This serves as a warning that the possible complications discussed in the "Chemical Evolution of Io's Surface" section of this paper are to be taken seriously.]

ACKNOWLEDGMENTS

This paper presents the results of one phase of research carried out at the Jet Propulsion Laboratory, California Institute of Technology, under Contract No. NAS 7-100 (185-50-72-06-00). We thank D. B. Nash, J. Conel, J. S. Lewis, R. Carlson, and C. Pilcher for helpful discussions and criticism. We also thank E. Olsen, who supplied the sample of Orgueil.

PICTURE OF GANYMEDE

Tom Gehrels
University of Arizona

Images of Ganymede (J3) taken by Pioneer 10 in blue and red light are presented.

Pioneers 10 and 11, using spin stabilized spacecraft with limited data link capability and with spin-scan imaging, have made reconnaissance missions for the more extensive flights with Mariner spacecraft. This is not to say, incidentally, that spin-scan imaging need be inferior to that with television, even in angular resolution, let alone in photometric accuracy (see Gehrels *et al.*, 1972; Russell and Tomasko, 1976).

The mission of Pioneer 10 and its thirteen experiments (there are 14 experiments connected with Pioneer 11) are described elsewhere (25 January 1974 *Science;* September 1974 *Journal of Geophysical Research;* Gehrels, 1974; Gehrels, 1976). One of the instruments is the imaging photopolarimeter (Gehrels *et al.*, 1974) which obtained the red and blue imaging data shown in Figure 18.1. Figure 18.2 shows the intensities in the blue and red scans. The width of the scan elements is 0.5 milliradian which is the size of the aperture in the focal plane of the telescope; the height is also 0.5 mrad (1 millisecond integration—or dwell-time at 5-rpm spin rate). The distance of Pioneer 10 from Ganymede was 7.8×10^5 km, and the resolution therefore is 390 km. There was, however, considerable overlap especially in the horizontal direction in Figure 18.2. The profiles of Figure 18.1 and of the scan numbers in Figure 18.2 have been approximately rectified to yield a circular appearance. Images of all the Galilean satellites produced by the Pioneer photopolarimeters form the Frontispiece of this book.

Lyot observed Ganymede visually (see Plate 40 of Dollfus, 1961b). With Plate 40 turned upside down, to have north up, the sub-spacecraft point (103°, $-18°$) is to the left of center, halfway to the right edge, and a little below the middle line. If the left half of Plate 40 (when upside down) is covered up, a general agreement of the present images with Lyot's drawing is seen. This helps to understand the rim-indents on the right in Figure 18.1 because dark areas are seen in these places in Lyot's drawing (near 0° longitude, $+15°$ and $-30°$ latitude). Incidentally, even though there is general agreement, I am rather dubious about the possibility that the finest details in Lyot's drawing could have been really observed; these details are one-thirtieth or so of the disc's diameter, or 0.06 arcsecs!

[406]

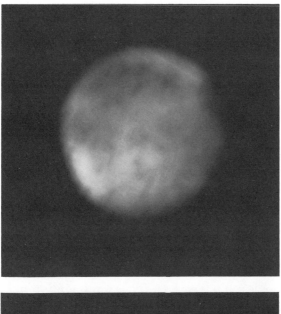

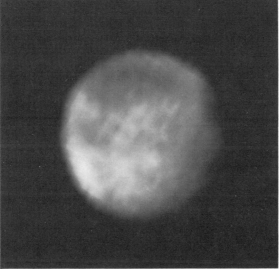

Fig. 18.1. Ganymede from spin-scan imaging by the imaging photopolarimeter of Pioneer 10 on December 3, 1973, 05^h 41^m 30^s–05^h 48^m 02^s U.T. (time interval of reception on Earth). Phase angle 37° (subsolar point is to the left of the sub-spacecraft point). The distance of the spacecraft from Ganymede was 7.8 × 10^5 km. On the top is the blue image (390–500 nm) and red (595–720 nm) is on the bottom. North (i.e., the direction perpendicular to the orbital plane) is up; the rising limb is on the left. Assuming that the rotation is synchronous and with axis perpendicular to the satellite orbit, the sub-spacecraft point is at −18° southern latitude and 103° longitude (zero longitude faces the Earth at superior geocentric conjunction; the longitude increases from right to left when north is up).

```
0  0  0  0  0  0  0  0  0  0  0  0  0  0  0  0  0  0  0  0  0  0  0  0  0  0  0  0  0  0  0  0  0  0  0

0  0  0  0  0  0  0  0  0  0  0  0  0  0  0  0  0  0  0  0  0  0  0  0  0  0  0  0  0  0  0  0  0  0  0

0  0  0  0  0  0  0  0  0  1  1  1  1  2  2  3  3  4  4  4  4  1  1  0  0  0  0  0  0  0  0  0  0  0  0

0  0  0  0  0  0  0  1  1  1  2  5  7  10 11 13 12 13 14 15 13 10  7  4  2  1  0  0  0  0  0  0  0  0  0

0  0  0  0  0  0  1  3  5  8  11 14 17 18 18 17 17 17 18 18 18 15 10  8  6  4  2  1  0  0  0  0  0  0  0

0  0  0  0  1  2  4  9  12 14 15 15 14 15 15 14 13 13 15 16 15 14 11 10  8  6  5  3  0  0  0  0  0  0  0

0  0  0  1  2  5  9  12 14 13 11 10 11 12 11 12 11 12 13 14 15 14 14 12  9  7  6  5  3  2  1  0  0  0  0

0  0  0  1  6  9  11 11 10 10 11 10 12 12 13 14 13 15 14 15 15 16 15 14 12 11  9  8  6  5  2  1  0  0  0

0  0  1  6  8  11 10 11 10 11 12 13 11 13 15 14 15 15 16 17 15 14 14 14 13 12 11 10  9  7  5  2  0  0  0

0  1  4  9  11 11 10 11 11 12 13 13 12 15 15 16 17 17 18 16 15 15 14 14 13 12 12 11 10  9  7  3  1  0  0

0  3  7  10 12 11 11 12 13 13 13 14 15 16 17 17 17 17 16 16 14 14 13 14 13 12 12 11 10 10  7  4  2  0  0

0  4  9  13 13 12 12 13 13 13 12 14 15 17 16 17 17 17 16 15 14 13 14 14 12 12 11 12 10  9  8  5  2  0  0

0  6  12 14 14 14 14 15 13 14 16 16 16 17 17 17 16 17 17 17 16 16 15 14 15 13 12 11 11 10  8  6  2  1  0

0  4  12 18 16 17 16 15 15 15 16 17 17 18 18 17 16 15 16 17 15 14 15 16 15 14 12 12 12 11  8  5  3  0  0

0  4  12 18 18 16 16 16 16 15 15 15 16 17 18 17 17 17 18 19 18 18 17 16 15 15 13 12 12 12 10  6  3  1  0

0  3  10 13 17 15 15 15 14 15 16 16 16 18 19 19 18 20 19 20 20 19 17 16 16 15 13 13 13 12  9  4  1  0  0

0  1  6  12 15 15 15 14 14 16 15 16 16 16 18 19 21 21 18 18 18 17 17 17 16 16 13 12 11  9  6  2  1  0  0

0  0  3  7  13 16 16 16 16 17 16 18 18 19 18 20 21 21 20 17 16 17 16 17 16 14 14 13  9  7  4  1  0  0  0

0  0  1  3  8  14 19 19 20 21 20 19 19 19 18 19 20 20 20 18 18 18 16 17 15 13 12 10  8  5  3  0  0  0  0

0  0  0  0  2  6  14 17 23 24 26 23 20 18 20 19 19 20 18 17 17 17 14 14 13 11 10  7  5  2  1  0  0  0  0

0  0  0  0  0  1  4  9  16 22 25 26 23 22 21 20 19 17 18 17 16 15 13 11  9  9  7  4  2  1  0  0  0  0  0

0  0  0  0  0  1  0  2  5  11 17 20 23 21 22 20 18 15 16 15 14 12 10  9  6  4  2  1  0  0  0  0  0  0  0

0  0  0  0  0  0  0  0  2  4  7  12 13 16 16 13 12 12 10  9  7  6  3  2  1  0  0  0  0  0  0  0  0  0  0

0  0  0  0  0  0  0  0  0  0  0  1  2  2  4  4  5  5  5  2  2  1  1  0  0  0  0  0  0  0  0  0  0  0  0

0  0  0  0  0  0  0  0  0  0  0  0  0  0  0  0  0  0  0  0  0  0  0  0  0  0  0  0  0  0  0  0  0  0  0
```

Fig. 18.2. Intensities on arbitrary but linear scales for Ganymede's blue image (left) and red (on the right). The numbers are in columns as obtained in scan lines that can be seen in Figure 18.1; north is rotated by 16° counterclockwise from straight up. A compression in the horizontal direction by about a factor 3 was used in order to present a round appearance.

Intensity calibrations have not as yet been made, but an approximate calibration is obtained as follows. It is seen in Figure 18.2 that the numbers are rather evenly distributed—about 15 for blue light and 22 or 23 for red—ignoring terminator effects and also ignoring the numbers near the rim where the aperture may not have been filled. With an average geometric albedo of about 40%, the maximum and minimum brightness levels seen in Figure 18.2 represent about 62% and 28% geometric albedo. Considering the resolution of 390 km, it appears that there is a gradual variation in the mixture of bright and less bright material.

We computed colors, red/blue. We made a work plot as a function of bright-

```
0  0  0  0  0  0  0  0  0  0  0  0  0  0  0  0  0  0  0  0  0  0  0  0  0  0  0  0  0  0  0  0  0  0  0  0  0  0
0  0  0  0  0  0  0  0  0  0  0  0  0  0  0  0  0  0  1  1  1  1  0  0  0  0  0  0  0  0  0  0  0  0  0  0  0  0
0  0  0  0  0  0  0  0  0  0  0  2  3  6  6  9  8 11 10 12  9  7  4  2  0  0  0  0  0  0  0  0  0  0  0  0  0  0
0  0  0  0  0  0  0  1  3  5  8 12 16 20 21 22 21 22 23 24 21 18 12 10  5  3  1  0  0  0  0  0  0  0  0  0  0  0
0  0  0  0  0  0  2  7 12 15 20 20 23 21 21 19 20 22 21 22 22 21 16 14 11  7  4  2  0  0  0  0  0  0  0  0  0  0
0  0  0  0  1  5  9 17 18 18 19 17 19 17 18 17 17 19 20 21 21 22 20 16 12 10  8  6  3  1  0  0  0  0  0  0  0  0
0  0  0  1  4 11 15 17 17 16 15 15 16 17 16 17 18 19 21 21 21 22 23 20 16 13 12  9  7  4  1  0  0  0  0  0  0  0
0  0  1  6 10 16 16 16 14 15 17 16 18 19 19 21 21 24 22 25 22 23 22 22 20 17 15 13 11  8  5  2  0  0  0  0  0  0
0  0  3 12 14 15 15 16 16 17 17 19 17 20 21 23 23 25 25 25 22 23 21 21 20 17 18 16 13 12  6  3  1  0  0  0  0  0
0  1  7 13 17 17 16 17 18 19 19 22 21 22 25 24 25 24 23 23 22 21 22 21 20 17 17 16 14 12  9  6  1  0  0  0  0  0
0  4 12 18 18 19 18 18 19 19 19 21 23 26 27 25 26 24 23 25 22 21 22 22 19 18 17 17 14 13  9  6  3  0  0  0  0  0
0  6 16 21 19 19 19 18 19 21 21 23 23 26 25 24 26 25 25 24 25 23 21 22 20 19 17 16 16 13  9  7  3  1  0  0  0  0
0  9 19 25 25 22 23 23 21 23 25 25 25 26 27 26 23 24 23 24 21 22 23 22 22 20 18 16 16 13 10  7  3  1  0  0  0  0
0  7 18 25 27 26 25 26 24 22 24 23 22 25 24 25 23 24 25 24 26 24 24 24 22 22 19 17 17 14 12  8  3  0  0  0  0  0
0  4 16 24 25 24 24 23 22 23 24 23 24 27 26 25 24 25 26 28 29 29 24 23 24 22 18 18 15 15 11  7  3  0  0  0  0  0
0  1  9 19 25 23 22 21 21 21 23 21 22 25 25 25 25 27 27 26 27 24 24 24 22 21 21 17 16 16  8  5  2  0  0  0  0  0
0  0  6 12 21 24 23 23 22 21 22 23 24 24 28 28 30 29 26 24 24 24 22 24 21 20 18 14 14 11  7  2  1  0  0  0  0  0
0  0  1  6 16 21 26 26 25 26 27 27 27 27 26 28 29 29 28 27 25 24 24 23 22 21 17 15 12  9  5  1  0  0  0  0  0  0
0  0  0  0  7 15 25 28 31 31 31 28 25 25 26 25 28 26 26 27 25 25 23 22 19 18 15 12  8  5  2  0  0  0  0  0  0  0
0  0  0  0  1  4 11 20 28 32 36 34 30 28 25 26 26 24 26 26 23 21 20 17 16 13 11  8  5  1  0  0  0  0  0  0  0  0
0  0  0  0  0  0  2  5 14 22 28 32 32 30 30 27 23 21 23 21 22 20 19 14 12  9  6  3  0  0  0  0  0  0  0  0  0  0
0  0  0  0  0  0  0  1  6 13 18 25 24 27 25 22 19 18 18 15 14 11  8  5  4  1  0  0  0  0  0  0  0  0  0  0  0  0
0  0  0  0  0  0  0  0  1  2  5  7 12 11 13 11 11  8  8  5  3  2  0  0  0  0  0  0  0  0  0  0  0  0  0  0  0  0
0  0  0  0  0  0  0  0  0  0  0  0  0  1  1  2  1  1  1  1  0  0  0  0  0  0  0  0  0  0  0  0  0  0  0  0  0  0
0  0  0  0  0  0  0  0  0  0  0  0  0  0  0  0  0  0  0  0  0  0  0  0  0  0  0  0  0  0  0  0  0  0  0  0  0  0
```

ness, $(R + B)/2$, and it shows a scatter diagram without correlation. Also a plot of R/B values over the surface shows no conspicuous color features, with the possible exception of the northernmost part of the dark area in the north that has R/B ~ 1.21 compared to the overall average of 1.44.

We have preliminary results on polarimetry of Ganymede obtained during the flyby of Pioneer 10. The polarization at quarter phase, in blue light, is only about 4%, which is much lower than that of the Moon (about 10%). Combined with ground-based polarimetry (Fig. 10.6), the curve looks quite similar to that of Figure 29B of Lyot (1929) obtained in the laboratory for powdered crystals of NaCl. This, however, is not a unique identification. Regolith areas (like lunar highlands) and frosted areas, randomly distributed over the surface of Ganymede, may explain the polarimetry just as well.

The ground-based infrared spectroscopy is quite convincing for the case of water frost (Chapt. 11, Johnson and Pilcher). But do these satellites have enough atmosphere to sustain the condensation of water frost? How unique is the identification? Could the spectroscopic observations be fitted with salts that have water molecules bound to them? Could there be a fairy-castle surface texture made up of clathrate hydrates?

ACKNOWLEDGMENTS

The satellite images for this chapter and for the Frontispiece of this book were produced by the imaging team of W. Swindell, J. W. Fountain, Y. P. Chen, and P. H. Smith. The intensities for the plots of Figure 18.2 were computed by M. S. Matthews. It is a pleasure to acknowledge the general support of the Imaging Photopolarimetry Team and of the NASA/Ames Pioneer Project Office.

DISCUSSION

CLARK CHAPMAN: Is there a smooth distribution of different albedos on Ganymede or is there a bimodal frequency histogram (corrected, of course, for the terminator shading)? A bimodal histogram might lend support to the two-component (water ice and silicate) surface model proposed, for instance, by Pilcher *et al.* (1972). Of course the water ice and silicate patches might occur on a spatial scale far lower than your resolution limit. But they might just as well be identical to the bright and dark patches shown in your imagery.

TOM GEHRELS: The histograms—omitting the numbers that are close to rim and terminator—show no bimodal frequency distribution.

DALE CRUIKSHANK: The Pioneer 10 images of Ganymede show a partial bright arc on the sunrise limb. Is this the same morning arc that Dollfus (1961b) reported from visual observations and which he attributed to an atmosphere?

TOM GEHRELS: Dollfus (1961b) states "Ganymede has shown whitening at the sunrise limb, covering permanent surface detail; this might be an indication of temporary light deposits on the ground or morning haze." The reader should judge for himself if he sees the effect here. But do consider that, with 37° phase angle and synchronous rotation, the Sun rose already some 17 hours before a feature is seen on the rising limb.

CARL PILCHER: It seems to me that a word in defense of water frost may be necessary to respond to the last comments in the chapter. The fact is that we see absorptions on Europa and Ganymede that can only be explained by the presence of substantial amounts of solid-phase water on the surfaces of those objects. The material may very well not be *pure* water—for example an impurity of 10 or 20% CH_4 or NH_3 couldn't be detected—but water *has* to be there. You are correct in saying that the water will evaporate and that a source to replenish the surface water is required, but, for an object like Ganymede with a density of 1.9 gm-cm^{-3}, all you need to do is to get some of the interior bulk water up to the surface.

This could happen via internal melting and degassing in the manner suggested for Io by Fanale *et al.* (Chapt. 17), or just by overturn of the surface layers following meteorite impact (Chapt. 25, Consolmagno and Lewis). However, you can't get away from the fact that the infrared spectra of Europa and Ganymede are dominated by condensed water.

TOM DUXBURY: I have made an approximate comparison of your observed surface brightness of Ganymede with a Lambertian surface on a pixel (i.e., picture element) by pixel basis for both the red and blue channels. Both channels indicate that the polar regions are significantly brighter than expected for a Lambertian surface. Possible explanations include Ganymede having a very rough surface with high albedo and having ice (or at least more ice) at the poles.

SATURN'S RINGS: A NEW SURVEY

Allan F. Cook and Fred A. Franklin
Harvard College Observatory and
Smithsonian Astrophysical Observatory

Observations of radar returns from Saturn's rings, together with radio interferometry of their absorption of radiation from the disk, combine to require an effective radius of ring particles of about 6 cm or larger. We suggest that the ring particles may also include, in addition to the known constituent ice, a mixture of the clathrated hydrate of methane and ammonia hydrate. The disappearance of the rings in emission longward of 1 mm in wavelength may, in the end, tell us more about the structure of individual ring particles than about their sizes. The interpretation of this phenomenon would be much improved by far-infrared and radio measurements of the absorption of methane clathrate and especially of ammonia hydrate near 90°K. A two-density model for ring particles is possible in which a matrix of low density ($\lesssim 0.4$ g-cm^{-3}) contains many nodules of higher density (~ 0.9 g-cm^{-3}) ice particles; in this case, radii nearly as large as the observed ring thickness would be possible. Improved resolution in radio observations at 21 cm or, if necessary, at longer wavelengths for narrow ring openings is perhaps the most useful method for determining upper limits on the particle size.

Since the publication of our previous survey of Saturn's rings (Cook *et al.*, 1973), a remarkable amount of new information concerning the rings has been obtained and analyzed. [See also the review by Pollack, 1975.] It is our plan here to consider interpretations of these new data only and to refer to material covered in the earlier review merely when necessary. We should like to begin by congratulating those who have contributed so much in the last few years (and by so doing, to supply further encouragement to them!).

A major concern of this review will be to obtain—or comment on—characteristic sizes (or limits on the sizes) of ring particles that are consistent with all present observations. It is well to recall that various determinations of particle radii may, inasmuch as they sample very different properties of ring particles, give very different results.

The organization of this review is to consider critically in successive sections each method of particle-size determination. In a final section, we shall check the consistency of the derived sizes with other ring properties. The radiometry of the rings is reviewed in some detail in Chapter 12 by Morrison; brightness temperatures of Ring B are listed in Table 12.5.

[412]

ABSORPTION FEATURES IN THE INFRARED

Pollack *et al.* (1973b) have fitted the infrared spectrum of the rings obtained by Kuiper *et al.* (1970) to a family of theoretical spectra of pure water ice as derived from a Mie scattering calculation. They determine a characteristic small-est dimension, i.e., a depth to the first internal reflection of $\sim 30 \mu$, and they identify this size with that of the smallest solid element of ice.

RADAR OBSERVATION IN BACKSCATTER

Goldstein and Morris (1973) have detected the rings in backscatter with the radar at Goldstone at a wavelength of 12.6 cm. They find, for an opaque Ring B and a half-covered Ring A, an equivalent Bond albedo for isotropic scattering of about 0.80 ± 0.08. [The ring dimensions are listed in Table 1.3.] But they actually measured the geometric albedo in direct backscatter; that is, the Bond albedo is to be divided by the isotropic phase integral, $q = 4$, whence the observed geometric albedo p is 0.20 ± 0.02, a factor of about 5 smaller than that at optical wavelengths for Ring B.

If we assume that the particles can be reasonably well represented by spheres, it is possible to use this geometric albedo to derive a lower limit to the particle size. A sphere with fine structure on a scale small compared to the wavelength of observation may be regarded as a uniform medium with a reduced index of refraction (van de Hulst, 1957, pp. 36–39). If the density is low, we employ the expression for Rayleigh scattering (van de Hulst, 1957, pp. 85–91) to find for p

$$\lim_{\epsilon \to 0} p = \frac{3}{4}\epsilon^2 x^4 \ , \quad \epsilon \equiv n - 1 \ , \quad x \equiv \frac{2\pi r}{\lambda} \ , \tag{1}$$

where n is the index of refraction, r the radius of a particle, and λ the wavelength. For large spheres, the geometric albedo is given by (van de Hulst, 1957, pp. 200–205)

$$\lim_{\substack{\epsilon \to 0 \\ \epsilon x \to \infty}} p = \epsilon^2 \ . \tag{2}$$

Equation (2) is a good approximation even for $\epsilon = 0.333$, for which van de Hulst (p. 232) quotes $p = 0.098$ while this expression gives 0.111. Other values in the range $1 < x < 5$ for $\epsilon = 0.333$ can be used to calculate p from van de Hulst (p. 153); for all these, $p < 0.06$.

We do not escape from this characteristically weak backscatter unless n lies between $\sqrt{2}$ and 2. In this case, the backscatter becomes intense (van de Hulst, 1957, pp. 130–152, 249–255), owing to the occurrence of a strong glory in the scattering pattern for sufficiently large particles. For example, at n = 1.55, corresponding approximately to a density of 0.6 g-cm^{-3}, van de Hulst's Figure 25 (p. 152) yields x between 3.6 and 4, with p rising to 0.24 at the latter value. For smaller x, the derived p is substantially less than 0.20. Accordingly, we reach our first conclusion: r $\lesssim$ 8 cm—smaller radii give too small a geometric albedo. A second conclusion is that loosely packed snow particles of arbitrary size (as long as they remain loosely packed) have densities too low to give an index of refraction as large as $\sqrt{2}$ in the radio region. Thus, we require densities similar to (or larger than) those found for a mixture of amorphous ice and clathrates of methane studied by Delsemme and Wenger (1970).

There will also be many solutions providing strong radar reflectivity for particles with large radii. In that case, the backscatter acquires its strength from a cylindrical return of illuminating radiation refracted through the front of the sphere and reflected from the axial point at the back to return along the cylinder of arrival. At an index of refraction of $\sqrt{2}$, this cylinder grazes the circumference of the sphere; at an index of refraction of 2, the cylinder has a vanishing radius about the axis of illumination. At large radii, internal absorption will weaken the return. Also, irregular deviations from a strict spherical shape will defocus the return and do so more effectively as the particle radius is increased. This is the strong backscatter by dielectric spheres pointed out by Pettengill and Hagfors (1974). Further discussion is put forth in Chapter 12 by Morrison.

The strength of the radar return thus imposes a high density. This can be taken to suggest a mixture of water ice, clathrated hydrates of methane, and probably ammonia hydrates, as suggested for the satellites of Jupiter and Saturn by Lewis (1971a). It also fits in with the drop in the brightness of the rings toward the ultraviolet found by Lebofsky et al. (1970), which, as they remark, requires the presence of something more than water ice.

The foregoing analysis appears to imply an upper limit associated with defocusing by the nonspherical shapes of irregular particles. The limit would occur very approximately at a radius of about a meter, unless the surface roughness is on a scale substantially less than the observed wavelength of 12.6 cm. Evasion of the one meter limit by postulating a many particle thick layer of cm-sized bodies is probably untenable because the rings have very probably collapsed, approaching a near monolayer in thickness (cf. Brahic, 1975). Thus, the "bright cloud" model of Pollack et al. (1973b) cannot be easily invoked here.

Can a collapsed bright cloud or snowbank be invoked instead? Snowbanks are also bright in reflection through multiple scattering. In the present case, this could be done only by filling a large particle of low bulk density with nodules or small spheres. The large particles could then have diameters up to the thickness of the rings deduced from their edge-on brightness determined by Focas and Dollfus (1969) and Kiladze (1969); that is, radii up to about 0.8 km would be possible.

This model, while possible, is more contrived than one of separate smaller particles, but it should be kept in mind.

THERMAL EMISSION FROM THE RINGS

Pollack *et al.* (1973b) have singled out the measurement of the disk at 1 mm by Rather *et al.* (1974) as being the most meaningful in the transition region between brightness temperatures nearly independent of wavelength (less than 1 mm) and invisibility of the rings in emission (greater than 2 mm). The result for the rings is a brightness temperature of $35 \pm 15°K$, which implies an emissivity of 0.4 ± 0.2 (see comments in Chapt. 12 by Morrison). We take the absorption per unit length of ice Ih from the discussion of Whalley and Labbé (1969) (their Fig. 2, with proportionality to the square of the temperature) by using the temperature of 90°K indicated by Pollack *et al.* (1973b). This absorption is 0.08 cm^{-1}, which yields a particle radius of 5 ± 3 cm for solid ice or 7 ± 4 cm for a density of 0.6 g-cm^{-3}. These figures may be upper limits because some of the probable other constituents may be more absorbing than ice. For this interpretation, the emissivity at 2 mm is 0.2 ± 0.1; that is, the rings are approaching invisibility at 2 mm.

We note that the model of low-density spheres, each filled with nodules of high density, would limit the penetration of any illumination at these wavelengths. This restriction would limit the absorption and, by Kirchhoff's law, limit the emissivity no matter how large the particles are. If the particles were constructed in this way from pure water ice, the radii found above would be representative of the linear dimension corresponding to an optical depth of unity in the scattered radiation, and the numerical values of the radii would not differ markedly from the radii of the nodules of higher density embedded in the low density matrix of the ring particles. The roughness associated with these nodules would be of small scale at the radar wavelength, so that isotropic reflection would be possible and would give a geometric albedo near 0.25, which is not in disagreement with the observed reflection.

We cannot expect pure water ice to be the sole constituent. We also note that experimental work on the clathrate hydrate of ethylene oxide indicates no change from ice Ih for absorption at long wavelengths such as 1 mm (Bertie *et al.*, 1973). This suggests that $CH_4 \cdot nH_2O$ is not likely to have a significantly different absorption. The crystalline hydrates of ammonia ($2NH_3 \cdot H_2O$, $NH_3 \cdot H_2O$, $NH_3 \cdot 2H_2O$) perhaps offer a better prospect for a different absorption.

INTERFEROMETRY OF SATURN ALONG THE POLAR AXIS

Briggs (1974b; 1974c) has done interferometry at the National Radio Astronomy Observatory in which he resolves Saturn along its polar axis to look for absorption of planetary radiation by the rings. He has worked at 3.7, 11.1, and 21.3 cm. His results require that particles of solid ice as ring particles have characteristic radii ⩾ 3 cm, which at a density of 0.6 g-cm⁻³ corresponds to 4 cm. Greater resolution and observations as the rings close up are very desirable at 21.3 cm, as absorption by the rings and limb darkening could not be separated by Briggs at this wavelength. Figure 12.10, Morrison, also considers Briggs' work.

A MODEL OF REPRESENTATIVE PARTICLES
TO FIT THE OBSERVATIONS

The two lower limits on the particle radius from the radar observations and the radio extinction measurements leave us with radii $\gtrsim$ 6 cm. The interpretation of the rapid decrease of the emission of the rings with increasing wavelength near 1 mm seems to us ambiguous; we have two alternatives. First, the absorption by the actual ring particles near this wavelength may be much larger than the value for pure ice of 0.08 cm⁻¹. This could lead to a characteristic diameter for the absorbing elements of ~ 1 mm or less. Presumably, such sizes are to be associated with surface structures on larger particles. Second, the absorption, even with impurities present, still may be not too much larger than 0.88 cm⁻¹, whence the particle radius is ~ 7 cm. Inasmuch as this is an upper limit, we would appear to have, in conjunction with the above lower limit, determined the particle size at radii of about 7 cm. There is, we feel, still a caveat even here. We refer to the possible conglomerate particles containing nodules of the order of 10 cm but of indefinite total size.

The strong radar return and the ultraviolet absorption of the rings combine to indicate a relatively high density for the particles ($\gtrsim$ 0.6 g cm⁻³). The particles are likely to be a mixture of water ice, clathrated hydrates of methane, and ammonia hydrates. Two future observations appear to us to be of special importance. First, we again call attention to the possibility of particles consisting of a low density medium and containing dense nodules. Radio observations at 21 cm with resolution greater than that achieved by Briggs (1974b) and a narrower ring opening could eliminate this possibility by setting the particle size at or near our present lower limit. The purpose would be to separate obscuration by the rings from limb darkening on the planet. Second, laboratory measurements in the infrared and through the radio region on the absorption of the methane clathrates and especially of the ammonia hydrates near 90°K would put more plausible limits on the radii of the particles, and measurements of their ultraviolet absorptions might constrain models of the composition of the particles.

IMPLICATIONS

In this final section, we wish to explore the consequences on other ring system parameters of the limiting values of particle radii determined earlier. These parameters are (1) the ring thickness, (2) the opposition effect, and (3) the ring mass.

If the average particle radius is of the order of a few (or even many) centimeters, then we must account for the observed ring thickness of $\sim$ 1 km. That is, what is the mechanism by which particles of this size range acquire, in the presence of collisional dissipation, sufficiently large vertical velocities to define a ring thickness of this amount? The valuable calculations of Brahic (1975) have amply borne out, and greatly extended, the inference of Jeffreys (1947) that partially inelastic collisions would reduce the ring thickness to a near monolayer in a time much less than a year. Calculations that both we and Bobrov (1970) have made argue that perturbations by the satellites cannot, in general, "pump up" a layer of particles in such a characteristic time, in order to produce a ring of the observed thickness. Once again, there appear to us to be two caveats. The first possible exception to this claim occurs at resonance, that is, in the neighborhood of Cassini's division, and it seems to us to be of some importance. It may well be that the measured thickness, which can be obtained only when the rings are (nearly) edge on (and the effective optical thickness is very large), merely appears to be a quantity characteristic of the entire ring but is, in fact, a local phenomenon occurring at or near resonance. The final answer as to whether— and if so, under what conditions—particle radii of a few centimeters are consistent with a ring thickness of $\sim$ 1 km must await (1) an extension of Brahic's numerical calculations that include perturbations by the inner satellites, Mimas in particular, and (2) an evaluation of the accommodation coefficient of macroscopic ice particles colliding at low relative velocities. The second possibility concerns a variable parameter, specifically, a variable accommodation coefficient. The possibility is indeed very hypothetical, but it still is conceivable that the accommodation coefficient is velocity dependent, with a positive slope of accommodation coefficient *vs.* velocity, so that the rings hover about a steady-state solution.

Only if the latter of these exceptions operates is the opposition effect—the non-linear brightness pip near zero phase angle—readily explained by mutual shadowing among particles (see Chapt. 9, Veverka).

If, on the other hand, at least some of the particles are larger, with radii approaching 100 m, then the thickness of the ring is easily, almost automatically, generated over a wide range of values of the accommodation coefficient (Brahic, 1975). We have inclined somewhat to the view of a population of large particles partly because of this reason and partly because large particles in Ring B correspond to a ring massive enough to account for the observed radial displacement of the center of the Cassini division from the location of the resonance at ½ Mimas' period. Lumme (1975) has carefully redetermined the position of the center and width of this division, with the result that the displacement just mentioned lies very close to 0.2 arcsec. Our calculations (Franklin *et al.,* 1971) that used this shift to derive a Ring B mass of $\sim 6 \times 10^{-6}$ of Saturn's were in the nature of a reconnaissance and were probably accurate to a factor of 2 or 3. [Greenberg (1976b) has reduced our estimate by a factor of $\sim$ 10.] The techniques developed by Brahic have opened the way to a more precise treatment. In our view, a major forward step in the analysis of ring properties will be the construction of a model that can produce the Cassini division. We envision a computer simulation that finally includes (1) the important perturbing satellites, (2) the oblateness of Saturn, (3) mutual partially inelastic collisions, and (4) self-gravitation of the ring particles. This test will determine whether these four effects can, as we have argued, produce the observed ring profile and, most especially, the width and location of Cassini's division.

Finally, with regard to the opposition effect for a (near) monolayer, we mention two papers that appeared since our last review. Hämeen-Anttila and Vaaraniemi (1974) have shown that such a model does account for the brightness variations of the rings with changing Saturnicentric declination of the Earth and Sun more satisfactorily than other models do, but that, for the monolayer, the opposition effect, clearly shown by the rings, is nearly absent. This lack is not a problem (see Cook et al., 1973), and we would further add the result (Franklin and Cook, 1974) that certain Saturnian satellites (i.e., single bodies that may be ice covered) also appear to exhibit opposition effects comparable in magnitude to that of the ring (cf. Chapt. 9, Veverka).

There remains the subject of the east-west asymmetries in the photometric and radiometric properties of the two-ring association. (Peale, Chapt. 6, discusses the possibility of synchronous rotation of ring particles.) We do not mean to discourage such observations (cf. Cruikshank and Pilcher, 1975) and, while the subject is fascinating, we have the feeling that the appropriate time for its review is still in the future.

There will be some who wish to consider a bimodal or some other more complicated distribution of particle sizes, or perhaps they might prefer particles even more complicated than our suggested two densities of icy matter. We are concerned here with finding the simplest model that fits the observations. In this context, we are loath to propose complications without proposing observations to confirm or deny their existence. We believe that we have maintained our self-discipline in this respect.

ACKNOWLEDGMENTS

We are pleased to thank Tobias Owen and E. Whalley for helpful discussions on the probable chemical composition of the rings and its consequences.

A portion of this work has been supported by the Jet Propulsion Laboratory, California Institute of Technology, under Contract No. 173720-31P with the Smithsonian Astrophysical Observatory, sponsored by the National Aeronautics and Space Administration.

DISCUSSION

THOMAS GOLD: The differential rotation of the rings is subject to a viscous type of decay that will drive the bulk of the material into the planet and a smaller amount outward (cf. Chapt. 7, Burns). The effective viscosity for an assembly of objects (moving under external gravitational influences) gets larger as the masses of the objects become larger. Does this not place a significant limit on the size of the pieces, to retain a reasonable lifetime for the rings?

FRED FRANKLIN: A. Brahic (1975) has shown that the upper limit to which you refer is of the order of several tens of meters. However, his calculations do not (yet) include the effect of the resonances that appear to determine the radial extent of the rings. Particles could—in fact, would—be at least temporarily trapped in such a resonance, and this process would markedly slow the inward or outward spiraling process that you mention (cf. Franklin *et al.*, 1971). Thus, it is conceivable that particles larger than the above characteristic figure could still populate the ring.

The presence of barriers established by resonances could also alter lifetimes derived for particles against the Poynting-Robertson effect (Chapt. 7, Burns). In this case one might expect (without the resonance barriers) that particles smaller than a few centimeters would be lost.

CLARK CHAPMAN: It is of interest for considering evolutionary processes in the rings (e.g., collisions) to know something of the size distribution: Is it Gaussian or is it a power law? If it is the latter, what does the term "representative size" mean and what limits can be set on the population index? Perhaps it is premature to ask some of these questions, but at least I would hope that people will be specific about what they mean. By "representative size," do you mean: More than half the visible cross section is due to particles with diameters within a factor of ten of the "representative size"?

ALLAN COOK: You are quite right: In our view, it *is* most premature to ask such questions when we are still groping for an average or characteristic particle size. A cursory glance at the current literature shows estimates that range between a few centimeters to a few hundred meters. It is wise to remember that the various methods employed for the determination of particle sizes may, in some sense, all be valid, that each one is sampling a different part of the size-distribution function or of the structural elements of a particle. The detailed knowledge that you seek is quite simply and definitely not now available.

GORDON PETTENGILL: J. Pollack (in Palluconi and Pettengill, 1974) suggested that one should attempt observations of occultations by the rings of discrete radio sources as a means of placing further limits on ring-particle sizes.

ALLAN COOK: The problem, though, is the paucity of sufficiently strong radio sources.

TITAN'S ATMOSPHERE AND SURFACE

Donald M. Hunten
Kitt Peak National Observatory

Titan's interior presumably is dominated by a melted NH_3-H_2O solution: the abundance of CH_4 in the atmosphere suggests its presence in the interior also. There also is evidence for H_2, although a high escape rate is implied. N_2, from NH_3 photolysis, could help in retarding this escape. Several lines of evidence suggest the presence of clouds and haze; both frozen CH_4 and organic polymers are plausible. The thermal-emission spectrum shows peaks at 8 and 12 μm, presumably due to CH_4 and C_2H_6 in a warm stratosphere, but no minimum at 17 μm due to a pressure-induced H_2 greenhouse. A decline in brightness temperature at longer wavelengths suggests a weak greenhouse effect from pressure-induced absorption in CH_4. A surface temperature near 125°K is suggested by the weight of the evidence, but a value as low as 90° is not excluded. This review is based on The Atmosphere of Titan, NASA SP-340 *(D. Hunten, ed., 1974).*

Titan, the largest satellite in the solar system, is more like a planet than some planets are. For example, it possesses a deeper atmosphere than that of Mars. Detection of Titan's atmosphere dates from the time of World War II (Kuiper, 1944, 1952), but most of our present knowledge has come since 1972. Trafton (1972a,b) has shown the amount of methane to be much greater than was previously suspected, and has given strong evidence for the presence of H_2. Infrared observers have disclosed a rich, informative thermal spectrum, which gives evidence on the composition and thermal structure of the atmosphere (Allen and Murdock, 1971; Morrison *et al.*, 1972; Gillett *et al.*, 1973; Joyce *et al.*, 1973; Low and Rieke, 1974b; cf. Morrison, Chapt. 12 herein). Titan's ultraviolet albedo is low, which implies the presence of an absorbing haze at high altitudes (Caldwell, 1975; Barker and Trafton, 1973a, 1973b). Its polarization curve speaks to the presence of clouds, perhaps of frozen methane, nearer the surface (Veverka, 1973a; Zellner, 1973; Veverka, Figs. 10.8–10.9). Interpretations have accompanied this flood of data, but perhaps most influential are the ideas of Lewis (1971a) that the interior of an icy body of Titan's size should be melted most of the way to the surface. These ideas are extended by Consolmagno and Lewis in Chapter 25 of this volume.

The present chapter is heavily based on the report of the "Titan Atmosphere Workshop" (Hunten, 1974), which contains a dozen summary papers and some additional material; however, the chapter is intended to be complete in itself, and lays emphasis on the atmosphere and on very recent developments.

Most of the discussion given here is based on a radius of 2500 km, even though it has been obvious that this result may have been biased by limb darkening (Hunten, 1972). The occultation radius of 2900 km (Elliot et al., 1975) may, on the other hand, refer to a level in the atmosphere as much as 200 or 300 km above the true surface.

Later chapters in this book dealing specifically with Titan are by Caldwell (Chapt. 21), who discusses the temperature inversion model, and by Andersson (Chapt. 22), who summarizes seventy-five years of photometry for Titan.

ATMOSPHERIC COMPOSITION

Spectroscopic estimates of abundances necessarily refer to the atmosphere visible above or within a cloud deck, for which there is considerable evidence. They depend also on the total pressure, which is unknown. Trafton (1972b) has given an elaborate analysis of this effect. If methane is the major gas, its abundance is 2 km-A (km-amagat). If there were 20 km-A of N_2, the methane would be only 0.1 km-A, but the total pressure at the cloud top would be raised from 20 to 350 mb. [*Note added in proof:* Lutz et al. (1976) find the methane abundance to be only 80 m-A, which implies the presence of 20–30 km-A of N_2 or Ne.]

Trafton (1972a, 1975a) has also reported a "possible detection" of H_2. Figure 20.1 shows one of his spectra, which is marginal by itself. But repeated measurements of the two lines to be expected seem to show them every time. Thus, the presence of absorption lines at the right wavelengths can hardly be doubted; if they are not due to H_2, they must be due to some other molecule. Indeed, there are plenty of unidentified absorptions in the spectrum, as discussed below; but it would require something of a coincidence for them to simulate H_2.

The reason one must be cautious about accepting the presence of H_2 is the large loss rate implied from such a light object as Titan. This loss, discussed further below, must almost certainly be balanced by an equal source, but no plausible source of the required magnitude has been suggested. Trafton's abundance of 5 km-A must therefore be treated with considerable reservation. Another possible complication is the pressure narrowing or broadening of the H_2 quadrupole lines by a foreign gas like N_2; our actual knowledge of these effects is confined to pure H_2 (Fink and Belton, 1969; Murray and Javan, 1972).

The thermal-emission spectrum (cf. Fig. 12.4) shows strong evidence for CH_4 and C_2H_6; but the abundance of the latter cannot be deduced without a good knowledge of the thermal structure.

For other possible gases we must turn to models of the formation and interior composition of Titan, since observational data are lacking. The work of Lewis (1971a) makes use of the mean densities of the satellites, along with the hypothesis that such bodies are accreted from the condensed fraction of the solar nebula. About 60% of the mass should be a solution of NH_3 in H_2O, and a

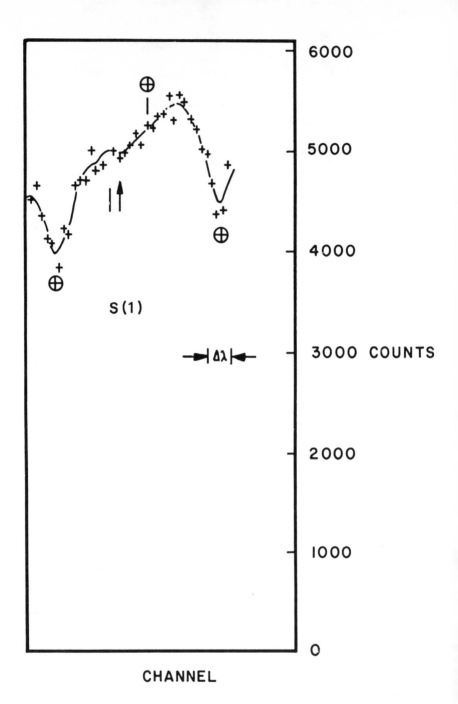

S(1)

$\Delta\lambda$

6000

5000

4000

3000 COUNTS

2000

1000

0

CHANNEL

Fig. 20.1. An example of one of Trafton's spectra, obtained in 1972, showing absorption at the wavelength (arrow) of the 3–0 S(1) line of H_2.

further 5% should be CH_4. The presence of the latter in the atmosphere fits this picture, which however suggests that there is far more methane remaining in the interior or on the surface. No H_2O, and very little NH_3, should be in the atmosphere at the prevailing temperatures. Photolysis of NH_3 and CH_4, and escape of H_2, could produce N_2, as well as a considerable range of other compounds, most of which should condense into aerosols or on the surface. Thus, the best candidate for a third atmospheric gas seems to be N_2. Noble gases might have been retained in small quantities. Cess and Owen (1973) have developed a greenhouse model based on a mixture of H_2 and Ne, although their rarity on Earth does not encourage this idea. Pollack (1973) has suggested that N_2 is improbable because it is not the thermodynamically stable form of nitrogen. To an aeronomer, this idea carries little weight. The stable form, NH_3, is rapidly photodissociated, and free hydrogen is rapidly lost from a light object like Titan. The main questions are whether the surface and lower atmosphere are warm enough to hold a significant quantity of NH_3, and whether the atmosphere is opaque enough to stop radiation below 2300 Å from reaching that region. One might also ask whether the nitrogen ends up primarily in compounds like methylamine, CH_3NH_2, and hydrazine, $(NH_2)_2$, or N_2. A considerable yield of the last seems inevitable; and once the hydrogen is gone, recycling of N_2 to NH_3 is most unlikely. Nitrogen has a bad reputation because it was postulated years ago for Mars and Venus simply by analogy with its abundance on Earth. Satellites in the outer solar system are a different matter altogether: ammonia is expected to be very abundant, and the process outlined above must almost certainly give plenty of N_2. We have therefore a strong *a priori* reason for expecting N_2 on Saturn's satellites.

Trafton (1974a) has reported the presence of additional absorptions that do not seem to be due to CH_4, though many of them are also seen in the spectrum of Uranus. More recently, however, he has proposed (Trafton, 1975b) that most of the lines can be explained by CH_4 absorption, the position on the curve of growth being different for Saturn and Titan. There remains some unidentified absorption in the range 10200–10700 Å. Low and Rieke (1974b) also find Titan to be unexpectedly dark at 1.65 μ. One may speculate that some of the photolysis products are responsible, but such possibilities are limited because most compounds must condense at the low temperatures of Titan's atmosphere. The best candidates are therefore C_2H_6, C_2H_4, and C_2H_2, and perhaps CH_3NH_2 if ammonia photolysis occurs. Not enough is known about the spectra of any of these compounds for an identification or rejection.

In the absence of an atmosphere, Lewis (1971a) would predict a surface of water ice containing CH_4 as a clathrate and NH_3 in solution. At a depth of a few tens of kilometers this medium should be melted (Chapt. 25, Consolmagno and Lewis). In the extreme case of a very deep atmosphere, it is conceivable that melting could extend all the way to the surface; the liquid CH_4 would then float on the H_2O-NH_3 solution. If the (p,T) relation passed through the critical point of methane, the atmosphere would merge into the ocean with no phase change, and could be regarded as having a surface pressure of some 1000 bars (cf. Lewis and Prinn, 1973). The relatively low temperature at 3.7 cm of Briggs (1974a), discussed below, rules out a hot surface, unless the atmosphere or ocean contains a microwave absorber such as ammonia. A more likely situation is a cold surface covered with a layer of photolysis products and their polymers. Such mixtures are usually dark in color, as is Titan.

CLOUDS AND HAZE

Two kinds of aerosol are to be expected in Titan's atmosphere: clouds of solid CH_4, and a photochemical haze (or smog). Veverka (1973a) and Zellner (1973) have published observations of Titan's polarization (cf. Chapt. 10, Veverka, Fig. 10.8), which can be obtained only to a phase angle of 6°. Despite this limitation, it is clear that negative polarization is absent, in striking contrast to observations of the Moon, Mars, Mercury, and many terrestrial solid surfaces. Positive polarization is shown by glassy surfaces and by atmospheric scattering (cf. Chapt. 10, Veverka). However, pure Rayleigh scattering by a gas is ruled out by the low albedo and the absence of the proper wavelength dependence. An absorbing aerosol is therefore suggested, and Veverka further suggests that it should be optically thick to hide the negative polarization from the surface. However, a dark, glassy surface would show positive polarization and could in principle account for part or all of the observations. While such surfaces are not widespread in the inner solar system, Titan may be a different story. Photolysis of its methane atmosphere is expected to yield various polymers, many of them dark—in other words, tar, though of much lower molecular weight than found in terrestrial parking lots. Thus, a dense cloud, though by no means improbable, is not uniquely required by Titan's polarization. The same statement applies to the strong limb darkening observed by Elliot *et al.* (1975). The most natural explanation is an optically deep atmosphere, whether clear or cloudy. But any surface that is not strongly back-scattering is also consistent.

Another line of evidence is the low ultraviolet albedo of Titan, observed at 2600 Å by Caldwell (1975) and above 3000 Å by Barker and Trafton (1973a, 1973b). The model by Danielson *et al.* (1973, 1974; cf. Chapt. 21, Caldwell) shows that an absorbing aerosol is required at stratospheric heights; otherwise the atmosphere would be too bright. One possibility is CH_4 ice, darkened by radiation; but a photochemical smog seems far more likely. Indeed, the stratosphere must be heated by the ultraviolet energy absorbed, and is probably too warm for methane condensation.

Trafton (1974a) has studied the absorption in manifolds of the $3\nu_3$ band of CH_4 near 1.1 μ. Although the line profiles are not resolved, he infers that the lines are "washed out," broader and shallower than would be expected for gaseous absorption alone. The best explanation is that cloud particles, mixed with the gas, are reflecting a continuous spectrum that adds to the line spectrum.

Thus, of the two kinds of aerosol, the photochemical haze at high altitudes is almost certainly present. CH_4 ice offers the most natural explanation of the spectral line profiles and the polarization, but does not seem to be absolutely required if the photochemical cloud is dense enough at low altitudes or if there is a smooth tarry surface.

ESCAPE AND RECYCLING

For an object as small and light as Titan, even temperatures as low as 100°K are not low enough to inhibit rapid thermal escape of hydrogen and helium. Jeans escape has been discussed briefly by Sagan (1973) and Trafton (1972a), both of whom found large fluxes and short lifetimes for H_2. A detailed treatment has been given by Hunten (1973a,b), covering not only Jeans escape but also blowoff and upward diffusion through a heavier gas. Recently Gross (1974) has examined the energy balance of the blowoff state and has concluded that the escape rate may be limited by the rate at which heat can be supplied to the gas. Loss times of 10^3 years or more are estimated. These times are much longer than those proposed by Hunten for pure H_2; but in any case, the stabilizing effect of diffusion is important, giving a loss time of around 10^6 years. The flux is still very large, and a correspondingly large source is needed to maintain a steady state.

If the molecular mixing ratio of H_2 to heavy gas is f_1, the "limiting" diffusive flux has the value

$$\phi_\ell \simeq \frac{b_1}{H_a} \left(\frac{f_1}{1 + f_1} \right) , \tag{1}$$

where H_a is the scale height of the background gas, and b_1 is the binary collision parameter, equal to the diffusion coefficient multiplied by the total number density. Through most of the atmosphere, the mixing ratio remains constant, but at great heights diffusive separation occurs. The H_2 density at the critical level adjusts itself to give a Jeans escape rate equal to eqn. (1). In one example (Hunten, 1973a), the critical level was at several Titan radii; the large area of the corresponding sphere is an important factor in matching the flow rates.

Trafton's work, discussed above, gives 5 km-A of H_2 and 2 km-A of CH_4, if there is no other gas present. The maximum value of f_1 is therefore 2.5, and eqn. (1) gives a flux of 1.3×10^{12} cm^{-2} sec^{-1}, or 1/300 of Titan's mass in the age of the solar system. If there were 20 km-A of N_2, the flux would still be 7×10^{11} cm^{-2} sec^{-1}. This flux, though far from a blowoff, is embarrassingly large, and no really credible source of the same magnitude has been suggested. Some of the possibilities follow:

Primordial: H_2 could be collected from the solar nebula only by adsorption or gravitation, both of which are far too small to give 1/300 of a Titan mass.

Photolysis of CH_4: The dissociation threshold is about 1600 Å; the global-mean flux of solar photons is therefore 1.6×10^9 cm^{-2} sec^{-1} (Ackerman, 1971). Further photochemistry (Strobel, 1974a) utilizes longer wavelengths and can give a yield of nearly 3 H_2 molecules per original dissociation. In addition, the atmosphere is sufficiently extended to appreciably increase the radius for ultraviolet absorption. From the downward fluxes of C_2H_2 and C_2H_6 implied by Strobel's Figures 1 and 2, the H_2 production rate is 9×10^9 cm^{-2} sec^{-1}. This result gives a reasonable lower limit of 0.5% to f_1 from eqn. (1) but would not support Trafton's abundance unless there were 1000 km-A of invisible gas such as N_2 or Ne.

Photolysis of NH_3: Absorption is found to 2300 Å, and the photon flux is 3×10^{11} cm^{-2} sec^{-1}. The quantum yield could approach 1.5. The biggest problem is that NH_3, with its small vapor pressure, must be confined to low altitudes, where the radiation cannot readily penetrate. In the most favorable case, the escape flux could be 4.5×10^{11} cm^{-2} sec^{-1}, which would require $f_1 = 0.14$, or 35 km-A of N_2.

Photolysis of H_2S: Sagan and Khare (1971) pointed out that H_2S absorbs up to 2700 Å; a global-mean photon flux of 2×10^{12} cm^{-2} sec^{-1} would thus be available. As with CH_4 and NH_3, the H_2 quantum yield may be 1 or greater; thus, the required source of H_2 may plausibly be met with H_2S, even with no extra background gas. The question then becomes whether H_2S is a plausible atmospheric constituent. In the models of Lewis (1971a), NH_3 is more abundant than H_2S and removes the latter from the atmosphere as NH_4HS. It is conceivable that this abundance ratio might somehow be reversed in Titan, or that nonequilibrium processes such as volcanism might release H_2S into the atmosphere. Certainly we do not know enough about Titan to rule out such ideas entirely.

Interior processes: Radiolysis from products of radioactive decay is estimated by Lewis to produce only 10^9 H atoms cm^{-2} sec^{-1} (Hunten, 1974, p. 116). Chemical action of hot ammonium hydroxide is conceivable but hard to quantify.

The sources that can be evaluated quantitatively are inadequate, and the potentially larger ones are totally speculative or require unpopular gases like H_2S. One must therefore conclude that an H_2 abundance as large as 5 km-A is improbable, even if there is a much larger amount of invisible, heavy gas. An amount one or two orders of magnitude less would cause much less embarrassment, but it would not lead to the observed absorptions.

McDonough and Brice (1973a,b) have pointed out that the H_2, once it has escaped from Titan, goes into orbit about Saturn and forms a moderately fat torus. Their estimated lifetime is long enough to support the idea that the gas might be recycled back to Titan and might thereby reduce the large escape rate to a more tolerable value. However, a diffusion-limited situation, such as outlined above, is highly resistant to such an effect. The first response of the atmosphere is to let H_2 accumulate at very high altitudes, so that the Jeans loss rate increases and the original net flux is restored. This behavior continues until the H_2 distribution all the way to the homopause (or turbopause) is essentially that of diffusive separation. For a rough treatment we may take the upper atmosphere as isothermal at $T = 100°K$; the barometric equation is then

$$n = n_h e^{\lambda - \lambda_h}, \tag{2}$$

where n is the H_2 number density, and the subscript h refers to the homopause. λ is a normalized potential energy:

$$\lambda = \frac{GMm}{kTr} = \frac{r}{H}, \tag{3}$$

where G is the gravitational constant, M and m the masses of Titan and of an H_2 molecule, k Boltzmann's constant, r the distance from the center of Titan, and H the scale height of H_2. The value of λ_h is 8.6. For very large distances, λ approaches zero, and the density approaches the finite value

$$n_\infty = n_h e^{-\lambda_h} = 2 \times 10^{-4} n_h. \tag{4}$$

If recycling is to have a major effect on Titan's H_2 budget, the density in the torus must be essentially n_∞.

We do not know the position of Titan's homopause, but a reasonable assumption is to take it at the same total number density, 10^{13} cm^{-3}, as on Earth (e.g., Hunten and Strobel, 1974). Since most of this gas is H_2 for the simplest Trafton model, we obtain an estimate of $N_\infty = 2 \times 10^9$ cm^{-3} from eqn. (4). If Titan's atmosphere is as strongly mixed as that of Mars (McElroy and McConnell, 1971), a value as small as 10^7 cm^{-3} is conceivable. This is still 10^3 times larger than the greatest value suggested by McDonough and Brice and changes their lifetime of 6 years (which is approximately the photoionization time) to a required value, estimated below, of 22,000 years. At a density of 10^7 cm^{-3}, the mean free path for gas-kinetic collisions between molecules is

$$L = \frac{1}{Qn} = 300 \text{ km}, \tag{5}$$

where Q is the cross section, 3×10^{-15} cm^2. Thus, such a torus is dominated by collisions, and the description by McDonough and Brice in terms of independent orbits is inappropriate. Instead, the gas may be regarded as residing in a potential well (Sullivan, 1973). Within the approximation of constant angular momentum, the restoring force per unit mass is always directed back to Titan's orbit, and has the value at a distance s

$$F' = -g_0 s/r_0 , \tag{6}$$

where g_0 is Saturn's gravity at the orbit. Solving the hydrostatic equation with the perfect gas law in the usual manner, we obtain for the pressure

$$p = p_0 \exp(-s^2/2r_0 H_0)$$
$$= p_0 \exp(-s^2/X_0^2) . \tag{7}$$

The scale height is $H_0 = kT/mg_0$, and the *scale radius* is $X_0 = (2r_0 H_0)^{1/2} = (2kTr_0/mg_0)^{1/2}$. Instead of the usual exponential distribution of pressure and density, we find a Gaussian. The minor diameter of the torus is of order $2X_0 = 4 \times 10^5$ km at 100°K or about 1/6 the major diameter. The effective cross-sectional area (which would contain all the gas at the central density) is πX_0^2. The corresponding volume is $2\pi^2 r_0 X_0^2 = 10^{33}$ cm^3 at 100°K; it is proportional to T.

In a torus with frequent collisions, there will be a high-velocity component of the energy distribution which is able to escape. The torus, just like the exosphere, is heated by absorption of solar ultraviolet below 800 Å. Unlike the exosphere, it lacks any significant means of losing heat. The heating efficiency is large, averaging about 0.86 according to Henry and McElroy (1969). Combining this with the solar flux, photon energy, and the cross section of H_2, we find a heating rate of about 5°K/Earth day, or 1800°K/year. S. Soter (personal communication, 1974) has pointed out that this estimate might be reduced, for a thin medium, by the escape of fast photoelectrons. For the high densities being discussed here, this effect is probably small, but, even if it were significant, the heating rate should be at least 500°K/year.

The Jeans formula does not apply to a torus, because the escape velocity depends strongly on direction. However, the error in using it is probably tolerable, because the loss rate depends so sharply on temperature. With the barometric equation and a collision cross section of 3×10^{-15} cm^2, we find the critical level at $s = 1.8 X_0$. Other parameters are:

Central density	2×10^7 cm^{-3}
Density at critical level	1×10^6 cm^{-3}
Area of critical level	3×10^{23} cm^2
Flux from Titan	10^{11} cm^{-2} sec^{-1}
Loss rate from Titan	8×10^{28} sec^{-1}
Required effusion velocity	0.3 cm sec^{-1}
Required temperature	500°K

(The loss rate is the product of density, area, and effusion velocity.)

From these estimates, it appears that the temperature of a dense torus would rise in less than a year to a value that would carry away the input from Titan. On the other hand, the filling time to the same density is 22,000 years. I conclude that a torus dense enough to recycle significant hydrogen to Titan cannot exist. A density around 10^3 cm^{-3}, in the range originally discussed by McDonough and Brice, is far less objectionable, but does not help Titan's hydrogen budget. Collisions still cannot be ignored: the mean free path is less than half the orbital circumference. Photoelectrons and fast ions might escape northward or southward, but would probably be returned to the torus by magnetic reflection, if Saturn's magnetic field is large enough. The heating efficiency thus probably remains high.

Pioneer 10 has observed Lyman α from a partial torus filling about ⅓ of Io's orbit (Carlson and Judge, 1974). This atomic hydrogen should be accompanied by a much larger density of H_2, unless it is produced in space by processes such as charge exchange.

THERMAL STRUCTURE

Temperatures can be discussed for four regions: (1) the solid (or liquid) surface; (2) the troposphere, in which the temperature gradient is negative due to some combination of dynamics and greenhouse effect; (3) the stratosphere-mesosphere, in which the temperature gradient is probably positive, though small or even negative at high altitudes; and (4) the thermosphere, exosphere, and orbiting torus.

On Earth, the number densities are 5×10^{18} cm^{-3} at the tropopause and 10^{14} cm^{-3} at the mesopause; similar or lower values might obtain on Titan. Figure 20.2 shows a temperature profile adopted here for reference.

It seems likely that the atmosphere is fairly deep and opaque; the dynamical regime derived by Leovy and Pollack (1973) is therefore probably able to maintain a fairly isothermal surface. For a Bond albedo of 0.20, the equilibrium temperature is 86°K, which is likely to be approximated at the tropopause. Some weak constraints on the surface temperature follow from the vapor pressure required to support methane in the atmosphere. The spectroscopic observations imply a methane pressure at the clouds of 20 mb or less; if N_2 were as abundant as suggested above, the partial pressure of CH_4 would be 1–2 mb. The vapor pressure of pure methane is 21 mb at 80°K and 350 mb at 100°K. For the clathrate hydrate (Delsemme and Wenger, 1970), the vapor pressures are 0.02 mb at 100°K and 20 mb at 145°K. It is probable (Lewis, 1972b) that methane was originally accumulated as the hydrate, and the corresponding vapor pressure curve might well be appropriate. But, as M. J. S. Belton (personal communication, 1974) has remarked, the ice crust, some tens of km deep, could be saturated with CH_4: the maximum CH_4/H_2O ratio is 1/5.75, with 1/6.9 more usual (Miller, 1961). Excess methane would then be released from the melted interior, and

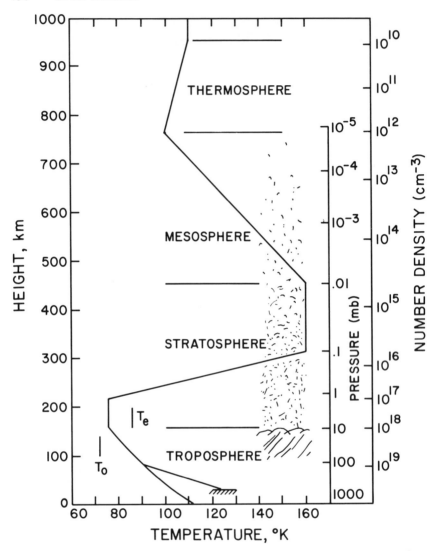

Fig. 20.2. A model atmosphere for Titan, based on a constant gravity of 140 cm sec^{-2} and a mean molecular weight of 16.

solid or liquid (the melting point is 89°K) would float on the ice and determine the vapor pressure. If liquid, it probably contains large amounts of photochemical polymers in solution, which would lower the vapor pressure slightly. Thus, a methane surface would probably be at or below 100°K, and an unsaturated clathrate surface in the 130–160° range. The lower curve in Figure 20.2 follows the methane vapor pressure relation up to 155 km (cf. Lewis and Prinn, 1973).

The following discussion is based on a radius of around 2550 km as assumed in most of the original papers. With a larger radius, the brightness temperatures will be less; but the value 2900 km (Elliot *et al.*, 1975) may not be appropriate at long wavelengths if it refers to a level high in the atmosphere. The atmospheric ''model'' of Figure 20.2 assumes a constant gravity of 140 cm sec^{-2} and a mean molecular weight of 16; significant quantities of H_2 would stretch the height scale considerably. This scale therefore must be regarded as a kind of geopotential height. By crude analogy with the Earth, the visible limb probably would be at about 200 km in Figure 20.2; the true geometrical height would be greater for a smaller gravity or a smaller molecular weight.

The radio brightness temperature at 8085 MHz is 115 ± 40°K (Briggs, 1974a) (see Morrison's discussion below). The corresponding surface temperature for a reasonable emissivity is 135 ± 45°K, which does not rule out a surface as cold as the equilibrium temperature. A surface temperature greater than some 200°K is ruled out unless hidden by some opacity such as that of ammonia.

The past few years have given us a wealth of significant data from radiometry in the thermal infrared (cf. Figures 12.4 and 21.1). Most of the results are summarized in Figure 20.3, taken from Low and Rieke (1974b), and Table 20.1. In addition, F. C. Gillett has kindly allowed me to quote the unpublished results of his most recent work, which gives improved definition of the C_2H_6 peak at 12.2 μ, and which confirms the lack of structure in the 10 μ region. The spectrum closely resembles that of Saturn (Gillett, 1975). Compare with other satellite results in Table 12.1.

The principal features visible in Figure 20.3 are: (1) a high brightness temperature at 8.0 μ, due to the 7.7 μ fundamental of CH_4; (2) a peak at 12 μ, probably due to the C_2H_6 fundamental; (3) neither a minimum nor a maximum at 17 μ, the peak of the H_2 pressure-induced absorption; and (4) a continued decline of brightness temperature to longer wavelengths.

In addition, Low and Rieke (1974b) estimate the brightness temperature in the 40–150 μ region from the energy balance, finding 64–74°K, depending on the Bond albedo.

Features (1) and (2) imply emission from a warm stratosphere, discussed below and by Caldwell in Chapter 21. Feature (3) rules out a significant greenhouse effect due to H_2, and feature (4) suggests a weak greenhouse effect by pressure-induced absorption in CH_4. All these interpretations were pointed out by Low and Rieke. (In principle, the minimum in the brightness temperature due to the 17 μ peak of H_2 opacity might be hidden by cloud opacity, but then the greenhouse would be due to the cloud, not the H_2.)

Brightness temperatures of 100°K and 125°K are seen in the 17 and 10 μ regions. One of these could be the surface temperature or the temperature of a dense cloud deck. If we adopt the 125° value, the opacity above 15 μ could be the pressure-induced absorption of CH_4. Pollack (1973) has given the required coefficients for pure methane by correcting laboratory data to low temperatures.

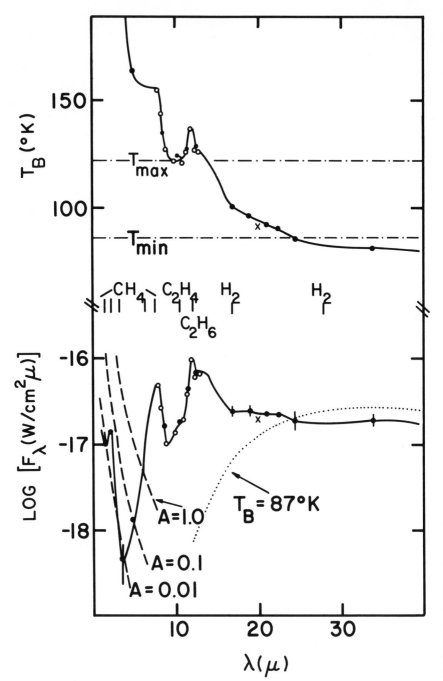

Fig. 20.3. Thermal spectrum of Titan, after Low and Rieke (1974b).

TABLE 20.1
Infrared Photometry of Titan

λ (μ)	Δλ (μ)	T_B (°K)	Ref.†
5.0	1.0	165*	L
8.0	0.12	158 ± 4	G
8.4	0.8	146 ± 3	G
8.8	1.0	136	L
9.0	0.14	130 ± 6	G
10.0	0.15	124 ± 3	G
10.3	1.3	125	L
10.6	5.0	124	L
11.0	0.17	123 ± 3	G
11.5	0.17	128 ± 2	G
11.6	0.8	128	L
12.0	0.18	139 ± 2	G
12.5	0.19	129 ± 2	G
12.6	1.0	129	L
13.0	0.20	128 ± 2	G
17.0	2.0	101	L
19.0	1.0	97	L
20.0	7.0	93 ± 2	M
21.0	8.0	93	L
22.5	5.0	91	L
24.5	1.0	86	L
34.0	12.0	82	L

*or an albedo ≤ 0.10.
†L, Low and Rieke (1974b); G, Gillett and Forrest (1974); M, Morrison *et al.* (1972).

At 17 and 25 μ they are $\beta = 1.5 \times 10^{-6}$ and 4×10^{-6} (cm-A-atm)$^{-1}$. For a scale height H, integration through the atmosphere gives an optical depth

$$\tau = \beta n_0 p_0 H/2 = \beta w p_0/2 = 515\, \beta w^2 \ , \tag{8}$$

where $w = n_0 H$ is the abundance in km-A, and the pressure due to 1 km-A of methane is 0.0103 atm. For $\tau = 1$ at the two wavelengths, 36 or 22 km-A is required, compared with the 2 km-A visible above the clouds (Trafton, 1972b). The corresponding surface pressures are 360 and 220 mb. The optical depth actually required to hide the flux from the surface would be somewhat greater than 1, but the required methane abundance increases only as the square root. These values do not seem excessive.

If an excess of N_2 is assumed, the total pressure requirement does not change much. The same abundance of nitrogen gives nearly twice the pressure because of its larger molecular weight, but the coefficient of pressure-induced absorption is probably smaller (cf. Pollack, 1973). The relation to Trafton's observations is also essentially unchanged, because they too give the product of (total pressure) times (methane abundance). Thus, a mixture of N_2 and CH_4 allows less CH_4 both above and below the cloud tops, but does not change the need for more gas by a factor of 10 in the lower region. The coefficient β for absorption by N_2 is much smaller than for CH_4 (Bosomworth and Gush, 1965). Such absorption would therefore be important only for a large excess of nitrogen.

If the surface temperature is 125°K, the temperature profile could lie between the two shown in Figure 20.2, which are saturated and dry adiabats. Clouds are shown opposite the upper portion of the saturated adiabat, but could extend lower.

The temperature at the tropopause should be somewhere between the observed 82°K brightness temperature at 34 μ (Table 20.1) and the Gold-Humphreys skin temperature, $2^{-1/4}T_e = 72°K$. Indeed, Low and Rieke (1974b) infer temperatures near the latter at very long wavelengths from the energy balance. At 72°, the vapor pressure of methane is 4 mb.

Danielson et al. (1973) were the first to point out that band emission from a warm stratosphere is likely to be important (see Chapt. 21, Caldwell). A lower limit to the temperature is given by the observed brightness temperature at 8 μ, 158°K, since a very small amount of CH_4 is opaque at this wavelength. The C_2H_6 band at 12.2 μ gives a lower temperature, and is thus probably not opaque. The model of Danielson et al. uses 160°K.

Although CH_4 and C_2H_6 contribute significant heat by their absorption of solar radiation (Wallace et al., 1974), the principal heat source for the stratosphere is probably ultraviolet absorption by a dark aerosol. The presence of this aerosol is required to explain the low ultraviolet albedo of Titan (Barker and Trafton, 1973b; Caldwell, 1975). Previously, Axel (1972) had applied similar reasoning to Jupiter. If the dust particles are small, they are poor infrared radiators, and in a vacuum they can become much hotter than the equilibrium temperature for a large body. In an atmosphere, their heat is primarily transferred to the gas, and radiated in the methane and ethane bands that are observed. There is also some optically thin continuous emission by the dust. In this kind of model (Danielson et al., 1973; Chapt. 21, Caldwell) the brightness temperature of 125°K around 10 μ is attributed to this dust emission, instead of the surface as assumed here. Their surface is much colder and could be located in the region of the temperature minimum in Figure 20.2.

Before we had as much detail in the infrared spectrum, several workers proposed another explanation of the high brightness temperatures near 8 μ (Allen and Murdock, 1971; Morrison et al., 1972; Hunten, 1972; Sagan, 1973; Pollack, 1973; Chapt. 12, Morrison). A warm surface, at about 160°K, was to give the short-wavelength emission; it was hidden at longer wavelengths by the pressure-induced absorption of H_2. Pollack's models suggested equal abundances of H_2 and CH_4, with a surface pressure of 440 mb. But they give a deep minimum in brightness temperature at 17 μ, where the H_2 absorption coefficient has a peak due to the broad S(1) line. Such a minimum is a direct consequence of H_2 absorption in a negative temperature gradient. The observed absence of the minimum (Fig. 20.3) is equally direct evidence that H_2 absorption is not involved in determining the thermal structure. The weaker S(0) line at 28.2 μ does not seem to be present either, though there are not enough data points to define it.

Though significant greenhouse absorption by H_2 seems to be ruled out, CH_4 is another matter, as discussed above. Indeed, many of Pollack's models are dominated by the CH_4 absorption above 30 μ, and give some idea of what this molecule can do. Quasi-polar absorption by CH_4 (Fox, 1975) may also be present in this region, but the lines are probably too narrow for a large effect on the mean opacity.

The observed band emissions probably arise from the lower stratosphere. At greater heights, the temperature might remain near 160°K, or could decrease if the mixing ratio of ultraviolet-absorbing dust decreases. A similar situation is found on Earth because of a decrease in the ozone mixing ratio. A reasonable range for the mesopause temperature is 100–160°K. Somewhat arbitrarily, the cooler value has been adopted for Figure 20.2. Strobel and Smith (1973) put the mesopause at a density near 5×10^{13} cm^{-3} for solar composition. With 10^4 times more methane, a density 10^2 times smaller, or 10^{12} cm^{-3}, would be appropriate.

In an ordinary thermosphere, the dominant balance is between heating from the ionosphere and downward conduction to the mesopause, a level dense enough to radiate all the energy from above. Strobel and Smith (1973) estimated a temperature contrast of 90°K for solar composition. For the more likely situation $CH_4/H_2 \geqslant 1$, Strobel (1974c) estimates less than 10°. Moreover, if H_2 is flowing outwards on its way to escape, it would cool adiabatically and might not be in thermal equilibrium with the CH_4 (cf. Gross, 1974). For most purposes, an adequate approximation is an isothermal "thermosphere" at the mesopause temperature. Another brief discussion is given by Strobel (1974a).

The torus might be considered as an outer exosphere with heat sources but no significant sinks; it would seem at least that conduction back to Titan is negligible for such a large object. Thus, as discussed above, a fairly large heating rate is expected, probably enough to be a major factor in limiting the density.

CHEMISTRY

Titan differs from the Jovian planets in (probably) having a cold surface to condense or dissolve its photochemical products. It does not have their cleansing action in which material carried to great depths is returned to thermodynamic equilibrium. Thus, it is probable that there is a deep layer of material at the surface. It may be dissolved in methane, or conceivably lie on top of it, with the methane reaching the atmosphere by means of fumaroles and volcanoes.

The photochemistry of two model Titanian atmospheres has been discussed by Strobel (1974a). The results are little different for pure CH_4 and equal (by number) mixtures of CH_4 and H_2. Abundances of 1 cm-A are found for C_2H_6 and C_2H_2, on the assumption that they flow downwards to the surface. The downward fluxes are (1.3 and 2.5) $\times$ 10^9 cm^{-2} sec^{-1}, which corresponds to an accumulation of around 30 kg cm^{-2} over the age of the solar system. The H_2 source, already described above, is 9 $\times$ 10^9 cm^{-2} sec^{-1}.

The production of heavier polymers has been briefly discussed by Strobel (1975). He is pessimistic about the efficiency of the process, estimating a total production over geologic time of less than 1 mole cm^{-2} in the hydrocarbon system. Alternatively, one can simply postulate a yield of a few percent, on the grounds that important reactions may not yet have been included. Many of these compounds are likely to condense, and probably correspond to the dark aerosol of Danielson et al. (1973). If the estimate of a few percent is valid, the accumulation on the surface could be 1 kg cm^{-2}.

Ammonia photochemistry has not been discussed specifically for Titan, but the work of Strobel (1973) can be adapted. The regions of active photochemistry for CH_4 and NH_3 are strongly separated because of the different abundances and the different wavelength regions that are important. Thus, reactions within the ammonia system tend to be emphasized over those between the ammonia and methane systems. A likely intermediate product is hydrazine, N_2H_4, which becomes a source of N_2 (McNesby and Okabe, 1964; Stief and DeCarlo, 1968).

Simulation experiments have been carried out at higher pressures by Khare and Sagan (1973) with mixtures of hydrocarbons, NH_3, and H_2S. They find a red-brown product whose absorption coefficient was used in the model of Danielson et al. Many organic compounds have been identified, especially when water is added to the product. Again, under Titanian conditions almost all these molecules would be frozen on the surface or dissolved in it. The hydrogen which is released would soon escape from the atmosphere.

REMARKS

Two major issues seem to stand out from this discussion. First, is H_2 really a major constituent of the atmosphere, more than the few percent to be expected from photolysis of CH_4 and NH_3? If so, what is the source? Second, is there a deep troposphere and a surface above 100°K, as sketched in Figure 20.2, or do all the features in the infrared spectrum arise from a warm stratosphere (cf. Chapt. 21, Caldwell)? The second question is particularly important to those who would like to measure the atmosphere and surface directly from an entry probe. The minimum atmosphere may not be deep enough to make such a mission attractive.

ACKNOWLEDGMENTS

I am indebted to D. F. Strobel for valuable discussions, and to F. C. Gillett and G. H. Rieke for access to unpublished data. Kitt Peak National Observatory is operated by the Association of Universities for Research in Astronomy, Inc., under contract with the National Science Foundation.

DISCUSSION

DAVID MORRISON: Note that Briggs' 4-cm brightness temperature, when corrected to the larger radius of Elliot *et al.* (1975), is only $87 \pm 26°K$, nearly as low as the equilibrium value. The corresponding thermometric temperature is probably about $100 \pm 30°K$. This may have important implications when considering possible greenhouse effects.

DONALD HUNTEN: An interesting point, but remember that the radio radius is probably less than the optical, so that the correction is probably considerably less.

THERMAL RADIATION FROM TITAN'S ATMOSPHERE

John Caldwell
Princeton University Observatory

The temperature inversion model of Titan's atmosphere is modified to include the thermal emission from a plausible amount of C_2H_2. Infrared photometry of Titan from 8 to 14μ agrees with model calculations for 0.5 cm-atm C_2H_6 and 1.0 cm-atm C_2H_2 and an optically thin dust layer in a temperature inversion region at $160°K$. The random band approximation used to calculate C_2H_6 opacity has been verified by laboratory transmission measurements. Measurements of the C_2H_2 emission at 13.7μ may provide an indirect check on the presence of H_2 on Titan.

Existing observations of Titan appear to be fully consistent with the inversion model if the surface radius equals 2700 km, if the atmospheric radius equals 2900 km, and if the surface temperature equals $78°K$. The energy balance depends sensitively on these three parameters.

Infrared observations (Low and Rieke, 1974b; Morrison, 1974d; Morrison and Cruikshank, 1974; see also Chapt.12, Morrison) over the past decade, and particularly over the past few years, have shown that Titan does not radiate even approximately like a black body, and have led to the unanimous conclusion that Titan's atmosphere significantly modifies its thermal emission. Brightness temperatures range from 120 to $160°K$ in the $8–14\mu$ atmospheric transmission window and are less than $100°K$ longward of 17μ. Titan's infrared spectrum is shown in Figures 12.4 and 20.3. Explanations of these observations have been widely divergent, however.

At one extreme, greenhouse models have been suggested by several groups. The details of these models have been extensively reviewed by Morrison and Cruikshank (1974) and are discussed by Hunten in Chapter 20. The greenhouse effect in a cold, reducing atmosphere relies largely on pressure-induced translational and rotational opacity of molecular hydrogen to block thermal emission from the surface and thus raise surface temperatures. The models require large amounts of either hydrogen (H_2) or some additional gas as a pressure broadening agent, and therefore imply that the total atmosphere on Titan is very massive. For example, Pollack (1973) calculates a minimum surface pressure of 0.4 atm, a minimum methane (CH_4) abundance of 30 km-atm, and a comparable H_2 abundance.

The opposite extreme is the temperature inversion model of Danielson *et al.* (1973), of which this paper is an extension. Since the ultraviolet albedo of Titan is extremely low (Caldwell, 1975), implying high altitude absorption of incident

[438]

solar radiation and thus high altitude heating, it was postulated that a temperature inversion occurs in the atmosphere.

Caldwell *et al.* (1973) predicted that this inversion would result in infrared emission peaks in Titan's spectrum from the following gases: CH_4 (7.7μ); ethylene, C_2H_4 (10.5μ); ethane, C_2H_6 (12.2μ) and acetylene, C_2H_2 (13.7μ). The latter three gases were expected as trace constituents resulting from the photolysis of CH_4, known to be present with a large abundance (Trafton, 1972b). Subsequently, Gillett *et al.* (1973) published a moderate resolution infrared spectrum of Titan which in fact showed emission peaks at 8 and 12μ. Strobel (1974a) concluded that C_2H_4 would be photodissociated before it could accumulate in detectable amounts. He also predicted a significant abundance of C_2H_2. The emission from C_2H_2 will be considered in detail in this paper.

A simple analysis (Danielson *et al.*, 1973) shows that the solar radiation absorbed by the high-altitude "dust" particles invoked to suppress Rayleigh scattering in the ultraviolet could quantitatively account for the observed thermal emission in the 8–14μ window by CH_4 and C_2H_6 molecules and by the dust particles themselves. The dust particles are also presumably produced by the photolysis of CH_4, but the details of this process are not well understood.

Photometric (Noland and Veverka, 1974; cf. Veverka, Chapt. 9 herein) and polarimetric observations (Zellner, 1973; Veverka, 1973a; cf. Veverka, Chapt. 10 herein) have been interpreted in terms of clouds in Titan's atmosphere. Since clouds are not likely to form in an atmosphere containing a large inversion, these observations require a different explanation in the inversion model. The explanation may be found in the scattering properties of the dust particles. For simplicity, the dust particles have been assumed to be purely absorbing (Danielson *et al.*, 1973). However, subsequent analysis (Danielson, private communication, 1974) suggests that the scattering and absorbing cross sections of the dust particles must be comparable at visible wavelengths. Moreover, the scattering by the dust exceeds the Rayleigh scattering of the CH_4 molecules. Hence the photometric and polarimetric properties of Titan in the inversion model are mainly determined by the physical characteristics of the atmospheric dust and by the surface. The latter is believed to be CH_4 snow contaminated by the dust particles which have settled out.

It is interesting to note that, although the greenhouse and inversion models are much different in all respects (cf. Chapt. 20, Hunten), they are not mutually incompatible. An inversion layer could conceivably exist above a greenhouse region. However, models of such a configuration would have to take into account the high altitude absorption of at least some of the solar incident radiation. Indeed, if the various arguments (reviewed by Morrison and Cruikshank, 1974; cf. Hunten, Chapt. 20) for thick clouds on Titan are correct, and if the amount of H_2 on Titan reported by Trafton (1972a, 1975a) is present, it will be necessary to consider such a juxtaposition.

It may also be noted that the question of the total mass of Titan's atmosphere is somewhat more than academic. One scenario discussed at the N.A.S.A. Titan workshop (Hunten, 1974) called for a ballistic probe to enter the atmosphere and accumulate data during the radio blackout period for later transmission to a flyby vehicle. The data would be transmitted only after atmospheric deceleration reduced the ionization field around the probe, at the about 20 mbar pressure level. In the inversion model of Danielson *et al.* (1973), the surface pressure is about 20 mbar.

PARAMETERS OF THE INVERSION MODEL

Radius

An accurate value of the radius of Titan is necessary for interpreting visual and infrared photometry and for calculating such quantities as surface gravity and mean density (cf. Chapt. 9, Veverka; Chapt. 12, Morrison). Values for the radius of about 2500 km (Morrison and Cruikshank, 1974) had recently been accepted, but Elliot *et al.* (1975) have now determined the radius to be 2900 ± 100 km from a lunar occultation. Lambertian limb darkening was employed in the reduction procedure by Elliot *et al.* Approximate calculations (Danielson, private communication, 1974) suggest that atmospheres with large absorption, such as Titan, may be approximated by a Lambert surface. Therefore, in this chapter, the new measurement by Elliot *et al.* (1975) will be accepted as the most probable value of the outermost radius at which the atmosphere of Titan scatters solar radiation. The effect on calculations of Titan's thermal properties due to uncertainties in this radius will be considered later.

Surface Temperature, Pressure, and Methane Abundance

In the inversion model, the above three quantities are intimately related. The surface temperature determines the vapor pressure of CH_4, and hence the column abundance, if CH_4 is the bulk constituent in the atmosphere. In the calculations of Danielson *et al.* (1973) a CH_4 abundance of about 2 km-atm was adopted following Trafton (1972b), implying a surface temperature of about 80°K and a surface pressure of about 20 mbar. It was shown by them that the latent heat of condensation of CH_4 was sufficient to maintain the entire surface of Titan at a temperature of 80°K, with a negligible fraction of the atmosphere condensing during the Titan night.

A more thorough evaluation of the energy balance given below indicates that these numbers are nearly unchanged. A baseline model is adopted in which the surface temperature is 78°K corresponding to a CH_4 abundance of at least 1.8 km-atm.

Briggs (1974d; cf. Chapt. 12, Morrison) has determined the surface brightness temperature to be 115 ± 35°K from radio wavelength interferometry. His measurements are virtually independent of the nature of Titan's atmosphere. However, Briggs used a radius of 2500 km in his reduction. If the true radius of Titan's surface is 2700 km, a brightness temperature of $100 \pm 30°K$ is implied by his measurements (cf. Morrison's comment following Hunten, in Chapt. 20). If the surface of Titan is covered with CH_4 frost, as assumed in the inversion model, the emissivity will be close to unity. In this case, Briggs' observations give the actual surface temperature. It is consistent with the inversion model.

Temperature of the Inversion Layer and Thermal Emission From Methane

The v_4 fundamental band of CH_4 at 7.7μ is so intense (Thorndike, 1947) that its center is optically thick in the Titan atmosphere. The band is seen in emission in Titan's spectrum. The narrow band photometry of Gillett et al. (1973) at 8μ in the wing of the band indicates a brightness temperature of about $160°K$. On the basis of this data point, a temperature of $160°K$ is assigned to the temperature inversion region.

Thermal Emission From the Dust

The concept of thermal emission by the ultraviolet-absorbing dust particles was introduced by Danielson et al. (1973) to account for the observed radiation near 10μ and near 20μ. Figure 21.1 includes the recent additional data points of Low and Rieke (1974b) and of Knacke et al. (1975) in the vicinity of 20μ.

Since the dust particles are expected to be small compared to infrared wavelengths, their thermal emissivities will vary as λ^{-1} if the index of refraction is independent of the wavelength λ. In this case, the thermal radiation of the dust will resemble blackbody emission at $160°K$ (the inversion temperature) modified by the emissivity factor. If the optical depth of the dust is normalized to make the dust emission agree with observations at 10μ ($\tau_{10\mu} \simeq 0.04$), the thermal emission at all wavelengths may be calculated. Figure 21.1 demonstrates the good agreement of such calculations with observations from 18 to 34μ, when the emission of the surface at $78°K$ is also included. Within the errors of the observations, some variation is possible in the λ^{-1} dependence of emissivity, in the optical depth of the dust at 10μ, and in the inversion temperature.

Ethane Abundance

C_2H_6 has a moderately strong band at 12.2μ. It is the v_9 fundamental, associated with the flexing mode of the molecule.

Photometry at 12.0μ by Gillett et al. (1973) indicates that Titan has an emission feature in this region. Similar photometry for Saturn (Gillett and Forrest, 1974), but with many more data points, shows a spectral emission feature centered at 12.2μ with a width matching model calculations of C_2H_6. Ridgway (1974) has also identified Q branches of the band in emission in high resolution scans of Jupiter. Furthermore, Strobel (1974a) has calculated that C_2H_6 will be a relatively stable product of the ultraviolet photolysis of CH_4, and hence will accumulate in a significant quantity in Titan's atmosphere. On the basis of this evidence, it is concluded that C_2H_6 is in emission in Titan's spectrum. Model calculations were therefore made to determine the C_2H_6 abundance and the total emission of the band.

C_2H_6 is a symmetric top molecule, and thus its perpendicular-type bands consist of a series of subbands, each with P, Q and R branches. The Q branches of the subbands are regularly spaced, and the line density in the Q branches makes them much more prominent than the P and R branches. For each subband, the P and R branches spread out beyond neighboring Q branches in other sub-bands (Herzberg, 1946, his Fig. 128).

Because of the large number of lines involved, a random band transmission model was used. Line positions and relative strengths were calculated following Herzberg (1946, Chapt. IV.2). Absolute strengths were found by normalizing to the total band intensity of Thorndike (1947). For convenience in calculation, the band was divided into thirteen arbitrary channels, each containing hundreds of individual lines. The calculated line strengths indicated that an inverse first power distribution of line strengths (Goody, 1964, eqn. 4.23) approximates the actual distribution in each channel. Mean absorptions were therefore calculated for each channel according to eqn. (4.36) from Goody (1964).

Laboratory observations to test the random band model are discussed in the Appendix to this chapter. With the reservations given there, the random band model of C_2H_6 opacity may be fitted to the observed data points to estimate an abundance. To calculate the emergent radiation, the following calculation was successively applied to arbitrary layers in the atmosphere (typically 20 were used), starting at the bottom:

$$I_{\lambda,e,i} = (I_{\lambda,e,i-1}) \, (\exp[-\tau_{\lambda,i}]) + (1 - \exp[-\tau_{\lambda,i}]) \, (B_{\lambda,i}) \,, \qquad (1)$$

where $I_{\lambda,e,i}$ is the emergent intensity from the ith layer, $I_{\lambda,e,i-1}$ is the emergent intensity from the next lower layer, $B_{\lambda,i}$ is the Planck function for the ith layer, and $\tau_{\lambda,i}$ is the optical depth of the ith layer, as calculated with the random band model. Eqn. (1) applies to the center of the disk. Emission was averaged over a hemisphere by assuming a secant variation of the line-of-sight optical depth.

Figure 21.1 shows the expected thermal emission from 0.5 cm-atm of C_2H_6 uniformly distributed through the atmosphere, assuming the entire atmosphere to be at 160°K. Figure 21.2 shows the same spectral range in more detail. The C_2H_6 abundance was adjusted to fit the observed data point of Gillett et al. (1973) at 12.0μ. Other data points on the C_2H_6 peak fit the model calculations well except for one point at 12.5 μ. Some of the discrepancy might be due to the overesti-mate of the optical thickness on the long wavelength side of the band, mentioned in the Appendix.

The data point at 13.0 μ appears to be inconsistent with the 12.2μ C_2H_6 emission feature; this point is the basis for the following discussion of acetylene.

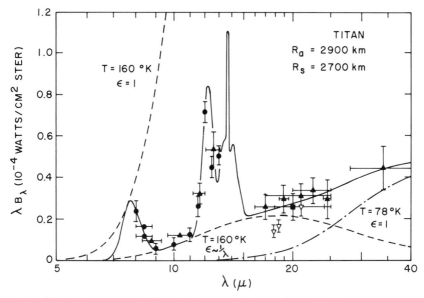

Fig. 21.1. Comparison of infrared photometric observations with the temperature
inversion model. Axes are drawn so that the area under the curves is proportional to
the energy traversing the atmospheric radius. Bandpasses for the closed circle data
points are similar to the width of the circles. The solid line represents the total
calculated emission from molecular bands and optically thin dust in a temperature
inversion region at 160°K, and a surface with unit emissivity at 78°K. The molecular abundances are 0.5 cm-atm C_2H_6 (12.2μ) and 1.0 cm-atm C_2H_2 (13.7μ). The
CH_4 peak (7.7μ) is only approximate. The predicted ratio of atmospheric emission
to surface emission was used to calculate the effective radius for each observational
data point. Closed circles are data from Gillett *et al.* (1973), closed triangles are
from Low and Rieke (1974b), open triangles are from Knacke *et al.* (1975), and the
closed square is from Morrison *et al.* (1972).

Acetylene Abundance and Thermal Emission

C_2H_2 has a strong band centered at 13.7 μ. It is the ν_5 fundamental, associated
with the bending mode of the linear molecule. If the wing of this band is
responsible for the high emission at 13.0 μ, the band will contribute significantly
to Titan's thermal emission. Strobel (1974a) has also predicted appreciable
amounts of C_2H_2 in Titan's upper atmosphere. Model calculations were therefore
undertaken to assess the influence of C_2H_2 on Titan's infrared spectrum.

The strength of the ν_5 band has been measured by Wingfield and Straley
(1955). Molecular constants were taken from Bell and Nielsen (1950) and from
Herzberg (1966). Relative line strengths were computed from the formulae of
Allen and Cross (1963, p. 104) and intensity alternation for odd and even J
values was incorporated, following Herzberg (1946). Difference bands detected
in the laboratory by Bell and Nielsen (1950) are expected to be negligible at
Titan's temperatures, and are ignored.

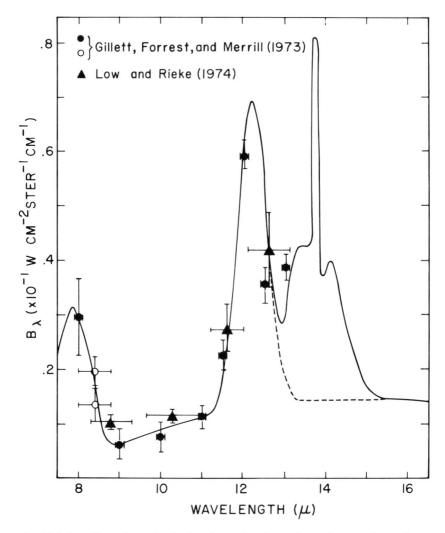

Fig. 21.2. Details of the molecular band emission from Titan. The solid line includes emission from optically thin dust, approximate CH_4 emission and calculated emission from 0.5 cm-atm C_2H_6 and from 1.0 cm-atm C_2H_2; the broken line shows the emission with no C_2H_2. Emission from C_2H_4 (10.5μ) is small, if any at all. CH_3 (16.5μ) has not been included in the calculations.

To estimate its effect on the spectrum of Titan, the band's opacity was calculated at typically 2500 wave numbers throughout the band, the exact number being model-dependent. Line profiles were assumed to be of the Voigt type (Doppler broadened Lorentzian) with no cut-off in the far wings; the Lorentz width at STP was assumed to be 0.1 cm^{-1}. Both these assumptions were made necessary by the lack of accurate data. The abundance of C_2H_2 was taken to be twice the C_2H_6 abundance, consistent with the photochemical model of Figure 1 in Strobel (1974a). The temperature inversion models considered in this paper all have pressures so low that individual lines are saturated and the precise abundance of C_2H_2 is therefore not critical in determining the opacity.

With the opacity determined as a function of wavelength, the calculation of thermal radiation was performed in the same manner as for C_2H_6. Figures 21.1 and 21.2 show the expected thermal emission from 1.0 cm-atm C_2H_2. The agreement with the observation at 13.0 μ is satisfactory. Further observations from 12 μ longward to the atmospheric cut-off will be required to establish the role of C_2H_2 in the atmosphere of Titan.

The details of the envelope of the C_2H_2 emission band depend on the resolution employed. In Figures 21.1 and 21.2, emission was smoothed over 10 wave numbers, comparable to the resolution of Gillett *et al.* (1973). The asymmetry of the band is largely due to the existence of R(0) and R(1) lines and the absence of P(0) and P(1) lines. Asymmetry of the Q branch itself also contributes to the overall asymmetry.

Positive identification of C_2H_2 may be possible through detection of individual lines of the R branch, as Ridgway (1974) did for Jupiter. The lines are optically thick at their centers, and well spaced. However, because of the low absolute flux from Titan, the Q branch, which has many strong but slightly spaced lines, might present a better target. At 13.7 μ, it is displaced from the center of the 15.0 μ CO_2 vibration-rotation band, but is obscured by the rotational opacity of H_2O. It may therefore require observations at aircraft or balloon altitudes.

Hydrogen Abundance

Trafton (1972a) discovered and subsequently confirmed (1975a) very weak absorption features in Titan's spectrum which coincide in wavelength with the S(0) and S(1) quadrupole lines of molecular hydrogen. He derives an abundance of the order of 5 km-atm from his observations. The features are so close to the detection limit, however, that H_2 has been excluded from present models.

A confirmation of the presence of C_2H_2 should provide an independent check on the amount of H_2 on Titan. Strobel's (1974a) models show that if an equal amount of H_2 is mixed with CH_4, C_2H_2 is less abundant than C_2H_6 by a factor of about two or more throughout the atmosphere. This reduced abundance of C_2H_2 (a factor of four smaller than employed in this paper) would lead to an emission peak which is probably too small to explain the spectrophotometry of Gillett *et al.* (1973) at 13.0μ. Failure to detect C_2H_2 on Titan would not constitute positive evidence for H_2, however, because C_2H_2 can also be destroyed by reactions not considered by Strobel. For example, polymerization of C_2H_2 (Garrison, 1947) could deplete its abundance and also possibly be the source of the "dust" particles mentioned above.

It may also be noted that in models of the atmosphere of the giant planets, where H_2 is the dominant species, Strobel (1974b) predicts that C_2H_2 will be less abundant than C_2H_6 by a factor of about 10^2, depending on mixing rates. Therefore, the absence of a prominent emission feature on Saturn longward of 13.0μ (Gillett and Forrest, 1974) is understandable. At higher resolution, C_2H_2 may be detectable on Saturn as it is on Jupiter (Ridgway, 1974).

Methyl Radical Thermal Emission

Strobel's (1974a) work indicates that CH_3 is the most abundant photolysis product in the highest layers in Titan's atmosphere. Furthermore, it has a fundamental (v_2) band at 16.5μ (Tan *et al.*, 1972) which is very favorably placed for thermal radiation efficiency. However, there is no indication of enhanced thermal emission in the 17.0μ broadband photometry of Low and Rieke (1974b). Also, the band strength is unknown. It was therefore excluded from model calculations. However, high resolution spectrophotometry, again possibly from aircraft altitudes, would be desirable to determine its influence, if any, on Titan's thermal radiation balance.

Upper Limit on Ethylene Abundance

The v_7 band of C_2H_4 at 10.54μ is 21 times stronger than the C_2H_6 band at 12.2μ (Thorndike *et al.*, 1947; Thorndike, 1947). The two bands are similar in line structure (Allen and Cross, 1963, p. 209). From Figure 21.2, any C_2H_4 emission peak appears to be at least about 30 times weaker than the C_2H_6 peak. Since observations by two different groups (Gillett *et al.*, 1973; Low and Rieke, 1974b), with independent calibration, are being compared, large relative inaccuracies may be introduced, but it may be concluded that C_2H_4 is at least two to three orders of magnitude less abundant than C_2H_6 in the thermal inversion region. This is in agreement with the predictions of Strobel (1974a).

THE ENERGY BALANCE

The energy balance on Titan depends sensitively on several factors. The first factor is the radius. If the true radius of Titan were 10% larger than the baseline value of 2550 km assumed by Danielson *et al.* (1973), the emitted intensities in graphs such as Figures 21.1 and 21.2 would be reduced by 20%. Moreover, the Bond albedo would be similarly reduced.

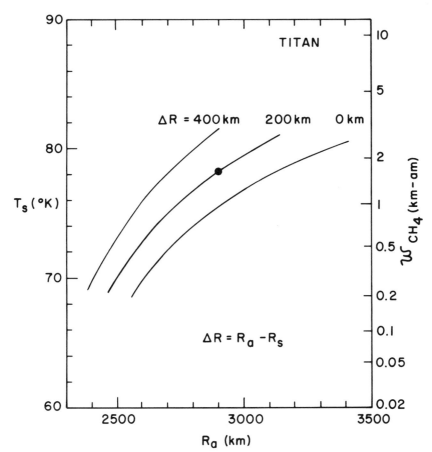

Fig. 21.3. The energy balance of Titan. The ordinate is the surface temperature necessary in the inversion model to balance emitted and absorbed radiation from Titan, for various atmospheric and surface radii. On the right, the column abundance of CH_4 (ω), corresponding to the surface temperature on the left, is shown. A value of $g = 125$ cm sec^{-1}, which is consistent with the adopted baseline model, was used to calculate ω.

A second factor is the depth of the atmosphere compared with the radius of the satellite's surface. Since the scale height of Titan's atmosphere is at least 60 km, it seems likely that the effective radius over which Titan absorbs solar radiation is at least 200 km larger than the radius of the solid surface. In this case, the emission from the solid surface is from an area which is about 13% smaller than the effective area for the absorption of solar radiation. Furthermore, the emitted radiation from the atmosphere (from molecules and dust) should also occur from an effective area which is significantly larger than the solid surface. The effective radius for the emitted atmospheric radiation is assumed to be the same as that for the absorbed solar radiation. In this chapter, a baseline model is adopted in which the atmospheric radius R_a = 2900 km (Elliot *et al.*, 1975) and the surface radius R_s = 2700 km.

A third factor is the surface temperature T_S. In the inversion model, it is closely related to the CH_4 abundance (ω_{CH_4}) in the atmosphere as illustrated in Figure 21.3. If T_S < 75°K, ω_{CH_4} is less than 0.9 km-am. This is about the lowest abundance which is compatible with Trafton's (1972b) determination of the CH_4 abundance on Titan. Hence T_S = 75°K is about the lowest surface temperature which is possible on Titan.

Figure 21.3 indicates how these three factors are interrelated in the inversion model. Although we have chosen R_a = 2900 km, $\triangle R = R_a - R_s$ = 200 km, and T_s = 78°K as a baseline, other combinations of these three parameters in the vicinity of the baseline point would also yield energy balance on Titan. Based on their infrared measurements, Low and Rieke (1974b) pointed out that the inversion model of Danielson *et al.* (1973) predicts that the emitted flux from Titan is substantially larger than that absorbed. Since R_a = 2550 km and $\triangle R$ = 0 were assumed by Danielson *et al.* (1973), it is evident from Figure 21.3 that T_s = 68°K ($\omega_{CH_4} \simeq$ 0.2 km-am) is required for energy balance. A surface temperature of 80°K, which was assumed by Danielson *et al.* (1973) leads to an energy imbalance of about 38%. This imbalance disappears, of course, if the new baseline parameters are adopted.

Figures 21.1 and 21.2 compare the observations of Titan with that predicted by the inversion model having the adopted baseline parameters. The agreement is very satisfactory.

ACKNOWLEDGMENTS

It is a pleasure to thank Darrel Strobel and Robert Danielson for many illuminating discussions, George Bird for his invaluable guidance in obtaining laboratory spectra, and Tobias Owen and Frank Low for supplying their infrared photometric data in advance of publication. This work was supported by NSF grant 39055 and by NASA grant NSG 7054.

APPENDIX

To test the random band model employed here, laboratory absorption experiments were performed on C_2H_6 with the cooperation of George Bird of Rutgers University. Examples of the laboratory transmission measurements are shown in Figure 21.4, where it may be seen that the random band model is in good agreement with the measurements, but has some discrepancies.

The first type of disagreement is that the actual band is slightly asymmetric.

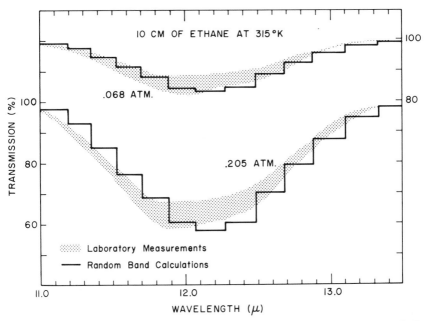

Fig. 21.4. Comparison of the random band transmission model used to calculate C_2H_6 emission with laboratory measurements. The calculated transmission is symmetric because the effects of the centrifugal distortion of the molecule have been omitted.

The random band model is symmetric because centrifugal distortion of the molecule has been ignored. This was made necessary by the lack of fundamental data on the C_2H_6 molecule. However, it is hoped that the appropriate molecular constants can be obtained from the Raman rotational spectrum of C_2H_6 in experiments currently being planned by Bird. It may be noted that the random band model used here predicts a greater optical depth than the measurements indicate on the long wavelength side of the band, and predicts a smaller optical depth on the short wavelength side.

The second area of disagreement between the observations and the calculations is that at the band center a smaller opacity is indicated by the calculations. This may result from a partial failure of the random band model there. Strong lines in the Q branches near the band center virtually coincide. Therefore, there may be more opacity concentrated in limited wavelength regions than the random band model assumes. In the wings of the band, energy level terms which are proportional to J^4 reduce this effect by spacing the lines farther apart.

The laboratory conditions on which Figure 21.4 was based differ from conditions on Titan by a factor of two or more in both temperature and pressure. Therefore, well calibrated transmission observations at low temperature and pressure would be desirable to check the random band model. This is doubly important because Bird (private communication, 1974) has suggested that there may be a ''hot band'' superimposed on the main band structure. The proposed hot band is another flexing model vibration with the lower level at the first torsional excitation level (275 cm^{-1} above the ground level). At laboratory temperatures, half the C_2H_6 molecules may be in torsionally excited states, whereas at 160°K, less than 20% are. Concentrating more absorption strength into fewer lines at low temperatures could reduce the randomness of the line spacings and thus introduce further errors in calculations assuming complete randomness.

The hot band could also be responsible for some of the asymmetry evident in Figure 21.4.

DISCUSSION

LAURENCE TRAFTON: Does the radius of Titan (as defined by the effective depth at which the monochromatic emission originates) vary with wavelength over the emission profile? If it does, then the derived spectrum would be exaggerated because of the variable emission area. This also would overestimate the degree of the temperature inversion.

JOHN CALDWELL: This effect has not been included in the above calculations. My intuitive feeling is that any uncertainties caused by ignoring it would be comparable to the effects of errors in the radius and errors in deriving the average transmission from the random band model. I believe that a more accurate treatment of Titan's limb emission would be justified in future calculations, but I'm quite sure that none of the major conclusions from the inversion model would be changed.

The temperature of the inversion region is only known approximately because of the difficulty of calculating opacity in the wing of the CH_4 band at 7.7μ. However, changing the nominal value of 160° would not have serious consequences for the model, because other parameters could be adjusted to offset the change. The inversion model described here does not give a unique representation of the infrared emission from Titan, but the parameters chosen are undoubtedly approximately correct, and the overall agreement between the observations and the adopted model is good.

VARIABILITY OF TITAN: 1896–1974

Leif E. Andersson
University of Arizona

22

Visual and photoelectric photometry of Titan for 1896–1900, 1922, 1951–56, and 1967–74 is discussed and the absolute magnitude is derived for each apparition covered. V(1,0) ranges from − 1.17 for 1951–56 to − 1.30 for 1973–74 and an uncertain − 1.36 for 1922. The coverage is too spotty to determine whether the variation is periodic or irregular, but a strict periodic variation with P = 29.5 yr (the orbital period of Saturn) is excluded. No attempt is made here to explain Titan's behavior.

Titan, the largest satellite of Saturn, shows no light variations attributable to axial rotation, unlike all other satellites for which sufficiently accurate photometry is available. This was first demonstrated convincingly by Harris (1961) and verified by all subsequent photoelectric studies with adequate coverage in orbital (and presumably rotational) phase. Furthermore, Titan has a very small variation with solar phase angle; the phase coefficient at 5500Å is approximately 0.005 mag/deg according to Noland *et al.* (1974b). Similar values were found by Blanco and Catalano (1971) and Andersson (1974). This is much less than for any other similarly sized or smaller body in the solar system (Chapt. 9,Veverka). These photometric properties of Titan are connected with the fact that it has an atmosphere, discovered by Kuiper (1944) and recently found to be extensive and complex (Chapt. 20, Hunten; Chapt. 21, Caldwell).

Although the short-term variations in the amount of light that Titan reflects are small, there remains the possibility of variations on a longer time scale. A brightening in recent years was suggested by Veverka (1974) and announced by Franklin (1974) on the basis of his observations in 1967 and 1972–74. [Other recent results are given by Noland *et al.* (1974a), Jones and Morrison (1975), Blanco and Catalano (1975), and Lockwood (1975a, 1975b).] I began observing Titan in 1970 September, as part of a program of UBV photometry of satellites of the Jovian planets, and, in view of an immediately obvious discrepancy with the photometry of Harris (1961), also made photometry of the comparison stars used by Wendell (1913) in his 1896–1900 visual photometry of Iapetus and Titan. Later, I added photometry of the stars used by Graff (1939) in 1922. In the meantime photometry by several observers (see references at bottom of Table 22.1) became available, allowing a relatively complete lightcurve in V for 1967–74 to be constructed.

The photometric system to which the various series of observations have been reduced is V of the UBV system since many of the observations have been made using this system and since the old visual photometry can be easily transformed to V.

AVAILABLE PHOTOMETRY OF TITAN

Titan's absolute magnitudes during known periods of observation are given in Table 22.1. The mean value for each apparition and series of observations is given through the 1969/1970 apparition; thereafter the magnitudes refer to time intervals of about two months. All series of observations since the middle of 1970 (excluding the last entry in the table) have observing dates in common with at least one other series, and an accurate mutual connection is possible without involving any magnitude transformations. Two of the recent series are originally on the UBV system: Andersson (1974) and Franklin and Cook (1974). The magnitudes found by Andersson are about $0^{m}\!.04$ brighter than those of Franklin and Cook, mostly due to a difference in the standard star magnitudes used. A compromise magnitude zero point midway between that of Andersson's photometry and that of Franklin and Cook is adopted; see also the next section of this chapter.

In the original publications the magnitudes are usually reduced to mean opposition distance, while Table 22.1 gives absolute magnitudes, i.e., reduced to unit distance from Sun and Earth. The difference between mean opposition magnitude and absolute magnitude is

$$V_{o} - V(1,0) = 9.555$$

for the commonly adopted $A = 9.540$ AU (semimajor axis of the orbit of Saturn). The magnitudes in Table 22.1 have been reduced to zero solar phase angle assuming a linear phase function with slope $dV/d\alpha = 0^{m}\!.004/deg$, the value derived by Andersson (1974). The means of the solar phase angles of the individual observations are given in the column labeled "$<\alpha>$". The number of nights entering in the magnitude mean is given under "n". The column entitled "ΔV" contains the correction that has been applied to the magnitudes in the original publication to put them on the V system; detailed justification for individual series of observations is given in the next section.

The typical uncertainty of the pre-photoelectric items in Table 22.1 is estimated at $0^{m}\!.04$. From the cases of independent transformations to the V system of simultaneous observations of Titan, it appears that the errors introduced by the transformations are typically about $0^{m}\!.02$. This is a reasonable estimate of the accuracy of the connection between the individual photoelectric series and the V system. However, the internal accuracy among the data in the interval 1970.7–1974.1 is better, about $0^{m}\!.01$, since the mutual connections are based on simul-

TABLE 22.1
Titan: Absolute Magnitude V (1,0)

Date	V(1,0)	$<\alpha>$	ΔV	n	Ref.*
1896.4	−1.27	1°.8	—	3	1
1897.5	−1.27	4.7	—	2	1
1898.5	−1.25	2.5	—	3	1
1899.6	−1.29	5.5	—	11	1
1900.6	−1.29	3.3	—	2	1
1922.2	−1.36	3.5	—	34	2
1951–1956	−1.17	2.5†	—	17	3
1967.92	−1.20	5.4	—	3	4
1968.84	−1.26	3.1	−0.08	7	5
1968.9	−1.20	4.4	—	6	6
1969.72	−1.27	4.1	−0.08	6	5
1969.87	−1.28	1.8	−0.08	3	5
1970.10	−1.27	6.0	−0.08	5	5
1970.74	−1.254	4.7	+0.020	7	7
1970.88	−1.268	1.5	+0.020	8	7
1971.74	−1.266	4.7	+0.020	8	7
1971.87	−1.260	1.9	+0.020	6	7
1972.04	−1.262	5.1	+0.020	7	7
1972.04	−1.260	5.0	+0.015	7	8
1972.17	−1.268	6.1	+0.020	7	7
1972.76	−1.261	5.8	−0.008	19	9
1972.92	−1.269	2.1	+0.015	8	8
1972.93	−1.266	1.6	−0.008	11	9
1972.94	−1.250	0.5	−0.026	7	10
1973.08	−1.272	5.0	+0.015	4	8
1973.13	−1.270	6.0	−0.026	4	10
1973.81	−1.302	5.8	−0.026	6	10
1973.82	−1.307	5.6	+0.015	4	8
1973.95	−1.282	1.1	+0.015	4	8
1974.04	−1.308	3.3	−0.026	7	10
1974.08	−1.295	4.2	+0.015	4	8
1974.23	−1.32	6.3	—	3	11

*References:
  1 Wendell (1913) and Andersson (1974)
  2 Graff (1939) and Andersson (1974)
  3 Harris (1961)
  4 Franklin (private communication, 1974)
  5 Blanco and Catalano (1971 and private communication, 1974)
  6 Veverka (1970)
  7 Andersson (1974)
  8 Jerzykiewicz (1973) and Lockwood (private communication, 1974) [see Lockwood, 1975a, 1975b]
  9 Noland et al. (1974b) [see Noland et al., 1974a]
  10 Franklin and Cook (1974)
  11 Blanco and Catalano (1974d) [see Blanco and Catalano, 1975]
†phase angle assumed for Harris' data.

TABLE 22.2
Other Titan Photometry

Reference	Years of Observation	No. of nights	Type of Photometry
Pickering (1879)	1877–78	18	visual
Guthnick (1910, 1914)	1905–08	92	visual
Graff (1924)	1921	20	visual
Widorn (1950)	1949	36	visual
Payne (1971)	1968–69	18	visual
McCord et al. (1971)	1968–69	?	narrowband, 0.3–1.1μ
Blair and Owen (1974)	1971–72	9	UBV
Younkin (1974)	1972	1	narrowband, 0.5–1.1μ

taneous observations rather than transformations of the respective magnitude systems. Therefore the magnitudes in this time interval are given to three decimal places in Table 22.1.

Table 22.2 lists other Titan photometry, which has not been utilized here because of low accuracy, uncertainty in the transformation to V, or unavailability of individual observations. Useful V(1,0) magnitudes may well be obtainable from some of these. It would be particularly valuable if a good Titan magnitude could be derived from the data of Guthnick (1910, 1914).

DISCUSSION OF INDIVIDUAL SERIES OF OBSERVATIONS

Wendell (1913)

Complete details of the photometry of Wendell's comparison stars and the rereduction of the Titan observations are given elsewhere (Andersson, 1974). It was found that the transformation equation for Wendell's visual magnitudes is

$$m_v = V + 0.30\,(B–V) + \text{constant.}$$

Not all of Wendell's observing nights (of which there were more than 60) are included in the annual means in Table 22.1. In some cases his residuals from the nightly mean are large, in others the magnitude of the comparison star is poorly determined according to Andersson's (1974) photometry. Also, comparisons with very blue or very red stars were considered less reliable. However, never in the five years does the mean of Wendell's observations differ by more than $0^{m}\!\!.04$ from the mean of the selected observations given in Table 22.1.

Graff (1939)

Details of the photometry of the comparison stars and of the rereduction of the Titan magnitudes are given by Andersson (1974). There seems to be both a magnitude scale error and a color term with respect to V in Graff's visual magnitudes:

$$V = 1.11m_v - 0.97 + \text{color term.}$$

The color term was evaluated for Titan with the help of three stars among the comparisons that have V and $B-V$ comparable to those of Titan. Graff assigned lower weight to a number of his observations that were made close to Titan's conjunctions with Saturn; these observations are not included in the average. The V(1,0) derived from Graff's data is the least certain of the items in Table 22.1; an error of $0^m.10$, while unlikely, cannot be ruled out.

Harris (1961)

Harris states that he measured Titan on 19 nights; however, only 17 data points are shown in his plot of V_0 vs. orbital phase. The observing dates are unknown, and only the limiting years (1951 and 1956) are given. Since Harris observed other Saturn satellites in 1952 and 1953, but no other satellites at all in 1954 or 1955, it is highly probable that the years of Titan observation were 1951, 1952, 1953, and 1956. The phase angles are of course unknown; $<\alpha> = 2^\circ.5$ was assumed for the reduction to zero phase. The error introduced by this assumption is at most $0^m.01$ (cf. Chapt. 9, Veverka).

Franklin (private communication, 1974)

No correction is applied to Franklin's 1967 observations.

Blanco and Catalano (1971)

Due to a reduction error, the published magnitudes of Titan require a correction of $-0^m.08$ (Blanco and Catalano, private communication, 1974). There is then a discrepancy with the 1968 observations by Veverka.

Veverka (1970)

No correction is applied. Only the six observations in Veverka's Figure 2 (p. 192) are used.

Andersson (1974)

All observations (43 nights) are used. The principal comparison star is $+17^\circ.703$ (van Bueren 23), for which V = 7.515 was adopted after connection to UBV standard stars on many nights. However, Johnson et al. (1962) give V = 7.54. Andersson has some nights of Iapetus photometry in common with Franklin and Cook, and his magnitudes are a few hundredths brighter. It is therefore likely that Andersson's photometry requires a small positive correction; +0.02 has been adopted.

Jerzykiewicz (1973); Lockwood (private communication, 1974)
[see Lockwood, 1975a, 1975b]

These are results from the Lowell Observatory Solar Variation Program, in which currently Titan, Uranus, and Neptune are measured in the b and y filters of the Strömgren system. The Titan series has observing dates in common with all other photometry since 1970 which are discussed in this chapter, except for that of Blanco and Catalano (1974d). Since exceptional care has been taken by the Lowell observers to insure the constancy of the photometric system from season to season, this photometry provides an excellent means of reducing to a common system the Titan magnitudes from 1970 to the present. Three dates in common with Andersson's photometry give

$$V_0 \text{ (Andersson)} - y_0 \text{ (Lowell)} = -0.005 \pm 0.005 \, (\sigma) \, .$$

Therefore, the correction of the Lowell magnitudes to the common system is +0.015; this may be taken as the definition of the common system.

Noland et al. (1974b)

This Strömgren-type photometry has three nights in common with the Lowell material.

$$y_0 \text{ (Noland } et\ al. \text{)} - y_0 \text{ (Lowell)} = 4.347 \pm 0.006$$

The magnitudes of Noland *et al.* are referred to the star 37 Tau. The correction to obtain magnitudes on the common system becomes +4.362. Noland *et al.* also give a Titan magnitude transformed to the V system, which is 0^m05 fainter than that implied by the correction derived above. However, a comparison of the V values for standard stars adopted by Noland *et al.* with other published V values for the same stars shows that their V system requires a zero point correction of about -0^m025. [See Noland *et al.* (1974a).]

Franklin and Cook (1974)

This BV photometry has five dates in common with the Lowell data.

$$V_0 \text{ (Franklin and Cook)} - y_0 \text{ (Lowell)} = +0.041 \pm 0.014$$

The resulting correction is −0.026.

Blanco and Catalano (1974d)

No correction is applied to this UBV photometry. Note that the published magnitudes include a different phase function than the one used in this paper [see Blanco and Catalano (1975).]

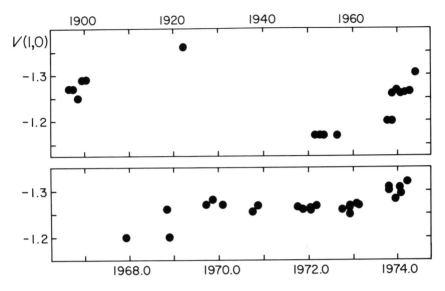

Fig. 22.1. *Top:* Mean absolute magnitude of Titan for each apparition. The discrepancy between the two points in 1968 is shown. The observations by Harris (1961) are indicated by points in 1951, 1952, 1953, and 1956. *Bottom:* Absolute magnitude of Titan, 1967–1974.

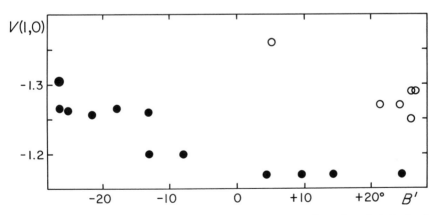

Fig. 22.2. Absolute magnitude of Titan, by apparition, versus Saturnicentric latitude of the Sun referred to the ringplane (B′). Filled symbols indicate photoelectric data.

TABLE 22.3
Titan: B–V Color

Year	B − V at $\alpha = 0$	$<\alpha>$	V(1,0)	n	Reference
1951–					
1956	1.295	2°.5*	−1.17	19?	Harris (1961)
1968.9	1.266	4.4	−1.20	6	Veverka (1970)
1968–	1.26	3.9	−1.27	21	Blanco and Catalano (1971
1970					and private communication 1974)
1970.8	1.270	3.0	−1.26	15	Andersson (1974)
1972.0	1.274	4.6	−1.26	28	Andersson (1974)
1973.1	1.258	6.0	−1.25	2	Franklin and Cook (1974)
1974.1	1.251	4.4	−1.26	2	Franklin and Cook (1974)
1974.2	1.254	6.3	−1.32	3	Blanco and Catalano (1974d)
					[see Blanco and Catalano, 1975]

*phase angle assumed for Harris' data.

SUMMARY

The data of Table 22.1 are plotted versus time in Figure 22.1. The coverage is too spotty to allow any statement about periodicity or total amplitude. The character of the variation since 1967, with a rise around 1969, a leveling, and another slight rise in 1973, seems to preclude a simple sinusoidal variation. An obvious periodicity to look for is Saturn's orbital period, 29.5 yr, since a connection might be expected with the distance from the Sun or with the Saturnicentric latitude of the Sun or of the Earth referred to the ring plane. Such a periodicity appears excluded by the magnitude discrepancy between the 1897 and 1956 observations, as well as the 1922 and 1951 data. However, if the 1922 data (which are quite uncertain) and the 1956 data (for which individual observations cannot be checked) are disregarded, Titan is on the average fainter when the rings (and therefore Titan's orbit and, presumably, its equator—see Chapt. 6, Peale) are seen edge-on (Fig. 22.2). This could be interpreted as indicating bright polar caps on Titan, or as a seasonal effect where Titan's albedo is higher at the times of solstices (as seen from Titan) than at the equinoxes. No conspicuous correlation with the 11-yr solar activity cycle is evident.

The B–V color according to different observers is given in Table 22.3. It was reduced to zero phase angle using $d(B-V)/d\alpha = 0.002$ mag/deg (cf. Chapt. 9, Veverka); the error resulting from the uncertainty in this phase coefficient is less than $0^{m}.01$. There is a suggestion of Titan becoming redder with fainter magnitude, but since color zero point errors in the different series of observations could conceivably amount to $0^{m}.02$ in some cases, the color-magnitude correlation requires more observational support.

While the uncertainties are large, the polarizations measured by Veverka (1970) seem to be larger than those found by Zellner (1973) when Titan was brighter (Chapt. 10, Veverka). Again, more observations are needed to establish whether a correlation exists.

ACKNOWLEDGMENTS

I thank G. W. Lockwood for permission to use unpublished Lowell observations, F. Franklin for details of his 1967 observations, and C. Blanco and S. Catalano for communicating the revision of their 1968–1970 photometry. This work is part of the author's Ph.D. dissertation at Indiana University.

PART V
SATELLITE ORIGIN

FORMATION OF THE OUTER PLANETS AND SATELLITES

23

A. G. W. Cameron
Harvard College Observatory

Recent models of Jupiter, Saturn, Uranus and Neptune suggest that all of these outer planets contain rocky and icy materials relative to hydrogen and helium in excess of solar composition. This points to a formation process in which condensed cores of the planets first collect together within the primitive solar nebula; when these cores become massive enough, substantial gravitational concentrations of gas in the solar nebula are formed in the vicinity of the planetary cores. In the cases of Jupiter and Saturn, it appears that the massive cir-cumplanetary envelope becomes unstable against hydrodynamic collapse onto the planetary core. It is probable that conservation of angular momentum will cause some of the collapsing envelope to go into orbit about the planetary core, forming a flattened disc. In the case of Uranus, it is likely that the large inclination of the spin axis of the planet was produced by a major collision during the terminal stages of assembly of the planet. If both proto-Uranus and the colliding body have associated with them a significant concentration of gas from the surrounding solar nebula, then it is probable that some of the gases would be placed into orbit about Uranus in the equatorial plane. This may spread out into a thin disc as a result of magnetic coupling and acceleration. For Jupiter, Saturn and Uranus solid condensates within their surrounding thin discs will be subject to the Goldreich-Ward instability mechanism, which will form condensed bodies of considerable size which are then able to accumulate to form the regular satellites associated with these planets.

STRUCTURE OF THE OUTER PLANETS

Since it is known that Jupiter and Saturn are predominantly composed of hydrogen and helium, the traditional exercise in constructing models of these planets has been to consider only these two substances in the construction of the models (Hubbard 1969, 1970). It is possible to construct quite reasonable models of Jupiter in this fashion, provided that the abundance of helium relative to hydrogen is raised to nearly double the proportions in the solar composition. In the case of Saturn, it does not appear possible to construct a good model with hydrogen and helium alone, since it is difficult to reproduce the gravitational moments of this planet by such a simple assumption.

In contrast, it has been known for some time that the bulk of Uranus and Neptune must be composed of icy materials, H_2O, NH_3, and CH_4, in addition to rocky materials. According to models constructed by Podolak and Cameron (1974), Uranus contains about 85% of its mass in the form of these substances, and Neptune about 75% of its mass. In these models the condition has been imposed that the hydrogen-to-helium ratio should be solar. When the same

[463]

assumption is made in the cases of Jupiter and Saturn, substantial amounts of chemically condensable materials are also found to be required for these planets. For Saturn, Podolak and Cameron found that about a third of the mass of planet should be in the form of rock and ice. In the case of Jupiter, they found that about one-sixth of the mass of the planet should be in the form of rock and ice (the ice being mostly mixed as H_2O vapor in the envelope), although, in this case, owing to the relatively small mass fraction of these constituents, the ratio of rock to ice is somewhat uncertain. For comparison it should be noted that the elements forming rocky and icy materials constitute only about 2% of the mass of the Sun.

As long as it was permissible to think of Jupiter and Saturn as having essentially solar composition, it was possible to suggest that these planets might arise from the operation of local gravitational instabilities within the solar nebula itself. For example, Urey (1966) has postulated that the solar nebula became gravitationally unstable against a break-up into bodies of which the rocky condensable component would have a mass comparable to that of the Moon. The total mass of such a body, as augmented by the icy constituents and by hydrogen and helium, would be comparable to that of the Earth. One would then suppose that Jupiter and Saturn were formed from the amalgamation of a large number of such bodies. One might further suppose that Uranus and Neptune were formed from similar bodies from which the bulk of the hydrogen and helium had escaped.

There are dynamical arguments against such a picture. The mass of a body which can be formed through a gravitational instability in the solar nebula depends on the temperature of the nebula. The higher the temperature, the larger the mass. However, whatever the temperature of the solar nebula, there will exist global instabilities which will have growth times much shorter than the local instabilities considered above, and they will involve very much larger amounts of material. These global instabilities involve bar-like deformations of the solar nebula, and the formation of rings (Hunter, 1965). Therefore, it is exceedingly unlikely that a situation would arise in which a very large number of local instabilities could have the preferred growth rates.

Furthermore, it is difficult to think of cosmogonic processes which would allow enrichment of helium relative to hydrogen in the material of the giant planets. If we think of some envelope of gas forming about a planetary core, separation of hydrogen from helium could only take place at some outer exospheric layer. A large-scale separation of hydrogen from helium would require the diffusion of hydrogen through the envelope toward the exospheric layer, from which it could thermally escape. The time scale for such diffusive separation would be very long compared to other characteristic time scales of the system, such as radiative cooling and mass addition. Thus, any such large extended envelopes are likely to contract into the body of the planet before any significant separation can occur. It is also unlikely that any solar wind effects can preferentially separate hydrogen from helium in the planetary atmosphere once it is formed, because such solar wind effects are likely to be more effective at the position of Jupiter than for the planets farther out, since the solar wind is stronger there. Yet Jupiter has the closest approach to the solar composition among the outer planets. These arguments form the basis upon which Podolak and Cameron (1974) constructed their models of the giant planets in which excess amounts of rocky and icy materials were included.

DYNAMICAL INSTABILITY IN THE SOLAR NEBULA

If one accepts the idea that the giant planets contain an excess of chemically condensable materials over solar composition, then a general mechanism of formation of the giant planets immediately suggests itself. We may start with the idea that the solar nebula at one time contained chemically condensed bodies of sufficient size (centimeters or larger) so that they could settle toward the midplane of the nebula to form a relatively thin disc (Cameron,1973). Goldreich and Ward (1973) have pointed out that such a thin disc becomes gravitationally unstable against the clumping of large numbers of such particles into bodies of significant size; this was an independent elaboration of a similar idea previously discussed by Safronov (1969). Chapter 26 by Safronov and Ruskol considers some of these characteristic time scales. As more recently developed by Ward (1974), two characteristic masses emerge from the analysis. The tendency of the bodies participating in the gravitational instability to clump together is halted at some stage through the operation of local conservation of angular momentum, which will cause the bodies to orbit about a common gravitating center. The larger characteristic mass is that of the entire mass of bodies participating in the gravitational instability. The smaller characteristic mass consists of a portion of the total amount of material which initially lay sufficiently close together so that it was able to contract to form a solid body before conservation of angular momentum halted the local contraction. The ultimate picture then is that the smaller mass m_1 is characteristic of a set of bodies which orbits about a mutual gravitating center, the entire cluster of bodies having a characteristic larger mass m_2. Ward (personal communication, 1974) has estimated that in the region of the solar nebula from which Jupiter was formed, m_1 and m_2 would be of the order 10^{16} and 10^{21} gm, respectively, in a low mass model of the solar nebula, and 10^{21} and 10^{28} gm, respectively, in a high mass model of the primitive solar nebula.

This instability process should operate quite quickly, but the amalgamation of the bodies of mass m_1 into bodies of mass m_2 would require a somewhat longer time scale in which gas drag effects have a chance to damp the relative orbital motions. Meanwhile, mutual gravitational perturbations will modify the orbits of the clusters of bodies, causing them to collide with one another, but it is not yet clear whether collisions among the clusters will take place on a longer or shorter time scale than that required for the amalgamation of bodies within each cluster to form a single body.

In this way one can expect massive planetary cores composed of chemically condensed material to form within the primitive solar nebula. As the planetary cores grow in mass, the local gas in the primitive solar nebula will become gravitationally concentrated toward the planetary cores, and eventually quite large amounts of mass in excess of the local background density of the primitive solar nebula can become attached to the planetary cores.

This process has been analyzed by Perri and Cameron (1974). They constructed numerical models in which an isentropic gas was concentrated toward a condensed planetary core and was in hydrostatic equilibrium, the density and temperature falling with increasing distance from the planetary core until a smooth match was made with the density and temperature of the background solar nebula at a distance where the gravitational potential gradient within the nebula was equal and opposite to that of the gravitating core together with its surrounding gas envelope. They then tested the hydrodynamic stability of the surrounding gas envelope against dynamical collapse onto the planetary core.

The results indicated that when the mass of the planetary core became large enough, the surrounding gas would become dynamically unstable against collapse onto the core. Perri and Cameron (1974) further demonstrated that the critical core mass at which instability occurred was very insensitive to the assumed position within the primitive solar nebula, and it was also very insensitive to the background pressure of the gas in the primitive solar nebula. However, the critical mass was quite sensitive to the adiabat which characterized the gas in the envelope. As the adiabat is lowered, the critical core mass decreases. Lowering of the adiabat corresponds to the cooling of the gas, and indeed, the cooling of the gas surrounding the planetary core will continually lower the adiabat as a function of time. The implication is that with continued increase in the mass of the core, and with continued cooling of the surrounding gas envelope, the time will come when the gas becomes dynamically unstable against collapse onto the core.

For a reasonable, low adiabat which might be achieved from a cooling of compressed gas from the primitive solar nebula, the critical core mass for dynamical collapse is still several tens of Earth masses. Since the excess condensable material in Jupiter amounts to 50 or 60 Earth masses, and since that in Saturn amounts to about 30 Earth masses, it appears very likely that dynamical collapse of the circumplanetary envelope onto those two objects could have taken place. The mass of the material participating in the collapse is comparable to the mass in the condensed core. After the collapse has taken place, it is no longer possible to achieve a hydrostatic continuation of this circumplanetary envelope into the gas of the surrounding solar nebula, and hence the planets will thereafter be gravitational sinks which will continue to absorb gas from a surrounding volume of space.

In the case of Uranus and Neptune, the mass of condensable material amounts to only 12 or 13 Earth masses. Thus, it is unlikely that these planets could have acquired gas from the surrounding solar nebula as a result of a hydrodynamic collapse process. Nevertheless, there would be a gravitational concentration of gas into a circumplanetary envelope about the cores of these objects as long as they were immersed in the primitive solar nebula, but this gas would continue to have a hydrostatic continuation into the surrounding solar nebula until such time as the T Tauri phase solar wind removed the surrounding primitive solar nebula. At that time, most of the circumplanetary envelope would be retained and would cool and contract to form the present hydrogen and helium atmospheres of these planets.

FORMATION OF CIRCUMPLANETARY DISCS

It has been shown that if gas from the primitive solar nebula is brought together from an extended radial interval in the primitive solar nebula, then conservation of angular momentum will require that the resulting planetary body be spinning rather rapidly in a prograde sense. This should already become manifest in the circumplanetary gas surrounding a planetary core in the sense described above. The situation is analogous to that in the evolution of a close binary system of stars, in which mass, transferred when one star expands and overflows its Roche lobe, goes into a prograde orbit about the companion star, and will form a gaseous disc surrounding that object. In the present instance we can regard the planetary cores as acquiring additional gas into their circumplanetary envelopes through a process in which the primitive solar nebula overflows its "Roche lobe" to go into orbit about its small binary companion. Thus the circumplanetary envelopes should have a rapid prograde rotation. Furthermore, the angular velocity of the outer parts of the circumplanetary envelope may possibly be increased as a result of angular momentum transfer in the envelope, resulting from the fact that the envelope can be expected to transport energy outwards by convection. The angular momentum transport in this case would result from turbulent viscosity.

Thus, when the dynamical collapse takes place onto the cores of Jupiter and Saturn, much of the collapsing gas can be expected to go into orbit about the resulting planet, forming relatively thin circumplanetary discs with differential rotation in the prograde sense.

In the cases of Uranus and Neptune, this is unlikely to take place. As we have seen, it is unlikely that a dynamical collapse process took place for either of these planets, and the circumplanetary envelopes formed about the planetary cores are unlikely to have been very extended initially. Thus, upon cooling and contraction, it is unlikely that they will be able to form circumplanetary discs.

However, in the case of Uranus we must also consider the process by which the planet acquired a spin axis tilted at 98° to the plane of its orbit about the Sun.

The tilting of planetary spin axes relative to the planes of their orbits is a process that probably depends upon the character of the largest collisions which occurred during the terminal stages of the accretion of these planets, as has been discussed by Safronov (1969). If there are a few collisions with massive bodies having a significant fraction of the planetary mass, then very large tilts of the rotation axis can result, but on a random basis.

Safronov has suggested that the largest collision involved in the formation of Uranus was with a body having 7% of the planetary mass. Actually, this number must be considered very uncertain, since we have no way of knowing the impact parameter involved for a single random event of that kind. Nevertheless, the implication is that the largest collision involved an incoming body with the mass of the order of an Earth mass or greater. Such a body probably had an impact parameter comparable to the radius of the proto-Uranus core.

Both proto-Uranus and the incoming body were thus probably large enough to have acquired a significant amount of hydrogen and helium in their envelopes, and hence as a result of the collision this surrounding gas would have to be set into rapid rotational motion in the equatorial plane of proto-Uranus with its tilted axis. Much of the gas was probably placed into orbit about Uranus, forming a small disc. It may have had a very different component of chemically condensable materials than the discs formed about Jupiter and Saturn, and any resulting satellites may thus also have a different bulk composition.

At this point we must discuss a feature of a largely gaseous planet which acquires a circumplanetary gaseous disc in orbit in the equatorial plane. Both of the methods of formation discussed here require that there should initially be a continuous fluid connection between the orbiting disc and the planetary atmosphere. Thus, we should expect Jupiter, Saturn, and Uranus to have been initially formed rotating fast enough so that the atmosphere in the equatorial plane became rotationally unstable. The present rates of rotation of these planets are significantly less than the implied initial rotation rate. Thus, it seems necessary that for these three cases some process operated to remove angular momentum from the planet. It is difficult to imagine a process which does not involve an initial planetary magnetic field. (De et al., Chapt. 24 herein, discuss processes, involving the primordial magnetic field, that they feel are critical in satellite formation.)

Since Jupiter and Saturn had to be formed before the primitive solar nebula had been completely dissipated, and since a dynamical collapse process was apparently involved, these two planets must have initially been formed at a very high temperature. In the case of Uranus, a major collision of the character described above would also produce a very hot atmosphere. Temperatures of at least several thousand degrees are implied from the amounts of gravitational potential energy released, and hence at least the inner portions of the circumplanetary discs are likely to be hot enough to have a significant ionic component. This provides the basis for a magnetic interaction, involving the initial planetary magnetic field, by means of which angular momentum can be transferred from the planetary rotation to the surrounding circumplanetary disc. This will cause the disc to separate from the planetary atmosphere and move to a somewhat larger orbiting distance. A significant reduction in the planetary rotation rate should result from this.

There should also exist purely fluid dynamical processes which can operate within the circumplanetary discs to cause outward transport of angular momentum within the discs. If energy transport by convection exists within the discs, then turbulent viscosity will produce such angular momentum transport. As discussed by Cameron and Pine (1973) for the primitive solar nebula, large-scale circulation currents should also be set up within such discs. In a region in convective equilibrium, such circulation currents result from the fact that a convecting region is only stable if it is co-rotating upon cylindrical surfaces, but such co-rotation is prevented by gas pressure gradients within the disc. In the case of a region in radiative equilibrium, the situation is more complicated, but circulation currents should be set up analogous to the Eddington-Sweet circulation currents which occur within the radiative envelopes of rotating stars. These circulation currents will also produce an outward transport of angular momentum, spreading the discs out toward larger radii.

FORMATION OF THE REGULAR SYSTEMS OF SATELLITES

Once a circumplanetary disc has cooled sufficiently for solid bodies to form by chemical condensation, it can be expected that the larger of these bodies will settle toward the mid-plane of the circumplanetary discs. At that point the Goldreich-Ward gravitational instability mechanism can once again operate, forming clusters of mass m_2 of bodies of mass m_1. Ward (personal communication, 1974) has estimated that for the circumplanetary disc about Jupiter, m_1 is of order 3×10^{18} gm and m_2 is nearly 10^{21} gm. The corresponding numbers for Saturn are nearly as large, and in the case of Uranus, m_1 is about 2×10^{16} gm, and m_2 is about 4×10^{18} gm. This instability mechanism should operate quite rapidly. As in the case of the solar nebula, the clusters of mass m_2 should be gradually consolidated into single bodies as a result of gas drag which damps the mutual orbiting motions within the cluster, and mutual gravitational perturbations among these clusters should lead to mutual collisions among them. (The idea of satellite clusters is developed in more detail by Safronov and Ruskol in Chapt. 26 in this volume.)

In this way we can expect the systems of regular satellites of Jupiter, Saturn, and Uranus to be formed. It can be expected that the circumplanetary discs will have greater surface densities, and hence greater temperatures, near the planet than at larger distances from the planet. In this way one can expect that the satellites formed closer to the planet should have a larger ratio of rock to ice than those formed farther away, as is implied by the regular decrease in the mean densities of the satellites with increasing distance away from the planet. The Goldreich-Ward instability mechanism may operate only on the rocky component of the material in the circumplanetary disc, and the icy component may be condensed within the disc and accumulated onto the satellites at a later time after the disc has undergone additional cooling.

Pollack and Reynolds (1974) have recently suggested that the radial variation in the composition of the regular satellites of Jupiter can be produced by the very high temperature associated with proto-Jupiter, which will prevent the condensation of ices in the vicinity of the inner satellites until a later stage than would be possible in the case of the outer regular satellites. An elaboration of their idea to all the Galilean satellites is presented by Fanale *et al.* in Chapter 17. This suggestion is clearly related to the one made just above, but I do not believe it is the whole story. The circumplanetary disc formed about Jupiter almost certainly had a high enough surface density so that it was opaque to the transmission of radiation. The surface density should also have been large enough so that adiabatic compression during the formation of the disc would produce initial temperatures high enough to vaporize the ices and possibly also the rocky material. This means that the inner part of the circumplanetary disc, which would have a higher surface density than the outer part, would require a longer time to cool, thus retarding the condensation of ices until a late stage in the accumulation of the rocky cores of the inner satellites. The high planetary temperature may act to further retard the cooling of the inner part of the circumplanetary disc.

All of this provides a logical framework within which the formation of Jupiter, Saturn, and Uranus, together with their regular systems of satellites, can be understood. [The dynamics of the Uranian satellites are presented by Greenberg (1975) and Singer (1975).] In the case of Neptune, no dynamic collapse of a surrounding gaseous envelope was likely to have taken place, and evidently no large collision took place at a sufficiently large impact radius to cause a significant axial tilt and the formation of a circumplanetary disc from which regular satellites could form. Furthermore, there is no indication in the case of Neptune that a mechanism is required to slow down the planetary rotation from near rotational instability at the equator to the present observed value.

In addition to the regular systems of satellites, the giant planets also have other satellites, generally with considerably larger orbits, and mostly with substantial orbital eccentricities and inclinations with respect to the equatorial plane. Some of these satellites have retrograde rotations. It seems likely that these irregular satellites have been independently formed within the solar nebula by the Goldreich-Ward gravitational instability mechanism and that they have subsequently been captured by the giant planets. A purely dynamical capture mechanism, applicable especially to Jupiter, has been suggested by Bailey (1971), but is criticized by Burns in Chapter 7, Morrison and Burns (1976) and

Greenberg (1976a). It is possible that other capture mechanisms exist, involving gas drag effects between the satellites and the gas of the primitive solar nebula in the vicinity of the giant planets, but these mechanisms have not yet been explored.

ACKNOWLEDGMENT

This work has been supported in part by grants from the National Science Foundation and the National Aeronautics and Space Administration.

DISCUSSION

B. R. DE: Dr. Cameron has invoked an interaction between the planetary magnetic field and the circumplanetary gas in connection with the Uranian satellites. He therefore accepts that the planets possessed magnetic fields during the satellite formation era. Consider then the Galilean satellite region of Jupiter which has a characteristic dimension $\ell \sim 10^{11}$ cm and a magnetic field of the order of $B \sim 10^{-3}$ gauss. In the protosatellite cloud that occupies this region, we may have a ratio between the numbers of ionized and un-ionized particles of $N_e/N_{total} \gtrsim 10^{-5}$, with the ionization caused by low energy cosmic rays, x-rays, starlight flux (ionizing particularly carbon) and natural radioactivity. This corresponds to an electrical conductivity $\sigma \gtrsim 10^{12}$ e.s.u. Now the Lundquist criterion states that if

$$L = \frac{B\ell\sigma}{c^2}\sqrt{\frac{\mu}{\rho}} \gg 1 ,$$

(c = velocity of light, ρ = gas density, μ = permeability), the magnetic field interacts strongly with the gas. Such an interaction results in electric currents flowing in the gas—and these currents further ionize the gas. The magnetic coupling between the planet and the gas causes energy to be fed into the gas at the expense of the rotational energy of the planet. If we put $N_{total} \sim 10^5 - 10^{10}$ cm^{-3}, we find $L \sim 10^8 - 10^6$, so the Lundquist criterion is easily satisfied.

The question then occurs as to whether or not hydromagnetic effects in the circumplanetary medium during the formation era can be avoided in a theory of satellite formation (cf. Chapt. 24, De *et al.*).

A. G. W. CAMERON: I expect gas densities *many* orders of magnitude higher than stated by Dr. De, so that the only effective ionization will be a short-lived thermal ionization which is a maximum near the planet. Under these conditions, magnetic coupling should be more important for slowing down the rotation of the planet than for extending the dimensions of the gaseous disc. I expect that internal hydrodynamic processes will be more important for the latter effect.

THE CRITICAL VELOCITY PHENOMENON AND THE ORIGIN OF THE REGULAR SATELLITES

24

Bibhas R. De, Hannes Alfvén,
and Gustaf Arrhenius
University of California, San Diego

The three well-developed satellite systems of Jupiter, Saturn, and Uranus may hold the key to the origin of the solar system. The spatial groupings in these satellite systems lead us to postulate a physical phenomenon that may have been operative during the formation era of the satellites. This is the now well-known and experimentally verifiable critical velocity phenomenon. A similar spatial grouping is present also in the planetary system, possibly indicating a process of formation of the planets similar to that of the satellites. It is suggested, however, that in trying to explore the genesis of the solar system, the primary emphasis should be laid on the satellite systems rather than the planetary system–since this approach puts important constraints on any theoretical model. A coherent series of processes that may have led to the formation of satellites (and, as a corollary, the planets) is outlined. The main thrust of the discussion is the crucial role of plasma phenomena in the formative processes of the solar system.

We shall discuss a physical concept leading to a theory of origin of the regular satellites, that is, the planetary satellites that are likely to have formed *in situ*. Thus, we exclude from our discussion the Moon (Lyttleton, 1967; Gerstenkorn, 1955, 1968, 1969; Singer, 1968, 1970a, 1972; Alfvén and Arrhenius, 1972; cf. Wood, Chapt. 27 herein), Triton (McCord, 1966), Phoebe, and the Jovian satellites VI through XIII (Bailey, 1971, 1972) in view of the likelihood that these satellites may be captured bodies. Capture processes are described in Chapter 7 by Burns. The regular satellites of Mars will be excluded because of the rudimentary character of this system (cf. Burns, 1972; Pollack, Chapt. 14 herein). We shall also exclude the regular satellite Nereid of Neptune because this provides us with only one member of a satellite system and because of the possibility that the Neptunian satellite system may have been modified due to the capture of Triton (McCord, 1966).

In the course of our discussion we shall recognize that because of the striking similarity between the satellite systems and the planetary system, it is both logically proper and aesthetically appealing to try to examine the premise that these two systems have an identical genesis. Hence we shall keep the planetary system in the back of our minds as we try to develop a theory for the satellite systems.

Note: For a more detailed development of the theory outlined here, the reader is referred to the monograph by Alfvén and Arrhenius (1976).

BAND STRUCTURES

Band Structure in the Solar System

We start by defining a quantity

$$\Gamma = \frac{M_c}{R_o} \ , \tag{1}$$

where M_c is the mass of a central body (a planet or the Sun) and R_o is the orbital radius of a secondary body (a satellite or a planet). Thus $G \, \Gamma$ is the specific gravitational potential energy of the orbiting secondary body, where G is the gravitational constant. In Figure 24.1 the quantity $\log \Gamma$ has been represented along the radius of a circle. In this diagram the radial distance represents the inward trajectory of an object falling freely towards the central body. In the left sector of this diagram we have drawn an arc for each of the satellites of the Jovian, the Saturnian and the Uranian systems. We notice that the satellites tend to fall in bands—one band (or possibly two) shortward of $\log \Gamma = 19$ and one band between $\log \Gamma = 19$ and 20. There is only one satellite (Amalthea) with $\log \Gamma$ greater than 20 and this may represent the single member of a third band. The planets Saturn and Uranus could not have any satellites in this band since the surfaces of these planets (as marked by broken lines in the diagram, Γ being now defined as the planetary mass divided by the planetary radius) lie outside this band.

In the right sector of the diagram we have included the planetary system in order to explore the similarities between this system and the satellite systems. Here again the planets fall in two bands—one containing the terrestrial planets and the other containing the Jovian planets. The terrestrial planets complement and confirm the suspected Amalthea band in the left sector. The Jovian band overlaps the two outer satellite bands. We have included the Moon and Triton along with the planets in view of the likelihood that these may be former planets.

Having observed this band structure, we must now ask if the grouping of secondary bodies in the satellite systems as well as in the planetary system into discrete ranges of Γ is physically significant. To answer this question, we shall look for a possible physical explanation of this grouping.

Band Structure for Cosmically Abundant Elements

Suppose that a cloud of gas containing atoms of mass m is falling freely towards a central body. Let v and R be the velocity of infall of the cloud and its distance from the central body at any instant of time. If we neglect the thermal velocity of the atoms compared to their velocity of infall, then the kinetic energy of an atom at any instant is

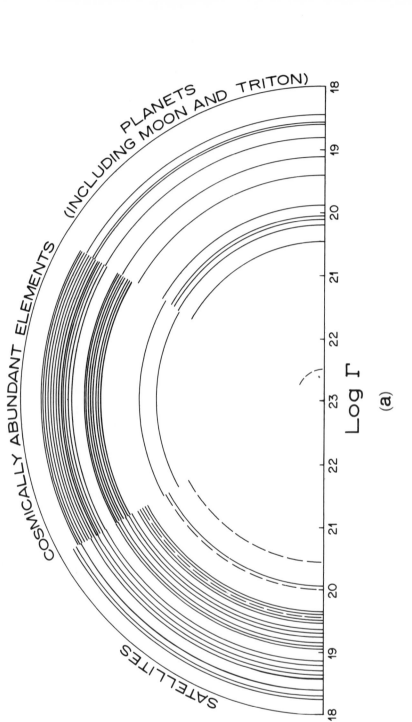

Fig. 24.1. (a) The band structure in the satellite systems, the planetary system and the cosmically abundant elements.

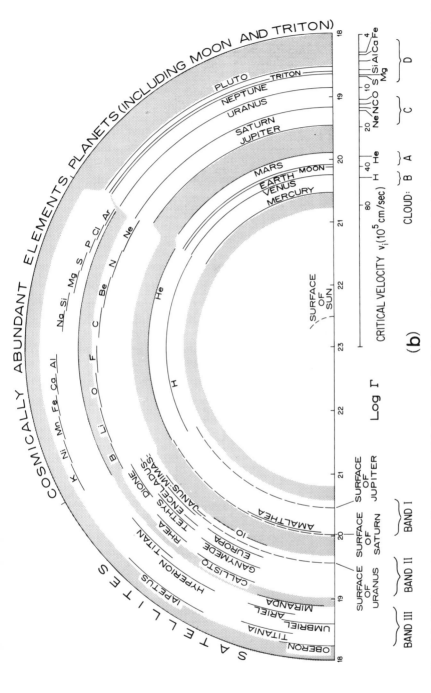

Fig. 24.1. (b) Key for the preceding diagram. This figure also includes a scale for the critical velocity for various elements, and the A, B, C and D clouds are indicated. The shading highlights the band structure.

[475]

$$\frac{1}{2}mv^2 = \frac{GM_c m}{R} . \qquad (2)$$

Suppose that when the atom reaches a distance $R = R_i$ ("the ionization distance") and a corresponding velocity $v = v_i$, its kinetic energy becomes equal to its ionization energy eV_i (e = electronic charge, V_i = ionization potential). Hence we can write

$$eV_i = \frac{1}{2}mv_i^2 = \frac{GM_c m}{R_i} . \qquad (3)$$

We may now define a parameter Γ_i for the infalling cloud (analogous to Γ for the satellites) by

$$\Gamma_i = \frac{M_c}{R_i} = \frac{eV_i}{Gm}$$

The parameter Γ_i for a cloud of infalling atoms is defined in terms of fundamental physical constants and is therefore itself a physical constant. In the middle sector of Figure 24.1 we have plotted the quantity $\log \Gamma_i$ for the first 23 cosmically abundant elements. We find that *these elements divide themselves into three distinct bands: Band I containing the elements in the first row of the periodic table, Band II containing the elements in the second row and Band III containing the elements in the third and the fourth rows.* Since this is a trend in a physical constant for the elements, it represents a fundamental property of matter.

Correspondence Between the Band Structures

On comparing the band structure for the elements in the middle sector of Figure 24.1 with the band structures in the left and the right sectors, we find that *the element bands coincide, substantially and without overlap, with the three satellite bands.*

In the planetary system, the terrestrial band coincides with Band I; the Jovian band coincides in part with Band II but partially overlaps Band III. The possibility arises therefore that the Jovian band has a correspondence to Band II but that it has shifted outward due to some reason that we should explore (for an explanation of this shift, see Alfvén and Arrhenius, 1974, 1975). The planet Pluto distinguishes itself from the Jovian sequence by having a higher density and orbital eccentricity, and hence it is possible that Pluto (and perhaps also Triton before its capture) has a correspondence to Band III. Neptune may have formed in a region of overlap between Bands II and III. In the following sections we shall use the band labels I, II and III to refer to the elements as well as to the satellites and the planets.

The correspondence between the band structures for the satellites and the planets on one hand and the elements on the other may be stated in the following way: Γ_i for the elements falls into ranges which are identical to Γ for the satellites and the planets. Hence we may conclude that from an observational point of view there exists an identity between the distributions of the parameters Γ_i and Γ: thus we may put $\Gamma_i \equiv \Gamma$. Since $\Gamma_i = M_c/R_i$ and $\Gamma = M_c/R_o$, this identity translates into the following statement: The average orbital distance R_o of the secondary bodies in a band is identical to the ionization distance R_i of a cloud containing one or more of the elements in that band.

It is this identity whose explanation we shall seek.

THE CRITICAL VELOCITY PHENOMENON

Concept of Critical Velocity

How was the matter that now forms the regular satellites derived? This material could not possibly have been ejected from the planets (see Alfvén and Arrhenius, 1974). There remain two other possibilities: (1) the planets captured heliocentric grains which later accumulated to form the satellites, somewhat as suggested in Chapter 26 by Safronov and Ruskol, or (2) the grains condensed in the neighborhood of the planets from gases accumulated by the gravitation of the planets, as mentioned in Chapter 23 by Cameron. The orbital angular momentum of the grains could then derive from the planetary spin by a mechanism described later. This would require the eventual satellites to orbit in the prograde sense in the equatorial plane of their respective planets, regardless of how this plane is oriented with respect to the ecliptic. (It is possible that the gravitationally accumulating gas already contained interstellar grains. Such grains would behave as a part of the gas, resulting in a colloidal plasma when the gas becomes ionized on reaching its critical velocity. Angular momentum would be supplied to this colloidal plasma hydromagnetically, so that these grains would eventually be found as accumulated larger bodies orbiting in the equatorial plane of the planet.) We shall pursue the second alternative here. We shall further make the assumption that, during the era of formation of the satellites, the planets possessed dipolar magnetic fields—the axes of the dipole being roughly aligned with the planetary spin axes. In the case of the formation of the planetary system, the same assumption can be made with regard to the Sun.

Returning now to the correspondence between the ionization distance R_i of a cloud and the orbital distance R_o of a group of secondary bodies, we note that such a correspondence would be expected if the infalling cloud is stopped at the distance R_i, forming a group of satellites at this distance. Such a stopping of the cloud can occur at the distance R_i if the gas becomes ionized and the magnetic field of the planet prevents further infall. Since the kinetic energy of the gas atoms equals their ionization energy at this location, the gas can, in principle, become ionized if there exists a mechanism which can convert the kinetic energy of the atoms into their ionization energy.

If such an ionization mechanism exists, then the velocity v_i, defined by eqn. (3), i.e., by

$$v_i = \left(\frac{2eV_i}{m}\right)^{1/2} , \qquad (5)$$

plays the role of a *critical velocity* such that when the velocity of a gas cloud reaches the value v_i appropriate to the gas, it becomes ionized (presumably by interacting with the background plasma and the magnetic field through which the cloud falls).

In this way, analysis of the orbital properties of the satellites and the planets, and its implications led to the concept of the critical velocity phenomenon (Alfvén, 1942–45, 1954), but an experimental demonstration of the phenomenon had not yet come.

The gas cloud we envisage as a precursor for a satellite group is a local cloud ("the source cloud") at a large distance (larger than the planet-to-outermost satellite distance) from the planet. This cloud is assumed to be initially at rest relative to the planet. When the cloud has reached the ionization distance R_i, the thermal velocity of the atoms is assumed to be much smaller than v_i, so that the atoms will be falling radially inward. (For a hydrogen cloud, the requirement is that the temperature of the cloud be much less than $10^5\,°K$ at this stage.)

There is assumed to be a thin "background" plasma medium surrounding the planet—the plasma being suspended in the magnetic field lines. The neutral gas atoms fall through this plasma. The density of the plasma is assumed to be so low that the mean free path of the atoms is much larger than the radial extent of the plasma medium. Thus there are very few collisions between the plasma particles and the atoms of the infalling gas. In spite of this, the infalling gas is supposed to be rapidly ionized when the relative velocity between this gas and the magnetized plasma exceeds the critical velocity. The original existence of a thin plasma around a planet—in fact the existence of a thin plasma anywhere in space—can be taken for granted.

Under these conditions, clouds of neutral gas falling toward a planet would be stopped in bands appropriate to the chemical composition of the cloud (or in bands appropriate to the dominant or "controlling" element in the clouds) and form condensates which eventually accrete to form the observed groups of satellites. For the formation of the planets, the same mechanism would be envisaged to take place around the Sun.

Applicability of the Critical Velocity Concept to the Satellite Formation Problem

When the preceding analysis was first made, it had three difficulties: (a) The ionization mechanism described above was not known to exist. (b) There was no theoretical basis for the hypothesis that masses of gas falling in toward the central bodies would have differing chemical compositions. (c) The chemical composition of the secondary bodies found in a band do not represent the chemical composition of the corresponding element band. Since practically nothing is known about the chemical composition of the satellites, let us turn to the planetary system for this discussion. For example, the terrestrial planets fall in a band which corresponds to the Γ values of H and He, but they contain very little of these elements. The band of the Jovian planets corresponds to C, N and O, but they were believed to consist mainly of H and He.

However, the above situation has changed drastically since the 1940s when it was first propounded because of both theoretical studies and empirical findings. Although we are still far from a final theory, it is fair to state that difficulty (a) has been eliminated by the discovery of the critical velocity phenomenon which is discussed in the next section. Difficulty (b) can be rationalized when we recognize that the separation of elements by plasma processes is a common phenomenon in space; such a separation process will be presented later. The apparent difficulty (c) is resolved by considerations given by Alfvén and Arrhenius (1976).

Experimental Verification of the Critical Velocity Phenomenon

In view of the fact that the theoretical treatment of plasma processes in space remains speculative unless supported by experiments, it was realized that further consideration of the applicability of the critical velocity concept to the problem of formation of satellites and planets depended on experimentally demonstrating the validity of this concept. As soon as the advance of thermonuclear technology made it possible, experiments were designed to investigate the interaction between a magnetized plasma and a non-ionized gas in motion relative to the plasma. These experiments have established beyond doubt the existence of the critical velocity phenomenon. A survey of these experiments has recently been given by Danielsson (1973), while a review of the various theories of the critical velocity phenomenon has been given by Sherman (1973).

The experiment most closely related to the cosmic situation was carried out by Danielsson (1969, 1970). The experimental arrangement is shown in Figure 24.2. A hydrogen plasma is generated and accelerated in an electrodeless plasma gun (a conical theta pinch). This plasma flows into a drift tube along a magnetic field. The direction of the magnetic field changes gradually from axial to transverse along the path of the plasma. A polarization electric field is developed and a plasma with a density of about 10^{11}–10^{12} cm^{-3} proceeds drifting across the magnetic field with a velocity up to 3×10^7 cm sec^{-1}. In the region of the transverse magnetic field the plasma penetrates into a small cloud of gas, which is released from an electromagnetic valve. This gas cloud has an axial depth of 5 cm and a density of 10^{14} cm^{-3} at the time of arrival of the plasma. The remainder of the system is under high vacuum. Under these conditions the mean free path

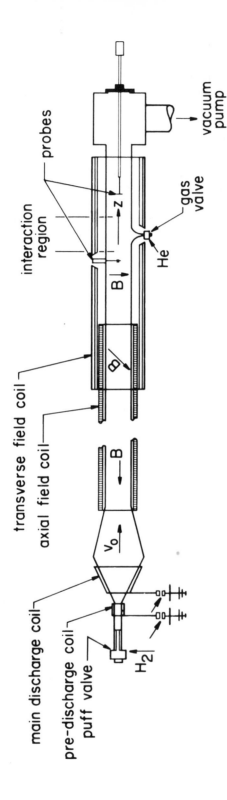

Fig. 24.2. The experimental arrangement of Danielsson (1969, 1970) for measuring critical velocity: The left part is a plasma gun, emitting a magnetized plasma with a velocity v_o. In a long drift tube, the longitudinal magnetization is changed to transverse magnetization. A thin cloud of gas is injected through the gas valve. If v_o is below the critical velocity, the plasma beam passes through the gas cloud with very little interaction since the mean free path in the gas cloud is large. If v_o is above the critical velocity, there is a strong interaction, decelerating the plasma beam and at the same time partially ionizing the gas cloud.

for direct binary collisions is much larger than 5 cm so that the interaction, according to common terminology, is collisionless.

The result of the experiment is that as long as the plasma velocity v_0 is below a certain value v_c, the plasma beam passes through the gas cloud with very little interaction, since the mean free path is large. If v_0 exceeds v_c, there is a strong interaction, bringing v_0 down to near v_c. At the same time the gas cloud becomes partially ionized. In general, the interaction leads to (a) local heating of the electrons, (b) ionization of the neutral gas and (c) deceleration of the plasma stream.

This and a number of other experiments have demonstrated the existence of a critical velocity v_c. Although under certain conditions there are deviations up to perhaps 50%, the general result is that v_c is the same as the velocity v_i given by eqn. (5).

THE FORMATION PROCESS

Chemical Composition of Infalling Clouds

Let us assume that the source cloud which contains all elements (having for instance abundances appropriate to some average "galactic" composition) is partially ionized to such an extent that all elements with ionization potentials higher than a certain value V_I are neutral, but all elements having ionization potentials lower than V_I are ionized. The Larmor radii of the electrons and ions are assumed to be much smaller, and the mean free paths much larger, than the dimensions of the source cloud. The neutral atoms will then begin to fall toward the central body.

Let V_I decrease slowly with time (e.g., through a magnetically induced cooling of the plasma as discussed by De, 1973b, in the case of solar prominences). When it has fallen below the ionization potential of helium, helium will recombine and a gas cloud dominated by helium will fall towards the gravitating central body. Helium reaches its critical velocity of 34.4×10^5 cm sec^{-1} at a Γ value of 0.9×10^{20}, that is, in the outer region of Band I in Figure 24.1. At this location the gas will become ionized and form a plasma cloud which will be referred to as the A cloud. When V_I decreases further and passes the ionization potential of hydrogen (which is nearly equal to the ionization potentials of oxygen and nitrogen), hydrogen, oxygen and nitrogen will start falling towards the central body. Since hydrogen is by far the most abundant element, we expect this infalling cloud to behave as hydrogen and stop at its Γ value of 1.9×10^{20}, i.e., in the inner region of Band I, forming what we shall call the B cloud. Although oxygen and nitrogen reach their respective critical velocities already in Band II, they will not be stopped in this band because of the quenching effect of the dominating component hydrogen on the acceleration of the electrons that would have ionized oxygen and nitrogen in Band II.

Next will follow an infall dominated by carbon which will be stopped at $v_i = 13.5 \times 10^5$ cm sec^{-1} and $\Gamma = 0.1 \times 10^{19}$ (Band II), forming the C *cloud*. Finally, the heavier elements (mainly silicon, magnesium and iron) will fall in to $\Gamma = 0.3 \times 10^{18}$ (Band III), producing the D *cloud* having a weighted mean critical velocity of 6.5×10^5 cm sec^{-1}.

In this way, clouds of gas having different compositions and falling towards the central body will be stopped in the different bands, thus removing difficulty (b) as mentioned earlier. Note that the formation of the plasma cloud in a band depends on the attainment of the critical velocity by the element or elements appropriate to this band. For less massive planets, the inner clouds cannot form due to inadequate acceleration of the infalling gas from the source cloud. We see from Figure 24.1 that Jupiter is massive enough to have an A cloud, but not a B cloud. The Galilean satellites of Jupiter formed in the Jovian C cloud. The inner Saturnian satellites formed in the Saturnian C cloud, while the outer satellites formed in the Saturnian D cloud. The satellites of Uranus formed in the Uranian D cloud.

In the planetary system, the Sun is massive enough to have a B cloud, in which the planets Mercury, Venus and Earth formed. Mars and perhaps also the Moon formed in the A cloud. The Moon probably should be included in the A cloud rather than the B cloud, considering the fact that the sizes of the Moon and Mars are much less disparate than the sizes of the Moon and the Earth for instance. However, from Figure 24.1 it seems likely that there was an overlap and possibly an intermixing between these two clouds around the Sun. The giant planets formed in the C cloud while Pluto and perhaps also Triton formed in the D cloud around the Sun.

Angular Momentum Transfer

One of the primary problems that any theory of the origin of the regular satellites must recognize is the necessity of imparting angular momentum to the satellite precursor medium in such a way that the eventual satellites would revolve around the planet in the prograde sense. The only plausible source of this angular momentum is the spin angular momentum of the planet.

The transfer of angular momentum from a rotating celestial body to a surrounding medium is a problem which has attracted much interest over the years. It has been concluded that an astrophysically efficient transfer can be produced only by hydromagnetic effects. A process of angular momentum transfer that is consistent with both the observations of particles and fields in space and the scheme of the present theory has been discussed elsewhere (Alfvén and Arrhenius, 1973; see also Fig. 24.4). The requirements for the magnetic field strengths of the central bodies for this process to be operative have been shown to be reasonable. The question of support of the plasma by electromagnetic force against the gravitation of the central body has also been discussed by Alfvén and Arrhenius (1973). It has been found that only a small fraction of the final mass of a secondary body can be supported by the electromagnetic force at any particular time. This means that the plasma density ρ_B at any time can only be a small fraction of the *distributed density* ρ_f (which is equal to the mass of final body divided by the space volume from which it derives; this volume may be taken to be a toroid with its large radius equal to the orbital radius of the secondary body and the small diameter equal to the sum of the half-distances to the immediate inner and the outer secondary bodies). Their analysis led to the conclusion that the instantaneous densities are less than the distributed densities by about 10^{-11} for the satellite systems to about 10^{-7} for the giant planets. Thus the *plasma densities we should consider are of the same order as the present number densities in the upper solar corona* (10^2–10^8 cm^{-3}). It should be observed that these values refer to the *average* densities. Since the plasma is necessarily strongly inhomogeneous due to the existence of current filaments that transfer the angular momentum (see Fig. 24.4), there may be localized dense regions with densities several orders of magnitude higher.

This is important because the time of condensation of a grain and its chemical and structural properties at its growth refer to the *local* condensation environment, and not to the entire primeval nebula. Assuming that the primordial components of meteorites were formed in the primeval circumsolar nebula, one can derive the properties of the medium from which they formed. The densities ($\sim 10^{10}$–10^{14} cm^{-3}) suggested in this way (Arrhenius, 1972; Arrhenius and De, 1973; De, 1973b) are much higher than ρ_B but still lower than ρ_f. These densities are likely to be characteristic of the localized dense regions where the major fraction of the condensation is likely to have taken place.

The localized dense regions are produced by the compression of the plasma into filaments which carry the electric current that causes the angular momentum transfer. The plasma medium in general resembles the solar corona, and the filaments embedded in this medium resemble the solar loop prominences, with the difference that the filaments have much larger dimensions, extending from the central body to the region of the eventual secondary bodies. For this reason, these filaments may be termed *superprominences*. Analogous to the solar prominences, the density in the superprominences is a few orders of magnitude larger than that of the surrounding medium, while the temperature in these regions is much lower. Typically, the temperature in the surrounding medium may be of the order of 10^5 °K, while the temperature in the superprominences may be 10^4 °K. The cooling of the plasma in these regions is induced by the compression of the plasma (see De, 1973a). Since high densities and low temperatures favor condensation, the major condensation of the plasma takes place in the superprominence regions.

Energy Source for the Plasma

The kinetic energy derived from hydromagnetic braking of a central body during angular momentum transfer is converted in part to the orbital energy of the plasma being accelerated and in part to thermal energy which is dissipated in the immediate vicinity (the "ionosphere") of the central body, in the plasma being accelerated and in the intervening medium.

Suppose that a central body is decelerated from angular velocity Ω_1 to Ω_2 while accelerating a mass M of a plasma cloud at a distance r from rest to an angular velocity ω. Then conservation of angular momentum requires that

$$\Theta\,(\Omega_1 - \Omega_2) = M\,r^2\,\omega \;, \tag{6}$$

where Θ is the moment of inertia of the central body. The energy released in this process is

$$W = \frac{1}{2}\,\Theta(\Omega_1^2 - \Omega_2^2) - \frac{1}{2}\,M\,r^2\omega^2 \;. \tag{7}$$

Putting $\Omega = (1/2)\,(\Omega_1 + \Omega_2)$ in the above two equations we have

$$W = M\,r^2\,(\Omega\omega - \omega^2) \;. \tag{8}$$

The velocity $r\omega$ will eventually equal the Kepler velocity, so that

$$M\,r^2\,\omega^2 = \frac{GM_cM}{r} \;, \tag{9}$$

where M now is the mass of the secondary body formed by the condensation of the plasma. Using eqn. (9) in eqn. (8), we have

$$W = \frac{GM_cM}{r}\;\left(\frac{\Omega}{\omega} - \frac{1}{2}\right)\; = \frac{GM_cM}{r}\;\left(\frac{\tau}{T_s} - \frac{1}{2}\right)\;, \tag{10}$$

where T_s is the spin period of the central body and τ is the orbital period of the secondary body at a distance r. This represents the energy released during the angular momentum transfer; it is used to sustain the plasma.

Condensation of Grains and Evolution of Grain Orbits

As the process of transfer of angular momentum proceeds, plasma condenses into grains preferentially in the superprominence regions. Although the plasma temperatures in such regions are of the order of several thousand degrees, the temperatures of solid grains condensing out of the plasma are typically of the order of 10^3 degrees (see, e.g., Arrhenius and De, 1973). Such a model of condensation under the condition of extreme temperature disequilibrium is necessitated by physical considerations (Arrhenius and De, 1973; also De, 1973b) and it helps us to understand many observations in meteoritic material (believed to be almost unaltered primordial condensates) that are not explainable in terms of physically untenable models with temperature equilibrium.

Furthermore, the flow of electric currents produces enhanced magnetic fields in the superprominence regions (in the same way as in the case of the solar prominences). If we assume meteoritic materials to have condensed from such a magnetized plasma in a circumsolar superprominence, we should expect to observe some imprint of the magnetic field in those meteoritic materials which are likely to preserve such an imprint. Indeed, the study of remanent magnetization in the carbonaceous chondrites indicates that the grains in these meteorites have condensed in an environment permeated by magnetic fields $\leq$ 0.1–1.0 oersted (Brecher, 1972; Brecher and Arrhenius, 1974, 1976). If we assume reasonably that the region of condensation of these materials is the asteroidal region and if we were to ascribe a magnetic field of the order of 0.1–1.0 oersted in this region to a primeval solar dipole field, it would require an unreasonably high dipole strength ($\sim 10^{40}$–10^{41} G cm^3). The superprominences with locally enhanced magnetic fields would thus be a very plausible condensation environment.

Having condensed from the plasma, a population of grains in orbit would interact mutually via inelastic collisions. Such collisions, provided they occur with a time scale much larger than the orbital period of the grains, do not lead to a spreading or scattering of the grain orbits, but rather they lead to a focusing of the orbits so that eventually the grains have orbits with orbital radii, eccentricities and inclinations all in very narrow ranges. Such a population of grains, evolved into a well-focused narrow orbital stream, is known as a *jet stream* (for detailed discussion see Alfvén and Arrhenius, 1970).

Within the jet streams, the grains collide and aggregate to form larger bodies. When one of these bodies has grown so large that it can gravitationally attract other bodies in the stream, it begins to sweep up all the matter in the jet stream, eventually to become a secondary body.

Formation of Planets: The Hetegonic Principle

We have stated at the outset that because of the striking similarity between the satellite systems and the planetary system, it is worthwhile to compare these two systems with a view to determining if both these systems originated as a result of identical series of processes. Throughout the development of the theory outlined here, we have included the planetary system—sometimes as a corollary, sometimes on an identical footing with the satellite systems. We have encountered no major difficulties in applying the processes discussed for the satellite systems to the planetary system. We can therefore put our entire analysis on a broader basis and aim at developing a general theory for the formation of secondary bodies

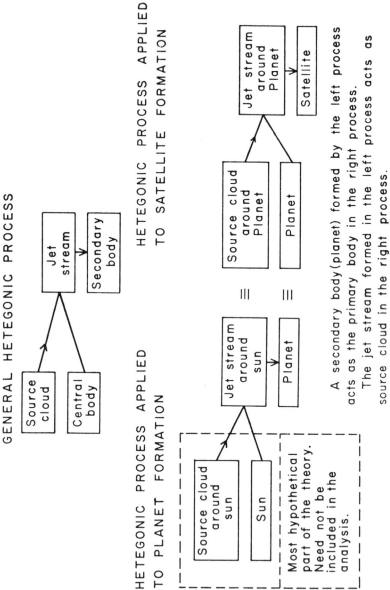

Fig. 24.3. Diagram showing how the speculative character of the theory is reduced by the hetegonic principle which implies that all theories should be applicable to both planetary and satellite systems. This eliminates the need to rely on hypotheses about the early Sun and ties the theory closer to observations.

around a central body, applicable to both the satellite systems and the planetary system. A theory based on this principle may be called a *hetegonic theory* and the principle that the satellite and the planetary systems have identical geneses may be called the *hetegonic principle* (the term hetegonic being derived from the Greek ἐταιροσ or ἐτησ meaning companion. Thus the concept refers to a general theory of formation of companion bodies.).

The hetegonic principle is a powerful principle because of the severe constraints it puts on any model. In spite of this it has usually been neglected in the formulation of solar system theories. Figure 24.3 shows how the principle is applied to the two similar series of processes leading to the formation of secondary bodies from a primeval dispersed medium. The chain of processes leading to the formation of planets around the Sun is repeated in the case of the formation of satellites around the planets, but in the latter case a small part (close to the planet) of the planetary jet stream provides the primeval cloud out of which the satellites form. There is only one basic chain for the processes discussed in this paper which applies to the formation of *both* satellites and the planets.

The consequence of this is that we can explore the processes leading to the formation of the planets without detailed assumptions regarding the properties of the early Sun. This is advantageous because these properties are very poorly known. If we adopt the hetegonic principle, we need not concern ourselves with the question of whether the Sun passed through a high luminosity Hayashi phase or whether the solar wind at some early time was much stronger than now. Neither of these phenomena could have influenced the formation of satellites (e.g., around Uranus) very much. The similarity between the planetary and the satellite systems shows that such phenomena have not played a major dynamic role.

A Model of the Supercorona

To summarize the theory outlined here, we can now sketch a model for the medium around a central body during the formation era. Because of the similarity of this medium to the present day solar corona and its much larger dimension, this medium may be called a *supercorona*.

The supercorona consists of regions of widely differing properties as shown in Figure 24.4.

LOW DENSITY PLASMA REGIONS

Most of the space in the supercorona outside the jet streams is filled with a low density plasma having densities perhaps in the range $10-10^5$ cm^{-3}. The supercorona is fed by injection of gas from a source cloud at a large distance ("infinity"). The transfer of angular momentum from the central body is achieved by processes in this plasma. This means that there is a system of strong electric currents producing filamentary structures or superprominences.

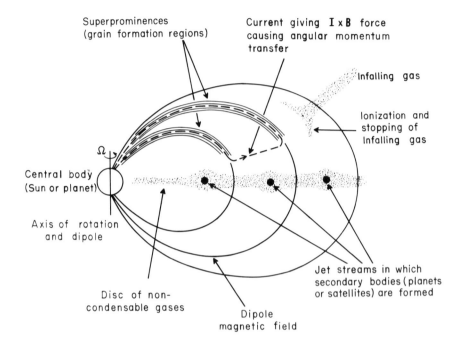

Superprominences
(grain formation regions)

Current giving I x B force
causing angular momentum
transfer

Infalling gas

Ionization and
stopping of
Infalling gas

Central body
(Sun or planet)

Axis of rotation
and dipole

Disc of non-
condensable gases

Jet streams in which
secondary bodies (planets
or satellites) are formed

Dipole
magnetic field

Fig. 24.4. A sketch of the series of hetegonic processes leading to the formation of secondary bodies around a spinning magnetized central body. The dipole magnet is located at the center of the central body and is aligned with the spin axis. A gas injected from infinity into the environment of the central body becomes ionized by interacting with the pre-existing magnetized plasma when the free-fall velocity of the gas exceeds the critical velocity. This ionized gas then remains suspended in the magnetic field. The rotation and the magnetic field together with the conducting plasma surrounding the central body constitute a *homopolar generator* (see e.g., Alfvén and Fälthammar, 1963, p. 11), producing an e.m.f. which drives a current in the plasma. This current I together with the magnetic field B gives rise to a force I x B which transfers angular momentum from the central body to the surrounding plasma. The current also produces prominence-like regions of gas (labelled *superprominences*) because of the filamentary nature of the current. These regions are denser and cooler than the surrounding medium and in these regions the condensation of grains takes place. Through inelastic collisions, the population of grains evolves into a number of jet streams while the non-condensable gases form a thin disc in the equatorial plane. (Not drawn to scale.)

FILAMENTARY STRUCTURES OR SUPERPROMINENCES

The primeval plasma medium structurally resembles the solar corona with embedded prominences produced by strong currents. These stretch from the surface of the central body out to the regions to which the angular momentum is being transferred. As in the solar corona, the filaments have a density which is orders of magnitude higher and a temperature which is much lower than those of the surrounding medium. As high density and low temperature favor condensation, most of the condensation takes place in the filaments. The grains condensed in these regions receive a tangential velocity which determines the Kepler orbit of the grains. Interaction among grains via inelastic collisions results in the formation of jet streams. As matter is depleted from the filaments by the condensation of grains, it is replenished by fresh plasma that is drawn into the filaments by electromagnetic forces (in a process analogous to the pinch effect).

JET STREAMS

Jet streams are made up of grains and gases, and occupy a toroidal volume around the central body. The small diameter of the toroid is only a few percent of the large diameter and hence the jet streams occupy about 10^{-3}–10^{-4} of the total volume of the supercorona. They are fed by injection of grains condensed in a network of filaments around them. The accretion of secondary bodies takes place in the jet streams.

NON-CONDENSABLE GAS CLOUDS

As the injected matter contains a large fraction of non-condensable gases—presumably they form the main constituent—there is an increasing supply of such gases in the filaments and in the interfilamentary medium. These gases accumulate near the equatorial plane, most of them concentrating in and around the jet streams. Hence the accretion in the jet stream may take place in a cloud of non-condensable gases. When an accreted body becomes so large that its gravitation becomes appreciable, it may capture an atmosphere from the gas supply of the jet stream.

It is likely that the jet stream cannot retain all the gas. Some of it may diffuse away, possibly forming a thin disc of gas which may leak into the central body or transfer gas from one jet stream to another. In Figure 24.4 the gas is supposed to form tori around the jet streams and flatten out to a disc elsewhere.

The behavior of non-condensable gases is necessarily the most hypothetical element in the model because we have very little information about them and most of it is essentially indirect.

CONCLUDING REMARKS

The theory we have outlined here is far from being complete. With our present knowledge of the solar system, it would be unrealistic to strive for completeness.

Yet we are able to arrange the various physically reasonable processes needed to explain the various phases of the evolution into a coherent sequence, which seems to constitute the basic infrastructure for a more detailed and more refined theory (for more details, see Alfvén and Arrhenius 1976). In evolving this sequence we have relied mainly on the wealth of information that has been gained about the behavior of particles and fields in space during recent years, as well as on the studies of pertinent problems in the laboratory. Further details and refinements of the theory must await an expansion of our knowledge of the solar system.

ACKNOWLEDGMENTS

This work was supported by the Planetology Program, Office of Space Science, under grant NASA NGL 05-009-110, the Lunar and Planetary Program Division under grant NASA NGL 05-009-002 and the Apollo Lunar Exploration Office under grant 05-009-154.

DISCUSSION

WILLIAM M. KAULA: If the satellite configurations are determined by hydromagnetic forces exerted by planetary magnetic fields on the proto-satellite material, then the satellite material must be heated to vaporization temperatures (more than about 1800°K) so that it can be ionized. This heating must occur *after* temperatures low enough for the planetary material to condense have occurred. It is difficult to imagine the off-again-on-again heating process involved.

B. R. DE: The protoplanetary plasma is heated hydromagnetically at the expense of the rotational energy of the Sun. The plasma condenses to grains, and the grains accrete to form the planets. The satellites form by an identical series of processes taking place around a planet when the planet has reached the last stage of its formation and has acquired a magnetic dipole. The heating of the plasma in this case takes place at the expense of the rotational energy of the planet. Thus the heating in the first instance refers to the circumsolar plasma (precursor for the planets) during the planet formation era, while the heating in the second instance refers to the circumplanetary plasma (precursor for the satellites) during the satellite formation era.

The condensation of the plasma can take place with the condensing solid having a temperature below 1800°K but the ambient plasma having a temperature of a few thousand degrees. The precursor plasma for the satellites or the planets may contain some interstellar grains in the colloidal state which behave as a part of the plasma. Some grains that may condense outside the region of secondary body formation may feed this plasma by the ablation process, discussed by Alfvén and Arrhenius (1976), while falling through it at high velocities. But we do not envisage a process where the original material is in the form of solids that need to be vaporized and ionized and thus brought to the plasma state—a process that you seem to presume in your question.

ROBERT MURPHY: If I understand your Fig. 24.1—and I don't think I do—wouldn't it imply that the densities of the satellites would increase as you move away from the planet?

BIBHAS R. DE: That would be the first impression from the diagram. However, a number of considerations enter into the question of correspondence between the chemical composition of a final secondary body (i.e., a satellite or a planet) and the elements that theoretically give rise to the band in which the body belongs. These considerations are discussed in some detail by Alfvén and Arrhenius (1976). The conclusion is that our model need not necessarily predict that the density of the secondary bodies should increase away from the central body (in fact, our model provides an explanation for an opposite trend both in the Galilean satellites and in the inner part of the planetary system; see the above reference). However, regardless of these considerations, the *spacing* of the final secondary bodies should correspond roughly to the theoretical bands. This is the main import of the diagram.

PRELIMINARY THERMAL HISTORY MODELS OF ICY SATELLITES

*Guy J. Consolmagno and
John S. Lewis
Massachusetts Institute of Technology*

We present here preliminary results of thermal history calculations for solar system bodies formed out of low-temperature condensates. Several diverse compositional and accretional models are examined, including integral and differential accretion modes of both chemically equilibrated and unequilibrated condensates. Critical sizes for the onset of melting and differentiation, and ages at the time of initial melting, are presented for several specific models.

The chemical composition and structure of satellites depends on the sequence of condensation and accretion of the solar nebula. The physical aspects of these processes are described in Chapter 23 by Cameron, Chapter 24 by De *et al.* and Chapter 26 by Safronov and Ruskol. In particular, one must consider whether the condensation sequence occurred under conditions of equilibrium or disequilibrium, and whether the accretion of the material occurred differentially, with each object accreting from one component of the condensation sequence, or integrally, with all components of the sequence contributing to all the objects. A general review of these phenomena is given in an article by Lewis (1974a). Consolmagno and Lewis (1976) extend the calculations presented here.

POSSIBLE MODELS FOR SATELLITES

The two chemical alternatives for each of the two accretion modes lead to four possible models for satellites. The condensation sequence for a nebula with solar composition has been found for both equilibrium and disequilibrium cases by Lewis (1972b). From this, the four models can be detailed.

Model 1: Equilibrium Condensation; Integral Accretion

At temperatures of 350°K, material similar to that of carbonaceous chondrite type Cl will have condensed from the solar nebula. This material is primarily iron and magnesium silicates, with carbon compounds, and hydrous silicates. The uncompressed density of such material would be $\sim$ 2.4 g cm^{-3}. If the condensation temperature dropped below $\sim$ 160°K, the satellite would be formed, according to this model, by a uniform mixture of roughly half water ice and half Cl material (by weight), with an overall density of 1.6–1.8 g cm^{-3}. Below $\sim$ 100°K, we would expect the ice phase to contain a hydrate of

[492]

$NH_3 \cdot H_2O$. Below $\sim$ 60°K, methane clathrate with composition $CH_4 \cdot xH_2O$ (x being about 6.7), formed by the complete reaction of H_2O ice with CH_4 gas, will dominate, lowering the density to approximately 1.5 g cm^{-3}. At temperatures below $\sim$ 20°K, methane ice and possibly ices of argon and the other rare gases would be included, lowering the total density of the body to 1 g cm^{-3}.

Note that in any body formed by this model at temperatures below 160°K, where ice will be an important constituent, there will be gravitational potential stored by the dense silicates which are distributed uniformly through the less dense ices.

Model 2: Equilibrium Condensation; Differential Accretion

Here the condensation is the same as in the previous model, but accretion occurs differentially, with each member of the condensation sequence forming its own type of body.

Under this model, a satellite which condensed at temperatures above 160°K would have the composition of carbonaceous chondrite type CI; one condensing between 160°K and 100°K would be made of water ice; one between 100°K and 60°K would be made of the ammonia hydrate $NH_3 \cdot H_2O$; below that, down to 30°K, methane clathrate hydrate bodies would be formed; and at any lower temperatures, bodies of methane or rare gas ices would be formed.

The chondritic satellite would have a density of $\sim$ 2.4 g cm^{-3}, while satellites made of any of the ices except methane would have densities of about 1 g cm^{-3}. The pure methane ice body would have a density of $\sim$ 0.6 g cm^{-3}.

Model 3: Disequilibrium Condensation; Integral Accretion

Here we assume that as each material condenses and accretes, it will cover previously condensed material, preventing it from reacting with uncondensed gases in the solar nebula. Thus, we have a planet being built up, layer by layer, without allowing chemical equilibrium to be achieved between the layers.

The core of such a body would consist of "dry rock," primarily iron and magnesium silicates. It would have a mean density greater than 3.7 g cm^{-3}. At 160°K, a layer of water ice with up to twice the mass of the rocky core would be deposited over it, lowering the density of the total body to 1.6 g cm^{-3}. Since this layer would contain all the water in the system, ammonia could not form a hydrate but would be free to react with H_2S gas, and at 140°K, NH_4SH ice would be deposited over the water ice. Assuming solar proportions of the elements in the nebula, additional ammonia would remain after the H_2S has been exhausted, and at temperatures below 100°K this would form a layer of NH_3 ice over the NH_4SH ice. The overall density of the body thus would be lowered to 1.5 g cm^{-3}. At temperatures below 20°K, CH_4 and rare gas ices would form a top layer on the body, lowering its density to $\sim$ 1 g cm^{-3}.

Thus, we would have a satellite already differentiated as it was formed. Since the densest material is already in the core, and the remaining ices are deposited roughly in order of decreasing density, there would be no significant gravitational potential energy stored in this system. However, since each layer was condensed out of chemical equilibrium with the previously condensed layers, there will be stored chemical potential energy.

Model 4: Disequilibrium Condensation; Differential Accretion

In this model, each material accretes differentially into its own body, as in Model 2, but the accretion is rapid enough that the material is never allowed to equilibrate with the remaining gases in the solar nebula, so that the disequilibrium condensation sequence is followed, as in Model 3.

Thus, there would be formed dry rock bodies, with densities greater than 3.7 g cm^{-3}; water ice bodies of density 1 g cm^{-3} if condensation temperatures fell below 160°K; NH$_4$SH ice bodies of density 1 g cm^{-3} below 100°K; and CH$_4$ and rare gas ice bodies of density 0.6 g cm^{-3} below 20°K.

THERMAL HISTORIES OF THE POSSIBLE MODELS

Work on thermal history modelling is still incomplete, but general comments can be made for each model. See further details in Consolmagno and Lewis (1976).

With the exception of the pure rock bodies, thermal histories for Models 2 and 4 will be uninteresting. Homogeneous, single component bodies lacking internal heat sources have no endogenous means to alter their structure with time.

The rock bodies do have heat sources, since decaying radionuclides are present in their minerals.

The case of a Cl chondrite material body, which could be formed both by Model 2 and by Model 1 if the temperature of condensation never fell below 160°K, has been suggested in Chapter 17 by Fanale et al. as a possible model for Io. They suggest heating due to radioactive nuclides would result in the release of water, carrying with it various dissolved salts, which would then be deposited on the surface.

Phobos, Deimos, and certain large asteroids may be other examples of a Cl chondritic composition body, or they may be "dry rock" material as formed by Model 4 or Model 3, again if the temperature of condensation never fell below 160°K. Review Chapter 14, Pollack, for other compositional clues.

Possible thermal histories for bodies formed by Model 3 have not been thoroughly examined, but several possibilities suggest themselves. One can expect a significant thermal gradient due to heat sources in the rocky core, which may lead to melting of the ice layer above it. Mixing of the ice and rocks, and of the NH$_4$SH and water ice, would lead to the release of chemical potential energy. Reactions upon melting of a mixture of NH$_4$SH and H$_2$O ices will yield a solution with a low freezing point, and also may release H$_2$S gas. If further melting of the water ice region occurs, it could possibly dissolve any H$_2$S gas released; if not, the gas would migrate toward the surface and react with the NH_3 ice layer if one is present, or be outgassed.

The time scale for any of these reactions would depend on the size, composition, and initial temperature of the body. More detailed modelling is needed to determine under what conditions these various scenarios might arise.

Model 1 has been the most thoroughly explored of these models, since it offers the most interesting possibilities. Lewis (1971b) made preliminary thermal models for the Cl + water ice and Cl + water ice + ammonia hydrate cases, assuming a steady state heat flow. This work is being expanded by the present authors, who are preparing computer simulations to predict the thermal evolution of satellites formed by this model (cf. Consolmagno and Lewis, 1976).

The Model 1 case divides itself into two classes. The first class is made of bodies with central pressure less than 2 kilobars, and radius less than 1000 km. This class includes the satellites Ariel, Umbriel, Oberon, Titania, Tethys, Iapetus, Dione, and Phoebe. If their composition is Cl chondrite material and water ice only, our preliminary results are shown in Figure 25.1. They indicate that the minimum sized object in which melting will occur has a radius of 650 km, and that melting will begin roughly 500 million years after accretion. However, as the heat production from radionuclides dies off exponentially, an enthalpy maximum is reached after about 1 billion years, and the planet starts to refreeze before melting has become significant. Even for a 1000 km radius object, it is not clear that significant melting will occur.

Heat released through gravitational infall is small, since the gravitational acceleration even at the surface of these small, light bodies is small, and approaches zero at the center, where melting first occurs. Thus, it does not appear that objects in this first class are significantly differentiated if they consist only of Cl chondrite material and water ice.

If ammonia is also present as a hydrate, the situation becomes more interesting. $NH_3 \cdot H_2O$ and H_2O ices have a eutectic melting point at only 173°K, and melting can therefore begin much earlier. Our results for this class are shown in Figure 25.2. In this case, even a 500 km radius object will begin to melt at the center, although again here the melting and differentiation will not be significant. But objects 700 km to 1000 km in radius can be significantly differentiated, with a 50–200 km core of silicates, a 300–400 km mantle consisting of a slurry of water ice and an ammonia-water solution, and a 300–400 km crust of unmelted ice. Again, melting begins at the center, and heat released from gravitational infall of silicates to form the core will be insignificant. Melting begins at roughly 250 million years, and reaches a maximum at 1 billion years.

Similar results may be expected for bodies containing methane clathrate. In such cases, CH_4 outgassing may occur. If CH_4 ice is initially present, a two-stage melting and differentiation process will occur.

The second class of objects, the larger satellites, must be considered separately because of the higher pressures in their interiors.

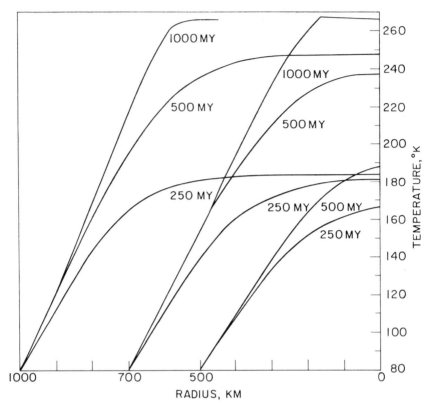

Fig. 25.1. Thermal profiles in bodies of 500, 700, and 1000 km radii with a composition of 35% silicates and 65% water ice by weight. The 500 km object reaches maximum temperatures after 500 million years and does not melt. The 700 km object may reach the melting point in the center after 1 billion years, but differentiation will not be significant. The 1000 km object will melt and differentiate after 1 billion years.

CALCULATIONS FOR LARGE SATELLITES

Above 2 kilobars, ice forms a number of high pressure phases. The relationships are complex, and are not well known at low temperatures. One must presume that the effects of pressure on the ammonia-water system will be even more complex. Since the high-pressure phase diagram of such a system is not known, we will concern ourselves at present only with a body made up of silicates and water ice.

We have found that icy bodies with a silicate weight fraction 0.25 or greater have central pressures which exceed 2 kilobars when the radius reaches 1000 km. With objects larger than this, that is, the Galilean satellites, Titan, Triton, and possibly Pluto, the high pressure forms of ice must be considered.

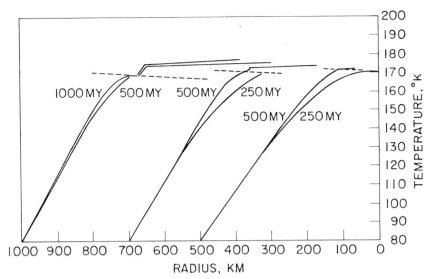

Fig. 25.2. Thermal profiles in bodies of 500, 700, and 1000 km radii with a composition of 35% silicates, 55% water ice, and 10% ammonia by weight. The eutectic melting temperature of the ammonia hydrate–water ice solution is shown by the dotted lines. We assumed a pressure dependence similar to the melting point depression of pure water. Above the dotted lines, convection occurs. The lines terminate at the core radius.

Simplified computer thermal models which ignore the different thermal properties of the various ice phases and which do not consider phase changes predict that for a 2500 km ice body, 40% silicate by weight, with typical Cl chondrite radioactive isotope abundances in the silicates, central temperatures will reach 600°K after 4.5 billion years. However, the heat required for phase changes in the ice and for melting the ice, and also the more efficient heat transfer from the center to the surface supplied by a convecting mantle, would tend to lower the central temperature by at least 100°K, and possibly much more. On the other hand, the infall of silicates from the mantle region will supply heat as gravitational potential energy is released, and will tend to concentrate the radioactive heat sources at the center, both effects resulting in a higher central temperature.

Whether these temperatures will lead to melting throughout the planet depends on the pressure structure within, which in turn both depends on and will determine the density of the material which makes up the planet. A fully melted body with 40% silicates by weight would have a density of only 1.4 g cm^{-3}. This is similar to the density for Callisto, but too low for Ganymede (cf. Table 1.4). For Ganymede to be totally melted and have its observed density, it would have to be made of 65–85% silicates by weight. However, we expect thermal evolution will have lowered these densities from their original state, since the phase changes for all forms of ice except ice I increase in volume with increasing temperature. Thus, to see how a satellite of this size would evolve with time, we must first get an idea of what the object looked like soon after accretion.

To use a likely case for a body comprised of water ice and silicate, we will examine the possible pressure structure of Callisto. Callisto has a radius of 2500 km ($\pm$ 150) and a mass of 91 $\times$ 10^{24} g, for a density of 1.4 g cm^{-3} (Morrison and Cruikshank, 1974; cf. Table 1.4). We will assume a composition of 40% silicates by weight, with the rest being water ice. We will consider two possible densities for the silicates, first a density of 3.7 g cm^{-3} (dry silicates from inhomogeneous accretion), and then one of 2.4 g cm^{-3} (hydrous silicates from homogeneous accretion).

Assuming the body is isothermal at 100°K, ice I will become ice II at 1.6 kilobars; ice II will become ice VI at 8 kilobars (this number is uncertain, and found by extrapolation of the II-VI phase boundary); and ice VI will become ice VIII at 13 kilobars. Thus, the problem is simply a matter of using the hydrostatic equation for pressure, dP= $-$ 4/3Gr$\pi\rho'\rho$dr (where ρ and ρ' are the densities of the object above and below the radius r) and solving for the radii at which the pressures of the phase transitions occur. Since the unmelted object will be more dense than the present object, we will assume an original radius of 2400 km for Callisto.

With a silicate density of 3.7 g cm^{-3}, we find the ice I-II transition occurs just 100 km below the surface, at 2300 km; the II-VI transition occurs at r = 1900 km; the VI-VIII transition at r = 1500 km. The central pressure will come to about 22 kilobars. The total mass of the satellite would equal 98 $\times$ 10^{24} g, slightly greater than the observed mass.

For the case where the silicate density is 2.4 g cm^{-3}, we find the first transition, ice I-II, at 2250 km, the II-VI transition at r = 1750 km, and the VI-VIII transition at r = 1300 km. The central pressure is 19 kilobars, the total mass is 87 $\times$ 10^{24} g, slightly less than the observed mass. It is also possible to vary the weight fractions, but any variation which still assumes reasonable densities for the silicate phase, and which gives a computed mass close to the observed figure would have to have a pressure structure similar to the ones computed above.

Assuming our structures are reasonable starting points for thermal evolution models, and that the heat sources are the same as in the smaller models, how will the internal structure of these bodies evolve with time?

A detailed computer model has not yet been completed. However, a discussion of the possible shapes such a model might take, and the likely results, can be fruitful.

The easiest way to picture such a model is to assume that the pressure at a given depth is constant for all time. Such a model is shown in Figure 25.3, assuming a silicate density of 2.4 g cm^{-3}. A series of possible temperature vs. radius curves, representing different times in the thermal evolution of the body, are superimposed on the H$_2$O phase diagram.

We can see the first result of heating on the body will be the growth of the ice VI region at the expense of the ice II and ice VIII regions. The latent heat of

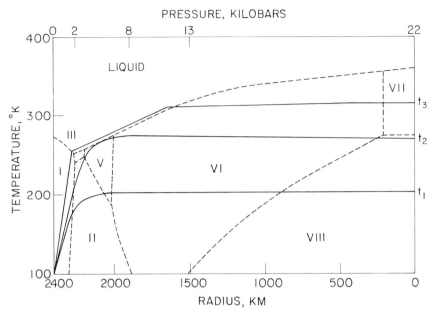

Fig. 25.3. Speculated thermal profiles for a homogeneous object of 2400 km radius with a composition 60% water ice and 40% carbonaceous chondrite type I material, by weight. The pressure scale is not linear. The thermal curves are superimposed over the pressure temperature phase diagram for water (adapted from Fletcher, 1970). Note that melting begins at shallow depths (always near the 2 kb level) if water ice is the only ice present. The time scale for first melting may be on the order of 1 billion years.

phase change in both cases is on the order of 250 cal mole^{-1}, so the temperature profile will be flattened along this region. Also, there is a volume change of $\sim 10\%$ involved, as the ice VIII expands to form ice VI, while the ice II contracts. There is no reason to expect that these changes should exactly cancel each other out, and there will probably be considerable seismic activity within the body as it expands or contracts. A similar effect, but on a smaller scale, will occur along the ice I-II boundary.

At about 205°K and 7 kilobars (roughly 400 km below the surface of a 2400 km object) ice V will form from ice II, requiring 290 cal mole^{-1}, and at 240°K and 2 kilobars, ice III will form from ice II, at the expense of 220 cal mole^{-1}. Again, flattening of the temperature curve and a notable volume change will occur.

Then, somewhere along the ice III or ice V boundaries with liquid water, the first melting will occur.

Note that, unlike the case of the smaller icy satellites, gravitational infall of silicate material should be a significant heat source. The 2500 km object has 125 times the mass of the 500 km object. The gravitational acceleration everywhere will be greater, and, in addition, the first melting occurs near the surface, where g is large, rather than at the center, where it is zero.

At 279°K, the remaining ice VIII will form ice VII. This will result in another flattening of the temperature curve, since the transition requires 260 cal mole^{-1}, but there is no associated volume change. In the meantime, the upper regions of the satellite are continuing to melt, overlaying unmelted regions with a "hot" layer of silicates. However, further into the satellite, regions with constantly higher pressures must melt. The increased pressures have associated with them increased melting temperatures, making melting difficult to achieve. For a satellite with the size of Callisto, having a central pressure of 20 kilobars, the melting point will be raised to 350°K. For a Ganymede-sized object, with central pressure of 35 kilobars, the melting point would be 450°K.

The effect of ammonia on an object of this size is unknown, but it will likely induce earlier melting of the ice.

Similar results can be expected when methane clathrate is added to the composition, with the additional effect of outgassing CH_4. When CH_4 ice itself is a part of the composition, it may form up to one-half the volume of the planet, giving an overall zero-pressure density of $\sim$ 1 g cm^{-3}. This is compatible with the new radius derived for Titan by Elliot *et al.* (1975).

CONCLUSIONS

The thermal history models considered herein accentuate the great ease of melting and differentiation of large icy satellites, and point the way toward several classes of models deserving of more detailed treatment. Satellites of the outer planets are found to be incapable of melting and differentiation if their radii are less than $\sim$ 500 km. The explorational significance of this predicted persistence of small bodies unaltered by endogenic processes is clear: such bodies merit close attention in any synoptic program of outer solar system exploration.

THE ACCUMULATION OF SATELLITES

V. S. Safronov and E. L. Ruskol
O. Yu. Schmidt Institute of Physics of the Earth
Moscow, USSR

An analysis of the formation processes of the regular satellites is given, following previous work (Ruskol, 1960, 1963, 1971a,b, 1972a,b; Kaula, 1971; Kaula and Harris, 1973) on the accretion of the Moon. The natural satellites are considered to coalesce from a circumplanetary swarm which itself is fed by heliocentric particles which are captured following inelastic collisions. Characteristic times are calculated for the important physical processes: the replenishment time of the swarm, the growth time of the mass of the swarm, the accumulation time for bodies in the swarm, and the orbital collapse time under the action of tides. During the capture of the swarm and its accumulation much material is lost to the planet's surface. Reasonable values for the mass of the satellite swarm can be found from this model.

The formation of the regular satellites of the planets is the principal subject of this chapter. The resemblance between the structure of satellite systems and the structure of the planetary system has often been pointed out. It has been interpreted as arising from a correspondence in the formation processes of planets and satellites. H. Alfvén (1971) and De *et al.* (Chapt. 24) emphasize that the theory of the origin of the planets must include a similar explanation for the origin of satellites. O. Yu. Schmidt (1959) suggested the formation of a swarm of solid particles and bodies revolving around the growing planet. The processes of the accumulation of these bodies and of the formation of the satellites is similar to the formation processes of the planets themselves. Black (1971) has found a common pattern in these systems and has even predicted the presence of an "asteroid belt" in the Saturnian satellite system between the orbits of Rhea and Titan.

The requirement of a similar origin makes the hypothesis of the formation of planets and satellites from gaseous condensations unacceptable (cf. De *et al.*, Chapt. 24). Most satellites are very small and could not be kept in a gaseous state due to an extremely rapid disintegration. On the other hand, the gaseous protoplanets of solar composition would be too massive and could not lose almost all their hydrogen and helium, which would comprise 98% of their masses (cf. Cameron, Chapt. 23). Hypotheses for the formation of satellites from highly rotating unstable planets or protoplanets are also untenable. In addition to the problem of chemical differences, the problem of slowing down an initially rapid planetary rotation arises.

The capture of satellites by the planets is a specific and a probably rare phenomenon (Chapt. 7, Burns). This mechanism can be applied only to single irregular satellites. But for groups of irregular satellites other explanations have been suggested. Colombo and Franklin (1971) have suggested that Jupiter's satellites (VI, VII, X, XIII) and (VIII, IX, XI, XII) were formed as a result of a collision of a satellite with an asteroid. Two collisions of satellites with asteroids were suggested earlier by Bronsten (1968).

The only reasonable mechanism for the formation of regular satellites seems to be their accumulation in a circumplanetary swarm of solid particles and bodies formed through the capture of heliocentric particles in the gravitational field of a growing planet following their inelastic collision near the planet. Based on this idea of O. Yu. Schmidt, a model for the formation of a circumterrestrial satellite swarm and of the subsequent formation of the Moon in this swarm has been developed by Ruskol (1960, 1962, 1963, 1971a, 1971b, 1972a, 1972b, 1973). Kaula in a detailed review (1971) and in a paper with Harris (1973) [see also Harris and Kaula (1975) and Kaula and Harris (1975)] has concluded that such a model is the most acceptable from a dynamical point of view and that the more exotic fission and capture models of the lunar origin are implausible (cf. Chapt. 27, Wood).

The most general features of the formation of a circumplanetary satellite swarm given by this model are described below.

CHARACTERISTIC TIMES FOR IMPORTANT PROCESSES

Time of Replenishment

The time of replenishment for a satellite swarm is an important quantity. The replenishment of the swarm occurs during the entire period of accumulation of the planet. The characteristic time τ_p for the planet to sweep out the material in its feeding zone (i.e., the time of the e-fold decrease of the mass of this material) is, according to Safronov (1969),

$$\tau_p = \frac{\delta R_p^3}{3(1 + 2\theta)\sigma_{10} R^2} P_1 , \tag{1}$$

where δ and R are the density and the radius of the planet, P_1 is the period of its revolution around the Sun, R_p is its final radius, σ_{10} is the initial surface density of the solid material in the zone of the planet, and θ is a nondimensional parameter of the order of several units. The latter is connected with the relative velocities of the bodies by the relation

$$v = \sqrt{\frac{GM}{\theta R}} , \tag{2}$$

where M is the mass of the planet. It is estimated that $\theta \approx 3$ to 5. Thus in the final stages of the Earth's growth, $\tau_P \approx 10^7$ years, and the Earth sweeps out 98% of the total mass in the zone during 6×10^7 to 10^8 years. τ_p increases with distance from the Sun due to the increase of P_1. For the outer planets (Uranus and Neptune) τ_p can reach one-half billion years.

Rate of Accumulation

The rate of the accumulation of bodies in the swarm is also a pertinent quantity. As the characteristic time τ_a for the accumulation of bodies (or particles) in the swarm we take the interval of the e-fold increase of the mass of the body due to its collisions and the resulting coagulation with bodies of smaller mass:

$$\tau_a = m/(dm/dt). \tag{3}$$

We assume an inverse power law for the distribution of the radii of bodies

$$N(r) = cr^{-n}, \qquad r_m < r < r_M . \tag{4}$$

The index "1" is attached to all quantities in the zone of the planet and the index "2" to similar quantities in the satellite swarm. Bodies in the swarm with radii less than r have a spatial density

$$\rho_2(r) = \rho_2(r/r_{M_2})^{4-n_2} , \tag{5}$$

where ρ_2 is the total density of the swarm and $n_2 < 4$. The rate of growth of a body is given by the well known equation $dm/dt = \xi \pi r^2 \rho(r)v$, where v is the relative velocity of the bodies and the factor ξ is several units. The latter is connected with the gravitational focusing factor $(1 + 2\theta$ for the largest bodies) and with the fact that the cross section for the collision of bodies of comparable sizes is between πr^2 and $4\pi r^2$. Then we find

$$\tau_a = \frac{4\delta r}{3\xi v_2 \rho_2} \left(\frac{r_{M_2}}{r} \right)^{4-n_2} . \tag{6}$$

For a swarm with relatively weak external perturbations we can take $\rho_2 v_2 = 4\sigma_2/P_2$, where σ_2 is the surface density of the swarm's material. It can be written in the form

$$\sigma_2 = cR_2^{-b}, \qquad R < R_2 < R_{L_1} , \tag{7}$$

where R_2 is the distance from the planet; it is bounded by the radius R of the planet and the distance R_{L_1} to the Lagrangian point L_1. Integrating $2\pi R_2 \sigma_2$ over this interval of R_2, we find the mass of the swarm μ. For an intermediate case,

$b = 2$, we have $\mu = 2\pi c \ln(R_{L_1}/R) \approx 30c$. Eliminating c by eqn. (7) and replacing $v_2\rho_2$ by σ_2 in eqn. (6) and then by μ, we find

$$\tau_a \approx \frac{M}{\mu} \left(\frac{R_2}{R}\right)^2 \frac{r_{M_2}}{R} \left(\frac{r}{r_{M_2}}\right)^{n_2-3} P_2 , \tag{8}$$

where P_2 is the period of revolution of the swarm around the planet at the distance R_2. One can assume $n_2 \approx 3.5$. The characteristic time τ_a for the accumulation of a body increases with its size. But even for the largest bodies ($r \sim r_{M_2}$), it is small in comparison with the time span τ_p for the replenishment of the swarm. So for $\mu/M \sim 10^{-3}$, $r_{M_2}/R \sim 10^{-1}$, $R_2/R \sim 30$, $r \sim r_{M_2}$, we find $\tau_a \sim 10^5 P_2$, that is, about several thousand years for the circumterrestrial swarm. τ_a is less for smaller bodies and for bodies closer to the planet. Thus $\tau_a \ll \tau_p$. This important fact means that various disturbances brought into the swarm by bodies from the planetary zone are almost wholly removed by collisions and the accumulation of bodies in the swarm. Therefore the distribution of sizes of bodies in the swarm is determined mainly by the accumulation process which led already in the early stages to the formation of large bodies—the embryos of future satellites.

Rate of Increase in Mass

The rate of increase in the mass of the swarm is now computed. In the same way as in the previous sections we can estimate the characteristic time τ_c during which bodies of mass m capture a mass equal to their own mass from the zone of the planet. We assume that trapping takes place when bodies of comparable size collide. Though bodies of the swarm with mass m also replenish smaller bodies, this replenishment is compensated to some extent by their being swept out through collisions with the larger bodies of the planetary zone. We assume also (as was done in the previous papers by Ruskol) that the replenishment takes place in the whole volume of the swarm and not only at its surface, that is, the swarm is relatively transparent for all bodies. The case of an optically thick swarm is considered by Harris and Kaula (1975). Actually the swarm is transparent for big bodies but is opaque for small bodies. Our model describes the early stages of the formation of the swarm. It can be shown that for $n_1 + n_2 < 7$ the model of the transparent swarm also should be preferred, but for $n_1 + n_2 > 7$ the model of the opaque swarm is definitely better. With our assumptions

$$\tau_c(r,R_2) = \frac{m}{W(R_2)\xi\pi r^2 m N_1(r)\Delta r v_1} = \frac{1}{W(R_2)\gamma\xi\pi r^2 N_1(r)rv_1} , \tag{9}$$

where the interval of radii of the trapped bodies is taken $\Delta r = \gamma r$ and $W(R_2)$ is the probability of capture in such collisions. Taking the inverse power law of eqn. (4) for $N_1(r)$ and using the relation,

$$\rho_1 = \int_{r_{m_1}}^{r_{M_1}} mN_1(r)dr \ ,$$

to avoid the constant c, we find at $n_1 < 4$

$$\tau_c(r,R_2) = \frac{\delta_1 r_{M_1}}{3(4 - n_1)W(R_2)\gamma\xi\sigma_1} \left(\frac{r}{r_{M_1}}\right)^{n_1 - 3} P_1 \ . \tag{10}$$

We see that small bodies are trapped more effectively. For large bodies $\tau_c(r,R_2)$ is much longer. From eqns. (1) and (10) it follows that $\tau_c(r,R_2) > \tau_p$ for $r = r_{M_1}$ if $r_{M_1}/R > (4 - n_1)\gamma\xi W/(1 + 2\theta) \sim 10^{-2}$. Averaging $\pi_c(r,R_2)$ over all values of the radii of bodies in the swarm, we find the characteristic time for an e-fold increase of the swarm's density ρ_2 is

$$\frac{\rho_2}{d\rho_2/dt} = \frac{\sigma_2}{d\sigma_2/dt} = \tau_c(R_2) = \frac{7 - n_1 - n_2}{4 - n_2} \tau_c(r_{M_2},R_2), \quad n_1 + n_2 < 7;$$

$$\tau_c(R_2) = \frac{1}{(4 - n_2)\ln(r_{M_2}/r_{m_2})} \tau_c(r_{M_2},R_2), \quad n_1 + n_2 = 7;$$

and $\hspace{10cm}$ (11)

$$\tau_c(R_2) = \frac{n_1 + n_2 - 7}{4 - n_2} \left(\frac{r_{m_2}}{r_{M_2}}\right)^{n_1 + n_2 - 7} \tau_c(r_{M_2},R_2), \quad n_1 + n_2 > 7 \ .$$

Taking the average value $\overline{W}$ of the capture probability in eqn. (11), we find the characteristic time τ_c for the increase of the whole mass of the swarm due to the trapping of the material.

Tidal Evolution of Orbits

Considered next is the rate of tidal evolution of satellite orbits (cf. Chapt. 7, Burns). When satellite embryos are large enough, those closest to the planet undergo appreciable tidal evolution. The rate of increase of their distance R_2 from the planet is given by the relation (MacDonald, 1964; Chapt. 7, Burns)

$$\frac{dR_2}{dt} \approx \frac{3k_2}{Q} \left(\frac{G}{M}\right)^{\frac{1}{2}} \frac{m}{R_2^{11/2}} R^5 \ , \tag{12}$$

where k_2 is the Love number of the second kind and Q is the specific dissipation function. The characteristic time τ_t for the increase of R_2 is

$$\tau_t = \frac{R_2}{dR_2/dt} \approx \left(\frac{4\pi\delta}{3}\right)^{5/3} \frac{R_2^{13/2}Q}{3k_2 G^{1/2} M^{7/6} m} \ . \tag{13}$$

The sizes of satellite systems are proportional to the radii R_{L_1} of the Hill gravitational spheres of their primaries. Taking $R_2 = \lambda R_{L_1} \propto \lambda M^{\frac{1}{3}} R_1$, one obtains

$$\tau_t \propto \frac{\lambda^{13/2} QMR_1^{13/2}}{k_2 m} . \tag{13'}$$

Due to the high degree of dependence of τ_t on the distance R_1 from the Sun, the tidal interaction is quite different for the terrestrial planets and for the giant planets. This difference is enhanced because of the high value of Q for the giant planets (Goldreich and Soter, 1966). Hence tidal evolution can be very significant in the satellite systems of the terrestrial planets (Burns, 1973; Ward and Reid, 1973) but is not significant in the region of the giant planets.

FORMATION OF A CIRCUMPLANETARY SATELLITE SWARM

As was mentioned above, the trapping of particles in the swarm results from their inelastic collisions. The higher an impact velocity, the greater is the energy loss in the collision. Hence the probability of capture for colliding bodies is a maximum (about ½) near the surface of the planet. A simple approximation for the decrease of this probability with distance from the planet is (Ruskol, 1963):

$$W(R_2) \approx \beta \frac{2\theta R}{R_2 + 2\theta R} , \quad R_2 < R_{L_1} \tag{14}$$

At the beginning of the swarm's formation, when its density ρ_2 was much less than ρ_1 (that in the zone of the planet), the swarm grew mainly due to mutual collisions of bodies in the planetary zone ("free-free" collisions). But, when ρ_2 became appreciably greater than ρ_1, the swarm began to grow mainly due to the trapping of bodies in the planetary zone as they collided with the bodies of the swarm ("bound-free" collisions).

The simultaneous consideration of free-free collisions and of bound-free collisions is somewhat complicated to calculate. The first are important only during the earlier stage. The trapping of the material in the swarm at this stage can be taken into account without appreciable error in the same way as was done during the second stage, that is, by bound-free collisions. It is sufficient for this to assume the initial density of the swarm is equal to the density ρ_1 of particles in heliocentric orbits. These free particles play the same role in the trapping of material as if they were the particles of the swarm.

For capture in a non-opaque swarm, collisions between bodies of comparable size are necessary. If the distribution of the radii of heliocentric and of planetocentric bodies is described by power laws as in eqn. (4) with indices n_1

and n_2, the bodies captured into the swarm (before their fragmentation) should have a power law distribution with the index $n_c = n_1 + n_2 - 3$. The number 3 occurs because the collision cross section is proportional to the square of the radius and the range Δr of the radii of captured bodies is proportional to the first degree of r. The value of n found theoretically for asteroids is about 3.5. It agrees with the values of n_1 found from observations of different objects in the solar system. With such values of n_1 and n_2 we have $n_c = 4$. In such a distribution a considerable part of the mass belongs to small particles. The fragmentation of trapped bodies can increase n_c even more. Therefore the trapping of new particles and bodies in the swarm must increase n_2. However, mutual collisions between particles of the swarm occur much more often than collisions of these particles with the particles in the zone of the planet ($\tau_a \ll \tau_c$). For this reason n_2 is determined in fact by the accumulation process in the swarm and the predominant trapping in the swarm of small particles does not really influence n_2:

$$n_2 \approx n_a + \frac{\tau_a}{\tau_c} (n_c - n_a) \approx n_a \ . \tag{15}$$

The main features of the swarm—its density, rate of accumulation and the sizes of its constituent bodies—depend significantly on the distance from the planet. Of great interest are the integral properties of the swarm: its total mass, which increases during planetary accumulation, and the mass which falls onto the planet from the dense innermost part of the swarm that spreads into the surface of the planet. While the mass of the planet grows, the distances to all the bodies of the swarm decrease (because of conservation of angular momentum, MR_2 is constant—Jeans' invariant). The radius of the planet increases and hence the inner part of the swarm joins the planet (see Gold's comment following Chapt. 19). There is also some additional contraction of the swarm because the angular momentum of the trapped bodies is smaller than that of particles in circular orbits at the capture distance.

Let μ_c be the whole mass which has been trapped in the circumplanetary swarm at the moment when the mass of the growing planet is M and let μ_p be the mass which has fallen onto the planet from the swarm. Then $\mu = \mu_c - \mu_p$ is the mass of the swarm. Let μ_{p_1} be the part of μ_p connected with the increase of M and σ_p be the surface density of the swarm near the planet. Then

$$dR/R = \tfrac{1}{3} \, dM/M, \qquad dR_2/R_2 = - \, dM/M \tag{16}$$

and

$$d\mu_{p_1} = 2\pi R \sigma_p (dR - dR_2) = \frac{8\pi}{3} R^2 \sigma_p \, dM/M \ . \tag{17}$$

Taking $\sigma_p = cR^{-b}$, according to eqn. (7), and then eliminating the constant c, we find

$$\frac{d\mu_{p_1}}{dM} = f_1(b) \frac{\mu}{M} \ , \tag{18}$$

where

$$f_1(b) = \frac{4}{3} \frac{2 - b}{(R_{L_1}/R)^{2-b} - 1} \text{ for } b \neq 2$$

$$f_1(b) = \frac{4}{3 \ln(R_{L_1}/R)} \quad \text{for } b = 2 \tag{19}$$

and

$$R_{L_1}/R = (\delta/3\rho^*)^{\frac{1}{3}}, \qquad \rho^* = 3M_\odot/4\pi R_1^{\ 3} \ . \tag{20}$$

We can try to estimate the mass μ_{p_2} which has fallen onto the planet from the swarm due to the small angular momentum of captured material (see Harris' comment following Chapt. 8). V. V. Radzievsky suggested that the average angular momentum relative to the planet of the material trapped at the distance R_2 must have been about $\omega_1 R_2^{\ 2}$, where ω_1 is the planet's angular velocity of revolution around the Sun (Ruskol, 1972a, 1972b). If this momentum was smaller than that for a circular orbit near the planet, $\sqrt{GMR} \propto R^2$, the material fell onto the planet. We assume that all bodies captured at a distance $R_2 < R_{2p}$, where

$$R_{2p} = \kappa R \ ,$$

fell onto the planet ($\kappa \approx 80$). It is seen from eqn. (11) that, if the radii of the largest bodies in the swarm did not depend on R_2, the captured mass was proportional to the mass of the swarm and

$$\frac{d\mu_{p_2}}{d\mu_c} = \frac{\mu(R_{2p})}{\mu} = \frac{R \int^{\kappa R} W(R_2)\sigma_2(R_2)R_2 dR_2}{R \int^{R_{L_1}} W(R_2)\sigma_2(R_2)R_2 dR_2} = f_2(b) \ , \tag{21}$$

where σ_2 and W are chosen according to eqns. (7) and (14).

The increase in the mass μ_c captured by the swarm is determined by the value of τ_c, which is found from eqns. (10) and (11) when the value for the probability of the capture $\overline{W}$ as averaged over the whole swarm is taken,

$$\frac{d\mu_c}{dt} = \frac{\mu}{\tau_c} \ . \tag{22}$$

In the same form we can write the equation for the growth of the planet

$$\frac{dM}{dt} = \frac{M}{\tau_M} . \qquad (23)$$

Then

$$\frac{d\mu_c}{dM} = \frac{\tau_M}{\tau_c} \frac{\mu}{M} = \alpha \frac{\mu}{M} . \qquad (24)$$

From eqns. (21) and (24) we find $d\mu_{p_2}/dM$ and, adding $d\mu_{p_1}/dM$ from eqn. (18) we obtain the equation for the whole mass $\mu_p = \mu_{p_1} + \mu_{p_2}$ which has fallen onto the planet from the swarm

$$\frac{d\mu_p}{dM} = \left[f_1(b) + f_2(b) \frac{\tau_M}{\tau_c} \right] \frac{\mu}{M} = \alpha_1 \frac{\mu}{M} , \qquad (25)$$

where

$$\mu = \mu_c - \mu_p . \qquad (26)$$

The equation for the change in μ is

$$\frac{d\mu}{dM} = \left[\{ 1 - f_2(b) \} \frac{\tau_M}{\tau_c} - f_1(b) \right] \frac{\mu}{M} = \alpha_2 \frac{\mu}{M} . \qquad (27)$$

This system of equations for μ, μ_c and μ_p contains one more unknown quantity, τ_M. The rate of growth of the planet is

$$\frac{dM}{dt} = \pi R^2 \rho_1 v_1 (1 + 2\theta) + \frac{d\mu_p}{dt} . \qquad (28)$$

Hence from eqns. (23), (25) and (28) one obtains

$$\tau_M = \tau_{M_0} (1 - \alpha_1 \frac{\mu}{M}) \approx \tau_{M_0} (1 - \alpha_{10} \frac{\mu}{M}) \qquad (29)$$

where τ_{M_0} and α_{10} are the values of τ_M and α_1 at $\mu = 0$:

$$\tau_{M_0} = \frac{4R\delta}{3(1 + 2\theta)\rho_1 v_1} , \quad \alpha_{10} = f_1(b) + f_2(b) \frac{\tau_{M_0}}{\tau_c} . \qquad (30)$$

The functions $f_1(b)$, $f_2(b)$ and $\overline{W}$ are listed in Table 26.1.

The system of eqns. (24)–(27) and (29) has its simplest solution when $\mu_p \ll M$

and $\tau_M \approx \tau_{M_0}$. The ratio τ_{M_0}/τ_c is proportional to the ratio $R/(r_{M_1}^{4-n_1} r_{M_2}^{n-3})$ and for $n_1 + n_2 > 7$ is also proportional to $(r_{M_2}/r_{m_2})^{n_1 + n_2 - 7}$. If these ratios are constant during the accumulation, τ_{M_0}/τ_c does not depend on M. The constancy of these ratios seems to be more acceptable than the constancy of r_{M_1}, r_{M_2}, r_{m_2} themselves. Supposing further that b is a constant, we have $\alpha_1 = $ const and $\alpha_2 = $ const and the system (24)–(27) is integrated immediately:

$$\mu = \mu_0 \left(\frac{M}{M_0}\right)^{\alpha_2} \quad , \quad \mu_p = \frac{\alpha_1}{\alpha_2} \mu \quad , \quad \mu_c = \frac{\alpha}{\alpha_2} \mu \quad . \tag{31}$$

For illustration we give in Table 26.2 the values of $\tau_{M_0}/\tau_c, \alpha_{10}, \alpha_{20}$ and μ_p/μ for $n_1 = 3.5$, $\sqrt{r_{M_1} r_{M_2}} = 0.1R$, $r_{M_2}/r_{m_2} = 10^{12}$ with different values of n_2, W and b.

The mass of the swarm increases with M when $\alpha_2 > 0$. When $\alpha_2 > 2/3$ the surface density of the swarm increases with M and at $\alpha_2 > 1$ its opacity increases. We see from Table 26.2 that suitable values of parameters for W constant are b = 2 and $n_2 = 3.5 \pm 0.1$, and for W taken from eqn. (14), b = 1.5 and $n_2 = 3.5 \pm 0.1$. The mass which has fallen onto the planet from the swarm is about five to ten times the mass of the satellites. From a consideration of planetary rotation it can be found that the last value is near the upper limit for the lunar swarm. For an opaque swarm suitable values of b and n_2 would be smaller.

At large values of n_2 the mass μ_p cannot be neglected in eqn. (28) and the simplest solution (31) is not correct. In this case it is sufficient to take a linear dependence of τ_M on μ (the second part of eqn. 29). Then a more correct solution can be easily found. However for larger n_2 (i.e., for $n_1 + n_2 > 7$) the opacity of the swarm should be also taken into account (cf. Harris and Kaula, 1975).

Therefore, using values of the parameters of the model which seem to be quite reasonable, it is possible to obtain a circumplanetary satellite swarm of sufficient mass which will develop into a satellite system. However, in order to confirm this general picture of the process, more detailed study is necessary. The parameters n_2, r_{M_2} and b should be estimated from the kinetics of the process itself. Then we can obtain a more definite judgment about the conditions during the formation and evolution of the swarm. A more detailed consideration of the accumulation of satellites would also include a study of the evolution of the orbits of growing bodies (see Harris' comment following Greenberg's Chapt. 8). Satellite embryos sufficiently massive and close to the planet move away due to tidal forces and avoid falling onto the planet. In some cases they could evolve into resonances with neighboring outer embryos (Greenberg et al., 1972; Colombo et al., 1974; Chapt. 8, Greenberg). However, we can say that such a capture did not occur during the tidal evolution of the Moon.

TABLE 26.1
Parameter Values for Various Swarm Models

The values of the parameters f_1 and f_2, which tell the manner in which the planet accumulates mass from the swarm (cf. eqns. 18 and 21), are given as functions of the negative exponential power in the density distribution law (eqn. 7) of the swarm. The average value $\overline{W}$ of the capture probability is also presented.

b	1.0	1.5	2.0	2.5	3.0	Model
$f_1(b)$	0.007	0.05	0.25	0.67	1.33	
$f_2(b)$	0.4	0.6	0.8	0.9	0.99	I, W = const
$f_2(b)$	0.71	0.88	0.97	0.98	0.99	II, W from eqn. (14)
$\overline{W}$	0.04	0.14	0.25	0.36	0.42	II, $\beta = \frac{1}{2}$

TABLE 26.2
Characteristic Numbers in the Development of Various Swarm Models*

n2	3	3.4	3.5	3.6	3.7	Model
τ_{M_0}/τ_c	0.7	2.	5.	20.	120.	I, $n_1 = 3.5$
α_{10}	0.8	1.8	4.2	16.	96.	W = const
α_{20}	0.11	0.15	0.75	3.8	24.	$\gamma\xi W = 1.6$
μ_p/μ		12.	5.6	4.2	4.0	b = 2
τ_{M_0}/τ_c	1.0	3.	7.	30.	180.	II, $n_1 = 3.5$
α_{10}	1.25	3.16	7.02	29.	175.	W from eqn. (14) $\beta = \frac{1}{2}$
α_{20}	−0.22	−0.16	−0.04	0.65	5.2	$\gamma\xi = 1$
μ_p/μ				45.	34.	b = 2
τ_{M_0}/τ_c	0.56	1.66	3.8	17.5	104.	
α_{10}	0.54	1.51	3.41	15.4	92.	II, b = 1.5
α_{20}	0.017	0.15	0.41	2.0	12.4	
μ_p/μ	32.	10.	8.3	7.5	7.4	

*Subscripts 1 refer to values in the planet's zone and 2 to those of the swarm zone; the zero subscript indicates an initial value. The parameter n is the power in the exponential distribution of particle sizes; for all models $n_1 \equiv 3.5$. The characteristic times are τ_M for the growth of the planet (eqn. 29) and τ_c for a characteristic particle in the swarm (eqn. 9). The rate of growth of the planet and the swarm are related to α_1 and α_2, respectively (see eqns. 25 and 27). The amount of matter which falls onto the planet relative to swarm mass is μ_p/μ.

The problem of the accumulation of satellites is also closely connected with the problem of their chemical composition (cf. Chapt. 25, Consolmagno and Lewis). Repeated collisions of particles with high impact velocities, especially in the inner part of the swarm, and the subsequent sweeping out of evaporated material by the solar wind will deplete the swarm of volatile substances. Hot giant planets could produce a considerable heating of the inner part of the swarm (Pollack and Reynolds, 1974; Chapt. 23, Cameron). The two effects have led to the higher densities of the inner satellites of the Jovian planets. On the other hand, a predominant selection of smaller particles into the swarm could produce a deficiency of metals in the satellites (as was probably the case for the Moon). This problem has not as yet been sufficiently studied and needs special investigation.

ORIGIN OF EARTH'S MOON

27

John A. Wood
Smithsonian Astrophysical Observatory

A brief review is given of the three most common hypotheses of the origin of the Moon: fission from the Earth, capture by the Earth and binary accretion. The last, which has the most adherents today, is considered most extensively.

When the United States and the Soviet Union embarked upon their recent programs of exploration and sampling of the Moon, it seemed reasonable to expect that the mode of origin of that body was soon to be revealed. The prevalent opinion was that the level of geological activity in the Moon had always been low, so many of its earliest properties were likely to be preserved to the present day. If samples of primordial undifferentiated lunar material had been included in the materials collected by the astronauts, it is possible that the process of lunar formation could have been read unambiguously in them. This was not to be the case, however. It turned out that the Moon had been subjected to violent processes in early times, and passed through a stage of intense geologic activity, which profoundly affected all the lunar materials that have been examined. None of the primordial Moon was preserved for our edification.

Consequently the origin of the Moon remains a mystery, and all the pre-Apollo hypotheses remain viable possibilities. The things we have learned about the early geochemical evolution of the Moon do provide important additional constraints on our ideas about lunar evolution, however. (The term "constraint" or "boundary condition" used in an interpretational context should, of course, be read with a proper degree of skepticism. What I present below are not observational facts but first-order interpretations placed on the observations. There is substantial agreement among lunar scientists that these interpretations are correct and strongly supported; but they *are* interpretations, not facts.)

MAJOR GEOCHEMICAL PROPERTIES OF THE MOON

There are substantial differences in chemical composition between the Earth and Moon. These are apparent qualitatively in the low levels of the more volatile elements in all lunar samples examined, compared with similar terrestrial materials, and also in the smallness or total absence of a metallic core in the Moon. [The moment of inertia and overall density of the Moon place an upper limit of

[513]

TABLE 27.1
Estimated Proportions of Element Groups in Earth and Moon
(Ganapathy and Anders, 1974)

Element group	Condensation temperature in solar nebula (°K)	Representative members	Abundance/ Cosmic abundance Earth	Moon
Refractory elements	>1400	Al, Ca, Ti, Ba, Sr, U, Th, Pt, Ir	~1	2.7
Silicates	1400–1200	Mg, Si	1*	1*
Metal	1400–1200	Fe, Ni, Co, Cr, Au	~1	0.25
Volatile elements	1300–600	S, Na, K, Cu, Zn, Te	0.25	0.05
Very volatile elements	<600	Cl, Br, Hg, Pb, In	0.02	0.0005

*Abundances normalized to 1.0 for magnesium silicates.

about 500 km on the radius of the lunar core, if it has one at all (Toksöz et al., 1974a, 1974b). This corresponds to < 2.5% of the volume of the Moon vs. 16% of the Earth's volume that is occupied by a high-density core.]

More quantitative estimates of the overall chemical composition of the Moon have been made by Ganapathy and Anders (1974) and Wänke et al. (1974) who analyzed lunar and terrestrial samples for critical pairs of elements that are known to behave differently during vaporization and condensation (the processes most likely to have established chemical differences between the planets and satellites) but not during subsequent melting and crystallization in the planets and satellites. Ganapathy and Anders' results are summarized in Table 27.1. These bulk abundances are estimated from observed crustal abundances using pairs of elements such as potassium and lanthanum to deduce the fractionation history.

Extensive melting occurred in the outer layers of the early Moon; at one time a magma layer at least 100 km deep completely enveloped the Moon. Flotation of plagioclase ($CaAl_2Si_2O_8$) crystals from the cooling magma produced the Moon's low-density crust ($\rho \sim 2.9$ g-cm^{-3}, ~ 60 km thick). Complete melting of the Moon at this stage cannot be excluded. The melt was not generated internally and delivered to the surface in small successive lava eruptions; the entire 100 km or more of magma had to be melted at one time in order to achieve a large-scale density fractionation. The melted layer existed at the surface, not inside the Moon, since the low-density floated layer is today exposed at the surface everywhere on the Moon except in the maria, where it has been covered by later lava eruptions.

The primary chemical differentiation of the Moon was accomplished at this time. An event of this scale would be reflected in the Rb-Sr (rubidium-strontium) model ages of rocks from the lunar crust. Since virtually all lunar materials collected—soils as well as highland and mare rocks—display Rb-Sr model ages in the range 4.3–4.6 × 10^9 years (cf. Wasserburg *et al.*, 1972), solidification of the lunar crust following the primary differentiation cannot have occurred later than this. The time scale of cooling and crustal solidification after melting on the scale noted is of the order of 10^8 years (the rate-determining factor is conductive heat loss through the solid crustal layer, once this has begun to form), so the high-energy process that melted the outer layers of the Moon must have operated in the first 1 or 2 × 10^8 years after the origin of the solar system.

Accretion of primordial material to the Moon must have all but ended by the time the cooling crust had grown thick enough to retain the newly added material near the surface, instead of letting it sink into the magma layer. The addition of several percent of the Moon's mass, in the form of chondritic material or any of the condensation components recognized by Ganapathy and Anders (1974), to the crust of the Moon would have caused conspicuous deviations from the mean composition that the crust displays, which is that of an igneous differentiate generated by internal planetary processes.

Severe bombardment of the lunar surface continued until ~ 4.0 × 10^9 years ago, however. The Rb-Sr internal isochron ages and argon gas retention ages of most highland rocks cluster about this value; apparently these are metamorphic ages in most cases, dating energetic events that caused internal re-equilibration of the rocks without necessarily promoting larger scale chemical fractionations as noted above.

HYPOTHESES OF LUNAR ORIGIN: FISSION

Many models for the origin of the Moon have been proposed, but all can be grouped under one of three basic mechanisms—fission, capture, and binary accretion—or occupy positions intermediate to these end members as shown in Figure 27.1. While all three models are still viable after Apollo, each has difficulties serious enough to make a skeptical observer doubt that we are very close to understanding the origin of the Moon. Some of the problems have to do with dynamics, others with chemistry. Several of the proposed mechanisms for placing the Moon in orbit about the Earth involve events of inherently low probability. (Dynamical aspects of the lunar origin are comprehensively reviewed by Kaula, 1971, and updated in Kaula and Harris, 1975.) The substantial differences in the overall chemical compositions of Moon and Earth noted above must be rationalized. A brief discussion of each of the three principal models follows. Orbital evolution of the Moon is not discussed (see Chapt. 7, Burns, for references to this area); most of the problems associated with orbital evolution are not unique to any one model of origin, so do not assist in identification of the correct model.

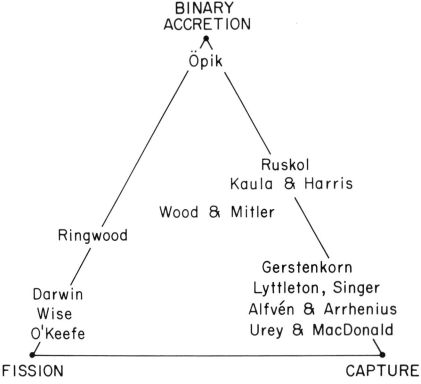

Fig. 27.1. Schematic representation of lunar origin ideologies.

The possibility that the Moon was once part of the Earth, and that rotational instability caused it to be spun off, was first proposed by Darwin (1880); more recently, the hypothesis has been further developed by Wise (1963, 1969) and O'Keefe (1966, 1969, 1970). For a fluid body with polytropic index less than 0.8 (a condition that applies to planets, as opposed to stars), the effect of rotational instability is to cause elongation in a direction perpendicular to the spin axis; the body progresses through the form of a Maclaurin spheroid, to that of a Jacobian ellipsoid, thence to a bowling-pin-shaped Poincaré figure; after which, rupture occurs at the narrow neck of the body, leaving the fragments in orbit about their common center of gravity (Fig. 27.2). It might be asked how the Earth could have formed in a state of rotational instability in the first place; one possibility is that it accreted as a homogeneous body rotating barely within the bounds of stability, and that subsequent internal differentiation into a dense core and lighter overlying mantle decreased its moment of inertia, thereby increasing its rotation rate (assuming conservation of angular momentum) to instability (Wise, 1963).

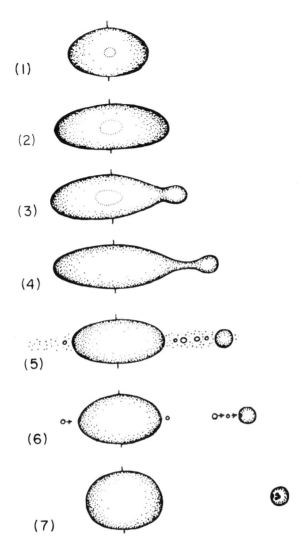

Fig. 27.2. Sequence of forms assumed by the Earth as it is spun to instability during core formation. Figure from Wise (1963).

The greatest virtue of this version of the fission model is that it accounts naturally for the first-order difference in chemistry between the Moon and Earth, namely, the fact that the Moon lacks a large dense core to match that of the Earth. If fission occurred after core separation in the Earth, the substance of the Moon would have been drawn from the Earth's metal-depleted mantle, and it would not have had the wherewithal to form a core of its own. Binder (1974,

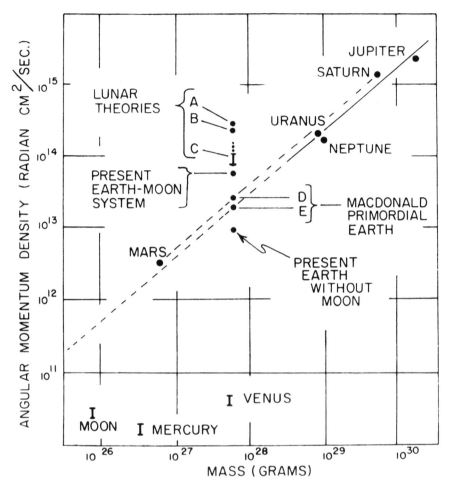

Figure 27.3. Relationship noted by MacDonald (1964) between the mass of a planet and its angular momentum density. Point A corresponds to Wise's (1963) fission model on this diagram; B represents minimum angular momentum density required to drive Earth to rotational instability and produce Moon by fission; C is the most favorable retrograde capture; D and E give possible early Earth spin periods which would satisfy MacDonald's relation. This illustration (from Wise, 1969) suggests that the angular momentum density of planets is proportional to $m^{5/6}$ (m = planetary mass). More recent proposals that the proportionality is $m^{2/3}$ instead (Burns, 1975; Fish, 1967; Hartmann and Larson, 1967) do not affect the text discussion.

1975) has explored the petrological consequences of melting and differentiation of a moon formed from terrestrial mantle material, and he concludes that the types of rock formed in such a process would correspond to the known lunar rock types.

The principal difficulty with the fission model lies in the excessive amount of angular momentum, corresponding to a rotation period no greater than 2.65 hours, needed to cause a moon to separate from the Earth. Special circumstances of formation have to be postulated to create an Earth spinning so much more rapidly than the other planets. MacDonald (1964; see, however, Burns, 1975) has noted a relationship between mass and angular momentum density that holds for most of the planets in the solar system (Fig. 27.3); for the primordial Earth to fission, it would have to possess ~ 6 × the angular momentum predicted by MacDonald's relationship [or have ~ 3 × that predicted by the more modern relationship, Burns, 1975], and twice the angular momentum of the present Earth-Moon system. Thus, a way must also be found to divest the Earth-Moon system of excessive angular momentum after the fission is accomplished. Mechanisms to effect these transferrals of angular momentum can be postulated, but they are speculative and diminish the credibility of the fission hypothesis.

The precipitation hypothesis of Ringwood (1970, 1972) is a variant of the fission model. Ringwood notes that the energy of accretion of the Earth is sufficient not only to melt, but also to boil the substance of our planet. The accreting Earth would be enveloped in a thick hot atmosphere of metal and oxide vapors. Such a system would have an effective polytropic index greater than 0.8, so if spun beyond the point of rotational stability, it would respond by shedding atmospheric gases at its equator, rather than by elongating and parting into discrete bodies as would a planetary body of low polytropic index (above). Ringwood proposes that this occurred in the case of the Earth and that, in fact, early spin rates exceeding rotational stability were the rule, not the exception, for planets of the solar system. As the hot atmospheric gases were spun out from the Earth's equator, they expanded and cooled, and the metal and oxide vapors condensed as dust grains in equatorial orbit about the Earth (Fig. 27.4). Eventually these accreted to form the Moon. The composition of the condensate available for lunar accretion would not have been identical to that of the Earth; the boiling-condensation mechanism amounts to a process of distillation, and in Ringwood's detailed model this operated to produce the observed differences in chemical composition between Earth and Moon.

Essentially the same objection applies to the precipitation hypothesis as to solid-body fission: An amount of angular momentum that appears excessive for planets of the solar system has to be assumed in order to promote separation of the substance of the Moon into orbit about Earth in the first place, and then a mechanism must be provided to dissipate a substantial amount of this angular momentum to bring the Earth-Moon system into its present state. (Proponents of both the fission and the precipitation hypotheses rely on the spinning off of gases from the Earth into interplanetary space to achieve this.)

In addition, the chemical differentiation proposed seems unlikely. On the face of it, a process of distillation would seem destined to raise the most volatile

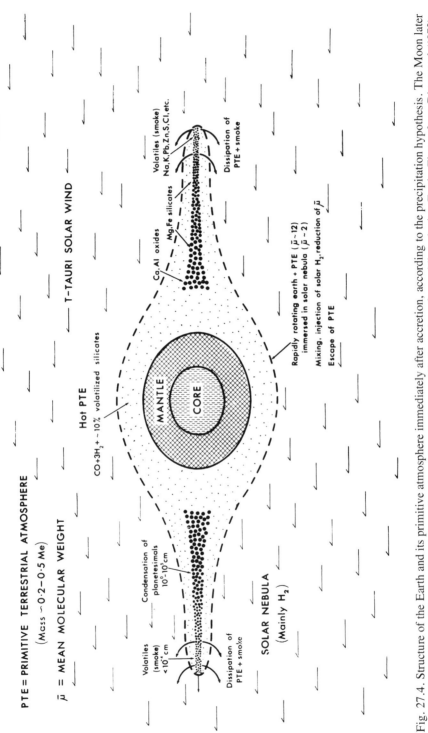

PTE = PRIMITIVE TERRESTRIAL ATMOSPHERE
(Mass ~ 0·2–0·5 Me)

$\bar{\mu}$ = MEAN MOLECULAR WEIGHT

T-TAURI SOLAR WIND

Hot PTE

$CO + 3H_2 + \sim 10\%$ volatilized silicates

Volatiles (smoke)
Na,K,Pb,Zn,S,Cl, etc.

Dissipation of
PTE + smoke

Ca,Al oxides

Mg,Fe silicates

MANTLE

CORE

Rapidly rotating earth + PTE ($\bar{\mu} \sim 12$)
immersed in solar nebula ($\bar{\mu} \sim 2$)

Mixing, injection of solar H_2, reduction of $\bar{\mu}$

Escape of PTE

Volatiles
(smoke)
<10^4 cm

Condensation of
planetesimals
$10^0 – 10^5$ cm

Dissipation of
PTE + smoke

SOLAR NEBULA
(Mainly H_2)

Fig. 27.4. Structure of the Earth and its primitive atmosphere immediately after accretion, according to the precipitation hypothesis. The Moon later accretes from solid particles condensed from the atmosphere and spun off at equator of a rapidly spinning Earth. Figure from Ringwood (1972).

elements and compounds to the outer limits of the system, where they could join the Moon, rather than the highly refractory compounds that are, in fact, most concentrated in the Moon (Table 27.1).

HYPOTHESES OF LUNAR ORIGIN: CAPTURE

A number of authors (e.g., Gerstenkorn, 1955, 1969; Lyttleton, 1967; Singer, 1968, 1970a, 1972; Urey and MacDonald, 1971) have proposed that the Moon was captured essentially intact from a heliocentric orbit into an orbit about the Earth. Others (Alfvén, 1963; Alfvén and Arrhenius, 1969, 1972) envisaged disruption of the Moon in the process, as it passed through Earth's Roche limit, followed by reaccumulation of the debris outside the Roche limit to form the present Moon.

Capture is not easy to accomplish (cf. discussion in Chapt. 7, Burns). In general, an object approaching the Earth from an independent heliocentric orbit describes a hyperbolic trajectory in the near vicinity of Earth, then passes into a new independent heliocentric orbit. If the object had essentially zero velocity relative to Earth while at a great distance from it (this could be realized only if the object originally pursued a heliocentric orbit almost identical to Earth's, but was substantially out of phase with Earth's position in its orbit), the object would describe a near-parabola about Earth when encounter occurred; but it would still adopt an independent heliocentric orbit when it receded from the Earth, and would not be captured.

For capture to occur, an object must be decelerated during its encounter with Earth; kinetic energy must be dissipated. If enough energy is dissipated, the object will not be able to escape Earth's gravitational field, but will fall into a closed orbit about Earth. The mechanisms of energy dissipation available are relatively feeble, however, so no great reduction of velocity can be accomplished. This means that only objects approaching in near-parabolic orbits, that is, objects originating in Earth-like heliocentric orbits, have any appreciable chance of being captured. Objects in eccentric orbits (reaching aphelion at Mars, for example) encounter the Earth at too high a velocity to be decelerated enough for capture. Thus, only capture from the Earth's own ''neighborhood'' in the solar system can be contemplated. This eliminates one of the great potential advantages of the capture hypothesis, namely, the opportunity to rationalize the differences in chemical composition between Earth and Moon by creating the Moon in some far-removed corner of the solar system.

The mechanism of energy dissipation that is usually invoked to accomplish lunar capture is tidal interaction between Earth and Moon (cf. Chapt. 7, Burns, for a discussion of the effects of tidal friction). The likelihood is small of an Earth-Moon encounter so perfectly staged that the tidal interaction could decelerate the Moon into a closed orbit. However, Kaula and Harris (1973, 1975) have pointed out that collisions between the prospective Moon and smaller objects already in orbit about the Earth are potentially a much more effective and realistic means of energy dissipation than is tidal interaction, by a factor of ~ 100. On these terms, lunar capture becomes a fairly plausible event. The process envisaged is no longer pure capture, however: A swarm of earlier captured orbiting objects is required to decelerate the protomoon when it arrives. (This process is described mathematically in Chapt. 26 by Safronov and Ruskol.) Most of these would eventually join the protomoon to form the Moon as we know it. Depending on whether the mass of one particular arriving "protomoon" greatly exceeds the mass of other members of the swarm or not, the process might better be termed binary accretion than capture. (The planetesimals that participate in binary accretion are, after all, assumed to have been captured from the solar nebula during accretion of the Earth.) Binary accretion will be discussed in later sections.

Öpik (1972) has pointed out that if a protomoon following a parabolic trajectory about the Earth were disrupted inside the Earth's Roche limit, the individual debris fragments would not subsequently pursue parabolic trajectories. All the fragments would be moving at the same velocity at the moment of disruption, but for the fragments closest to the Earth (i.e., those derived from the Earth-facing side), this would amount to less than the parabolic velocity, while fragments from the farthest-out face of the protomoon would be moving at greater than parabolic velocity. Debris from the inner half of the protomoon would therefore be captured into closed geocentric orbits (Fig. 27.5), and this would happen without recourse to tidal or collisional decelerations. For protomoon approaches at greater than the parabolic velocity, less than half of the debris is captured. For values of geocentric V_∞ greater than about 2.5 km sec^{-1}, nothing would be captured.

Wood and Mitler (1974) proposed that many such encounters contributed an orbiting swarm of debris particles to the Earth, from which the Moon eventually accreted. The appeal of this concept stems from the selective retention in orbit of a particular portion of each passing protomoon. If the protomoons had been melted and gravity-differentiated before encounter, and if the latter occurred in the right velocity range, then debris from the silicate mantles of the protomoons would be systematically captured and metallic iron core material rejected into heliocentric orbits. This would provide ideal raw material for the accretion of our metal-poor Moon. Unfortunately, the process is very inefficient. Without making special assumptions about the sizes and orbits of the protomoons, it becomes necessary to process an enormous mass of material through Earth's Roche limit to obtain the makings of our Moon; it then becomes very difficult to prevent some of the large amount of rejected material from joining the Moon and nullifying the chemical fractionation achieved.

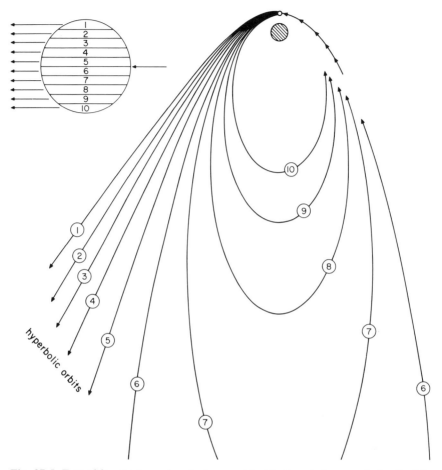

Fig. 27.5. Fate of fragments produced when an object is disrupted upon entering Earth's Roche limit at parabolic velocity. Fragments from 10 arbitrary layers in the object (upper left), each layer approximately equidistant from the center of mass of the Earth, are followed: 1-5 are lost in hyperbolic orbits relative to a terrestrial reference frame; 6-10 are captured in elliptical orbits. Figure from Wood and Mitler (1974).

HYPOTHESES OF LUNAR ORIGIN: BINARY ACCRETION

Although fission and capture cannot be excluded as possible modes of origin for Earth's Moon, binary accretion currently appears to be the most promising mechanism (editorial opinion of the author). This model requires the least in the way of special assumptions and has been the most carefully developed in recent years.

Formation of the Moon by this means cannot be separated from the more fundamental question of the accretion of the planets. There are currently two strongly divergent bodies of thought regarding the formation of the planets: One might be termed the American school, which pictures a massive ($\sim$ 1 solar mass) nebula in which rapid ($\sim 10^3$ years), inefficient accretion produced the terrestrial planets (e.g., Cameron, 1973; Chapt. 23, herein); accretion on such a short time scale would cause extensive heating and melting in the planets and Earth's Moon. By contrast the Russian school, grounded in the thought of O. Yu. Schmidt (1959), envisages comprehensive accretion from a nebula of more modest scale ($\sim$ 0.1 solar mass). Under these circumstances accretion would take much longer ($\sim 10^8$ years), and accretional energy could not in most cases be retained efficiently enough to cause initial melting of the terrestrial planets (e.g., Safronov, 1972). A critical comparison of these models of planetary formation is beyond the scope of this chapter. The binary accretion mechanism of Moon formation will be sketched for both contexts.

In the small-nebula (Russian) model, accretion of small solid particles in nearly circular orbits in the solar nebula at first proceeded rapidly, producing a hierarchy of subplanets up to several hundreds of kilometers in dimension in the zone of what is now the terrestrial planets, in a time of the order of 10^3 years. As the subplanets achieved substantial size, however, they began to interact gravitationally and perturbed one another into eccentric orbits. After this, there was a competition between disruptive high-velocity collisions and constructive slower encounters between the more massive objects and small particles. In the end, after $\sim 10^8$ years, the particles had been assembled into the array of terrestrial planets we know today. The gases of the solar nebula were dissipated during an epoch of T Tauri activity in the Sun after $\sim 10^7$ years, so much of the accretional process occurred in the absence of gas (Safronov, 1972).

Particles accreting to the growing Earth approached in both the direct and the retrograde senses, relative to the Earth (Giuli, 1968). The same would apply to particles that made close approaches but did not join the Earth. Impacts between prograde and retrograde particles would have checked the motions of both, and left some fraction of them in direct orbit about the Earth (assuming there was a net excess of direct angular momentum, as is indicated by the rotation of the Earth). These would have assisted in the deceleration of later particles that passed near the Earth, so that some of them too were captured. In this way, a swarm of particles in geocentric orbit was established, from which the Moon was to grow. Mutual collisions among the members of the swarm tended to circularize their orbits and bring them into a common plane.

Ruskol (1971a, 1973) estimates that our Moon did not nucleate in or join the circumterrestrial swarm until the Earth had achieved $\sim$ 0.5 of its present mass; earlier than that, the population of protoplanetary objects at large in the solar system was numerous enough to endanger the existence of a satellite nucleus, and make likely its destruction by collision. Once nucleation of the Moon was achieved, the time needed to sweep up the circumterrestrial swarm would have been relatively short, of the order of 10^3 years; but since the time scale of planetary formation was $\sim 10^8$ years, and material continued to be added to the circumterrestrial swarm over this period, the assembly of the Moon was similarly protracted. Ruskol estimates that a moon could not have grown by more than $\sim$ 300 km in the circumterrestrial swarm, and so invokes the capture of a nucleus of $\sim$ 1500-km radius or, better, several nuclei of $\sim$1000-km radius, into the circumterrestrial swarm from heliocentric orbits. The nuclei were decelerated during close Earth encounters by collisions with objects already orbiting in the swarm. In Chapter 26, Safronov and Ruskol discuss similar processes for the formation of the regular satellites.

Accretion over 10^8 years precludes melting of the outer layers of the Moon by straightforward accretional heating, as has been postulated by many geochemists. Otherwise, the Russian model is not seriously in conflict with the geochemical constraints mentioned earlier. (The chemical composition of the Moon in this context is discussed below.) Ruskol points out that substantial heating of the Moon would have occurred if several submoons ultimately joined one another; most of the potential energy of accretion of the Moon would be released at this stage, not in the accretion of the individual submoons. If the submoons were already warm from 10^8 years of ^{235}U decay, or possibly from electrical heating by the T Tauri solar wind (Sonett *et al.*, 1968), this final assembly of components might well have caused the melting and primary differentiation of the Moon that are indicated in the lunar samples. Melting would have occurred in the interiors of coalescing submoons rather than at their surfaces, but this is not a serious objection; a part-melted, part-solid melange of lunar material would have quickly achieved a density stratification in which a light magma layer surrounded a core of denser unmelted rock.

In the large-nebula (American) model, accretion of the terrestrial planets occurred rapidly and inefficiently from particles revolving in near-circular orbits within the gaseous nebula (Cameron, 1973; Cameron and Pine, 1973; Chapt. 23, Cameron). Within a very short time, $\sim 10^3$ years, the inner portion of the nebula was dissipated (by being drawn into the growing Sun, not by T Tauri emission). Unaccreted dust grains (of aggregate mass much greater than the mass of accreted planets or subplanets) were removed with the gases. Whether there was a lengthy postnebula phase of activity in which subplanets were perturbed into eccentric orbits, then gradually collected into a small number of planets in near-circular orbits, is not specified. Growth of a moon by binary accretion would have followed along the general lines discussed in the Russian model, except the duration of the process was greatly shortened. Under these circumstances, melting of the outer hundreds of kilometers of the Moon is easily accomplished by accretional heating.

CHEMICAL FRACTIONATION IN THE
BINARY ACCRETION MODELS

The Russian and American models differ drastically in the amount of chemical fractionation they allow to accompany planetary formation. The Russian model involves complete cooling and exhaustive condensation of the nebular gases before planetary accretion, so all the chemical components listed in Table 27.1 would have been present as dust particles in the inner solar system and available for accretion. Only the ices of hydrogen, carbon, nitrogen, and oxygen compounds would have been prevented from condensing by the heat of the Sun. Differences in physical behavior among the various condensate species have to be invoked to rationalize the differences in chemical composition between planets.

Similarly, the Russian model has to rely on differences in the physical properties of the minerals to explain the chemical differences between Moon and Earth. Ruskol (1971b), citing Orowan (1969), notes that silicate grains (being brittle) would tend to fracture on collision, while impacting metal particles would ductile-weld together. The resulting fine silicate dust would be more easily captured into the circumterrestrial swarm than would heavier metal aggregations, because of the larger surface-to-mass ratio the silicate particles present to objects already in the swarm, which have the potential of slowing and capturing them. Consequently the circumterrestrial swarm, and ultimately the Moon, would be enriched in silicates and depleted in metals. Volatile elements would be systematically lost during high-energy collisions in the swarm. Neither of these selection factors would greatly affect material accreting directly to the Earth. However, these processes have difficulty explaining the apparent enrichment of refractory silicates and oxides over magnesium silicates in the Moon (Table 27.1).

In the American model, the cooling time and dissipation time of gases in the inner nebula are comparable (Cameron and Pine, 1973). While complete cooling and condensation of the gases probably occurred before they were dissipated, the opportunity does exist for planets or subplanets to accrete larger proportions of the earliest formed, most refractory condensates than they do of the later, low-temperature minerals. Lewis (1972a) explains the density differences of the terrestrial planets in this way; Mercury, closest in to the Sun, received the greatest proportion of refractory elements and metallic iron, Venus the second greatest, and so forth. Similar processes are thought to produce the density differences in the Galilean satellites (cf. Pollack and Reynolds, 1974; Chapt. 17, Fanale et al.; Chapt. 23, Cameron).

Proponents of the American model can also resort to fractionation of condensates formed at different temperatures in the nebula to account for the chemical differences of Earth and Moon. D. Anderson (1972) has proposed that the Moon is composed almost wholly of high-temperature condensates. Table 27.1 indicates that the Moon is systematically enriched in high-temperature condensates, and depleted in low-temperature materials, relative to Earth. [However, the position of metallic iron in this sequence is equivocal. The temperature ranges for equilibrium condensation of metallic iron and magnesium silicates, the two most abundant components in Earth and Moon, are practically coincident. It is hard to see how the fractionation of equilibrium condensates alone can have wrought any significant difference in the metal/magnesium silicate ratio between Earth and Moon. Blander and Katz (1967) have argued that metallic iron in fact tends to supersaturate in the gas phase, finally condensing at low temperatures under disequilibrium conditions. If metallic iron is actually a low-temperature condensate, its depletion in the Moon along with other low-temperature components is more easily understood. But this amounts to having one's cake and eating it too: If metallic iron behaves as a low-temperature component in the formation of the Moon, then its superabundance in Mercury cannot be ascribed to its high temperature of condensation.]

Leaving the metallic iron discrepancy aside, how can the Moon have avoided collecting late-condensing low-temperature material in the same proportions as the Earth? Any type of material joining the Earth from heliocentric orbit also had a calculable probability of joining the Moon. The difference in gravitational capture cross sections of the nearly grown Earth and Moon is often held responsible for the exclusion of late-arriving material from the latter (cf. Chapt. 26, Safronov and Ruskol). However, the effect is not strong enough. In the limit as $V_\infty \to 0$ for particles accreting from heliocentric orbits, the Earth/Moon ratio of accretion rates per unit surface area is 22.3, about adequate to account for the compositional discrepancies between Earth and Moon in Table 27.1. But the numbers in Table 27.1 are thought to represent the bulk compositions of Earth and Moon, not just the composition of a shallow surface layer on each, and in this case, the relevant parameter is the accretion rate per unit volume of Earth and Moon, not per unit surface area. In addition, material joining the Earth-Moon system late can be expected to arrive with high values of V_∞, not at low velocities. When these circumstances are taken into account, Earth's advantage over the Moon in gravitational capture cross section disappears (Fig. 27.6). In fact, the advantage can swing to the Moon.

The advantage the Earth has in being able to hold onto material that impacts it at high velocities is probably more important. In a hypervelocity impact, most of the projectile material is incorporated in a spray of debris that is jetted hydrodynamically from the point of impact in the earliest stages of the event. The jetted material moves at velocities comparable to the original projectile velocity (Gault et al., 1963). Thus, late projectile material entering the Earth-Moon system with high V_∞ is probably largely rejected by the Moon at velocities sufficient to escape the Earth-Moon system, but is systematically trapped by the

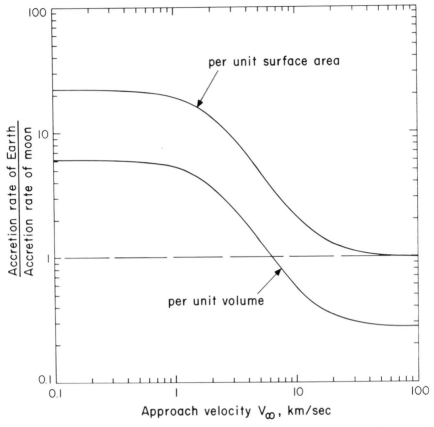

Fig. 27.6. The relationship (ratio) of accretion rates of Earth and Moon, as a function of the geocentric velocity at infinity of a population of approaching particles. Present masses and dimensions of Earth and Moon are assumed.

larger gravitational field and decelerating atmosphere of the early Earth. The tendency of volatile elements to be rejected from the Moon is even more pronounced if they arrive in a potentially explosive cometary matrix (Wasson, 1971).

Clearly, there are still difficulties in understanding the chemical differences of Earth and Moon in terms of a binary accretion model. Possibly the most serious objection to binary accretion of any school, however, transcends chemistry. Binary accretion explains the existence of the Moon as a natural consequence of the processes operating in the early solar system. No special occurrences are invoked; this makes binary accretion the most aesthetically appealing model. But if Moon formation is so inevitable, then why don't all the planets have satellites, with dimensions comparable to Earth's Moon? One is now forced to make special assumptions about all the other planets: Mercury's dynamical history was too much dominated by the Sun; Venus's moon (if it had one) was spun down to its surface by tidal interactions with a planet rotating more slowly than the period of the satellite (or even in a direction opposite to the satellite's orbit) (McCord, 1968; Burns, 1973; Counselman, 1973; Ward and Reid, 1973); there were fundamental differences in the process of satellite formation for Jovian planets. Mars is probably the worst embarrassment; there is no obvious reason why it should not have accumulated a moon comparable to Earth's. Perhaps in the end it would be more plausible to make a special assumption for only one planet, and invoke fission or capture for Earth's Moon.

ACKNOWLEDGMENT

This work was supported in part by grant NGL 09-015-150 from the National Aeronautics and Space Administration.

DISCUSSION

GREGORY WILLIAMS: The evidence referred to certainly establishes the necessity of major differentiation having occurred, and the K-La ratio indeed suggests that the sorting (in the fluid state) of the various components of the Moon's material from each other took place at a single site. However, the arguments presented do not substantiate the contention that the site of the differentiation was *necessarily* the Moon. The material could as well have been pre-processed in a protoplanet. In this regard it is to be recalled that the products of a collisionally disrupted planet tend to be injected into a narrow range of orbits (because of the comparatively low velocity of the ejected material) rather than contributing to the general solar system mix.

WILLIAM M. KAULA: It should be pointed out that the enrichment of refractory silicates estimated for the whole Moon depends on a considerable extrapolation. The enrichment of 2.7 is derived from assuming that (i) the heat flows measured at the Apollo 15 and Apollo 17 sites are representative of the whole Moon, and hence measure the total uranium content; and (ii) chondritic ratios of refractory silicates to uranium prevail.

REFERENCE SECTION

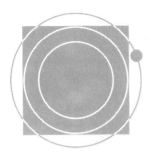

TABULATIONS OF SATELLITE POSITIONAL OBSERVATIONS, AND OF THEIR DISCUSSION

P. K. Seidelmann
U.S. Naval Observatory

In 1972 serious discussion began concerning the status of, and the need to improve, the ephemerides of the natural satellites (Chapt. 3, Aksnes), both for the spacecraft missions being planned to Jupiter and Saturn, and for the ground-based astronomers.

Table A1 compiles discussions of observations of natural satellites up through 1972. Since then emphasis has been placed on obtaining as complete an observational coverage of the different satellites as possible and on evaluating the quality of the observations which can be obtained with the various telescopes; Table A2 indicates the number of plates of observational coverage that were obtained during 1973–74. The accuracy of the observations obtained by the Jet Propulsion Laboratory at Table Mountain Observatory is uncertain. Harvard's Agassiz station will be used only for problem cases and under special circumstances, since the observations from McDonald Observatory and from the U.S. Naval Observatory should provide sufficient coverage and accuracy for most cases. These observations, augmented by those from the University of Arizona, provide adequate observational data with a few exceptions. The exceptions are Amalthea (J5) and Janus (S10), which are going to require special attention. In these cases special observational techniques must be utilized to obtain observations of the satellites which are close to the primary (cf. Pascu, Chapt. 5, herein). A different technique, with a special filter, is planned to try to obtain more accurate observations of the faint outer satellites of Jupiter.

The U.S. Naval Observatory will act as a collection agent for positional observations of the satellites so that observational data and information are available from one source. R. Millis, at Lowell Observatory, has been collecting the observational data for mutual events of the Galilean satellites. People with either type of observation are urged to forward them to the appropriate place to assure the availability of observations for comparison with theories.

Theoretical work on the Galilean satellites is being pursued at the Bureau des Longitudes and at the Jet Propulsion Laboratory, while theoretical work on the Saturnian satellite system is underway at the University of Texas and the Jet Propulsion Laboratory. For the other satellite ephemerides it is felt that the theoretical development is not the limiting factor concerning the accuracy of the

[533]

ephemerides, but rather the limiting factors are the distribution and accuracy of the observations (Chapt. 5, Pascu) and the derived constants necessary for the ephemerides. Some of these theories are discussed by Aksnes (Chapt. 3, herein), and Kovalevsky and Sagnier (Chapt. 4, herein).

It is hoped that by this coordinated effort improved ephemerides of all the natural satellites might be derived by 1980.

DISCUSSION

DALE CRUIKSHANK: In the course of photometric observations of eclipses of the Galilean satellites made for studies of the post-eclipse brightening effect (Veverka, Chapt. 9), observers over the past 12 years have accumulated and published data on the (O − C) values of half-intensity times. Are these data of any potential value for improvement of satellite theories, and, if so, how accurate must the timing be?

JAY H. LIESKE: I would answer an enthusiastic yes. Some people may be disappointed with the large divergence among the timing (i.e., longitude) corrections for the Galilean satellites (especially for J3) and wonder about the worthwhile nature of obtaining more data when, even with present data, we seem to be in such a sorry state. I would reply that at the present time we can *only* adjust the longitudes for the Galilean satellites because we do not have the partial derivatives which would enable us to change orbital parameters, such as eccentricities, or physical parameters, such as masses and oblateness parameters. In the near future we hope to have such partial derivatives available and at that time the data should prove to be extremely worthwhile.

Sampson said 10 sec was the precision needed for the eclipse times. Since we do not have partial derivatives available now, we cannot readily study the sensitivity of the data. It also would be worthwhile to have the time history of an event available in addition to the timing of the mid-point, since such data will yield information which also will be useful once partial derivatives become available.

S. FERRAZ-MELLO: The publication of the available photoelectric data on the timing of eclipses may be useful; however, much care must be taken with its use. We know that these observations have systematic errors which are not due to the observation but to the observed phenomenon and which must be related to the atmosphere of Jupiter.

BRIAN O'LEARY: This coordinated effort in improving the ephemerides of planetary satellites could be expanded to assist in predicting stellar occultations by satellites (cf. O'Leary, Chapt. 13, herein). In particular, by taking astrographic plates of the sky ahead of the motions of the outer planets, it would be possible to identify candidate stars for satellite occultations.

TABLE A.1

Articles Discussing Observations of Satellite Positions

Satellite	Reference	Accuracy
Jupiter		
V	van Woerkom (1950)	
	Sudbury (1969)	
I–IV	Sampson (1910)	
	Marsden (1966)	
	Kovalevsky (1959)	
VI	Bobone (1937a)	
	Mulholland (1965)	
VII	Bobone (1937b)	
X	Wilson (1939)	residuals C.A. 8″.0
	Lemechova (1961)	
	Herget (1968b)	residuals C.A. 0″.6
XII	Herrick (1952)	C.A. 0″.5
	Herget (1968b)	C.A. 0″.5
XI	Herget (1968b)	
	Herget (1968a)	C.A. 1″.5
VIII	Herget (1968b)	residuals — 2″.15 rms
IX	Bec (1969)	4″.0 m.e.
	Herget (1968a)	4″.3 p.e.
	de Polavieja and Edelman (1972)	
	Herrick (1952)	
Saturn		
X Janus		
I Mimas	G. Struve (1933)	
	Kozai (1957)	
II Enceladus	G. Struve (1933)	
	Kozai (1957)	
III Tethys	G. Struve (1933)	
	Kozai (1957)	
IV Dione	G. Struve (1933)	
	Kozai (1957)	0″.15 p.e.
V Rhea	G. Struve (1933)	
	Kozai (1957)	0″.14 p.e.
VI Titan	G. Struve (1933)	
	Kozai (1957)	0″.14 p.e.
VII Hyperion	Woltjer (1928)	
	Message (1960)	
VIII Iapetus	G. Struve (1933)	
IX Phoebe	Ross (1905)	
	Elmabsout (1970)	

TABLE A.1 (Continued)
Articles Discussing Observations of Satellite Positions

Satellite	Reference	Accuracy
Uranus		
V Miranda	Dunham (1971)	long. at epoch $\pm$ 0˝28
I Ariel	Harris (1949)	
	Dunham (1971)	$\pm$ 0˝024
II Umbriel	Harris (1949)	
	Dunham (1971)	$\pm$ 0˝053
III Titania	Harris (1949)	
	Dunham (1971)	$\pm$ 0˝040
IV Oberon	Harris (1949)	
	Dunham (1971)	$\pm$ 0˝030
Neptune		
I Triton	Eichelberger and Newton (1926)	
	Gill and Gault (1968)	
II Nereid	van Biesbroeck (1957)	O $-$ C's of 1˝0

Other References for Saturn's Satellites
    Laves (1938)
    Jeffreys (1953, 1954)
    Garcia (1970, 1972)

Theoretical Investigations
    Charnow *et al.* (1968)
    Ferraz-Mello (1966)
    Kovalevsky (1959)

TABLE A.2
Observations of Natural Satellites 1973–74

Mars

Phobos, Deimos   48 plates—Pascu (USNO)

100 obs. Mariner 9—Duxbury (JPL) (see Born and Duxbury, 1975)

Jupiter

V

I–IV   40 mutual occultations and eclipses—Millis (Lowell)

15 ordinary eclipses—Millis

Mutual events data bank—Millis

17 plates—Aksnes (SAO)

40 plates—Duxbury

24 plates—Pascu/Fiala

some—Ianna (McCormick)

VI   6 plates—Aksnes →

VII

X

XII

XI

VIII   few—Shelus (McDonald) →   some—Roemer (Arizona) →

IX   few—Shelus →

TABLE A.2 (Continued)
Observations of Natural Satellites 1973–74

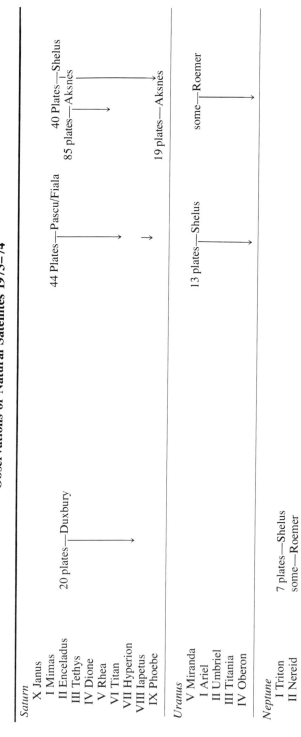

Saturn
X Janus
I Mimas
II Enceladus
III Tethys
IV Dione
V Rhea
VI Titan   20 plates——Duxbury
VII Hyperion
VIII Iapetus
IX Phoebe

Uranus
V Miranda
I Ariel
II Umbriel
III Titania
IV Oberon

Neptune
I Triton   7 plates——Shelus
II Nereid   some——Roemer

44 Plates——Pascu/Fiala   40 Plates——Shelus
                              85 plates——Aksnes

                              19 plates——Aksnes

13 plates——Shelus   some——Roemer

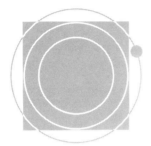

REFERENCES AND BIBLIOGRAPHY

Ackerman, M. 1971. Ultraviolet solar radiation related to mesospheric processes. In *Mesospheric Models and Related Experiments* (G. Fiocco, ed.), pp. 149–59. Dordrecht, Holland: D. Reidel.

Adams, J. B. 1974. Uniqueness of visible and near infrared diffuse reflectance spectra of pyroxenes and other rock-forming minerals. In *Infrared and Raman Spectroscopy of Lunar and Terrestrial Minerals* (C. Karr, ed.). New York: Academic Press.

Adams, J. B., and McCord, T. B. 1971. Alteration of lunar optical properties: Age and composition effects. *Science* 171:567–71.

Aggarwal, H. R., and Oberbeck, V. R. 1974. Roche limit of a solid body. *Astrophys. J.* 191:577–88.

Aksnes, K. 1974a. Mutual phenomena of Jupiter's Galilean satellites, 1973–74. *Icarus* 21:100–11.

_____. 1974b. An ephemeris for Neptune II (Nereid). *IAU Circ.* 2665.

Aksnes, K., and Franklin, F. A. 1974. Reduction techniques and some results from occultations of Europa by Io. *Bull. Amer. Astro. Soc.* 6:382.

_____, and _____. 1975a. Results of 1973 occultations of Europa by Io. Presented at IAU Colloquium No. 28. *Astron J.* 80:56–63.

_____, and _____. 1975b. DeSitter's theory "melts" Europa's polar cap. *Nature* 258:503–5.

Aksnes, K., and Marsden, B. G. 1974. Jupiter XIII. *IAU Circ.* 2711.

Alden, H. L. 1943. Observations of the satellite of Neptune. *Astron. J.* 50:110–11.

Alden, H. L., and O'Connell, W. C. 1928. Photographic measures of the satellites of Saturn in 1926. *Astron. J.* 38:52–55.

Alfvén, H. 1942–45. On the cosmogony of the solar system. *Stockholms Observatoriums Annaler,* Bd. 14, No. 2; Bd. 14, No. 5; Bd. 14, No. 9.

_____. 1954. *On the Origin of the Solar System.* London: Oxford University Press.

_____. 1963. The early history of the Moon and the Earth. *Icarus* 1:357–63.

_____. 1971. Plasma physics, space research, and the origin of the solar system. *Science* 172:991–94.

Alfvén, H., and Arrhenius, G. 1969. Two alternatives for the history of the Moon. *Science* 165:11–17.

_____, and _____. 1970. Structure and evolutionary history of the solar system, I. *Astrophys. Space Sci.* 8:338–421.

_____, and _____. 1972. Origin and evolution of the Earth-Moon system. *The Moon* 5:210–30.

_____, and _____. 1973. Structure and evolutionary history of the solar system, III. *Astrophys. Space Sci.* 21:117–76.

_____, and _____. 1974. Structure and evolutionary history of the solar system, IV. *Astrophys. Space Sci.* 29:63–159.

[539]

————, and ————. 1976. *Evolution of the Solar System*. NASA SP–345. Washington, D.C.: U.S. Gov't. Printing Office.

Alfvén, H. and Fälthammar, C.-G. 1963. *Cosmical Electrodynamics* (2nd ed.). London: Oxford University Press.

Allan, R. R. 1969. Evolution of Mimas-Tethys commensurability. *Astron. J.* 74:497–506.

Allen, C. W. (ed.). 1963. *Astrophysical Quantities* (2nd ed.). London: Univ. of London, The Athlone Press.

Allen, D. A. 1970. The infrared diameter of Vesta. *Nature* 227:158–59.

Allen, D. A., and Murdock, T. L. 1971. Infrared photometry of Saturn, Titan, and the rings. *Icarus* 14: 1–2.

Allen, H. C., Jr., and Cross, P. C. 1963. *Molecular Vib-rotors: The Theory and Interpretation of High Resolution Infrared Spectra*. New York: John Wiley.

Anderson, D. L. 1967. The anelasticity of the mantle. *Geophys. J. Roy. Astro. Soc.* 14:135–64.

————. 1972. The origin of the Moon. *Nature* 239:263–65.

————, and Archambeau, C. B. 1964. The anelasticity of the Earth. *J. Geophys. Res.* 69:2071–84.

Anderson, J. D. 1971. Feasibility of determining the mass of an asteroid from a spacecraft flyby. In *Physical Studies of Minor Planets* (T. Gehrels, ed.), pp. 577–83. NASA SP–267. Washington, D.C.: U.S. Gov't. Printing Office.

Anderson, J. D.; Null, G. W.; and Wong, S. K. 1974a. Gravitational parameters of the Jupiter system from the doppler tracking of Pioneer 10. *Science* 183:322–23.

————; ————; and ————. 1974b. Gravity results from Pioneer 10 doppler data. *J. Geophys. Res.* 79:3361–64.

Andersson, L. E. 1972. Photometry of Jupiter VI and Phoebe (Saturn IX). *Bull. Amer. Astro. Soc.* 4:313.

————. 1974. *A Photometric Study of Pluto and Satellites of the Outer Planets*. Ph.D. dissertation. University of Indiana, Bloomington, Indiana.

Andersson, L. E., and Burkhead, M. S. 1970. Observations of the satellite Jupiter VI. *Astron. J.* 75:743.

Andoyer, H. 1915. Sur le calcul des ephemerides des quatres anciens satellites de Jupiter. *Bull. Astron.* 32:177–224.

Arlot, J.–E. 1975. A comparison of some photographic observations of the Galilean satellites with Sampson's theory. Presented at IAU Colloquium No. 28. *Cel. Mech.* 12:39–50.

Armstrong, K. R.; Harper, D. A., Jr.; and Low, F. J. 1972. Far-infrared brightness temperatures of the planets. *Astrophys. J.* 178:L89–L92.

Arrhenius, G. 1972. Chemical effects in plasma condensation. In *From Plasma to Planet*, Nobel Symp. 21 (A. Elvius, ed.), pp. 117–32. New York: John Wiley.

Arrhenius, G., and De, B. R. 1973. Equilibrium condensation in a solar nebula. *Meteoritics* 8:297–313.

Artem'ev, A. V. 1969. Planetary rotation induced by elliptically orbiting particles. *Solar System Research* 3:15–21.

Arvesen, J. C.; Griffin, R. N.; and Pearson, B. D. 1969. Determination of extraterrestrial solar spectral irradiance from a research aircraft. *Applied Optics* 8:2215–22.

Aumann, H. H.; Gillespie, C. M., Jr.; and Low, F. J. 1969. The internal powers and effective temperatures of Jupiter and Saturn. *Astrophys. J.* 157:L69–L72.

Aumann, H. H., and Kieffer, H. H. 1973. Determination of particle sizes in Saturn's rings from their eclipse cooling and heating curves. *Astrophys. J.* 186:305–12.

Avramchuck, V. V. 1964. Polychromatic polarimetry of some lunar regions. In *Physics of the Moon and Planets* (I. Koval, ed.), pp. 1–10. Trans. 1966. Jerusalem: Israel Program for Scientific Translations Ltd. NASA TT F–382.

Axel, L. 1972. Inhomogeneous models of the atmosphere of Jupiter. *Astrophys. J.* 173:451–68.

Bailey, J. M. 1969. The Moon may be a former planet. *Nature* 223:251–53.

_____. 1971. Origin of the outer satellites of Jupiter. *J. Geophys. Res.* 76:7827–32.

_____. 1972. Studies on planetary satellites. Satellite capture in the three-body elliptical problem. *Astron. J.* 77:177–82.

Barker, E. S., and Trafton, L. M. 1973a. The reflectivity of Titan from 3000–4350 Å. *Bull. Amer. Astro. Soc.* 5:305.

_____, and _____. 1973b. Ultraviolet reflectivity and geometrical albedo of Titan. *Icarus* 20:444–54.

Bartholdi, P., and Owen, F. 1972. The occultation of Beta Scorpii by Jupiter and Io. II - Io. *Astron. J.* 77:60–65.

Bass, M. N. 1971. Montmorillonite and serpentine in the Orgueil meteorite. *Geochim. Cosmochim. Acta* 35:139–48.

Baum, W. A., and Code, A. D. 1953. A photometric observation of the occultation of σ Arietis by Jupiter. *Astron. J.* 58:108–12.

Bec, A. 1969. Détermination de la masse de Jupiter par l'étude du mouvement de son neuvième satellite. *Astron. Astrophys.* 2:381–87.

Becklin, E. E.; Neugebauer, G.; Hansen, O.; and Kieffer, H. 1973. Stellar flux calibration at 10 and 20 μ using Mariner 6, 7 and 9 results. *Astron. J.* 78:1063–66.

Beckmann, P. 1968. *The Depolarization of Electromagnetic Waves.* Boulder, Colorado: Golem Press.

Beletskii, V. V. 1972. Resonance rotation of celestial bodies and Cassini's laws. *Cel. Mech.* 6:356–58.

Bell, E. E., and Nielsen, H. H. 1950. The infra-red spectrum of acetylene. *J. Chem. Phys.* 18:1382–94.

Bell, R. J. 1972. *Introductory Fourier Transform Spectroscopy,* p. 23. New York: Academic Press.

Berge, G. L., and Muhleman, D. O. 1973. High-angular-resolution observations of Saturn at 21.1 cm wavelength. *Astrophys. J.* 185:373–81.

_____, and _____. 1974. The brightness temperature of Callisto at 3.71 cm wavelength. *Bull. Amer. Astro. Soc.* 6:385. See also *Science* 187:441–43.

Berge, G. L., and Read, R. B. 1968. The microwave emission of Saturn. *Astrophys. J.* 152:755–64.

Bergstralh, J. T.; Matson, D. L.; and Johnson, T. V. 1975. Sodium D-line emission from Io: Synoptic observations from Table Mountain Observatory. Presented at IAU Colloquium No. 28. *Astrophys. J.* 195:L131–L135.

Berlage, H. P. 1948. The disc theory of the origin of the solar system. In *Proc. Koninkl. Ned. Akad. Wetenschap.* 51:796.

Bertie, J. E.; Othen, D. A.; and Solinas, M. 1973. The infrared spectra of ethylene oxide hydrate and hexamethylenetetramine hydrate at 100°K. In *Physics and Chemistry of Ice* (E. Whalley, S. J. Jones and L. W. Gold, eds.), pp. 61–65. Ottawa: Royal Society of Canada.

Bhatnagar, P. L., and Whipple, F. L. 1954. Accretion of matter by a satellite. *Astron. J.* 59:121–24.

Binder, A. B. 1974. On the origin of the Moon by rotational fission. In *Lunar Science V,* p. 63. Houston: Lunar Science Institute. See also *The Moon* 11:53–76.

———. 1975. On the petrology and structure of a gravitationally differentiated Moon of fission origin. *The Moon* 13:431–73.

Binder, A. B., and Cruikshank, D. P. 1964. Evidence for an atmosphere on Io. *Icarus* 3:299–305.

———, and ———. 1966. Photometric search for atmospheres on Europa and Ganymede. *Icarus* 5:7–9.

Black, D. C. 1971. On the equivalence of the planet-satellite formation process. *Icarus* 15:115–19.

Blair, G. N., and Owen, F. N. 1973. UBV photometry of the inner satellites of Saturn. *Bull. Amer. Astro. Soc.* 4:313.

———, and ———. 1974. The UBV orbital phase curves of Rhea, Dione, and Tethys. *Icarus* 22:224–29.

Blanco, C., and Catalano, S. 1971. Photoelectric observations of Saturn's satellites Rhea and Titan. *Astron. Astrophys.* 14:43–47.

———, and ———. 1974a. Relation between light variations of solar system satellites and their interaction with interplanetary medium. In *Exploration of the Planetary System,* IAU Symposium No. 65 (A. Woszczyk and C. Iwaniszewski, eds.), pp. 533–38. Dordrecht, Holland: D. Reidel.

———, and ———. 1974b. On the photometric variations of the Saturn and Jupiter satellites. *Astron. Astrophys.* 33:105–11.

———, and ———. 1974c. Mutual eclipses of Jupiter's satellites. *Astron. Astrophys.* 33:303–9.

———, and ———. 1974d. *IAU Circ.* No. 2679.

———, and ———. 1975. On the secular variations of Io and Titan. Presented at IAU Colloquium No. 28. *Icarus* 25:585–87.

Blander, M., and Katz, J. L. 1967. Condensation of primordial dust. *Geochim. Cosmochim. Acta* 31:1025–34.

Block, L. 1972. Potential double layers in the ionosphere. *Cosmical Electrodyn.* 3:349–76.

Bobone, J. 1937a. Tablas del VI satellite de Jupiter. *Astron. Nachr.* 262:321–46.

———. 1937b. Tablas del VII satellite de Jupiter. *Astron. Nachr.* 263:401–12.

Bobrov, M. S. 1970. Physical properties of Saturn's rings. In *Surfaces and Interiors of Planets and Satellites* (A. Dollfus, ed.), pp. 377–461. New York: Academic Press.

Born, G. H. 1974. Mars physical parameters as determined from Mariner 9 observations of the natural satellites and doppler tracking. *J. Geophys. Res.* 79:4837–44.

Born, G. H. and Duxbury, T. C. 1975. Phobos and Deimos ephemerides from Mariner 9 TV data. Presented at IAU Colloquium No. 28. *Cel. Mech.* 12:77–88.

Bosomworth, D. R., and Gush, H. P. 1965. Collision-induced absorption of compressed gases in the far infrared, part II. *Can. J. Phys.* 43:751–69.

Bowell, E. L. G. 1971. Polarimetric studies. In *Geology and Physics of the Moon* (G. Fielder, ed.), pp. 125–33. New York: American Elsevier.

Bowell, E.; Dollfus, A.; and Geake, J. E. 1972. Polarimetric properties of the lunar surface and its interpretation. Part 5: Apollo 14 and Luna 16 lunar samples. *Proc. Third Lunar Science Conference,* pp. 3103–26. Houston: Lunar Science Institute.

Bowell, E., and Zellner, B. 1974. Polarizations of asteroids and satellites. In *Planets, Stars and Nebulae* (T. Gehrels, ed.), pp. 381–404. Tucson: Univ. of Arizona Press.

Brahic, A. 1975. A numerical study of a gravitating system of colliding particles: Applications to the dynamics of Saturn's ring, and to the formation of the solar system. Presented at IAU Colloquium No. 28. *Icarus* 25:452–58.

Brecher, A. 1972. Memory of early magnetic fields in carbonaceous chondrites. In *On the Origin of the Solar System* (H. Reeves, ed.), pp. 260–73. Paris: Centre Nationale de la Recherche Scientifique.

Brecher, A., and Arrhenius, G. 1974. The paleomagnetic record in carbonaceous chondrites: Natural remanence and magnetic properties. *J. Geophys. Res.* 79:2081–106.

———, and ———. 1976. The paleomagnetic record in carbonaceous chondrites: Modeling of natural remanence and paleofield intensities. Submitted to *Phys. Earth Planet. Interiors*.

Briggs, F. H. 1973. Observations of Uranus and Saturn by a new method of radio interferometry of moving sources. *Astrophys. J.* 182:999–1011.

———. 1974a. The radio brightness of Titan. *Icarus* 22:48–50.

———. 1974b. The microwave properties of Saturn's rings. *Astrophys. J.* 189:367–77.

———. 1974c. *Radio Interferometer Studies of Solar System Objects*. Ph.D. dissertation. Cornell University, Ithaca, New York.

———. 1974d. Radio observations of Titan. In *The Atmosphere of Titan* (D. M. Hunten, ed.), p. 143. NASA SP–340. Washington, D.C.: U.S. Gov't. Printing Office.

Brinkmann, R. T. 1973. Jovian satellite-satellite eclipses and occultations. *Icarus* 19:15–29.

———. 1976. On the inversion of mutual occultation light curves. *Icarus* 27:69–89.

Brinkmann, R. T., and Millis, R. L. 1973. Mutual phenomena of Jupiter's satellites in 1973–74. *Sky and Telescope* 45:93–95.

Bronshten, V. A. 1968. On the origin of the irregular satellites of Jupiter (In Russian). *Astron. Vestnik* 2:No. 1, 29–36.

Brouwer, D. 1924. *Bull. Astron. Inst. Netherlands* 2:119.

———. 1946. The motion of a particle with negligible mass under the gravitational attraction of a spheroid. *Astron. J.* 51:223–31.

———. 1959. Solution of the problem of artificial satellite theory without drag. *Astron. J.* 64:378–97.

Brouwer, D., and Clemence, G. M. 1961a. *Methods of Celestial Mechanics*. New York: Academic Press.

———, and ———. 1961b. Orbits and masses of planets and satellites. In *Planets and Satellites* (G. P. Kuiper and B. M. Middlehurst, eds.), pp. 31–94. Chicago: University of Chicago Press.

Brouwer, D., and van Woerkom, A. J. J. 1950. *Astro. Papers, Amer. Ephem. Naut. Almanac* 13:Part 2.

Brown, E. W. 1896. *Introductory Treatise on Lunar Theory.* London: Cambridge University Press.

Brown, R. A. 1974. Optical line emission from Io. In *Exploration of the Planetary System*, IAU Symposium No. 65 (A. Woszczyk and C. Iwaniszewski, eds.), pp. 527–31. Dordrecht, Holland: D. Reidel.

Brown, R. A., and Chaffee, F. H., Jr. 1974. High resolution spectra of sodium emission from Io. *Astrophys. J.* 187:L125–L126.

Brown, R. A.; Goody, R. M.; Murcray, F. J.; and Chaffee, F. H., Jr. 1975. Further studies of line emission from Io. *Astrophys. J.* 200:L49–L53.

Brown, R. A., and Yung, Y. L. 1976. Io, its atmosphere and optical emissions. In *Jupiter* (T. Gehrels, ed.). Tucson: University of Arizona Press.

Brownlee, R. R., and Cox, A. N. 1961. Early solar evolution. *Sky and Telescope* 21:No. 5, 2–6.

Burns, J. A. 1968. Jupiter's decametric radio emission and the radiation belts of its Galilean satellites. *Science* 159:971–72.

————. 1972. Dynamical characteristics of Phobos and Deimos. *Rev. Geophys. Space Phys.* 10:463–83.

————. 1973. Where are the satellites of the inner planets? *Nature Physical Science* 242:23–25.

————. 1975. The angular momenta of solar system bodies: Implications for asteroid strengths. *Icarus* 25:545–54.

————. 1976a. The consequences of the tidal slowing of Mercury. *Icarus* 28:453–58.

————. 1976b. An elementary derivation of the perturbation equations of celestial mechanics. *Am. J. Phys.* 44:944–9.

Burns, J. A., and Safronov, V. S. 1973. Asteroid nutation angles. *Mon. Not. Roy. Astron. Soc.* 165:403–11.

Burns, J. A.; Soter, S.; and Lamy, P. L. 1976a. An improved derivation of the Poynting-Robertson equation. *Eos* 57:152.

————; ————; and ————. 1976b. Forces on small particles due to solar radiation. *Icarus* 31:000–00.

Burns, R. G. 1970. *Mineralogical Applications of Crystal Field Theory*. London: Cambridge University Press.

Burton, H. E. 1929. Elements of the orbits of the satellites of Mars. *Astron. J.* 39:155–64.

Byl, J., and Ovenden, M. W. 1975. On the satellite capture problem. *Mon. Not. Roy. Astron. Soc.* 173:579–84.

Caldwell, J. 1975. Ultraviolet observations of small bodies in the solar system by OAO–2. Presented at IAU Colloquium No. 28. *Icarus* 25:384–96.

Caldwell, J. J.; Larach, D. R.; and Danielson, R. E. 1973. The continuum albedo of Titan. *Bull. Amer. Astro. Soc.* 5:305.

Cameron, A. G. W. 1963. Formation of the solar nebula. *Icarus* 1:339–42.

————. 1972. Orbital eccentricity of Mercury and the origin of the Moon. *Nature* 240:299–300.

————. 1973. Accumulation processes in the primitive solar nebula. *Icarus* 18:407–50.

Cameron, A. G. W., and Pine, M. R. 1973. Numerical models of the primitive solar nebula. *Icarus* 18:377–406.

Carlson, R. W. 1975. Atmospheres of the outer planet satellites. In *Exploration of the Outer Solar System* (AIAA Conference). Cambridge, Mass: MIT Press.

Carlson, R. W.; Bhattacharyya, J. C.; Smith, B. A.; Johnson, T. V.; Hidayat, B.; Smith, S. A.; Taylor, G. E.; O'Leary, B. T.; and Brinkmann, R. T. 1973. An atmosphere on Ganymede from its occultation of SAO–186800 on 7 June 1972. *Science* 182:53–55.

Carlson, R. W., and Judge, D. L. 1974. Pioneer 10 ultraviolet photometer observations at Jupiter encounter. *J. Geophys. Res.* 79:3623–33.

————, and ————. 1975. Pioneer 10 UV photometer observations of the Jovian hydrogen torus: The angular distribution. Presented at IAU Colloquium No. 28. *Icarus* 24:395–99.

Carlson, R. W.; Matson, D. L.; and Johnson, T. V. 1975. Electron impact ionization of Io's sodium emission cloud. *Geophys. Res. Ltrs.* 2:469–72.

Carr, M. H. 1974. Tectonism and volcanism of the Tharsis region of Mars. *J. Geophys. Res.* 79:3943–49.

Cassidy, W. A., and Hapke, B. 1975. Effects of darkening processes on surfaces of airless bodies. Presented at IAU Colloquium No. 28. *Icarus* 25:371–83.

Cess, R., and Owen, T. 1973. Titan: The effect of noble gases on an atmospheric greenhouse. *Nature* 244:272–73.

Chapman, C. R.; McCord, T. B.; and Johnson, T. V. 1973. Asteroid spectral reflectivities. *Astron. J.* 78:126–40.

Chapman, C. R.; Morrison, D.; and Zellner, B. 1975. Surface properties of asteroids: A synthesis of asteroid spectrophotometry, radiometry, and polarimetry. *Icarus* 25:104–30.

Chapman, C. R.; Pollack, J. B.; and Sagan, C. 1969. An analysis of the Mariner 4 cratering statistics. *Astron. J.* 74:1039–51.

Charnow, M. L.; Musen, P.; and Maury, J. L. 1968. Application of Hansen's method to the Xth satellite of Jupiter. *J. Astronautical Sci.* 15:303–12.

Chebotarev, G. A. 1968. Bjull. Inst. Theor. Astro.-Leningrad 11:341.

Clark, Jr., S. P. 1966. *Handbook of Physical Constants* (Geological Society of America Memoir 97). New York: Pub. Geol. Soc. America.

Clark, Jr., S. P.; Turcotte, D. L.; and Nordmann, J. C. 1975. Accretional capture of the Moon. *Nature* 258:219–20.

Cloutier, P. A.; Anderson, H. R.; Park, R. J.; Vondrak, R. R.; Spiger, R. J.; and Sandel, B. R. 1970. Detection of geomagnetically aligned currents associated with an auroral arc. *J. Geophys. Res.* 75:2595–2600.

Coffeen, D. L. 1965. Wavelength dependence of polarization. IV. Volcanic cinders and particles. *Astron. J.* 70:403–13.

Coffeen, D. L., and Hansen, J. E. 1974. Polarization studies of planetary atmospheres. In *Planets, Stars and Nebulae* (T. Gehrels, ed.), pp. 518–81. Tucson: University of Arizona Press.

Cohen, C. J.; Hubbard, E. C.; and Oesterwinter, C. 1973. Planetary elements for 10,000,000 years. *Cel. Mech.* 7:438–48.

Colombo, G. 1965. Rotational period of the planet Mercury. *Nature* 208:575.

——. 1966. Cassini's second and third laws. *Astron. J.* 71:891–96.

Colombo, G., and Franklin, F. A. 1971. On the formation of the outer satellite groups of Jupiter. *Icarus* 15:186–91.

Colombo, G.; Franklin, F. A.; and Shapiro, I. I. 1974. On the formation of the orbit-orbit resonance of Titan and Hyperion. *Astron. J.* 79:61–72.

Colombo, G., and Shapiro, I. I. 1966. Rotation of the planet Mercury. *Astrophys. J.* 145:296–307.

Comrie, L. J. 1931. Phenomena of Saturn's satellites. *Mem. Brit. Astron. Assoc.* 30:97–106.

Conel, J. E. 1974. Solution of the equation of heat conduction with time-dependent sources: Programmed application to planetary thermal history. JPL Tech. Mem. 33–718.

Consolmagno, G. J., and Lewis, J. S. 1976. Structural and thermal models of icy Galilean satellites. In *Jupiter* (T. Gehrels, ed.), pp. 1035–51. Tucson: University of Arizona Press.

Cook, A. F., and Franklin, F. A. 1964. Rediscussion of Maxwell's Adams prize essay on the stability of Saturn's rings. *Astron. J.* 69:173–200.

——, and ——. 1970. An explanation of the light curve of Iapetus. *Icarus* 13:282–91.

Cook, A. F.; Franklin, F. A.; and Palluconi, F. D. 1973. Saturn's rings: A survey. *Icarus* 18:317–37.

Counselman, III, C. C. 1973. Outcomes of tidal evolution. *Astrophys. J.* 180:307–14.

Crawford, R. T. 1938. The tenth and eleventh satellites of Jupiter. *Pub. Astro. Soc. Pac.* 50:344–47.

Cruikshank, D. P. 1974. Atmospheres of the Galilean satellites, Triton and Pluto. Presented at IAU Colloquium No. 28.

———. 1975. Radiometric studies of Trojan asteroids and Jovian satellites 6 and 7. *Bull. Amer. Astro. Soc.* 7:377–78.

———. 1976a. The diameters of Triton and Pluto. In preparation.

———. 1976b. Radii and albedos of four Trojan asteroids and Jovian satellites 6 and 7. *Icarus* 30:Jan. 77.

Cruikshank, D. P., and Morrison, D. 1973. Radii and albedos of asteroids–1, 2, 3, 4, 6, 15, 51, 433 and 511. *Icarus* 20:477–81.

———, and ———. 1976. The Galilean satellites of Jupiter. *Sci. Am.* 234(5):108–16.

Cruikshank, D. P., and Murphy, R. E. 1973. The post-eclipse brightening of Io. *Icarus* 20:7–17.

Cruikshank, D. P., and Pilcher, C. B. 1974. The spectrum of Saturn's rings: 3000–8000 cm⁻¹. Presented at IAU Colloquium No. 28.

Cuzzi, J. N., and Dent, W. A. 1975. Saturn's rings: The determination of their brightness, temperature and opacity at centimeter wavelengths. *Astrophys. J.* 198:223–27.

Cuzzi, J. N., and van Blerkom, D. V. 1974. Microwave brightness of Saturn's rings. *Icarus* 22:149–58.

Danielson, R. E.; Caldwell, J. J.; and Larach, D. R. 1973. An inversion in the atmosphere of Titan. *Icarus* 20:437–43.

———; ———; and ———. 1974. An inversion in the atmosphere of Titan. In *The Atmosphere of Titan* (D. M. Hunten, ed.). NASA SP–340, pp. 92–109. Washington, D.C.: U.S. Gov't. Printing Office.

Danielson, R. E., and Tomasko, M. G. 1971. High resolution images of Io. *Bull. Amer. Astro. Soc.* 3:243. See also unsigned article in *Sky and Telescope* 42:10.

Danielsson, L. 1969. On the interaction between a plasma and a neutral gas. Research Report 69-17, Division of Plasma Physics, Royal Institute of Technology, Stockholm.

———. 1970. Experiment on the interaction between a plasma and a neutral gas. *Phys. Fluids* 13:2288–94.

———. 1973. Review of the critical velocity of gas-plasma interaction, Part I: Experimental observations. *Astrophys. Space Sci.* 24:459–85.

Darwin, G. H. 1879a. On the bodily tides of viscous and semi-elastic spheroids, and on the ocean tides on a yielding nucleus. *Phil. Trans. Roy. Soc.* (London) 170:1–35.

———. 1879b. On the precession of a viscous spheroid and on the remote history of the Earth. *Phil. Trans. Roy. Soc.* (London) 170:447–538.

———. 1880. On the secular change in elements of the orbit of a satellite revolving about a tidally distorted planet. *Phil. Trans. Roy. Soc.* (London) 171:713–891.

Davilà, H.; Dèbarbat, S.; and Journet, A. 1975. Observations of Jupiter's Galilean satellites using astrolabes. Presented at IAU Colloquium No. 28. *Cel. Mech.* 12:51–55.

Davis, D. R. 1974. Secular changes in Jovian eccentricity: Effect on the size of capture orbits. *J. Geophys. Res.* 79:4442–43.

De, B. R. 1973a. *Some Astrophysical Problems Involving Plasmas and Plasma-Solid Systems.* Ph.D. dissertation. University of California, San Diego.

———. 1973b. On the mechanism of formation of loop prominences. *Solar Phys.* 31:437–47.

Delaunay, C. E. 1860, 1867. Mémoires de l'Académie des Sciences de Paris, 28:883 and 29:931.

Delsemme, A. H., and Wenger, A. 1970. Physico-chemical phenomena in comets. I. Experimental study of snows in a cometary environment. *Planet. Space Sci.* 18:709–15.

de Polavieja, M. G., and Edelman, C. 1972. Détermination de la masse de Jupiter à l'aide du mouvement du son neuvième satellite JIX par un filtre de Kalman-Bucy. *Astron. Astrophys.* 16:66–71.

Deprit, A.; Henrard, J.; and Rom, A. 1971. Analytical lunar ephemeris: Delaunay's theory. *Astron. J.* 76:269–72.

Dermott, S. F. 1968. On the origin of commensurabilities in the solar system—I. The tidal hypothesis and II. The orbital period relation. *Mon. Not. Roy. Astron. Soc.* 141:349–76.

————. 1969. On the origin of commensurabilities in the solar system—III. The resonant structure of the solar system. *Mon. Not. Roy. Astron. Soc.* 142:143–49.

————. 1973. Bode's law and the resonant structure of the solar system. *Nature Phys. Sci.* 244:18–21.

de Sitter, W. 1915. Determination of the mass of Jupiter. *Cape Annals* 12:part I, 153.

————. 1924. *Bull. Astron. Inst. Neth.* 2:99.

————. 1931. Jupiter's Galilean satellites. *Mon. Not. Roy. Astron. Soc.* 91:706–38.

————. 1938. On the system of astronomical constants. *Bull. Astron. Inst. Neth.* 8:213–31.

de Vaucouleurs, G. 1964. The physical ephemerides of Mars. *Icarus* 3:236–47.

Dicke, R. H. 1966. The secular acceleration of the Earth's rotation and cosmology. In *The Earth-Moon System* (B. G. Marsden and A. G. W. Cameron, eds.), pp. 98–164. New York: Plenum Press.

Divine, N. 1974. Model atmospheres of Titan. Presented at IAU Colloquium No. 28. See "Titan Atmosphere Models," JPL Tech. Mem. 33–672.

Dollfus, A. 1955. Study of planets by means of the polarization of their light. NASA TT F–188.

————. 1956. Polarization de la lumière renvoyée par les corps solides et les nuages naturels. *Ann. d'Ap.* 19:83–113.

————. 1961a. Polarization studies of planets. In *Planets and Satellites* (G. P. Kuiper and B. M. Middlehurst, eds.), pp. 343–99. Chicago: University of Chicago Press.

————. 1961b. Visual and photographic studies of planets at the Pic du Midi. In *Planets and Satellites* (G. P. Kuiper and B. M. Middlehurst, eds.), pp. 534–71. Chicago: University of Chicago Press.

————. 1962. The nature of the surface of planets and the Moon (in French). *Handbuch der Physik* 54:180–239.

————. 1967. The discovery of Janus, Saturn's tenth satellite. *Sky and Telescope* 34:136–37.

————. 1970. Diamètres des planètes et satellites. In *Surfaces and Interiors of Planets and Satellites* (A. Dollfus, ed.), pp. 46–139. New York: Academic Press.

————. 1971. Physical studies of asteroids by polarization of the light. In *Physical Studies of Minor Planets* (T. Gehrels, ed.), pp. 95–116. NASA SP–267. Washington, D.C.: U.S. Gov't. Printing Office.

————. 1975. Optical polarimetry of the Galilean satellites of Jupiter. Presented at IAU Colloquium No. 28. *Icarus* 25:416–31.

Dollfus, A., and Bowell, E. 1971. Polarimetric properties of the lunar surface and its interpretation. Part I. Telescopic observations. *Astron. Astrophysics* 10:29–53.

Dollfus, A., and Focas, J. H. 1966. Polarimetric study of the planet Mars. Final scientific report AF–61(052)–508–AF. Cambridge Research Lab.

Dollfus, A.; Focas, J. H.; and Bowell, E. 1969. La planète Mars: La nature de sa surface et les propriétés de son atmosphère, d'après la polarization de sa lumière. II: La nature du Sol de la planète Mars. *Astron. Astrophys.* 2:105–21.

Dollfus, A. and Murray, J. B. 1974. La rotation, la cartographie et la photometrie des satellites de Jupiter. In *Exploration of the Planetary System.* (A. Woszczyk and C. Iwaniszewski, eds.), pp. 513–25. Dordrecht, Holland: D. Reidel.

Du Fresne, E. R., and Anders, E. 1962. On the chemical evolution of the carbonaceous chondrites. *Geochim. Cosmochim. Acta* 26:1085–1114.

Duncombe, R. L.; Klepczynski, W. J.; and Seidelmann, P. K. 1973a. The masses of the planets, satellites, and asteroids. *Fundamentals of Cosmic Physics* 1:119–65.

Duncombe, R. L.; Seidelmann, P. K.; and Klepczynski, W. J. 1973b. Dynamical astronomy of the solar system. *Annual Review Astron. Astrophys.* 11:135–54.

Dunham, D. 1971. Motions of the satellites of Uranus. *Bull. Amer. Astro. Soc.* 3:415. See also *The Motions of the Satellites of Uranus.* Ph.D. dissertation. Yale University, New Haven, Conn.

Duxbury, T. C. 1973. Mapping Phobos. *Bull. Amer. Astro. Soc.* 5:307.

———. 1974. Phobos: Control network analysis. *Icarus* 23:290–99.

Duxbury, T. C.; Johnson, T. V.; and Matson D. L. 1974. Preliminary evaluation of some mutual events: Implications for modeling. *Bull. Amer. Astro. Soc.* 6:382.

———; ———; and ———. 1975. Galilean satellite mutual occultation data processing. Presented at IAU Colloquium No. 28. *Icarus* 25:567–84.

Eckert, W. J., and Eckert, D. A. 1967. The literal solution of the main problem of the lunar theory. *Astron. J.* 72:1299–1308.

Eckhardt, D. H. 1967. Lunar physical libration theory. In *Measure of the Moon* (Z. Kopal and C. L. Goudas, eds.), pp. 40–51. Dordrecht, Holland: D. Reidel.

Egan, W. G. 1967. Polarimetric measurements of simulated lunar surfaces. *J. Geophys. Res.* 72:3233–46.

———. 1969. Polarimetric and photometric simulation of the Martian surface. *Icarus* 10:223–37.

Egerov, V. A. 1959. *Iskusstvennyye Sputniki Zemli #3.* Acad. Sci. USSR. NASA TT–F9 (1960).

Eichelberger, W. S. 1911. *Publs. U.S. Nav. Observ.*, 2nd ser., 6:1.

Eichelberger, W. S., and Newton, A. 1926. The orbit of Neptune's satellite and the pole of Neptune's equator. *Astron. Papers Amer. Eph.* 9:275–337.

Eichhorn, H. 1971. The behavior of magnitude dependent systematic errors. In *Conference on Photographic Astrometric Technique* (H. Eichhorn, ed.), pp. 241–47. NASA CR–1825.

Elliot, J. L.; Veverka, J.; and Goguen, J. 1975. Lunar occultation of Saturn. I. The diameters of Tethys, Dione, Rhea, Titan and Iapetus. Presented at IAU Colloquium No. 28. *Icarus* 26:387–407.

Elmabsout, B. 1970. Méthode semi-numérique de résolution du problème de Hill— Application à Phoebe. *Astron. Astrophys.* 5:68–83.

Epstein, E. E.; Dworetsky, M. M.; Montgomery, J. W.; Fogarty, W. G.; and Schorn, R. A. 1970. Mars, Jupiter, Saturn, and Uranus: 3.3-mm brightness temperatures and a search for variations with time or phase angle. *Icarus* 13:276–81.

Epstein, S., and Taylor, H. P., Jr. 1970. The concentration and isotopic composition of hydrogen, carbon and silicon in Apollo 11 lunar rocks and minerals. In *Proc. Apollo 11 Lunar Sci. Conf. 2* (A. A. Levinson, ed.), pp. 1085–96. New York: Pergamon Press.

———, and ———. 1972. O^{18}/O^{16}, Si^{30}/Si^{28}, C^{13}/C^{12} and D/H studies of Apollo 14 and 15 samples. *Lunar Science,* vol. 3 (C. Watkins, ed.), pp. 236–38. Houston: Lunar Science Institute.

Everhart, E. 1973. Horseshoe and Trojan orbits associated with Jupiter and Saturn. *Astron. J.* 78:316–28.

Fanale, F. P., and Cannon, W. A. 1974. Surface properties of the Orgueil meteorite: Implications for the early history of solar system volatiles. *Geochim. Cosmochim. Acta* 38:453–70.

Fanale, F. P.; Johnson, T. V.; and Matson, D. L. 1974a. The surface composition of Io. *Bull. Amer. Astro. Soc.* 6:384.

_____; _____; and _____. 1974b. Io: A surface evaporite deposit? *Science* 186: 922–24.

Fauth, P. 1940. *Jupiter Beobachtungen zwischen 1910 und 1938/39*. Berlin: G. Schönfelds Verlagsbuchhandlung.

Feibelman, W. A. 1967. Concerning the "D" ring of Saturn. *Nature* 214:793–94.

Ferraz-Mello, S. 1966. Recherches sur le mouvement des satellites Galiléans de Jupiter. *Bull. Astron.* 3: Ser. 1, 287–330.

_____. 1975. The problem of the Galilean satellites of Jupiter. Presented at IAU Colloquium No. 28. *Cel. Mech.* 12:27–37.

Ferrin, I. R. 1974. Saturn's rings. I. Optical thickness of rings A, B, D and structure of ring B. *Icarus* 22:159–74.

_____. 1975. Saturn's rings. II. Condensations of light and optical thickness of Cassini's division. *Icarus* 26:45–52.

Fink, U., and Belton, M. J. S. 1969. Collision narrowed curves of growth for H_2 applied to new photoelectric observations of Jupiter. *J. Atmos. Sci.* 26:952–62.

Fink, U.; Dekkers, N. H.; and Larson, H. P. 1973. Infrared spectra of the Galilean satellites of Jupiter. *Astrophys. J.* 179:L155–L159.

Fink, U., and Larson, H. P. 1975. Temperature dependence of the water ice spectrum between 1 and 4 microns: Application to Europa, Ganymede and Saturn's rings. *Icarus* 24:411–20.

Fink, U.; Larson, H. P.; and Gautier, III, T. N. 1976. New upper limits for atmospheric constituents on Io. *Icarus* 27:439–46.

Fish, F. F., Jr. 1967. Angular momenta of the planets. *Icarus* 7:251–56.

Fish, R. A.; Goles, G. G.; and Anders, E. 1960. The record in the meteorites. III. On the development of meteorites in asteroidal bodies. *Astrophys. J.* 132:243–58.

Fletcher, N. H. 1970. *The Chemical Physics of Ice*. London: Cambridge University Press.

Focas, J. H., and Dollfus, A. 1969. Propriétés optiques et épaisseur des anneaux de Saturne observés par la tranche en 1966. *Astron. Astrophys.* 2:251–65.

Fox, Kenneth. 1975. Estimates of quasi-polar absorption by methane in the atmosphere of Titan. Presented at IAU Colloquium No. 28. *Icarus* 24:454–59.

Franklin, F. A. 1974. *IAU Circular* 2628.

Franklin, F. A., and Colombo, G. 1970. A dynamical model for the radial structure of Saturn's rings. *Icarus* 12:338–47.

Franklin, F. A.; Colombo, G.; and Cook, A. F. 1971. A dynamical model for the radial structure of Saturn's rings. II. *Icarus* 15:80–91.

Franklin, F. A., and Cook, A. F. 1965. Optical properties of Saturn's rings. II. Two-color phase curves of the two bright rings. *Astron. J.* 70:704–20.

_____, and _____. 1974. Photometry of Saturn's satellites: The opposition effect of Iapetus at maximum light and the variability of Titan. *Icarus* 23:355–62.

Franz, O. G. 1975. A photoelectric color and magnitude of Mimas. *Bull. Amer. Astro. Soc.* 7:388.

Franz, O. G., and Millis, R. L. 1970. A search for an anomalous brightening of Io after eclipse. *Icarus* 14:13–15.

550   REFERENCES AND BIBLIOGRAPHY

_____, and _____. 1973. UBV photometry of Enceladus, Tethys and Dione. *Bull. Amer. Astro. Soc.* 5:304.

_____, and _____. 1974. A search for posteclipse brightening of Io in 1973. II. *Icarus* 23:431–36.

_____, and _____. 1975. Photometry of Dione, Tethys and Enceladus on UVB system. Presented at IAU Colloquium No. 28. *Icarus* 24:433–42.

Frey, H. 1975. Post-eclipse brightening and non-brightening of Io. *Icarus* 25:439–46.

Gaffey, M. J. 1974. *A Systematic Study of the Spectral Reflectivity Characteristics of the Meteorite Classes with Applications to the Interpretation of Asteroid Spectra for Mineralogical and Petrological Information.* Ph.D. dissertation. Massachusetts Institute of Technology, Cambridge, Mass.

Ganapathy, R., and Anders, E. 1974. Bulk compositions of the Moon and Earth, estimated from meteorites. In *Proc. Fifth Lunar Science Conf.: Geochim. Cosmochim. Acta, Suppl.* 5: 1181–206.

Garcia, H. A. 1970. *The Determination of the Mass of Saturn and Certain Zonal Harmonic Coefficients by Means of Photographic Astrometry of Five Saturnian Satellites.* Ph.D. dissertation. Georgetown University, Washington, D.C.

_____. 1972. The mass and figure of Saturn by photographic astrometry of its satellites. *Astron. J.* 77:684–91.

Garrison, W. M. 1947. On the polymerization of unsaturated hydrocarbons by ionizing radiation. *J. Chem. Phys.* 15:78–79.

Gatley, I.; Kieffer, H.; Miner, E.; and Neugebauer, G. 1974. Infrared observations of Phobos from Mariner 9. *Astrophys. J.* 190:497–503.

Gault, D. E., and Moore, H. J. 1965. Scaling relationships for microscale to megascale impact craters. *Proc. 7th Hypervelocity Impact Symposium* 6:341–51.

Gault, D. E.; Shoemaker, E. M.; and Moore, H. J. 1963. Spray ejected from the lunar surface by meteoroid impact. *NASA Tech. Note D–1767.*

Gault, D. E., and Wedekind, J. A. 1969. The destruction of tektites by micrometeoroid impact. *J. Geophys. Res.* 74:6780–94.

Gautier, T. N.; Fink, U.; Larson, H. P.; and Treffers, R. R. 1976. Infrared spectra of the satellites of Saturn: Identification of water ice on Iapetus, Rhea, Dione, and Tethys. Div. Planet. Sci. Annual Meeting (abstract), *Bull. Amer. Astro. Soc.* 8:464.

Geake, J. E.; Dollfus, A.; Garlick, G. F. J.; Lamb, W.; Walker, C.; Steigmann, G.; and Tetulaer, C. 1970. Luminescence, electron paramagnetic resonance, and optical properties of lunar material. *Science* 167:717–20.

Gehrels, T. 1956. Photometric studies of asteroids. V. The light curve and phase function of 20 Massalia. *Astrophys. J.* 123:331–338.

_____. 1970. Photometry of asteroids. In *Surfaces and Interiors of Planets and Satellites* (A. Dollfus, ed.), pp. 319–76. London: Academic Press.

_____. 1974. The convectively unstable atmosphere of Jupiter. *J. Geophys. Res.* 79:4305–07.

_____, ed. 1976. *Jupiter.* Tucson: University of Arizona Press.

Gehrels, T.; Coffeen, T.; and Owings, D. 1964. Wavelength dependence of polarization. III. The lunar surface. *Astron. J.* 69:826-52.

Gehrels, T.; Coffeen, D.; Tomasko, M.; Doose, L.; Swindell, W.; Castillo, N.; Kendall, J.; Clements, A.; Hämeen-Anttila, J.; KenKnight, C.; Blenman, C.; Baker, R.; Best, G.; and Baker, L. 1974. The imaging photopolarimeter experiment on Pioneer 10. *Science* 183:318–20.

Gehrels, T.; Suomi, V. E.; and Krauss, R. J. 1972. The capabilities of the spin-scan imaging technique. In *Space Research XII* (S. A. Bowhill, L. D. Jaffee and M. J. Rycroft, eds.), pp. 1765–70. Berlin: Akademie-Verlag.

Gerstenkorn, H. 1955. Über Gezeitenreibung beim Zweikörper Problem. *Z. Astrophys.* 36:245–74.

———. 1968. A reply to Goldreich. *Icarus* 9:394–95.

———. 1969. The earliest past of the Earth-Moon system. *Icarus* 11:189–207.

Gill, D. 1913. Determination of the mass of Jupiter and the elements of the orbits of the older satellites. In *History and Description of the Cape Observatory*, pp. 84–101. Bellevue, Edinburgh: Neill and Co., Ltd.

Gill, J. R., and Gault, B. L. 1968. A new determination of the orbit of Triton, pole of Neptune's equator, and mass of Neptune. *Astron. J.* 73:S95.

Gillett, F. C. 1975. Further observations of the 8-13 micron spectrum of Titan. *Astrophys. J.* 201:L41–L43.

Gillett, F. C., and Forrest, W. J. 1974. The 7.5- to 13.5-micron spectrum of Saturn. *Astrophys. J.* 187:L37–L40.

Gillett, F. C.; Forrest, W. J.; and Merrill, K. M. 1973. 8-13μ observations of Titan. *Astrophys. J.* 184:L93–L95.

Gillett, F. C.; Merrill, K. M.; and Stein, W. A. 1970. Albedo and thermal emission of Jovian satellites I-IV. *Astrophys. Lett.* 6:247–49.

Gindilis, L. M.; Divari, N. B.; and Reznova, L. Y. 1969. Solar radiation pressure of interplanetary dust. *Soviet Astron.-AJ* 13:114–19.

Giuli, R. T. 1968. On the rotation of the earth produced by gravitational accretion of particles. *Icarus* 8:301–23.

Gold, T. 1972. Quoted by Burns (1972), p. 476.

———. 1975a. Remarks on the paper, 'The tidal loss of satellite-orbiting objects and its implications for the lunar surface,' by Mark J. Reid. *Icarus* 24:134–35.

———. 1975b. Resonant orbits of grains and the formation of satellites. Presented at IAU Colloquium No. 28. *Icarus* 25:489–91.

Goldreich, P. 1963. On the eccentricity of satellite orbits in the solar system. *Mon. Not. Roy. Astron. Soc.* 126:257–68.

———. 1965a. Inclination of satellite orbits about an oblate precessing planet. *Astron. J.* 70:5–9.

———. 1965b. An explanation of the frequent occurrence of commensurable mean motions in the solar system. *Mon. Not. Roy. Astron. Soc.* 130:159–81.

———. 1966a. History of the lunar orbit. *Rev. Geophys.* 4:411–39.

———. 1966b. Near-commensurate satellite orbits in the solar system. In *Theory of Orbits in the Solar System and Stellar Systems* (G. Contopoulos, ed.), pp. 268–70. New York: Academic Press.

———. 1972. Tides and the Earth-Moon system. *Sci. Amer.* 226(4):43–52.

Goldreich, P., and Peale, S. J. 1966. Spin-orbit coupling in the solar system. *Astron. J.* 71:425–38.

———, and ———. 1967. Spin-orbit coupling in the solar system. II: The resonant rotation of Venus. *Astron. J.* 72:662–68.

———, and ———. 1968. Dynamics of planetary rotations. *Ann. Rev. Astron. Astrophys.* 6:287–320.

———, and ———. 1970. The obliquity of Venus. *Astron. J.* 75:273–83.

Goldreich, P., and Soter, S. 1966. Q in the solar system. *Icarus* 5:375–89.

Goldreich, P., and Ward, W. R. 1973. The formation of planetesimals. *Astrophys. J.* 183:1051–61.

Goldstein, R. M., and Morris, G. A. 1973. Radar observations of the rings of Saturn. *Icarus* 20:260–62.

———, and ———. 1975. Ganymede: Observations by radar. *Science* 188:1211–12. See also *Bull. Amer. Astro. Soc.* 7:387.

Golitsyn, G. S. 1975. Another look at atmospheric dynamics on Titan and some of its general consequences. *Icarus* 24:70–75.

Goody, R. M. 1964. *Atmospheric Radiation. I. Theoretical Basis.* Oxford: Clarendon Press.

Gorgolewski, S. 1970. Possible detection of thermal radio emission at 3.5 mm from Callisto. *Astrophys. Lett.* 7:37.

Graboske, H. C., Jr.; Pollack, J. B.; Grossman, A. S.; and Olness, R. J. 1975. The structure and evolution of Jupiter: The fluid contraction stage. *Astrophys. J.* 199:265–81.

Gradie, J., and Zellner, B. 1973. A polarimetric survey of the Galilean satellites. *Bull. Amer. Astro. Soc.* 5:404.

Graff, K. 1924. Über den Lichtwechsel der Saturntrabanten Titan, Rhea, Tethys, Dione und Enceladus im Frühjahr 1921. *Astron. Nachr.* 220:321–23.

———. 1939. Der Lichtwechsel der Saturntrabanten Titan und Japetus im Jahre 1922. *Sitz. ber. Akad. Wiss. Wien,* Klasse IIa, 148:49–57.

Gray, D. E. (ed.) 1972. *American Institute of Physics Handbook,* pp. 2:104–2:118. New York: McGraw-Hill.

Grebenikov, E. A. 1958. The analytical theory of motion of Iapetus. *Astron. Zh.* 35:904–16. Trans. in *Soviet Astronomy-AJ* 2:850–62.

———. 1959. The perturbed motion of Saturn's eighth satellite Iapetus. *Astron. Zh.* 36:361–69. Trans. in *Soviet Astron.-AJ* 3:353–61.

Greenberg, R. J. 1973a. Evolution of satellite resonances by tidal dissipation. *Astron. J.* 78:338–46.

———. 1973b. The inclination-type resonance of Mimas and Tethys. *Mon. Not. Roy. Astron. Soc.* 165:305–11.

———. 1974. Outcomes of tidal evolution for orbits with arbitrary inclination. *Icarus* 23:51–58.

———. 1975. The dynamics of Uranus' satellites. *Icarus* 24:325–33.

———. 1976a. The motions of satellites and asteroids: Natural probes of Jovian gravity. In *Jupiter* (T. Gehrels, ed.), pp. 122–32. Tucson: University of Arizona Press.

———. 1976b. The location of Cassini's division in Saturn's rings. *Bull. Amer. Astro. Soc.* 8:433, 8:462.

Greenberg, R. J.; Counselman, C. C.; and Shapiro, I. I. 1972. Orbit-orbit resonance capture in the solar system. *Science* 178:747–49.

Greenberg, R. J., and Whitaker, E. A. 1974. Miranda: Her eccentricity and inclination. *Bull. Amer. Astro. Soc.* 6:207.

Greene, T. F.; Shorthill, R. W.; and Despain, L. G. 1971. Boeing Research Lab., DI-30-141 93-1.

Greene, T. F.; Vermilion, J. R.; Shorthill, R. W.; and Clark, R. N. 1975. The spectral reflectivity of selected areas on Europa. Presented at IAU Colloquium No. 28. *Icarus* 25:405–15.

Greenstein, J. L., and Arpigny, C. 1962. The visual region of the spectrum of Comet Mrkos (1957d) at high resolution. *Astrophys. J.* 135:892–905.

Gromova, L. V.; Moroz, V. I.; and Cruikshank, D. P. 1970. The spectrum of Ganymede in the region 1–1.7 microns. *Astron. Circular (USSR),* No. 569, p. 6 (in Russian).

Grosch, H. R. J. 1948. The orbit of the eighth satellite of Jupiter. *Astron. J.* 53:180–87.

Gross, S. H. 1974a. The atmospheres of Titan and the Galilean satellites. *J. Atmos. Sci.* 31:1413–20.

———. 1974. The atmosphere of Titan. *Rev. Geophys. Space Phys.* 12:435–46.

Grossman, A. S.; Graboske, H. C.; Pollack, J. B.; Reynolds, R. T.; and Summers, A. 1972. An evolutionary calculation of Jupiter. *Phys. Earth Planet. Interiors* 6:91–98.

Guérin, P. 1970. The new ring of Saturn. *Sky and Telescope* 40:88.

Gurnett, D. A. 1972. Sheath effects and related charged-particle acceleration by Jupiter's satellite Io. *Astrophys. J.* 175:525–33.

Guthnick, P. 1910. Ergebnisse aus photometrischen Messungen der Saturntrabanten. 1. Über den Lichtwechsel des Japetus. *Berlin Erg.* No. 14.

_____. 1914. Die veränderlichen Satelliten des Jupiter und Saturn. *Astron. Nachr.* 198:233–56.

Haas, W. H. 1950. Four independent simultaneous drawings of Ganymede. *Sky and Telescope* 9:59.

Hagihara, Y. 1961. The stability of the solar system. In *Planets and Satellites* (G. P. Kuiper and B. M. Middlehurst, eds.), pp. 95–158. Chicago: University of Chicago Press.

_____. 1962. Recommendations on notation of the earth potential. *Astron. J.* 67:108.

_____. 1972. *Celestial Mechanics,* vol. 2. Cambridge, Mass.: MIT Press.

Hall, A. 1878. Observations and orbits of the satellites of Mars with data for ephemerides in 1879. Washington, D.C.: U.S. Gov't. Printing Office.

_____. 1885. The orbit of Iapetus. Washington Observations for 1882, Appendix I, pp. 1–82.

_____. 1893. Observations of Mars, 1892. *Astron. J.* 12:185–88.

Hall, A., Jr.; Lawson, E. A.; and Bower, E. C. 1926. Corrections to the elements of the satellites of Mars. *Astron. J.* 37:69–74.

Hall, J. S., and Riley, L. A. 1974. A photometric study of Saturn and its rings. *Icarus* 23:144–56.

Hämeen-Antilla, K. A.; Laakso, P.; and Lumme, K. 1965. The shadow effect in the phase curves of lunar type surfaces. *Ann. Acad. Scien. Fennicae,* Series A, No. 172.

Hämeen-Antilla, K. A., and Vaaraniemi, P. 1974. A theoretical photometric function of Saturn's rings. *Aarne Karjalainen Obs., Univ. of Oulu Report* 18:15. Oulu, Finland. Also *Icarus* 25 (1975):470–78.

Hansen, O. L. 1972. *Thermal Radiation from the Galilean Satellites Measured at 10 and 20 Microns.* Ph.D. dissertation. California Institute of Technology, Pasadena, California.

_____. 1973. Ten micron eclipse observations of Io, Europa, and Ganymede. *Icarus* 18:237–46.

_____. 1974. 12-micron emission features of the Galilean satellites and Ceres. *Astrophys. J.* 188:L31–L33.

_____. 1975. Infrared albedos and rotation curves of the Galilean satellites. *Icarus* 26:24–29.

_____. 1976. Thermal emission spectra of 24 asteroids and the Galilean satellites. *Icarus* 27:463–71.

Hapke, B. W. 1963. A theoretical photometric function for the lunar surface. *J. Geophys. Res.* 68:4571–86.

_____. 1966. An improved theoretical lunar photometric function. *Astron. J.* 71:333–39.

Hapke, B. W.; Cohen, A. J.; Cassidy, W. A.; and Wells, E. N. 1970. Solar radiation effects on the optical properties of Apollo 11 samples. In *Proc. Apollo 11 Lunar Sci. Conf. 3* (A. A. Levinson, ed.), pp. 2199–212. New York: Pergamon Press.

Harris, A. W. 1975. Collisional breakup of particles in a planetary ring. *Icarus* 24: 190–92.

Harris, A. W., and Kaula, W. M. 1975. A co-accretional model of satellite formation. Presented at IAU Colloquium No. 28. *Icarus* 24:516–24.

Harris, D. L. 1949. *The Satellite System of Uranus*. Ph.D. dissertation. University of Chicago.

———. 1961. Photometry and colorimetry of planets and satellites. In *Planets and Satellites* (G. P. Kuiper and B. M. Middlehurst, eds.), pp. 272–342. Chicago: University of Chicago Press.

Hartmann, W. K. 1975. Lunar "cataclysm": A misconception? *Icarus* 24:181–87.

Hartmann, W. K., and Davis, D. R. 1975. Satellite-sized planetesimals. Presented at IAU Colloquium No. 28. *Icarus* 24:504–15.

Hartmann, W. K.; Davis, D. R.; Chapman, C. R.; Soter, S.; and Greenberg, R. 1975. Mars: satellite origin and angular momentum. Presented at IAU Colloquium No. 28. *Icarus* 25:588–94.

Hartmann, W. K., and Larson, S. M. 1967. Angular momenta of planetary bodies. *Icarus* 7:257–60.

Hendershott, M. C. 1972. The effects of solid earth deformations on global ocean tides. *Geophys. J.* 29:389–402.

Hénon, M. 1970. Numerical exploration of the restricted problem. VI. Hill's case: Non-periodic orbits. *Astron. Astrophys.* 9:24–36.

Henry, R. J. W., and McElroy, M. B. 1969. The absorption of extreme ultraviolet solar radiation by Jupiter's upper atmosphere. *J. Atmos. Sci.* 26:912–17.

Heppenheimer, T. A. 1975. On the presumed capture origin of Jupiter's outer satellites. *Icarus* 24:172–80.

Herget, P. 1967. Private communication, quoted by J. Kovalevsky (1970).

———. 1968a. Outer satellites of Jupiter. *Astron. J.* 73:737–42.

———. 1968b. Ephemerides of comet Schwassmann-Wachmann I and the outer satellites of Jupiter. *Publ. Cincinnati Observatory* 23:1–118.

Herrick, S. 1952. Jupiter IX and Jupiter XII. *Pub. Astro. Soc. Pac.* 64:237–41.

Herring, J. R., and Licht, A. L. 1959. Effect of solar wind on the lunar atmosphere. *Science* 130:266.

Herzberg, G. 1946. *Molecular Spectra and Molecular Structure. II. Infrared and Raman Spectra of Polyatomic Molecules*. Princeton: Van Nostrand Co.

———. 1966. *Molecular Spectra and Molecular Structure. III. Electronic Spectra and Electronic Structure of Polyatomic Molecules*. Princeton: Van Nostrand Co.

Herzog, A. D., and Beebe, R. F. 1975. Interpretations of surface features of Europa obtained from occultations by Io. *Icarus* 26:30–36.

Hill, C. W. 1878. *Amer. J. Math.* 1:5.

Hodges, Jr., R. R.; Hoffman, J. H.; and Johnson, F. S. 1974. The lunar atmosphere. *Icarus* 21:415–26.

Holden, E. S. 1881. Investigations of the objective and micrometers of the 26 inch equatorial. Washington Observations for 1877, Appendix I, 1–44.

Hopfield, J. J. 1966. Mechanism of lunar polarization. *Science* 151:1380–81.

Hori, G. 1960. The motion of an artificial satellite in the vicinity of the critical inclination. *Astron. J.* 65:291–300.

Hubbard, R. F.; Shawhan, S. D.; and Joyce, G. 1974. Io as an emitter of 100 keV electrons. *J. Geophys. Res.* 79:920–28.

Hubbard, W. B. 1969. Thermal models of Jupiter and Saturn. *Astrophys. J.* 155:333–44.

———. 1970. Structure of Jupiter: Chemical composition, contraction, and rotation. *Astrophys. J.* 162:687–97.

_____. 1974. Tides in the giant planets. *Icarus* 23:42–50.

Hunt, G. R., and Salisbury, J. W. 1970. Visible and near-infrared spectra of minerals and rocks. I. Silicate minerals. *Mod. Geol.* 1:283–300.

Hunten, D. M. 1972. The atmosphere of Titan. *Comm. Astrophys. Space Phys.* 4:149–154.

_____. 1973a. The escape of H_2 from Titan. *J. Atmos. Sci.* 30:726–32.

_____. 1973b. The escape of light gases from planetary atmospheres. *J. Atmos. Sci.* 30:1481–94.

_____, ed. 1974. *The Atmosphere of Titan.* NASA SP-340. Washington, D.C.: U.S. Gov't. Printing Office.

Hunten, D. M., and Strobel, D. F. 1974. Production and escape of terrestrial hydrogen. *J. Atmos. Sci.* 31:305–17.

Hunter, C. 1965. Oscillations of self-gravitating disks. *Mon. Not. Roy. Astron. Soc.* 129:321–43.

Hunter, R. B. 1967. Motions of satellites and asteroids under the influence of Jupiter and the Sun. *Mon. Not. Roy. Astron. Soc.* 136:245–65.

Irvine, W. M. 1966. The shadow effect in diffuse reflection. *J. Geophys. Res.* 71:2931–37.

Irvine, W. M., and Lane, A. P. 1973. Photometric properties of Saturn's rings. *Icarus* 18:171–76.

Janssen, M. A. 1974. Short wavelength radio observations of Saturn's rings. In *The Rings of Saturn* (F. D. Palluconi and G. H. Pettengill, eds.), pp. 83–96. NASA SP-343. Washington, D.C.: U.S. Gov't. Printing Office.

Jefferys, W. H. 1968. Perturbation theory for strongly perturbed systems. *Astron. J.* 73:522–27.

_____. 1970. A Fortran-based list processor for Poisson series. *Cel. Mech.* 2:474–80.

Jeffreys, H. 1947. The effect of collisions on Saturn's rings. *Mon. Not. Roy. Astron. Soc.* 107:263–67.

_____. 1953. On the masses of Saturn's satellites. *Mon. Not. Roy. Astron. Soc.* 113:81–96.

_____. 1954. Second-order terms in the figure of Saturn. *Mon. Not. Roy. Astron. Soc.* 114:433–36.

_____. 1961. Effects of tidal friction on eccentricity and inclination. *Mon. Not. Roy. Astron. Soc.* 122:339–43.

_____. 1970. *The Earth* (5th ed.). Cambridge: Cambridge University Press.

Jerzykiewicz, M. 1973. Solar variation and atmospheric transparency program—Summary of results for first two years of the program. Unpublished report, Lowell Observatory.

Johnson, H. L. 1965. The absolute calibration of the Arizona photometry. *Comm. Lunar Planet. Lab.* 3:73–77.

Johnson, H. L.; Mitchell, R. I.; and Iriarte, B. 1962. The color-magnitude diagram of the Hyades cluster. *Astrophys. J.* 136:75–94.

Johnson, T. V. 1969. *Albedo and Spectral Reflectivity of the Galilean Satellites of Jupiter.* Ph.D. dissertation. California Institute of Technology, Pasadena, California.

_____. 1971. Galilean satellites: Narrowband photometry 0.30 to 1.10 microns. *Icarus* 14:94–111.

Johnson T. V., and Fanale, F. P. 1973. Optical properties of carbonaceous chondrites and their relationship to asteroids. *J. Geophys. Res.* 78:8507–18.

Johnson, T. V., and McCord, T. B. 1970. Galilean satellites: The spectral reflectivity 0.30–1.10 microns. *Icarus* 13:37–42.

————, and ———. 1971. Spectral geometric albedo of the Galilean satellites, 0.3 to 2.5 microns. *Astrophys. J.* 169:589–94.

Johnson, T. V., and McGetchin, T. R. 1973. Topography on satellite surfaces and the shape of asteroids. *Icarus* 18:612–20.

Johnson, T. V.; Matson, D. L.; and Carlson, R. W. 1976. Io's atmosphere and ionosphere: New limits on surface pressure from plasma models. *Geophys. Res. Ltrs.* 3:293–96.

Johnson, T. V.; Veeder, G. J.; and Matson, D. L. 1975. Evidence for frost on Rhea's surface. Presented at IAU Colloquium No. 28. *Icarus* 24:428–32.

Jones, H. S. 1939. The rotation of the Earth, and the secular accelerations of the Sun, Moon, and planets. *Mon. Not. Roy. Astron. Soc.* 99:541–48.

Jones. T. J., and Morrison, D. 1974. A recalibration of the radiometric/photometric method of determining asteroid sizes. *Astron. J.* 79:892–95.

————, and ———. 1975. The secular brightening of Titan. *Bull. Amer. Astro. Soc.* 7:384.

Jordan, J. F., and Lorell, J. 1975. Mariner 9: An instrument of dynamical science. *Icarus* 25:146–65.

Joyce, R. R.; Knacke, R. F.; and Owen, T. 1973. An upper limit on the 4.9-micron flux from Titan. *Astrophys. J.* 183:L31–L34.

Judge, D. L., and Carlson, R. W. 1974. Pioneer 10 observations of the ultraviolet glow in the vicinity of Jupiter. *Science* 183:317–18.

Kanaev, I. I. 1970. Observations of the satellites of Mars made with the special light reducing shutter. *Comm. Pulkova Obs.* 185:117–18.

Kaula, W. M. 1964. Tidal dissipation by solid friction and the resulting orbital evolution. *Rev. Geophys.* 2:661–85.

————. 1966. *Theory of Satellite Geodesy.* Waltham, Mass.: Blaisdell.

————. 1968. *An Introduction to Planetary Physics: The Terrestrial Planets.* New York: John Wiley.

————. 1971. Dynamical aspects of lunar origin. *Rev. Geophys. Space Phys.* 9:217–38.

————. 1975a. Mechanical processes affecting differentiation of proto-lunar material. *Soviet-Amer. Conf. Cosmochem. Moon and Planets.* In press.

————. 1975b. The seven ages of a planet. *Icarus* 26:1–15.

Kaula, W. M., and Bigeleisen, P. E. 1975. Early scattering by Jupiter and its collision effects in the terrestrial zone. *Icarus* 25:18–33.

Kaula, W. M., and Harris, A. W. 1973. Dynamically plausible hypotheses of lunar origin. *Nature* 245:367–69.

————, and ———. 1975. Dynamics of lunar origin and orbital evolution. *Rev. Geophys. Space Phys.* 13:363–71.

Kawata, Y., and Irvine, W. M. 1974. Models of Saturn's rings which satisfy the optical observations. In *Exploration of the Planetary System* (A. Woszczyk and C. Iwaniszewska, eds.), pp. 441–64. Dordrecht, Holland: D. Reidel.

————, and ———. 1975. Thermal emission from a multiple scattering model of Saturn's rings. Presented at IAU Colloquium No. 28. *Icarus* 24:472–82.

Keihm, S. J., and Langseth, M. G. 1975. Lunar microwave brightness temperature observations re-evaluated in the light of Apollo program findings. *Icarus* 24:211–30.

Kelly, M. C.; Mozer, F. S.; and Fahleson, U. V. 1971. Electric fields in the nighttime and daytime auroral zone. *J. Geophys. Res.* 76:6054–66.

KenKnight, C. E. 1972. Autocorrelation methods to obtain diffraction-limited resolution with large telescopes. *Astrophys. J.* 176:L43–L45.

KenKnight, C. E.; Rosenberg, D. L.; and Wehner, G. K. 1967. Parameters of the optical properties of the lunar surface powder in relation to solar wind bombardment. *J. Geophys. Res.* 72:3105–30.

Kerr, F. J., and Whipple, F. L. 1954. On the secular accelerations of Phobos and Jupiter V. *Astron. J.* 59:124–27.

Khare, B. N., and Sagan, C. 1973. Red clouds and reducing atmospheres. *Icarus* 20:311–21.

Kieffer, H. 1970. Spectral reflectance of CO_2-H_2O frosts. *J. Geophys. Res.* 75:501–9.

Kieffer, H. H., and Smythe, W. D. 1974. Frost spectra: Comparison with Jupiter's satellites. *Icarus* 21:506–12.

Kiladze, R. I. 1969. Observations of Saturn's rings at the moment of the Earth's transit through their plane (in Russian). *Abastumani Astrophys. Obs. Bull. No.* 37:151–64.

Kliore, A.; Cain, D. L.; Fjeldbo, G.; Seidel, B. L.; and Rasool, S. I. 1974. Preliminary results on the atmosphere of Io and Jupiter from Pioneer 10 S-Band occultation experiment. *Science* 183:323–24.

Kliore, A. J.; Fjeldbo, G.; Seidel, B. L.; Sweetnam, D. N.; Sesplaukis, T. T.; Woiceshyn, P. M.; and Rasool, S. I. 1975. Atmosphere of Io from Pioneer 10 radio occultation measurements. Presented at IAU Colloquium No. 28. *Icarus* 24:407–10.

Knacke, R. F.; Owen, T.; and Joyce, R. R. 1975. Infrared observations of the surface and atmosphere of Titan. Presented at IAU Colloquium No. 28. *Icarus* 24:460–64.

Knopoff, L. 1964. *Q. Rev. Geophys.* 2:625–60.

Koutchmy, S. 1975. Saturn's rings: A photometric study of ring C. *Icarus* 25:131–35.

Koutchmy, S., and Lamy, P. L. 1975. Study of the inner satellites of Saturn by photographic photometry. Presented at IAU Colloquium No. 28. *Icarus* 25:459–65.

Kovalevsky, J. 1959. Méthode numérique de calcul de perturbations générales. Application au VIIIe satellite de Jupiter. *Bull. Astron.* 23:1–89.

———. 1966. Sur la théorie du mouvement d'un satellite a fortes inclinaison et excentricité. In *The Theory of Orbits in the Solar System and in Stellar Systems* (G. Contopoulos, ed.), pp. 326–44. New York: Academic Press.

———. 1967. *Introduction to Celestial Mechanics.* Dordrecht, Holland: D. Reidel.

———. 1970. Détermination des masses des planètes et satellites. In *Surfaces and Interiors of Planets and Satellites* (A. Dollfus, ed.), pp. 2–44. New York: Academic Press.

Kowal, C. T.; Aksnes, K.; Marsden, B. G.; and Roemer, E. 1975. The thirteenth satellite of Jupiter. *Astron. J.* 80:460–64.

Kozai, Y. 1957. On the astronomical constants of Saturnian satellite system. *Ann. Tokyo Astron. Obs.,* 2nd ser. 5:73–106.

Kuiper, G. P. 1944. Titan: A satellite with an atmosphere. *Astrophys. J.* 100:378–83.

———. 1951. On the origin of the irregular satellites. *Proc. Natl. Acad. Sci.* 37:717–21.

———. 1952. Planetary atmospheres and their origin. In *The Atmospheres of the Earth and Planets,* rev. ed. (G. P. Kuiper, ed.), pp. 306–405. Chicago: Univ. of Chicago Press.

———. 1956. On the origin of the satellites and the Trojans. In *Vistas in Astronomy,* vol. 2 (A. Beer, ed.), pp. 1631–66. New York: Pergamon Press.

———. 1957. Infrared observations of planets and satellites. *Astron. J.* 62:245.

———. 1961a. Limits of completeness. In *Planets and Satellites* (G. P. Kuiper and B. M. Middlehurst, eds.), pp. 575–91. Chicago: University of Chicago Press.

———. 1961b. Quoted by D. L. Harris, *Planets and Satellites* (G. P. Kuiper and B. M. Middlehurst, eds.), p. 289. Chicago: University of Chicago Press.

_____. 1972. The Lunar and Planetary Laboratory and its telescopes. *Comm. Lunar Planetary Lab.* 9:199–247.

_____. 1973. Comments on the Galilean satellites. *Comm. Lunar Planet. Lab.* 10:28–34.

Kuiper, G. P.; Cruikshank, D. P.; and Fink, U. 1970. The composition of Saturn's rings. *Sky and Telescope* 39:14, 80.

Kupo, I.; Mekler, Y.; and Eviatar, A. 1976. Detection of ionized sulfur in the jovian magnetosphere. *Astrophys. J.* 205:L51-3.

Lagus, P. L., and Anderson, D. L. 1968. Tidal dissipation in the Earth and planets. *Phys. Earth Planet. Interiors.* 1:505–10.

Lambeck, K. 1975. Effects of tidal dissipation in the oceans on the Moon's orbit and the Earth's rotation. *J. Geophys. Res.* 80:2917–25.

Lambeck, K.; Cazenave, A.; and Balmino, G. 1974. Solid earth and ocean tides estimated from satellite orbit analyses. *Rev. Geophys. Space Phys.* 12:421–34.

Lamy, Ph.L. 1974. Interaction of interplanetary dust grains with the solar radiation field. *Astron. Astrophys.* 35:197–207.

_____. 1975. *On the Dynamics of Interplanetary Dust Grains.* Ph.D. dissertation. Cornell University, Ithaca, N. Y.

Lamy, P. L., and Burns, J. A. 1972. Geometrical approach to torque free motion of a rigid body having internal energy dissipation. *Am. J. Phys.* 40:441–45.

Laplace, P. S. 1839. *Celestial Mechanics,* vol. 4 (N. Bowditch, trans.), pp. 126–44. Reprinted 1966. New York: Chelsea Publ. Co.

Laves, K. 1938. The theory of motions of the satellites of Saturn. *Vierteljahrsschr. d. Astronom. Gesellschaft* 71:310–416.

Lebofsky, L. A. 1972. Reflectivities of ammonium hydrosulfides: Application to interpretation of reflection spectra of outer solar system bodies. *Bull. Amer. Astro. Soc.* 4:362.

_____. 1973a. Spectral reflectivities of ices. *Bull. Amer. Astro. Soc.* 5:307.

_____. 1973b. *Chemical Composition of Saturn's Rings and Icy Satellites.* Ph.D. dissertation. Massachusetts Institute of Technology, Cambridge.

_____. 1975. Stability of frosts in the solar system. *Icarus* 25:205–17.

Lebofsky, L. A., and Fegley, Jr., M. B. 1976. Chemical composition of icy satellites and Saturn's rings. *Icarus* 28:379–88.

Lebofsky, L. A.; Johnson, T. V.; and McCord, T. B. 1970. Saturn's rings: Spectral reflectivity and compositional implications. *Icarus* 13:226–30.

Lee. T. 1972. Spectral albedos of the Galilean satellites. *Comm. Lunar Planet. Lab.* 9:179–80.

Lemechova, E. N. 1961. New system of elements and tables of the motion of Jupiter's X satellite. *Bull. Inst. Theor. Astron.* (Akad. Nauk, USSR) 8:103–33 and 8:512.

Leovy, C. B., and Pollack, J. B. 1973. A first look at atmospheric dynamics and temperature variations on Titan. *Icarus* 19:195–201.

_____, and _____. 1975. Further comment on Titan's atmospheric scaling. *Icarus* 24:76–77.

Levallois, J. J., and Kovalevsky, J. 1971. *Géodésie Générale,* Tome 4. Paris: Eyrolles.

Levin, A. E. 1931. Mutual eclipses and occultations of Jupiter's satellites. *Mem. Brit. Astron. Assoc.* 30:149–83.

Lewis, J. S. 1971a. Satellites of the outer planets: Their physical and chemical nature. *Icarus* 15:174–85.

_____. 1971b. Satellites of the outer planets: Thermal models. *Science* 172:1127–28.

_____. 1972a. Metal/silicate fractionation in the solar system. *Earth Planet. Sci. Lett.* 15:286–90.

_____. 1972b. Low temperature condensation from the solar nebula. *Icarus* 16:241–52.

_____. 1973. Chemistry of the outer solar system. *Space Sci. Rev.* 14:401–11.

_____. 1974a. The chemistry of the solar system. *Sci. Amer.* 230(3):50–65.

_____. 1974b. The temperature gradient in the solar nebula. *Science* 186:440–43.

_____. 1974c. Composition, structure and differentiation histories of satellites. Presented at IAU Colloquium No. 28. See Consolmagno, G., and Lewis, J. S., Ch. 25.

Lewis, J. S., and Prinn, R. G. 1973. Titan revisited. *Comm. Astrophy. Space Phys.* 5:1–7.

Lieske, J. H. 1973. On the 3-7 commensurability between Jupiter's outer two Galilean satellites. *Astron. Astrophys.* 27:59–65.

_____. 1974. A method of revitalizing Sampson's theory of the Galilean satellites. *Astron. Astrophys.* 31:137–39.

_____. 1975. Computer-developed construction of analytical expressions for the coordinates and partial derivatives of the Galilean satellites. Presented at IAU Colloquium No. 28. *Cel. Mech.* 12:5–17.

Lindberg, L.; Witalis, E.; and Jacobsen, C. 1960. Experiments with plasma rings. *Nature* 185:452–53.

Lockwood, G. W. 1975a. The secular and orbital brightness variations of Titan, 1972–1974. *Astrophys. J.* 195:L137–L139.

_____. 1975b. Planetary brightness changes: Evidence for solar variability. *Science* 190:560–62.

Lorell, J., and Shapiro, I. I. 1973. Mariner 9 celestial mechanics experiment: A status report. *J. Geophys. Res.* 78:4327–29.

Low, F. J. 1965. Planetary radiation at infrared and millimeter wavelengths. *Bull. Lowell Obs.* 5:184–87.

Low, F. J., and Rieke, G. H. 1974a. The instrumentation and techniques of infrared photometry. In *Methods of Experimental Physics,* vol. 12, Part A (N. Carleton, ed.), pp. 415–62. New York: Academic Press.

_____, and _____. 1974b. Infrared photometry of Titan. *Astrophys. J.* 190:L143–L145.

Lowan, A. N. 1933. On the cooling of a radioactive sphere. *Phys. Rev.* 44:769–75.

Lumme, K. 1975. Shape of the Cassini division in different colors. Presented at IAU Colloquium No. 28. *Icarus* 24:483–91.

Lutz, B. L.; Owen, T.; and Cess, R. D. 1976. Laboratory band strengths of methane and their application to the atmospheres of Jupiter, Saturn, Uranus, Neptune, and Titan. *Astrophys. J.* 203:541–51.

Lyot, B. 1929. Research on the polarization of light from planets and from some terrestrial substances. *Ann. Obs. Meudon 8.* Also NASA TT F-187 (1964), Washington, D.C.

_____. 1953. L'aspect des planètes au Pic du Midi dans une lunette de 60 cm d'ouverture. *Astronomie* 67:3–21.

Lyttleton, R. A. 1936. On the possible results of an encounter of Pluto with the Neptunian system. *Mon. Not. Roy. Astron. Soc.* 97:108–15.

_____. 1967. Dynamical capture of the Moon by the Earth. *Proc. Roy. Soc. (London)* A296:285–92.

McAdoo, D. C., and Burns, J. A. 1973. Further evidence for collisions among asteroids. *Icarus* 18:285–93.

_____, and _____. 1974. Approximate axial alignment times for spinning bodies. *Icarus* 21:86–93.

McCord, T. B. 1966. Dynamical evolution of the Neptunian system. *Astron. J.* 71:585–90.

————. 1968. The loss of retrograde satellites in the solar system. *J. Geophys. Res.* 73:1497–1500.

McCord, T. B.; Adams, J. B.; and Johnson, T. V. 1970. Asteroid Vesta: Spectral reflectivity and compositional implications. *Science* 168:1445–47.

McCord, T. B., and Chapman, C. R. 1975a. Asteroids: Spectral reflectance and color characteristics. *Astrophys. J.* 195:553–62.

————, and ————. 1975b. Asteroids: Spectral reflectance and color characteristics. II. *Astrophys. J.* 197:781–90.

McCord, T. B., and Johnson, T. V. 1970. The spectral reflectivity of the lunar surface (0.30 to 2.50μ) and implications for remote mineralogical analysis. *Science* 169:855–58.

McCord, T. B.; Johnson, T. V.; and Elias, J. H. 1971. Saturn and its satellites: Narrow-band spectrophotometry (0.3-1.1μ). *Astrophys. J.* 165:413–24.

McCord, T. B., and Westphal, J. A. 1971. Mars: Narrow-band photometry from 0.3 to 2.5 microns of surface regions during the 1969 apparition. *Astrophys. J.* 168:141–53.

McCuskey, S. W. 1963. *Introduction to Celestial Mechanics.* Reading, Mass.: Addison-Wesley.

MacDonald, G. J. F. 1959. Calculations on the thermal history of the Earth. *J. Geophys. Res.* 64:1967–2000.

————. 1964. Tidal friction. *Rev. Geophys.* 2:467–541.

————. 1966. The origin of the Moon: Dynamical considerations. In *The Earth-Moon System* (B. Marsden and A. G. W. Cameron, eds.), pp. 165–210. New York: Plenum Press.

McDonough, T. R. 1975. A theory of the Jovian hydrogen torus. Presented at IAU Colloquium No. 28. *Icarus* 24:400–06.

McDonough, T. R., and Brice, N. M. 1973a. New kind of ring around Saturn? *Nature* 242:513.

————, and ————. 1973b. A Saturnian gas ring and the recycling of Titan's atmosphere. *Icarus* 20:136–45.

McElroy, M. B., and McConnell, J. C. 1971. Dissociation of CO_2 in the Martian atmosphere. *J. Atmos. Sci.* 28:879–84.

McElroy, M. B., and Yung, Y. L. 1975. The atmosphere and ionosphere of Io. *Astrophys. J.* 196:227–50.

McElroy, M. B.; Yung, Y. L.; and Brown, R. A. 1974. Sodium emission from Io: Implications. *Astrophys. J.* 187:L127–L130.

McGovern, W. E. 1971. An upper limit to hydrogen and helium concentrations on Titan. In *Planetary Atmospheres* (C. Sagan, T. C. Owen and H. J. Smith, eds.), pp. 394–400. Dordrecht, Holland: D. Reidel.

McNamara, D. H. 1964. Narrowband photometry of stars, planets and satellites. North American Aviation, Inc., Space and Information Systems Division, *SID 64-78,* Accession No. 52156–64.

McNesby, J. R., and Okabe, H. 1964. Vacuum ultraviolet photochemistry. *Advances in Photochemistry* 3:157–240.

Macy, W. W., Jr., and Trafton, L. M. 1975a. Io's sodium emission cloud. Presented at IAU Colloquium No. 28. *Icarus* 25:432–38.

————, and ————. 1975b. A model for Io's atmosphere and sodium cloud. *Astrophys. J.* 200:510–19.

Marsden, B. G. 1966. *The Motions of the Galilean Satellites of Jupiter.* Ph.D. dissertation. Yale University, New Haven, Conn.

————. 1975. Probable new satellite of Jupiter. *IAU Circ.* 2855.

Marsden, B. G., and Cameron, A. G. W., eds. 1966. *The Earth-Moon System.* New York: Plenum Press.

Mason, B. 1958. *Principles of Geochemistry* (2nd ed.). New York: John Wiley.

———. 1973. The carbonaceous chondrites: A selective review. *Meteoritics* 6:60–71.

Masursky, H.; Batson, R. M.; McCauley, J. F.; Soderblom, L. A.; Wildey, R. L.; Carr, M. H.; Milton, D. J.; Wilhelms, D. E.; Smith, B. A.; Kirby, T. B.; Robinson, J. C.; Leovy, C. B.; Briggs, G. A.; Young, A. T.; Duxbury, T. C.; Acton, C. H.; Murray, B. C.; Cutts, J. A.; Sharp, R. P.; Smith, S.; Leighton, R. B.; Sagan, C.; Veverka, J.; Noland, M.; Lederberg, J.; Levinthal, E.; Pollack, J. B.; Moore, J. T.; Hartmann, W. K.; Shipley, E. N.; de Vaucouleurs, G.; and Davies, M. E. 1972. Mariner 9 television reconnaissance of Mars and its satellites: Preliminary results. *Science* 175:294–305.

Matson, D. L. 1971. Infrared observations of asteroids. In *Physical Studies of Minor Planets* (T. Gehrels, ed.), pp. 45–50. NASA SP-267. Washington, D.C.: U.S. Gov't. Printing Office.

Matson, D. L.; Johnson, T. V.; and Fanale, F. P. 1974. Sodium D-line emission from Io: Sputtering and resonant scattering hypothesis. *Astrophys. J.* 192:L43–L46.

Mendis, D. A., and Axford, W. I. 1974. Satellites and magnetospheres of the outer planets. *Ann. Rev. Earth Planet. Sci.* 2:419–74.

Mertz, L., and Coleman, I. 1966. Infrared spectrum of Saturn's rings. *Astron. J.* 71:747.

Message, P. J. 1960. Satellite theory. *IAU Trans.* 10:111.

Miller, S. L. 1961. The occurrence of gas hydrates in the solar system. *Proc. Nat'l. Acad. Sci.* 47:1798–1808.

Millis, R. L. 1973. UBV photometry of Iapetus. *Icarus* 18:247–52.

———. 1974. Mutual occultations and eclipses of the Galilean satellites. *Bull. Amer. Astro. Soc.* 6:382.

Millis, R. L., and Thompson, D. T. 1975. UBV photometry of the Galilean satellites. *Icarus* 26:408–19.

Millis, R. L.; Thompson, D. T.; Harris, B. J.; Birch, P.; and Sefton, R. 1974. A search for posteclipse brightening of Io in 1973. I. *Icarus* 23:425–30.

Minton, R. B. 1973. The polar caps of Io. *Comm. Lunar Planet. Lab.* 10:35–39.

Mitler, H. E. 1975. Formation of an iron-poor Moon by partial capture, or: Yet another exotic theory of lunar origin. *Icarus* 24:256–68.

Mogro-Campero, A. 1975. Angular momentum transfer to the inner Jovian satellites. *Nature* 258:692–93.

Molchanov, A. M. 1968. The resonant structure of the solar system. *Icarus* 8:203–15. See also critiques in *Icarus* 11 (1969):88–113.

Moroz, V. I. 1961. On the infrared spectra of Jupiter and Saturn (0.9–2.5μ). *Astron. Zh.* 38:1080–81; trans. in *Soviet Astron.-AJ* 5:827–31.

———. 1965. Infrared spectrophotometry of satellites: The Moon and the Galilean satellites of Jupiter. *Astron. Zh.* 42:1287; trans. in *Soviet Astron.-AJ* 9:999–1006.

———. 1967. *Physics of Planets.* Nauka, Moscow. NASA TT F-515.

Morrison, D. 1973a. Determination of radii of satellites and asteroids from radiometry and photometry. *Icarus* 19:1–14.

———. 1973b. Rotational variations of Io and Europa: Photometric and radiometric observations. *Bull. Amer. Astro. Soc.* 5:404.

———. 1974a. Albedos and densities of the inner satellites of Saturn. *Icarus* 22:51–56.

———. 1974b. Infrared radiometry of the rings of Saturn. *Icarus* 22:57–65.

———. 1974c. Radiometric diameters and albedos of 40 asteroids. *Astrophys. J.* 194:203–12.

_____. 1974d. Infrared photometry and spectrophotometry of Titan. In *The Atmosphere of Titan* (D. M. Hunten, ed.), pp. 12–16. NASA SP-340. Washington, D.C.: U.S. Gov't. Printing Office.

_____. 1975. Radiometry and photometry of the Galilean satellites. In preparation.

Morrison, D., and Burns, J. A. 1976. The Jovian satellites. In *Jupiter* (T. Gehrels, ed.), pp. 991–1034. Tucson: University of Arizona Press.

Morrison, D., and Cruikshank, D. P. 1973. Thermal properties of the Galilean satellites. *Icarus* 18:224–36.

_____, and _____. 1974. Physical properties of the natural satellites. *Space Sci. Rev.* 15:641–739.

Morrison, D.; Cruikshank, D. P.; and Murphy, R. E. 1972. Temperatures of Titan and the Galilean satellites at 20μ. *Astrophys. J.* 173:L143–L146.

Morrison, D.; Cruikshank, D. P.; Murphy, R. E.; Martin, T. Z.; Berry, J. G.; and Shipley, J.P. 1971. Thermal inertia of Ganymede from 20-micron eclipse radiometry. *Astrophys. J.* 167:L107–L111.

Morrison, D., and Hansen, O. L. 1975. Radiometry of two eclipses of Callisto. In preparation.

Morrison, D.; Jones, T. J.; Cruikshank, D. P.; and Murphy, R. E. 1975. The two faces of Iapetus. Presented at IAU Colloquium No. 28. *Icarus* 24:157–71.

Morrison, D.; Morrison, N. D.; and Lazarewicz, A. 1974. Four-color photometry of the Galilean satellites. *Icarus* 23:399–416.

Morrison, D., and Simon, T. 1973. Broad-band 20-micron photometry of 76 stars. *Astrophys. J.* 186:193–206.

Moulton, F. R. 1914a. *An Introduction to Celestial Mechanics* (2nd ed.). New York: Macmillan Co.

_____. 1914b. On the stability of direct and retrograde satellites. *Mon. Not. Roy. Astron. Soc.* 75:40–57.

Mozer, F. S., and Fahleson, U. V. 1970. Parallel and perpendicular electric fields in an aurora. *Planetary Space Sci.* 18:1563–71.

Mukai, T., and Mukai, S. 1973. Temperature and motion of grains in interplanetary space. *Pub. Astro. Soc. Japan* 25:481–88.

Mukai, T.; Yamamoto, T.; Hasegawa, H.; Fujiwara, A.; and Koike, C. 1974. On circumsolar grain materials. *Pub. Astron. Soc. Japan* 26:445–58.

Mulholland, J. D. 1965. *A Theory of the Sixth Satellite of Jupiter.* Ph.D. dissertation. Univ. of Cincinnati, Cincinnati, Ohio.

Munk, W. H., and MacDonald, G. J. F. 1960. *The Rotation of the Earth.* London: Cambridge University Press.

Murphy, R. E. 1973. Temperatures of Saturn's rings. *Astrophys. J.* 181:L87–L90.

_____. 1974. Variations in the infrared brightness temperature of Saturn's rings. In *The Rings of Saturn* (F. D. Palluconi and G. H. Pettengill, eds.), pp. 65–72. NASA SP-343. Washington, D.C.: U.S. Gov't Printing Office.

Murphy, R. E., and Aksnes, K. 1973. Polar cap on Europa. *Nature* 224:559–60.

Murphy, R. E.; Cruikshank, D. P.; and Morrison, D. 1972. Radii, albedos, and 20 micron brightness temperatures of Iapetus and Rhea. *Astrophys. J.* 177:L93–L96.

Murray, B. C.; Belton, M. J.; Danielson, G. E.; Davies, M. E.; Gault, D. E.; Hapke, B.; O'Leary, B.; Strom, R. G.; Suomi, V.; and Trask, N. 1974. Mercury's surface: Preliminary description and interpretation from Mariner 10 pictures. *Science* 185:169–79.

Murray, B. C.; Ward, W. R.; and Yeung, S. C. 1973. Periodic insolation variations on Mars. *Science* 180:638–40.

Murray, B. C.; Westphal, J. A.; and Wildey, R. L. 1964b. The eclipse cooling of Ganymede. *Astrophys. J.* 141:1590–92.

Murray, B. C.; Wildey, R. L.; and Westphal, J. A. 1964a. Observations of Jupiter and the Galilean satellites at 10 microns. *Astrophys. J.* 139:986–93.

Murray, J. B. 1975. New observations of surface markings on Jupiter's satellites. Presented by A. Dollfus at IAU Colloquium No. 28. *Icarus* 25:397–404.

Murray, J. R., and Javan, A. 1972. Effects of collisions on Raman line profiles of hydrogen and deuterium gas. *J. Molec. Spectrosc.* 42:1–26.

Nagy, B. 1966. Investigations of the Orgueil carbonaceous meteorite. *Geologisku Foreningens i Stockholm Fordhandlingar* 88:235–72.

Nash, D. B., and Conel, J. E. 1973. Vitrification darkening of rock powders: Implications for optical properties of the lunar surface. *The Moon* 8:346–64.

Nash, D. B.; Matson, D. L.; Johnson, T. V.; and Fanale, F. P. 1975. Production of Na-D line emission from rock specimens by proton bombardment: Implications for Io emissions. *J. Geophys. Res.* 80:1875–79.

Newburn, R. L., and Gulkis, S. 1973. A survey of the outer planets: Jupiter, Saturn, Uranus, Neptune, Pluto and their satellites. *Space Sci. Rev.* 14:179–271.

Newcomb, S. 1875. The Uranian and Neptunian systems. *Wash. Obs. for 1873, App. I.,* 1–74.

———. 1891. On the motion of Hyperion. *Astron. Papers Amer. Ephem. III, 347.*

———. 1895. The elements of the four inner planets and the fundamental constants of astronomy. *Supplement to the American Ephemeris and Nautical Almanac for 1897,* pp. 1–202.

Newton, R. R. 1972. *Medieval Chronicles and the Rotation of the Earth.* Baltimore: Johns Hopkins Press.

Nicholson, S. B. 1939. Discovery of the tenth and eleventh satellites of Jupiter and observations of these and other satellites. *Astron. J.* 48:129–32.

Noland, M. 1975. *Photometric Properties of the Satellites of Mars and Saturn.* Ph.D. dissertation. Cornell University, Ithaca, N.Y.

Noland, M., and Veverka, J. 1974. The phase coefficient of Titan. *Bull. Amer. Astro. Soc.* 6:386–87.

———, and ———. 1975. The phase coefficient of Titan. In preparation. See Noland (1975).

———, and ———. 1976a. Predicted lightcurves of Phobos and Deimos. *Icarus* 28:401–4.

———, and ———. 1976b. Photometric function of Phobos and Deimos. I. Disc-integrated photometry. *Icarus* 28:405–14.

———, and ———. 1976c. Photometric function of Phobos and Deimos. II. Surface photometry of Phobos. *Icarus* 30:Jan. 77.

———, and ———. 1976d. Photometric function of Phobos and Deimos. III. Surface photometry of Deimos. *Icarus* 30:Jan. 77.

Noland, M.; Veverka, J.; and Goguen, J. 1974a. New evidence for the variability of Titan. *Astrophys. J.* 194:L157–L158.

Noland, M.; Veverka, J.; Morrison, D.; Cruikshank, D. P.; Lazarewicz, A. R.; Morrison, N. D.; Elliot, J. L.; Goguen, J.; and Burns, J. A. 1974b. Six-color photometry of Iapetus, Titan, Rhea, Dione, and Tethys. *Icarus* 23:334–54.

Noland, M.; Veverka, J.; and Pollack, J. B. 1973. Mariner 9 polarimetry of Phobos and Deimos. *Icarus* 20:490–502.

Nolt, I. G.; Radostitz, J. V.; Donnelly, R. J.; Murphy, R. E.; and Ford, H. C. 1974. Thermal emission from Saturn's rings and disk at 34μm. *Nature* 248:659–60.

Null, G. W. 1969. A solution for the mass and dynamical oblateness of Mars using Mariner-IV doppler data. *Bull. Amer. Astro. Soc.* 1:356.

Null, G. W.; Anderson, J. D.; and Wong, S. K. 1974. Gravitational coefficients of Jupiter and its Galilean satellites determined from Pioneer 10 spacecraft data. Presented at IAU Colloquium No. 28. See J. D. Anderson *et al.* (1974a; 1974b).

Oetking, P. 1966. Photometric studies of diffusely reflecting surfaces with applications to the brightness of the Moon. *J. Geophys. Res.* 71:2505–13.

Öhman, Y. 1955. A tentative explanation of negative polarization in diffuse reflection. *Ann. Stockholm Obs.* 18:1.

O'Keefe, J. A. 1966. The origin of the Moon and the core of the Earth. In *The Earth-Moon System.* (B. G. Marsden and A. G. W. Cameron, eds.), pp. 224-233. New York: Plenum Press.

———. 1969. Origin of the Moon. *J. Geophys. Res.* 74:2758–67.

———. 1970. The origin of the Moon. *J. Geophys. Res.* 75:6565–74.

O'Leary, B. T. 1972. Frequencies of occultations of stars by planets, satellites, and asteroids. *Science* 175:1108–12.

O'Leary, B. T.; Marsden, B. G.; Dragon, R.; Hauser, E.; McGrath, M.; Backus, P.; and Roboff, H. 1976. The occultation of K Geminorum by Eros. *Icarus* 28:133–46.

O'Leary, B. T., and Rea, D. G. 1968. The opposition effect on Mars and its implications. *Icarus* 9:405–28.

O'Leary, B. T., and van Flandern, T. C. 1972. Io's triaxial figure. *Icarus* 17:209–15.

O'Leary, B. T., and Veverka, J. 1971. On the anomalous brightening of Io after eclipse. *Icarus* 14:265–68.

Öpik, E. J. 1972. Comments on lunar origin. *Irish Astron. J.* 10:190–238.

Orowan, E. 1969. Density of the Moon and nucleation of planets. *Nature* 222:867.

Ovenden, M. W. 1972. Bode's law and the missing planet. *Nature* 239:508–9.

Ovenden, M. W.; Feagin, T.; and Graf, O. 1974. On the principle of least interaction action and the Laplacean satellites of Jupiter and Uranus. *Cel. Mech.* 8:455–71.

Owen, F. N., and Lazor, F. J. 1973. Surface color variations of the Galilean satellites. *Icarus* 19:30–33.

Owen, T. C. 1965. Saturn's rings and the satellites of Jupiter: Interpretation of the infrared spectrum. *Science* 149:974–75.

Owen, T. C., and Cess, R. D. 1975. Methane absorption in the visible spectra of the outer planets and Titan. *Astrophys. J.* 197:L37–L40.

Palluconi, F. D., and Pettengill, G. H., eds. 1974. *The Rings of Saturn.* NASA SP-343. Washington, D.C.: U.S. Gov't. Printing Office.

Pannella, G. 1972. Paleontological evidence on Earth's rotational history since early Precambrian. *Astrophys. Space Sci.* 16:212–37.

Parkinson, T. D. 1975. Excitation of the sodium D line emission observed in the vicinity of Io. *J. Atmos. Sci.* 32:630–33.

Pascu, D. 1967. Untitled letter about Martian satellites. *Sky and Telescope* 34:22.

———. 1972. *The Motions of the Satellites of Mars from Photographic Observations Made in 1967, 1969, and 1971.* Ph.D. dissertation. The University of Virginia, Charlottesville.

———. 1973a. Photographic photometry of the Martian satellites. *Astron. J.* 78:794–98.

———. 1973b. Astrometric techniques for photographic observations of natural satellites. *Bull. Amer. Astron. Soc.* 5:362.

Pascu, D. 1975. An analysis of photographic observations of the Martian satellite photographic observations of 1967. Presented at IAU Colloquium No. 28. *Icarus* 25:479–83.

Pauliny-Toth, I. K.; Witzel, A.; and Gorgolewski, S. 1974. The brightness temperatures of Ganymede and Callisto at 2.8 cm wavelength. *Astron. Astrophys.* 34:129–32.

Payne, R. W. 1971. Visual photometry of Titan. *Brit. Astro. Assn.* 81:123–29.

Peale, S. J. 1966. Dust belt of the earth. *J. Geophys. Res.* 71:911–33. See also *J. Geophys. Res.* 72:1124–27.

———. 1968. Evidence against a geocentric contribution to the zodiacal light. *J. Geophys. Res.* 73:3025–33.

———. 1969. Generalized Cassini's Laws. *Astron. J.* 74:483–89.

———. 1972. Determination of parameters related to the interior of Mercury. *Icarus* 17:168–73.

———. 1973. Rotation of solid bodies in the solar system. *Rev. Geophys. Space Phys.* 11:767–93.

———. 1974. Possible histories of the obliquity of Mercury. *Astron. J.* 79:722–44.

———. 1975. Dynamical consequences of meteorite impacts on the Moon. *J. Geophys. Res.* 80:4939–46.

———. 1976. Excitation and relaxation of the precession, wobble, and libration of the moon. *J. Geophys. Res.* 81:1813–27.

Peale, S. J., and Gold, T. 1965. Rotation of the planet Mercury. *Nature* 206:1240–41.

Peek, B. M. 1958. *The Planet Jupiter.* London: Faber and Faber.

Perri, F., and Cameron, A. G. W. 1974. Hydrodynamic instability of the solar nebula in the presence of a planetary core. *Icarus* 22:416–25.

Perrine, C. D. 1902. Photographic observations of the satellite of Neptune. *Lick Obs. Bull.* 39:70–71.

Persson, H. 1963. Electric field along a magnetic line of force in a low-density plasma. *Phys. Fluids* 6:1756–59.

———. 1966. Electric field parallel to the magnetic field in a low-density plasma. *Phys. Fluids* 9:1090–98.

Peters, C. F. 1973. Accuracy analysis of the ephemerides of the Galilean satellites. *Astron. J.* 78:951–56.

———. 1975. Eclipses of natural planetary satellites. Presented at IAU Colloquium No. 28. *Cel. Mech.* 12:99–110.

Peterson, C. 1975. An explanation for Iapetus' asymmetric reflectance. Presented at IAU Colloquium No. 28. *Icarus* 24:499–503.

Petrescu, M. G. 1935. Positions de Jupiter obtenues à l'équatorial photographique de l'Observatoire de Bucarest. *Jnl. des Obs.* 18:179–83.

———. 1938, 1939. Positions photographiques des satellites de Jupiter obtenues à l'équatorial de l'Observatoire de Bucarest. *Jnl. des Obs.* 21:13–16 and *Jnl. des Obs.* 22:8–10.

Pettengill, G. H., and Dyce, R. B. 1965. A radar determination of the rotation of the planet Mercury. *Nature* 206:1240.

Pettengill, G. H., and Hagfors, T. 1974. Comments on radar scattering from Saturn's rings. *Icarus* 21:188–90.

Pettit, E. 1961. Planetary temperature measurements. In *Planets and Satellites* (G. P. Kuiper and B. M. Middlehurst, eds.), pp. 400–28. Chicago: University of Chicago Press.

Pickering, E. C. 1879. Satellites of Saturn. *Ann. Harvard Coll. Obs.* 11: Part II, 247–70.

———. 1907. Eclipses of Jupiter's satellites 1878–1903. *Ann. Harvard Coll. Obs.* 52: Part 1, 1–149.

Pierce, D. A. 1975. Observations of Saturn's satellites 1789–1972. *Pub. Astro. Soc. Pac.* 87:785–87.

Pilcher, C. B.; Chapman, C. R.; Lebofsky, L. A.; and Kieffer, H. H. 1970. Saturn's rings: Identification of water frost. *Science* 167:1372–73.

Pilcher, C. B.; Ridgway, S. T.; and McCord, T. B. 1972. Galilean satellites: Identification of water frost. *Science* 178:1087–89.

Podolak, M., and Cameron, A. G. W. 1974. Models of the giant planets. *Icarus* 22:123–48.

Podosek, F. A. 1970. Dating of meteorites by the high-temperature release of iodine-correlated Xe^{129}. *Geochim. Cosmochim. Acta* 34:341–65.

Poincaré, H. 1957. *Méthodes Nouvelles de la Mécanique Céleste* (2nd ed.). New York: Dover.

Pollack, J. B. 1973. Greenhouse models of the atmosphere of Titan. *Icarus* 19:43–58.

————. 1975. The rings of Saturn. *Space Sci. Rev.* 18:3–93.

Pollack, J. B.; Grossman, A. S.; Moore, R.; and Graboske, H. C., Jr., 1976a. The formation of Saturn's satellites and rings as influenced by Saturn's contraction history. *Icarus* 29:35–48.

————; ————; ————; and ————. 1977. A calculation of Saturn's gravitational contraction history. *Icarus* 30:111–28.

Pollack, J. B., and Reynolds, R. T. 1974. Implications of Jupiter's early contraction history for the composition of the Galilean satellites. *Icarus* 21:248–53.

Pollack, J. B., and Sagan, C. 1969. An analysis of Martian photometry. *Space Sci. Rev.* 9:243–99.

Pollack, J. B.; Summers, A.; and Baldwin, B. 1973b. Estimates of the size of the particles in the rings of Saturn and their cosmogonic implications. *Icarus* 20:263–78.

Pollack, J. B.; Veverka, J.; Noland, M.; Sagan, C.; Hartmann, W. K.; Duxbury, T. C.; Born, G. H.; Milton, D. J.; and Smith, B. A. 1972. Mariner 9 television observations of Phobos and Deimos. *Icarus* 17:394–407.

Pollack, J. B.; Veverka, J.; Noland, M.; Sagan, C.; Duxbury, T. C.; Acton, C. H., Jr.; Born, G. H.; Hartmann, W. K.; and Smith, B. A. 1973a. Mariner 9 television observations of Phobos and Deimos, 2. *J. Geophys. Res.* 78:4313–26.

Pollack, J. B.; Veverka, J.; Noland M.; Lane, A. L.; Barth, C. A.; and Rhoads, J. 1976b. Measurements of the spectral reflectivity of Phobos and Deimos. Compositional implications. In preparation.

Porter, J. G. 1960. The satellites of the planets. *J. Brit. Astron. Assn.* 70:33–59.

Price, M. J. 1973. Optical scattering properties of Saturn's ring. *Astron. J.* 78:113–20.

————. 1974. Optical scattering properties of Saturn's ring. II. *Icarus* 23:388–98.

————. 1975. Saturn's rings and Pioneer 11. *Icarus* 24:492–98.

————. 1976. Infrared thermal models for Saturn's ring. *Icarus* 27:537–44.

Price, M. J., and Baker, A. 1975. Illumination of Saturn's ring by the ball. I. Preliminary results. *Icarus* 25:136–45.

Przibran, K. 1956. *Irradiation Colours and Luminescence* (transl. J. E. Coffyn). London: Pergamon Press Ltd.

Rapaport, M. 1973. Sur l'étude analytique des satellites de Saturne, Encelade-Dione. *Astron. Astrophys.* 22:179–86.

Rather, J. D. G.; Ulich, B. L.; and Ade, P. A. R. 1974. Planetary brightness temperature measurements at 1.4 mm wavelength. *Icarus* 23:448–53.

Reid, M. 1973. The tidal loss of satellite-orbiting objects and its implications for the lunar surface. *Icarus* 20:240–48.

————. 1974. Tidal friction and the formation of satellites. Presented at IAU Colloquium No. 28.

______. 1975. On the gravitational stability of satellite-orbiting objects: A reply to T. Gold. *Icarus* 24:136–38.

Renz, F. 1898. Positionen der Jupiterstrabanten. *Mem. Acad. Sci. St. Petersbourg* 7:1–172.

Rice, F. O. 1956. Colors on Jupiter. *J. Chem. Phys.* 24:1259.

Ridgway, S. T. 1974. Jupiter: Identification of ethane and acetylene. *Astrophys. J.* 187:L41–L43.

Rieke, G. H. 1975a. The temperature of Amalthea. *Icarus* 25:333–34.

______. 1975b. The thermal radiation of Saturn and its rings. *Icarus* 26:37–44.

Ringwood, A. E. 1970. Origin of the Moon: The precipitation hypotheses. *Earth Planet. Sci. Lett.* 8:131–40.

______. 1972. Some comparative aspects of lunar origin. *Phys. Earth Planet. Interiors* 6:366–76.

Robertson, H. P. 1937. Dynamical effects of radiation in the solar system. *Mon. Not. Roy. Astron. Soc.* 97:423–38.

Roemer, E., and Lloyd, R. E. 1966. Observations of comets, minor planets and satellites. *Astron. J.* 71:443–57.

Rose, L. E. 1974. Orbit of Nereid and mass of Neptune. *Astron. J.* 79:489–90.

Ross, F. E. 1905. Investigations on the orbit of Phoebe. *Ann. Harvard Coll. Obs.* 53:101–42.

Roy, A. E. 1965. *The Foundation of Astrodynamics*. New York: Macmillan Co.

Roy, A. E., and Ovenden, M. W. 1954. On the occurrence of commensurable mean motions in the solar system. *Mon. Not. Roy. Astron. Soc.* 114:232–41.

______, and ______. 1955. On the occurrence of commensurable mean motions in the solar system–II. The mirror theorem. *Mon. Not. Roy. Astron. Soc.* 115:296–309.

Rubincam, D. P. 1975. Tidal friction and the early history of the Moon's orbit. *J. Geophys. Res.* 30:1537–48.

Ruskol, E. L. 1960. Origin of the Moon. I. Formation of a swarm of bodies around the Earth. *Soviet Astronomy-AJ* 4:657–68.

______. 1962. The origin of the Moon. In *The Moon.*, Proc. IAU Symposium 14 (Z. Kopal and Z. Kadla, eds.), pp. 149–55. New York: Academic Press.

______. 1963. Origin of the Moon. II. The growth of the Moon in the circumterrestrial swarm of satellites. *Soviet Astronomy-AJ* 7:221–27.

______. 1971a. The origin of the Moon. III. Some aspects of the dynamics of the circumterrestrial swarm. *Astron. Zh.* 48:819–29. Also in *Soviet Astronomy-AJ* 15 (1972):646–54.

______. 1971b. On the possible differences in the bulk chemical composition of the Earth and the Moon forming in the circumterrestrial swarm. In *The Moon,* IAU Symposium 47 (S. K. Runcorn and H. C. Urey, eds.), pp. 426–28. Dordrecht, Holland: D. Reidel. Also *Astron. Zh.* 48:1336–38; translated in *Soviet Astron.-AJ* 15 (1972):1061–63.

______. 1972a. On the initial distance of the Moon forming in the circumterrestrial swarm. In *The Moon,* IAU Symposium 47 (S. K. Runcorn and H. C. Urey, eds.), pp. 402–4. Dordrecht, Holland: D. Reidel.

______. 1972b. The role of the satellite's swarm in the origin of the rotation of the Earth (in Russian). *Astron. Vestn.* 6:91–95.

______. 1973. On the model of the accumulation of the Moon compatible with the data on the composition and the age of lunar rocks. *The Moon* 6:190–201.

______. 1975a. On the origin of the Moon. *Proc. Soviet Amer. Conf. Cosmochem. Moon and Planets*. In press.

———. 1975b. *The Origin of the Moon* (in Russian). Moscow: Nauka. English translation: NASA TT-F-16623.

Russell, E. E., and Tomasko, M. G. 1976. Spin-scan imaging application to planetary missions. In *Chemical Evolution of the Giant Planets* (C. Ponnamperuma, ed.). New York: Academic Press.

Russell, H. N. 1916. On the albedo of the planets and their satellites. *Astrophys. J.* 43:173–95.

Safronov, V. S. 1969. *Evolution of the Protoplanetary Cloud and Formation of the Earth and Planets*. Moscow: Nauka. National Tech. Inform. Serv., Document N-72-26718, Springfield, Va. (Translation, 1972).

———. 1972. Accumulation of the planets. In *On the Origin of the Solar System* (H. Reeves, ed.), pp. 89–113. Paris: Centre National de la Recherche Scientifique.

Sagan, C. 1971. The solar system beyond Mars: An exobiological survey. *Space Sci. Rev.* 11:827–66.

———. 1973. The greenhouse of Titan. *Icarus* 18:649–56.

Sagan, C., and Khare, B. N. 1971. Long-wavelength ultraviolet production of amino acids on the primitive Earth. *Science* 173:417–20.

Sagnier, J.-L. 1975. A new theory of the motions of the Galilean satellites of Jupiter. Presented at IAU Colloquium No. 28. *Cel. Mech.* 12:19–25.

Sampson, R. A. 1909. A discussion of the eclipses of Jupiter's satellites 1878–1903. *Ann. Harvard Coll. Obs.* 52:part 2, 149–344.

———. 1910. *Tables of the Four Great Satellites of Jupiter*. London: Wesley.

———. 1921. Theory of the four great satellites of Jupiter. *Mem. Roy. Astron. Soc.* 63:1–270.

Sandner, W. 1965. *Satellites of the Solar System.* (Translated from German by A. Helm.) New York: American Elsevier.

Scattergood, T. W.; Lesser, P.; and Owen, T. 1975. Production of organic molecules in the outer solar system by proton irradiation: Laboratory simulations. Presented at IAU Colloquium No. 28. *Icarus* 24:465–71.

Schilling, G. F. 1964. On exospheric drag as the cause of the supposed secular accelerations of Phobos. *J. Geophys. Res.* 69:1825–29.

Schmidt, O. Yu. 1959. *The Origin of the Earth.* London: Lawrence and Wishart. Original publication by USSR Academy of Science, Moscow (1950).

Scott, R. F., and Robertson, F. I. 1968. Soil mechanics surface sampler. In *Surveyor Project Final Report Part II. Science Results*, pp. 195–206. Jet Propulsion Laboratory Report 32-1265.

Seidelmann, P. K.; Klepczynski, W. J.; Duncombe, R. L.; and Jackson, E. S. 1971. Determination of the mass of Pluto. *Astron. J.* 76:488–492.

Shankland, T. J. 1970. Pressure shift of infrared absorption bands in minerals and effect on radiation heat transport. *J. Geophys. Res.* 75:409–413.

Shapiro, I. I. 1963. The prediction of satellite orbits. In *The Dynamics of Satellites* (M. Roy, ed.), pp. 257–312. New York: Academic Press.

Shapiro, I. I.; Lautman, D. A.; and Colombo, G. 1966. Earth's dust belt: Fact or fiction? I. Forces perturbing dust particle motion. *J. Geophys. Res.* 71:5695–704.

Sharpless, B. P. 1939. Observations of the satellites of Mars in 1939. Unpublished. U.S. Naval Obs. archives.

———. 1945. Secular accelerations in the longitudes of the satellites of Mars. *Astron. J.* 51:185–86.

Shelus, P. J.; Abbot, R. I.; and Mulholland, J. D. 1975. Observations of outer planet satellites. Presented at IAU Colloquium No. 28. *Cel. Mech.* 17:57.

Sherman, J. C. 1973. Review of the critical velocity of gas-plasma interaction. Part II: Theory. *Astrophys. Space Sci.* 24:487–510.

Shklovskii, I. S., and Sagan, C. 1966. *Intelligent Life in the Universe.* San Francisco: Holden-Day Inc.

Shnirev, G. D.; Grechushnikov, B. N.; and Moroz, V. I. 1964. Investigation of the infrared spectrum of Saturn by the method of Fourier transformation. *Astron. Circ.* No. 302, 1.

Shor, V. A. 1975. The motion of the Martian satellites. Presented by A. T. Sinclair at IAU Colloquium No. 28. *Cel. Mech.* 12:61–75.

Sill, G. T. 1973. Reflection spectra of solids of planetary interest. *Comm. Lunar Planet. Lab.* 10:1.

Sinclair, A. T. 1972a. The motions of the satellites of Mars. *Mon. Not. Roy. Astron. Soc.* 155:249–74.

_____. 1972b. On the commensurabilities amongst the satellites of Saturn. *Mon. Not. Roy. Astron. Soc.* 160:169–87.

_____. 1974a. On the commensurabilities amongst the satellites of Saturn. II. *Mon. Not. Roy. Astron. Soc.* 166:165–79.

_____. 1974b. A theory of the motion of Iapetus. *Mon. Not. Roy. Astron. Soc.* 169:591–605.

_____. 1975a. The orbital resonance amongst the Galilean satellites of Jupiter. *Mon. Not. Roy. Astron. Soc.* 171:59–72.

_____. 1975b. The orbital resonance amongst the Galilean satellites. Presented at IAU Colloquium No. 28. *Cel. Mech.* 12:89–96.

Singer, S. F. 1968. The origin of the Moon and its geophysical consequences. *Geophys. J. Roy. Astron. Soc.* 15:205–26.

_____. 1970a. Origin of the Moon by capture and its consequences. *Trans. Am. Geophys. Union* 51:637–41.

_____. 1970b. How did Venus lose its angular momentum? *Science* 170:1196–98. See also comment by B. M. French 1971, *Science* 173:169–70. Reply by S. F. Singer 1971, *Science* 173:170.

_____. 1971. The Martian satellites. In *Physical Studies of Minor Planets* (T. Gehrels, ed.), pp. 399–405. NASA SP-267. Washington, D.C.: U.S. Gov't. Printing Office.

_____. 1972. Origin of the Moon by tidal capture and some geophysical consequences. *The Moon* 5:206–9.

_____. 1975. When and where were the satellites of Uranus formed? Presented at IAU Colloquium No. 28. *Icarus* 25:484–88.

Sinton, W. M. 1961. Recent radiometric studies of the planets and the Moon. In *Planets and Satellites* (G. P. Kuiper and B. M. Middlehurst, eds.), pp. 429–41. Chicago: University of Chicago Press.

_____. 1972. A near-infrared view of the Uranus system. *Sky and Telescope* 44:304–5.

_____. 1973. Does Io have an ammonia atmosphere? *Icarus* 20:284–96.

_____. 1974. A filter radiometer test for an ammonia atmosphere on Io. *Bull. Amer. Astro. Soc.* 6:383.

Smith, B. A. 1970. Phobos: Preliminary results from Mariner 7. *Science* 168:828–30.

Smith, B. A.; Cook, A. F.; Feibelman, W. A.; and Beebe, R. F. 1975. On a suspected ring external to the visible rings of Saturn. Presented at IAU Colloquium No. 28. *Icarus* 25:466–69.

Smith, B. A., and Smith, S. A. 1972. Upper limits for an atmosphere on Io. *Icarus* 17:218–22.

Smith, D. W. 1975. Mid-eclipse brightening of the Galilean satellites. *Icarus* 25: 447–51.

Smith, J. C., and Born, G. H. 1976. Secular acceleration of Phobos and Q of Mars. *Icarus* 27:52–54.

Smythe, W. 1975. Spectra of hydrate frosts. Presented at IAU Colloquium No. 28. *Icarus* 24:421–27.

Sonnett, C. P.; Colburn, D. S.; and Schwartz, K. 1968. Electrical heating of meteorite parent bodies and planets by dynamo induction from a pre-main sequence T-Tauri solar wind. *Nature* 219:924.

Soter, S. 1971. *The Dust Belts of Mars*. Cntr. Radiophys. Space Res. Report No. 462. See also Ph.D. dissertation, Cornell University, Ithaca, New York.

Soter, S., and Harris, A. 1976. The equilibrium shape of Phobos and other small bodies. *Icarus* 30:Jan. 77.

Soulié, G.; Broqua; Dupouy; Rousseau; Teulet; and Pourteau, L. 1968. Positions de grosses planètes et de leurs satellites et de la lune. *Jnl. des Obs.* 51:315–28.

Stacey, F. D. 1969. *Physics of the Earth*. New York: John Wiley.

Steavenson, W. H. 1948. Observations of the satellites of Uranus. *Mon. Not. Roy. Astron. Soc.* 108:186–88.

———. 1964. The satellites of Uranus. *J. Brit. Astron. Assn.* 64:54–59.

Stebbins, J. 1927. The light-variations of the satellites of Jupiter and their application to measures of the solar constant. *Lick Obs. Bull.* 13:1–11.

Stebbins, J., and Jacobsen, T. S. 1928. Further photometric measures of Jupiter's satellites and Uranus, with tests for the solar constant. *Lick Obs. Bull.* 13:180–95.

Stief, L. J., and DeCarlo, V. J. 1968. Vacuum-ultraviolet photochemistry. VIII. *J. Chem. Phys.* 49:100–105.

Stockwell, J. N. 1873. *Smithsonian Contrib.* 18:9.

Stöffler, D.; Gault, D. E.; Wedekind, J.; and Polkowski, G. 1975. Experimental hyper-velocity impact into quartz sand: Distribution and shock metamorphism of ejecta. *J. Geophys. Res.* 80:4062–77.

Strobel, D. F. 1973. The photochemistry of NH$_3$ in the Jovian atmosphere. *J. Atmos. Sci.* 30:1205–9.

———. 1974a. The photochemistry of hydrocarbons in the atmosphere of Titan. *Icarus* 21:466–70.

———. 1974b. Hydrocarbon abundances in the Jovian atmosphere. *Astrophys. J.* 191:L47–L49.

———. 1974c. Temperature of the thermosphere. In *The Atmosphere of Titan* (D. Hunten, ed.), p. 144. NASA SP-340. Washington, D.C.: U.S. Gov't. Printing Office.

———. 1975. Aeronomy of the major planets: Photochemistry of ammonia and hydrocarbons. *Rev. Geophys. Space Phys.* 13:372–82.

Strobel, D. F., and Smith, G. R. 1973. On the temperature of the Jovian thermosphere. *J. Atmos. Sci.* 30:718–25.

Strömgren, B. 1963. Problems of internal constitution and kinematics of main sequence stars. *Quart. J. Roy. Astron. Soc.* 4:8–36.

Struve, G. 1924–33. Neue Untersuchungen im Saturnsystem. *Veröff. der Univ. Sternwarte zu Berlin–Babelsburg* 6:Part I (1924), 1–16; Part 4 (1930), 1–61; Part 5 (1933), 1–44.

———. 1928. Preliminary results of a comparison between visual and photographic observations of the satellites of Saturn, made in Johannesburg, in 1926. *Astron. J.* 39:9–12.

Struve, H. 1888. Beobachtungen der Saturnstrabanten. *Obs. Pulkova,* Suppl. 1:1–132.

———. 1898a. Beobachtungen der Saturnstrabanten. *Public. Obs. Central Nicholas,* Series II, 11:1–337.

———. 1898b. Beobachtungen der Marstrabanten im Washington. Pulkowa, und Lick Observatory. *Mem. Acad. Sci., St. Petersbourg* 8:1–73.

———. 1903. Suggestions concerning future observations of the satellites of Uranus. *Pub. Astro. Soc. Pac.* 15:183–86.

———. 1906. Bestimmung der Sacularbewegung des V Jupitermondes. *Sitz. Berlin* 44:790–810.

———. 1911. Über die Lage der Marsachse und die Konstanten in Marssystem. *Sitz. ber. Königlich Preuss. Akad. der Wiss. für 1911,* 1056–83.

———. 1913. Bahnen der Uranustrabanten, Erste Abteilung: Oberon und Titania. *Abhand. Königlich Preuss. Akad. der Wiss.,* 1–109.

———. 1915. Bestimmung der Halbmesser von Saturn aus Verfinisterungen seiner Monde. *Sitz. ber. Königlich Preuss. Akad. der Wiss.* 47:805–23.

Stumpff, K. 1948. Concerning the albedos of planets and the photometric determination of the diameters of asteroids (in German). *Astron. Nachr.* 276:108–28.

Sudbury, P. V. 1969. The motion of Jupiter's fifth satellite. *Icarus* 10:116–43.

Sullivan, R. J. 1973. Titan: A model for a toroidal gas cloud surrounding its orbit. Preprint.

Sverdrup, H. V.; Johnson, M. W.; and Fleming, R. H. 1942. *The Oceans: Their Physics, Chemistry and General Biology.* New York: Prentice-Hall.

Szebehely, V. G., and Giacaglia, G. E. O. 1964. On the elliptical restricted problem of three bodies. *Astron. J.* 69:230–35.

Tan, L. Y.; Winer, A. M.; and Pimentel, G. C. 1972. Infrared spectrum of gaseous methyl radical by rapid scan spectroscopy. *J. Chem. Phys.* 57:4028–37.

Taylor, G. E. 1972. The determination of the diameter of Io from its occultation of β Scorpii C on May 14, 1971. *Icarus* 17:202–8.

Taylor, G. E.; O'Leary, B. T.; van Flandern, T. C.; Bartholdi, P.; Owen, F.; Hubbard, W. B.; Smith, B. A.; Smith, S. A.; Fallon, F. W.; Devinney, E. J.; and Oliver, J. 1971. Occultation of Beta Scorpii C by Io on May 14, 1971. *Nature* 234:405–6.

Thorndike, A. M. 1947. The experimental determination of the intensities of infrared absorption bands. III. Carbon dioxide, methane and ethane. *J. Chem. Phys.* 15: 868–74.

Thorndike, A. M.; Wells, A. J.; and Wilson, E. B., Jr. 1947. The experimental determination of the intensities of infrared absorption bands. II. Measurements on ethylene and nitrous oxide. *J. Chem. Phys.* 15:157–65.

Tisserand, F. 1896. *Traité de Mécanique Céleste.* Paris: Gauthier-Villars. Reprinted as 2nd ed. in 1960.

Toksöz, M. N.; Dainty, A. M.; and Solomon, S. C. 1974a. A summary of lunar structural constraints. In *Lunar Science V,* pp. 801–3. Houston: Lunar Science Institute.

Toksöz, M. N.; Dainty, A. M.; Solomon, S. C.; and Anderson, K. R. 1974b. Structure of the Moon. *Rev. Geophys. Space Phys.* 12:539–67.

Toksöz, M. N.; Solomon, S. C.; and Minear, J. W. 1972. Thermal evolution of the Moon. *The Moon* 4:190–213.

Trafton, L. M. 1972a. On the possible detection of H_2 in Titan's atmosphere. *Astrophys. J.* 175:285–93.

———. 1972b. The bulk composition of Titan's atmosphere. *Astrophys. J.* 175: 295–306.

———. 1973. Interpretation of Titan's infrared spectrum in terms of a high-altitude haze layer. *Bull. Amer. Astro. Soc.* 5:305.

_____. 1974a. Titan: Unidentified strong absorptions in the photometric infrared. *Icarus* 21:175–87.

_____. 1974b. Titan's spectrum and atmospheric composition. In *The Atmosphere of Titan* (D. Hunten, ed.), pp. 17–41. NASA SP-340. Washington, D.C.: U.S. Gov't. Printing Office.

_____. 1974c. The source of Neptune's internal heat and the value of Neptune's tidal dissipation factor. *Astrophys. J.* 193:477–80.

_____. 1975a. Near-infrared spectrophotometry of Titan. Presented at IAU Colloquium No. 28. *Icarus* 24:443–53.

_____. 1975b. The morphology of Titan's methane bands. I. Comparison with a reflecting layer model. *Astrophys. J.* 195:805–14.

_____. 1975c. Detection of a potassium cloud near Io. *Nature* 258:690–92.

_____. 1976. A search for emission features in Io's extended cloud. *Icarus* 27:429–37.

Trafton, L.; Parkinson, T.; and Macy, W., Jr. 1974. The spacial extent of sodium emission around Io. *Astrophys. J.* 190:L85–L89.

Trainor, J. H.; Teegarden, B. J.; Stilwell, D. E.; McDonald, F. B.; Roelof, E. C.; and Webber, W. R. 1974. Energetic particle population in the Jovian magnetosphere: A preliminary note. *Science* 183:311–13.

Ulich, B. L., and Conklin, E. K. 1975. Brightness temperature measurements at 3mm wavelength. *Bull. Amer. Astro. Soc.* 7:391.

_____. 1976. Observations of Ganymede, Callisto, Ceres, Uranus, and Neptune at 3.33 mm wavelength. *Icarus* 27:183–89.

Uphoff, C.; Roberts, P. H.; and Friedman, L. D. 1974. Orbit design concepts for Jupiter orbiter missions. Presented at AIAA Mechanics and Control of Flight Conference, Anaheim, California, August 5–9, 1974. In *Jnl. Spacecraft Rockets* 13(1976):348–55.

Urey, H. C. 1952. *The Planets: Their Origin and Development.* New Haven, Conn.: Yale University Press.

_____. 1966. Chemical evidence relative to the origin of the solar system. *Mon. Not. Roy. Astron. Soc.* 131:199–223.

Urey, H. C., and MacDonald, G. J. F. 1971. Origin and history of the Moon. In *Physics and Astronomy of the Moon*, 2nd ed. (Z. Kopal, ed.), pp. 213–89. New York: Academic Press.

Vanasse, G. A., and Sakai, H. 1967. Fourier spectroscopy. *Progress in Optics*, vol. 6 (E. Wolf, ed.), pp. 259–330. Amsterdam: North-Holland Publishing Co.

van Biesbroeck, G. 1946. The fifth satellite of Jupiter. *Astron. J.* 52:114.

_____. 1957. The mass of Neptune from a new orbit of its second satellite, Nereid. *Astron. J.* 62:272–74.

van de Hulst, H. C. 1957. *Light Scattering by Small Particles.* New York: John Wiley.

van de Kamp, P. 1967. *Principles of Astrometry.* San Francisco: W. H. Freeman Co.

van den Bosch. 1927. Ph.D. dissertation, Utrecht University. Unpublished.

van Flandern, T. C. 1975. A determination of the rate of change of G. *Mon. Not. Roy. Astron. Soc.* 170:333–42.

van Woerkom, A. J. J. 1950. The motion of Jupiter's fifth satellite, 1892–1949. *Astron. Papers Am. Ephem. 13*, part 1:7–77.

Vermilion, J. R.; Clark, R. N.; Greene, T. F.; Seamans, J. F.; and Yantis, W. F. 1974. Low resolution map of Europa from four occultations by Io. *Icarus* 23:89–96.

Veverka, J. 1970. *Photometric and Polarimetric Studies of Minor Planets and Satellites.* Ph.D. dissertation. Harvard University, Cambridge, Mass.

_____. 1971a. Polarization measurements of the Galilean satellites of Jupiter. *Icarus* 14:355–59.

_____. 1971b. Asteroid polarimetry: A progress report. In *Physical Studies of Minor Planets* (T. Gehrels, ed.), pp. 91–94. NASA SP-267. Washington, D.C.: U.S. Gov't. Printing Office.

_____. 1971c. The physical meaning of phase coefficients. In *Physical Studies of Minor Planets* (T. Gehrels, ed.), pp. 79–90. NASA SP-267. Washington, D.C.: U.S. Gov't. Printing Office.

_____. 1971d. The meaning of Russell's law. *Icarus* 14:284–85.

_____. 1973a. Titan: Polarimetric evidence for an optically thick atmosphere? *Icarus* 18:657–60.

_____. 1973b. Polarimetric observations of 9 Metis, 15 Eunomia, 89 Julia, and other asteroids. *Icarus* 19:114–17.

_____. 1973c. The photometric properties of natural snow and of snow-covered planets. *Icarus* 20:304–10.

_____. 1974. Photometry and polarimetry. In *The Atmosphere of Titan* (D. M. Hunten, ed.), pp. 42–57. NASA SP-340. Washington, D.C.: U.S. Gov't. Printing Office.

Veverka, J., and Noland, M. 1973. Asteroid reflectivities from polarization curves: Calibration of the "slope-albedo" relationship. *Icarus* 19:230–39.

Veverka, J.; Noland, M.; Sagan, C.; Pollack, J. B.; Quam, L.; Tucker, R. B.; Eross, B.; Duxbury, T. C.; and Green, W. 1974. A Mariner 9 atlas of the moons of Mars. *Icarus* 23:206–89.

von Weizsäcker, C. F. 1944. Über die entstehung des Planetsystems. *Z. Astrophys.* 22:319–55.

Vu, D. T., and Sagnier, J.-L. 1974. Mouvement des satellites Galilean de Jupiter: Mise sur ordinateur de la theorie de Sampson. Groupe de Recherches de Geodesie Spatiale, *Bull. 11.*

Wallace, L.; Prather, M. J.; and Belton, M. J. S. 1974. The thermal structure of the atmosphere of Jupiter. *Astrophys. J.* 193:481–93.

Wamsteker, W. 1972. Narrow-band photometry of the Galilean satellites. *Comm. Lunar Planet. Lab.* 9:171–77 and 10:70.

_____. 1974. The surface composition of Io. Presented at IAU Colloquium No. 28. See Wamsteker *et al.* (1974).

Wamsteker, W.; Kroes, R. L.; and Fountain, J. A. 1974. On the surface composition of Io. *Icarus* 23:417–24.

Wänke, H.; Palme, H.; Baddenhausen, H.; Drebus, G.; Jagoutz, E.; Kruse, H.; Spettel, B.; and Teschke, F. 1974. Composition of the Moon and major lunar differentiation processes. In *Lunar Science V.*, pp. 820–22. Houston: Lunar Science Institute.

Ward, W. R. 1974a. Some remarks on the accretion problem. *Proc. Internat. Mtg. Planetary Physics and Geology*, Rome.

_____. 1974b. Climatic variations on Mars. 1. Astronomical theory of insolation. *J. Geophys. Res.* 79:3375–86.

_____. 1975a. Past orientation of the lunar spin axis. *Science* 189:377–79.

_____. 1975b. Tidal friction and generalized Cassini's laws in the solar system. *Astron. J.* 80:64–70.

Ward, W. R., and Reid, M. J. 1973. Solar tidal friction and satellite loss. *Mon. Not. Roy. Astron. Soc.* 164:21–32.

Wasserburg, G. J.; Turner, G.; Tera, F.; Podosek, F. A.; Papanastassiou, D. A.; and Huneke, J. C. 1972. Comparison of Rb-Sr, K-Ar, and U-Th-Pb ages: Lunar chronology and evolution. In *Lunar Science III* (C. Watkins, ed.), pp. 788–90. Houston: Lunar Science Institute.

Wasserman, L. H.; Elliot, J. L.; Veverka, J.; and Liller, W. 1976. Galilean satellites: Observations of mutual occultations and eclipses in 1973. *Icarus* 27:91–107.

Wasson, J. T. 1971. Volatile elements on the Earth and the Moon. *Earth Planet. Sci. Lett.* 11:219–25.

Watson, K.; Murray, B. C.; and Brown, H. 1963. The stability of volatiles in the solar system. *Icarus* 1:317–27.

Weidenschilling, S. J. 1975. Mass loss from the region of Mars and the asteroid belt. *Icarus* 26:361–66.

———. 1976. Accretion of the terrestrial planets. II. *Icarus* 27:161–70.

Wendell, O. C. 1913. Observations of satellites of Saturn. *Ann. Harvard Coll. Obs.* 69, part II:218–25.

Wetherill, G. W. 1974. Problems associated with estimating the relative impact rates on Mars and the Moon. *The Moon* 9:227–31.

Wetterer, M. K. 1971. *The Moons of Jupiter.* New York: Simon and Schuster.

Whalley, E., and Labbé, H. J. 1969. Optical spectra of orientationally disordered crystals. III. Infrared spectra of the sound waves. *J. Chem. Phys.* 51:3120–27.

Whitaker, E., and Greenberg, R. J. 1973. Eccentricity and inclination of Miranda's orbit. *Comm. Lunar Planet. Lab.* 10:70–80.

Widorn, T. 1950. Der Lichtwechsel des Saturnsatelliten Japetus im Jahre 1949. *Sitz. ber. Österr. Akad. Wiss.,* Abt. IIa, 159:189–99.

Wilkins, G. A. 1965. Orbits of the satellites of Mars. *Tech. Note No. 1.* Sussex, England: H. M. Nautical Almanac Office.

———. 1966. A new determination of the elements of the orbits of the satellites of Mars. In *The Theory of Orbits in the Solar System and in Stellar Systems,* IAU Symposium 25 (G. Contopoulos, ed.), pp. 271–73. Dordrecht, Holland: D. Reidel.

———. 1967. The determination of the mass and oblateness of Mars from the orbits of its satellites. *Tech. Note No. 10.* Sussex, England: H. M. Nautical Almanac Office. Also in *Mantles of the Earth and Terrestrial Planets* (S. K. Runcorn, ed.), pp. 77–84. New York: Interscience.

———. 1968. The problems of the satellites of Mars. *Tech. Note No. 16.* Sussex, England: H. M. Nautical Almanac Office. Published (1970) *Inst. Naz. Alta. Matematica, Sump. Math.,* vol. 3, Bologna.

———. 1969. Motion of Phobos. *Nature* 224:789.

Wilkins, G. A., and Sinclair, A. T. 1974. The dynamics of the planets and their satellites. *Proc. Roy. Soc. (London)* A336:85–104.

Williams, J. G.; Slade, M. A.; Eckhardt, D. H.; and Kaula, W. M. 1973. Lunar physical librations and laser ranging. *The Moon* 8:469–83.

Wilson, R. H., Jr. 1939. Revised orbit and ephemeris for Jupiter X. *Pub. Astro. Soc. Pac.* 51:241–42.

Wingfield, E. C., and Straley, J. W. 1955. Measurement of bond moments in C_2H_2 and C_2D_2 from infrared intensities. *J. Chem. Phys.* 23:731–33.

Wise, D. U. 1963. An origin of the Moon by rotational fission during formation of the Earth's core. *J. Geophys. Res.* 68:1547–54.

———. 1969. Origin of the Moon from the Earth: Some new mechanisms and comparisons. *J. Geophys. Res.* 74:6034–46.

Woltjer, J. 1928. The motion of Hyperion. *Ann. van de Sterrewacht te Leiden* 16, Part 3, 1–139.

Wood, J. A., and Mitler, H. E. 1974. Origin of the Moon by a modified capture mechanism, *or* half a loaf is better than a whole one. In *Lunar Science V,* pp. 851–53. Houston: Lunar Science Institute.

Woolard, E. W. 1944. The secular perturbations of the satellites of Mars. *Astron. J.* 51:33–36.

Wrixon, G. T., and Welch, W. J. 1970. The millimeter wave spectrum of Saturn. *Icarus* 13:163–72.

Yoder, C. F. 1973. *On the Establishment and Evolution of Orbit-Orbit Resonances.* Ph.D. dissertation. University of California, Santa Barbara.

———. 1975. Establishment and evolution of satellite-satellite resonances. Presented at IAU Colloquium No. 28. *Cel. Mech.* 12:97.

Younkin, R. L. 1974. The albedo of Titan. *Icarus* 21:219–29.

Yung, Y. L., and McElroy, M. B. 1975. Ganymede: Possibility of an oxygen atmosphere. *Bull. Amer. Astro. Soc.* 7:387.

Zadunaisky, P. E. 1954. A determination of new elements of the orbit of Phoebe, ninth satellite of Saturn. *Astron. J.* 59:1–6.

Zellner, B. H. 1972a. Minor planets and related objects. VIII. Deimos. *Astron. J.* 77:183–85.

———. 1972b. On the nature of Iapetus. *Astrophys. J.* 174:L107–L109.

———. 1973. The polarization of Titan. *Icarus* 18:661–64.

Zellner, B. H., and Capen, R. C. 1974. Photometric properties of the Martian satellites. *Icarus* 23:437–44.

Zellner, B.; Gehrels, T.; and Gradie, J. 1974. Minor planets and related objects. XVI. Polarimetric diameters. *Astron. J.* 79:1100–10.

Zellner, B., and Gradie, J. 1975. A polarimetric survey of the Galilean satellites. In preparation.

Zellner, B.; Wisniewski, W. Z.; Andersson, L.; and Bowell, E. 1975. Minor planets and related objects. XVIII. UBV photometry and surface composition. *Astron. J.* 80:986–95.

Zmuda, A. J.; Heuring, F. T.; and Martin, J. H. 1967. Dayside magnetic disturbances at 1100 kilometers in the auroral oval. *J. Geophys. Res.* 72:1115–17.

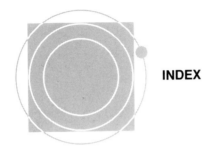

INDEX

separation of effects, 177–78, 185, 364; with orbital longitude, 172, 177–79, 191, 200–6, 368; with solar phase 172, 174–77, 368–75. *See also* individual satellite names, Lightcurves; Photometry
Brinkmann, R. T., 11, 14, 31, 36, 229, 236, 271, 302, 303, 306–8, 381
Bronshten, V. A., 142, 502
Brouwer, D., 16, 29, 33, 45, 49, 54, 91, 107, 116, 138, 140, 142, 164
Brown, E. W., 54
Brown, H., 393
Brown, R. A., 256, 399–401
Burkhead, M. S., 235
Burns, J. A., 3, 5, 8, 18, 19, 21, 33, 58, 59, 85, 88, 91–94, 96, 100, 113, 115–16, 126, 132–36, 141–42, 145, 147, 151, 159, 161, 163–64, 167, 176–77, 181–82, 191, 193–200, 203, 205–7, 226, 236, 252, 260–65, 271, 282, 285, 328, 334, 340–44, 355, 364, 419, 451, 453, 456, 470, 472, 501, 505, 506, 515, 518–19, 521, 529, 539
Burns, R. G., 234
Burton, H. E., 39
Byl, J., 141

Cain, D. L., 393, 401, 405
Caldwell, J. J., 8, 22, 194, 228–29, 232, 259–61, 272, 278, 280, 420, 421, 424, 431, 434, 436, 438–41, 446, 448, 451
Callisto (J4): color variations, 251, 371; differences between leading and trailing sides, 177, 186, 189, 223, 225, 251–53, 371, 377; eclipse measurements, 287–90; ice crust, 389; interior model, 497–500; lightcurve, 177–78, 251, 371; opposition effect, 176–78, 189; photometry, 184–90; polarimetry, 223–25, 371; radio detection, 278; reflectance from 0.3 to 1.1 μm, 236–39; reflectance from 1.0 to 2.5 μm, 239–44; surface, 189, 223, 225, 241, 288–89; temperature, 278; thermal inertia, 288. *See also* Galilean satellites
Cameron, A. G. W., 23, 114, 127, 141, 145, 255, 343, 380, 381, 383, 463–71, 477, 492, 501, 512, 524–26
Camichel, H., 10
Cannon, W. A., 383, 398

Capen, R. C., 15–16, 182, 184–85, 235, 330–31, 338
Capture: critique of Bailey's mechanism, 142; of Martian satellites, 343; of material, 477, 502–8, 521–22, 526; of Moon (*see* Moon, origin of); of outer satellites, 141–43, 470–72; into resonances, 166, 168, 510. *See also* Accumulation of satellites
Carbonaceous chondrites: crystal structure, 383–84, 392; formation in solar nebula, 380–81, 485, 492, 494; leaching of samples, 390–92; minerals, 391–92, 399, 492; paleomagnetic fields in, 485
Carlson, R. W., 11, 236, 271, 302, 303, 306–8, 381, 401, 402, 405, 429
Carme (J11). *See* Jupiter's outer satellites
Carr, M. H., 320, 342
Cassini, G. D., 4, 27, 28
Cassini's division. *See* Saturn's rings
Cassini's laws, 103–4
Cassini spin states, 105–7: selection of, 109, 111
Castillo, N., 208, 230, 406
Catalano, S., 177, 184, 187, 189, 193–94, 196–97, 200, 202–5, 208, 236, 259, 363–68, 370–72, 374, 376, 378, 451, 453, 455–56, 458
Cazenave, A., 121–22, 125, 131
Cess, R. D., 421, 423
Chaffee, F. H., Jr., 256, 399
Chapman, C. R., 134, 235, 283, 285, 315, 343, 344, 345, 395, 410, 419
Charnow, M. L., 536
Chebotarev, G. A., 141
Chemical composition, 20: condensation models, 492–500; effect of collisions during accumulation on, 512; from density, 17; diagnostic absorption features (frosts, gases, rocks), 234, 235; spectral identification, 233; surface composition, 232–68. *See also* individual satellite names; Spectrophotometry
Circulatory solution. *See* Perturbation solution
Circumplanetary cloud, 466–69, 477, 506–12: evolution of, 145–46
Clark, R. N., 14, 37, 208, 271, 396
Clark, S. P., Jr., 144, 384, 388
Clemence, G. C., 16, 29, 45, 49, 91, 107, 116, 138, 140, 163
Clements, A., 208, 230, 406
Clouds. *See* Titan
Code, A. D., 311
Coffeen, D. L., 174, 208, 210, 229, 230, 406